SUSTAINABLE ENERGY TECHNOLOGIES FOR SEAWATER DESALINATION

SUSTAINABLE ENERGY TECHNOLOGIES FOR SEAWATER DESALINATION

MARC A. ROSEN
AIDA FARSI
Faculty of Engineering and Applied Science, Ontario Tech University, Oshawa, Ontario, Canada

ELSEVIER
ACADEMIC PRESS
An imprint of Elsevier
elsevier.com/books-and-journals

Academic Press is an imprint of Elsevier
125 London Wall, London EC2Y 5AS, United Kingdom
525 B Street, Suite 1650, San Diego, CA 92101, United States
50 Hampshire Street, 5th Floor, Cambridge, MA 02139, United States
The Boulevard, Langford Lane, Kidlington, Oxford OX5 1GB, United Kingdom

British Library Cataloguing-in-Publication Data
A catalogue record for this book is available from the British Library

Library of Congress Cataloging-in-Publication Data
A catalog record for this book is available from the Library of Congress

ISBN: 978-0-323-99872-7

For Information on all Academic Press publications visit our website at https://www.elsevier.com/books-and-journals

Publisher: Charlotte Cockle
Acquisitions Editor: Lisa Reading
Editorial Project Manager: Howi M. De Ramos
Production Project Manager: Anitha Sivaraj
Cover Designer: Matthew Limbert

Typeset by Aptara, New Delhi, India

Dedication

To our families, friends, and colleagues for their support and inspiration.

Contents

About the Authors

Marc A. Rosen, PhD is a Professor at Ontario Tech University (officially University of Ontario Institute of Technology) in Oshawa, Ontario, Canada, where he served as founding Dean of the Faculty of Engineering and Applied Science. With over 70 research grants and contracts and 900 technical publications, Dr. Rosen is an active teacher and researcher in sustainable energy, desalination, thermodynamics, and environmental impact. Much of his research has been carried out for industry. Dr. Rosen has served as President of the Engineering Institute of Canada and of the Canadian Society for Mechanical Engineering. He has acted in many professional capacities, including Editor-in-Chief of various journals and a Director of Oshawa Power and Utilities Corporation. Dr. Rosen has worked for such organizations as Imatra Power Company in Finland, Argonne National Laboratory near Chicago, the Institute for Hydrogen Systems near Toronto, and Ryerson University in Toronto, where he served as Chair of the Department of Mechanical, Aerospace and Industrial Engineering. Dr. Rosen has received numerous awards and honors and is a fellow of numerous societies, including the Engineering Institute of Canada, Canadian Academy of Engineering, Canadian Society for Mechanical Engineering, American Society of Mechanical Engineers, International Energy Foundation, and Canadian Society for Senior Engineers.

Aida Farsi, PhD is a postdoctoral fellow at Ontario Tech University, Oshawa, Ontario, Canada, in the Faculty of Engineering and Applied Science. Dr. Farsi is a member of the leadership team of the ASME Ontario Chapter and presently serves as Treasurer. She is an active researcher in sustainable energy technologies, clean hydrogen production methods, desalination systems, and energy system analysis and modeling. She has received several honors and awards. Dr. Farsi has been an active member of Women in Engineering at Ontario Tech University and has been a technical reviewer for many high-impact journals.

Preface

Of all of the Earth's natural water resources, only 1% is available as freshwater. The main natural freshwater sources include lakes, rivers, and aquifers, and all of these are under pressure due to climate change and rising global demands for freshwater. Industrialization and population growth are the primary reasons for the increasing freshwater demands. Although many measures have been employed for satisfying freshwater demands, it is unlikely that these measures will be able to meet the increasing needs in the future for freshwater for all countries and in all sectors (agricultural, industrial, municipal, etc.). Desalination of seawater and brackish water is a proven and viable technique for producing freshwater from saline water. But conventional desalination plants are heavily dependent on fossil fuel resources, which are diminishing and which when combusted contribute to climate change, one of the causes of freshwater scarcity. This book describes desalination processes and technologies, with a focus on sustainable approaches and technologies for freshwater production. Given that sustainability and sustainable energy are relatively new areas, it is particularly important that their applications in seawater desalting processes receive increased attention.

Sustainable Energy Technologies for Seawater Desalination constitutes a comprehensive reference on desalination using sustainable energy resources, covering both the main technology options as well as integrated and combined systems. To provide a more holistic appreciation, energy, exergy, economic, and optimization methods are considered throughout. Comprehensive mathematical modeling is provided of the performance of thermal or membrane desalination processes combined with solar, wind, geothermal, and nuclear energy resources. Factors affecting the design, analysis, and optimization of renewable energy-based desalination systems are also considered. Analyses of desalination and sustainable energy technologies are described, utilizing energy and exergy methods, complementing the material on optimization procedures considering economic factors. Finally, the book addresses the future potential and main challenges of sustainable freshwater production systems as emerging technologies. A primary aim of this book is to provide a unique textbook and resource containing organized and synthesized knowledge about the analysis of desalination systems driven by sustainable energy technologies that can guide students, researchers, and engineers in understanding and applying key concepts and cutting edge developments in the field.

The main features of this book follow:

✓ A comprehensive textbook and resource is provided on sustainable freshwater production.

✓ The modelling, analysis, and optimization of renewable energy-based desalination processes are described using energy and exergy methods, economic assessments, and various optimization procedures.

✓ Numerous examples and case studies are incorporated to illustrate the material covered and practical applications.

✓ Up-to-date information is presented desalination using sustainable energy resources along with recent developments and advances.

✓ A coherent approach is followed throughout, starting from fundamentals and basics and culminating with advanced processes, systems and applications.

Sustainable Energy Technologies for Seawater Desalination is structured so as to build a coherent understanding for the reader while progressing through the book, as outlined here. In Chapters 1 and 2, introductory material is covered, like the status of desalination and its historical development, the working principles of various desalination technologies, the potential for sustainable freshwater production and emerging trends in alternative freshwater supply strategies. In Chapter 3, sustainability, sustainable energy, and sustainable energy methods and technologies are described. Also, a sustainable energy evaluation is presented for a particular energy system, the net-zero energy building. In Chapter 4, energy and exergy methods are reviewed. Theoretical and practical aspects of thermodynamics most relevant to energy and exergy methods are described and general applications and implications of exergy analysis results are discussed and contrasted with those of energy analysis. Models for widely-used desalination techniques are presented in Chapter 5, to help develop an understanding of desalination technologies and their performance, design, analysis, and optimization. In Chapter 6, the first and second laws of thermodynamics are applied to a generalized desalination process to establish a benchmark for the least energy required for desalination. In Chapter 7, a consistent definition of exergy efficiency is provided for a general desalination system through which the thermodynamic performance of desalination systems can be compared. Also, irreversibilities associated with desalting processes are assessed via the entropy generation in desalination systems. Various configurations of sustainable desalination systems are presented, compared, and discussed in Chapter 8, in terms of operational conditions, performance, and efficiency. In Chapter 9, the economics of freshwater production is examined, in part through a techno-economic analysis of desalination systems and renewable energy technologies. In addition, exergo-economic, and exergo-environmental assessments are presented to address the proper allocation of economic resources to optimize the design and operation of desalination systems and to evaluate the environmental impact of desalination systems using renewable and nonrenewable energy. Finally, in Chapter 10, a structure for optimization of combined or stand-alone desalination systems using renewable energy sources is presented and applied to determine the optimum design parameters based on the objective functions considered.

Many case studies and illustrative examples are incorporated to help the reader understand and carry out practical applications of sustainable energy technologies for seawater desalination. We hope this book allows seawater desalination technologies to be applied more widely in a sustainable manner, using both conventional systems and renewable and advanced technologies. By applying thermodynamics, economics and optimization methods to sustainable desalination technologies, it is believed that they can be made more clean, sustainable, and efficient means for freshwater production.

Marc A. Rosen
Aida Farsi

Acknowledgments

Many people and entities assisted, directly or indirectly, in bringing this book to fruition, and we are grateful to all of them.

We particularly thank past and present colleagues and students for thoughtful discussions and debates, which helped influence aspects of this book. That assistance helped in enhancing the book's coverage and content and in making it more useful as a textbook and resource on desalination using sustainable energy resources.

We are also thankful to companies and other organizations that contributed documents, data, and illustrations used in preparing this book. The materials allowed for coverage of recent developments and sound practicality, as well as real-world and industrial relevance.

Last, but not least, we thank our families and friends who have provided great motivation and support, as well as understanding and patience. In particular, Marc A. Rosen thanks his wife, children, parents, siblings, and other family members who have been a great and continuous source of inspiration and support. Furthermore, Aida Farsi warmly thanks her fiancé, Alireza, parents, and siblings, for their love, support, guidance, and motivation.

Marc A. Rosen
Aida Farsi

CHAPTER ONE

Introduction to desalination and sustainable energy

1.1 Resources and the need for seawater desalination

Although water is the one of the most abundant substances on the planet, only 0.5% of natural water resources are freshwater. The global need for freshwater is increasing dramatically, while natural readily available freshwater resources, such as rivers, lakes and aquifers, are diminishing. Numerous freshwater supply options have been employed in the past, but they cannot meet the rising freshwater demands in all sectors (municipal, industrial, electrical, irrigation, etc.). Desalination of brackish water and seawater has received a great deal of attention over the years as a viable means of satisfying freshwater demands around the world. Population growth, urbanization, and industrialization appear to be the main drivers in recent years of the need for desalination, as a reliable technique for producing freshwater from a relatively low-quality water resource.

This book addresses seawater desalination processes, with a focus on sustainable approaches to providing freshwater. This chapter, in particular, describes the status of desalination systems, including the number of desalination plants installed around the world and their freshwater production capacities as well as other important factors such as feed water type, geographical parameters, and desalting technologies. This chapter also provides insights into the historical development of desalination, and the potential for driving desalination systems using sustainable energy resources. Using sustainable energy in seawater desalination systems has the potential to permit continuous freshwater production with low impacts on the environment.

Water covers roughly three-fourths of the surface of the earth and constitutes the most abundant substance on the planet (USGS, 2021a). In Table 1.1, water resources are classified based on salinity (total dissolved solids). Only 2.5% of the world's water is freshwater and brackish water, and these resources are primarily found in lakes and rivers, polar regions, and ground water (USGS, 2021a). Of the 2.5% fresh/brackish water resources,

Sustainable Energy Technologies for Seawater Desalination
DOI: https://doi.org/10.1016/B978-0-323-99872-7.00002-4

Table 1.1 Classification of water sources in terms of total dissolved solids.

Water source	Total dissolved solids (g/kg)
Drinking water	< 0.5
Freshwater	< 1
Brackish water	1–15 (diluted brackish water)
	15–35 (concentrated brackish water)
Seawater	35–45

Data from (WHO, 2008; Kucera, 2019).

about 70% is frozen in the form of glaciers, snows, and ices (USGS, 2021a). About 29.75% of freshwater and brackish water resources are found underground, usually in deep aquifers, while rivers and lakes contribute only 0.25% of overall freshwater and brackish water resources (USGS, 2021a). Fig. 1.1 shows a breakdown of water resources around the world.

Population growth and industrialization are significantly increasing freshwater demand and this effect is expected to grow in the future. In addition, climate change and its effects are becoming more concerning and accelerating as industrialization and population grow, resulting in worsening droughts in many parts of the world. Pollution of water resources due to the activities of people and civilizations is another reason of freshwater shortages.

The uneven distribution of freshwater resources exacerbates the problem of water scarcity. Some regions such as Scandinavia, northern Russia, northern North America (Canada, Alaska), and central and southern coastal regions of South America have large freshwater resources (NRC, 2004). For example, Canada has 20% of the global freshwater resource (Government of Canada, 2018), yet it has less than 0.5% of total world population (Worldometer, 2021).

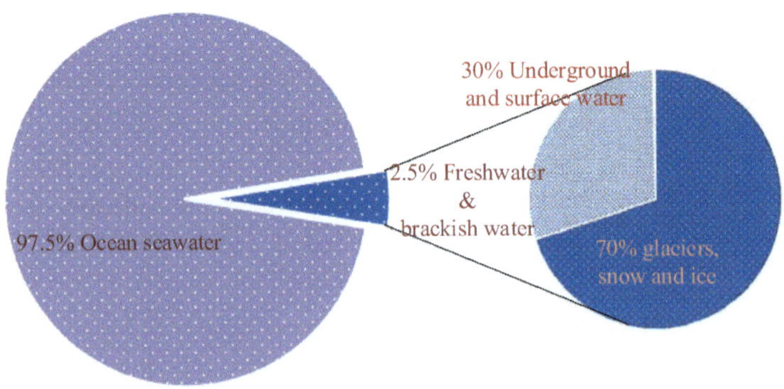

Fig. 1.1 ***Distribution of world's water resources.***

Around 70% of global water consumption is used for agriculture, 20% for industry, and 10% for domestic applications (The World Bank 2017).

There are several methods to measure water stress (e.g., the Falkenmark Indicator) with which current and projected future water stress levels can be described. The Falkenmark indicator is one of the widely used indexes for measuring water scarcity in an area. With this method, water scarcity is defined in terms of total water resources that are available to the population of an area. In fact, the water scarcity is measured through the amount of renewable water resources available for each person per year (Falkenmark, 1989). Renewable water supplies are water supplies that are continuously replenished. Renewable water sources include precipitation from the atmosphere (e.g., rain, snow) and water quantities transferred through streams, rivers, and shallow ground water. Note that renewable water resources do not include deep aquifers of water that are formed by precipitates over centuries. If the amount of renewable water for each person in a country is below 1700 m^3 per year, that country is experiencing water stress. Also, if the amount of renewable water available for each person is below 1000 m^3, the country is experiencing water scarcity and below 500 m^3, absolute water scarcity (White, 2012).

Fig. 1.2 shows the global water stress for two years: 2019 and 2030. It is evident that the worldwide water scarcity is expected to increase over this time period. According to UN medium population projections (GRID-Arendal, 2009), 48 countries will encounter water stress by 2025, corresponding to 2.8 billion people (or 35% of the global population of 7.9 billion in 2021) facing water crises. The majority of these countries are in North Africa, south and middle East Asia, and the southern United States. Also, it is predicted that, by 2050, the number of countries facing water stress will reach 54 by 2025, corresponding to 4 billion people (GRID-Arendal, 2009).

Approaches such as transferring freshwater from the water-rich areas to the water-poor areas, conservation, recycling/reuse of freshwater sources, and freshwater production using desalination can be employed to reduce water scarcities. Among these, desalination is recognized as a viable approach to meet freshwater demand around the world.

In terminology, the word "desalination" means remove salts from (MWD, 2021); however, desalination is conventionally defined as removing minerals and dissolved solids from water. The purpose of a desalination process is to clean saline water and thereby to provide water with an allowable limit of dissolved solids (around 500 ppm or less based on the World

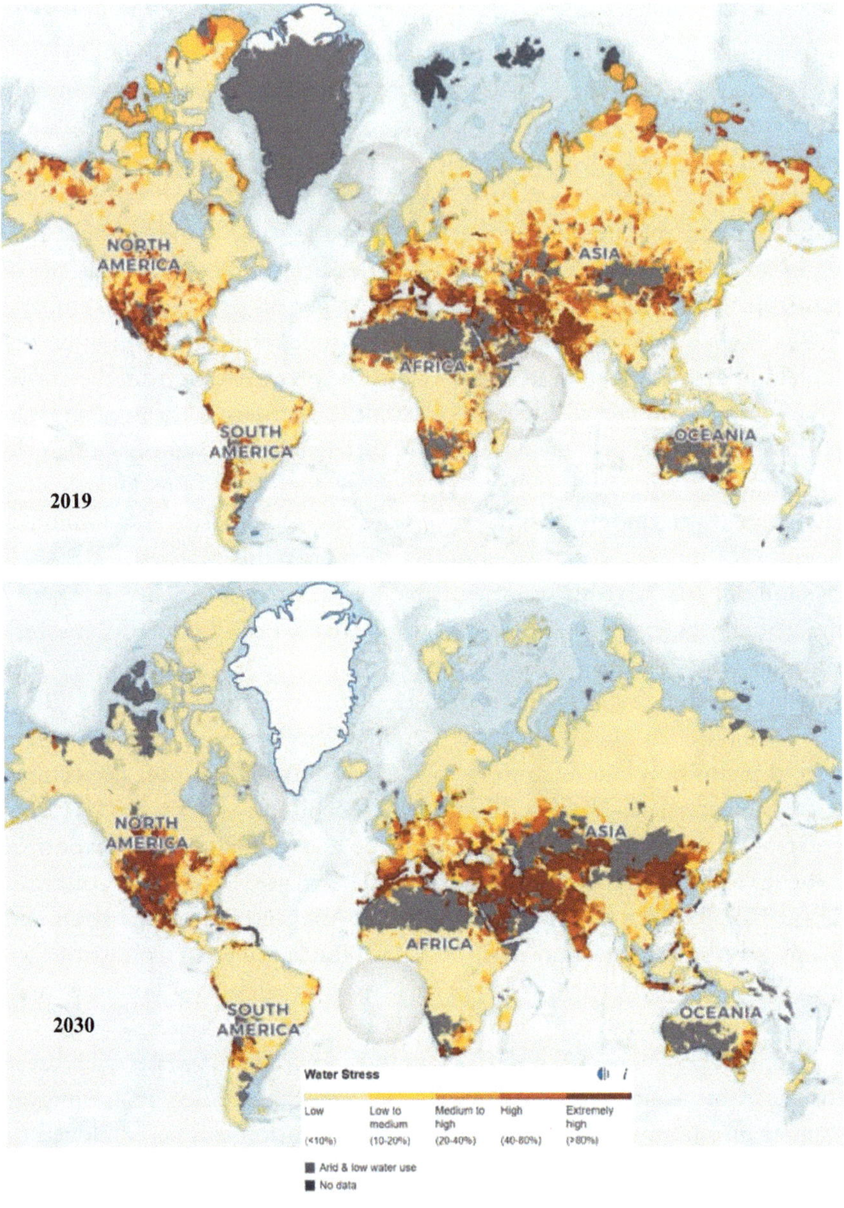

Fig. 1.2 ***Global water stress in 2019 and 2030.*** *Source: (WRI Aqueduct, 2019) available at https://www.wri.org/applications/aqueduct/water-risk-atlas (accessed on 24 August 2021).*

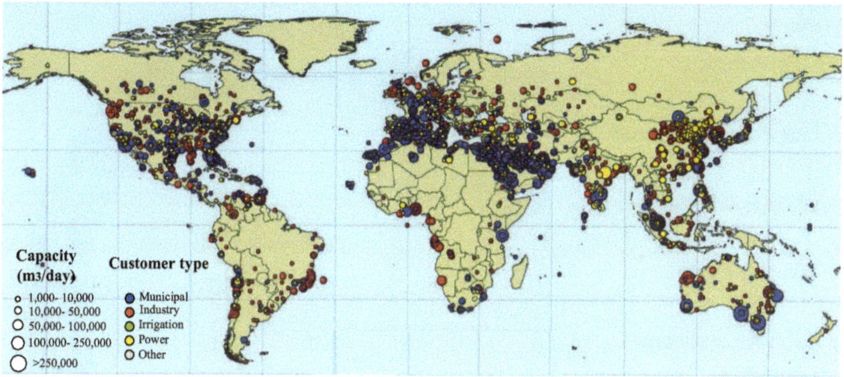

Fig. 1.3 ***Global share of operational desalination plants, with capacities and customer types.*** *Source: (Jones et al., 2019).*

Health Organization (WHO, 2017)). Water can be classified into freshwater, brackish water, and seawater based on its salinity. Total dissolved solids are 500–1000 ppm for freshwater, 100–35000 ppm for brackish water, and 35,000–45,000 ppm for seawater (USGS, 2021b). The salinity of the feed water input to a desalination system varies depending on the feed water source, which can be saline aquifer, seawater, or wastewater.

Fig. 1.3 shows global desalination plants and their capacities and user sectors. It can be seen that the many of the desalination plants are installed in the United States, China, North Africa, the Middle East, and Australia. The majority of desalination facilities are located on coastlines, and desalination plants installed on and around coastlines are typically larger than inland desalination facilities. Most of the desalination plants dedicated to municipal applications are concentrated in the Middle East and North Africa (MENA) region, while most desalination plants for nonmunicipal applications are found in North America, Western Europe, East Asia, and Pacific regions. Fig. 1.4 shows the numbers and capacities of desalination plants, broken down by region. Around half of global freshwater is produced in the MENA region, with Saudi Arabia contributing 15% of global freshwater production, the United Arab Emirates 10%, and Kuwait 3.7%. The East Asia and Pacific regions have the next highest share of global freshwater production, due to large capacities in China (7.5%) and the United States (11.2%). About half of the freshwater production in Western Europe is associated with Spain, which is responsible for 5.7% of global freshwater production capacity. It is noteworthy that most of the desalination facilities are located in countries with relatively high incomes.

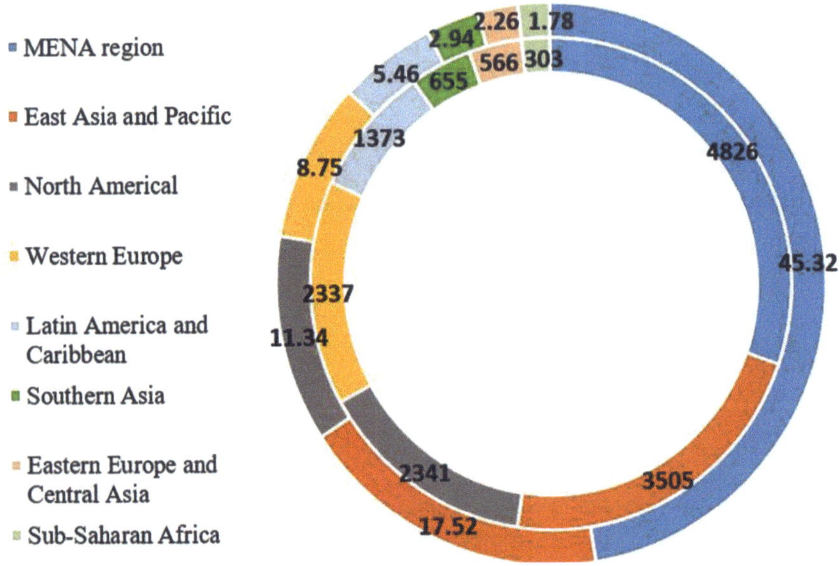

Fig. 1.4 ***Data on freshwater production facilities worldwide, broken down by geographical area: freshwater production capacity (in million m^3/day) (outside ring); and number of freshwater production facilities (inside ring).*** *Data from Jones et al. (2019).*

The municipal sector is the largest freshwater user, consuming 62.3% of the total freshwater produced by desalination plants, while the industrial sector consumes 30.2% of the produced freshwater (Jones et al., 2019). The electrical and irrigation sectors, respectively, consume 4.8% and 1.8% of the total freshwater production from desalination plants (Jones et al., 2019).

Among desalination technologies, reverse osmosis (RO) systems are the most widely used, accounting for about 69% of installed desalination facilities. The next most common thermal desalination technologies are multistage flash (MSF) and multieffect desalination (MED), contributing 18% and 7%, respectively, of the total number of operational desalination plants around the world (Jones et al., 2019). In general, industrial desalination systems can be categorized as thermal (or phase change) processes such as MSF, MED, and mechanical vapor compression (MVC) and membrane-based (or single phase) process such as RO. Fundamentals and models of industrial desalination technologies are presented and discussed in Chapters 2 and 5.

1.2 History of desalination

In this section, the history is presented of the development and commercialization of currently used desalination systems. The history of desalination technologies is divided into three categories: MSF, MED, and MVC; solar stills and humidification–dehumidification (HDH); and membrane-based desalination systems, including RO and membrane desalination (MD).

1.2.1 History of MSF, MED, and MVC desalination systems

Historically, one of the first practices of desalination was on ships, which carried on to as late as 1800 (El-Dessouky and Ettouney, 2002). Single-stage stills were used in a batch mode to evaporate seawater with heat supplied by cock stoves. In such systems, condensation heat was not recovered. Instead sponges were used to absorb the material that evaporated.

The evaporation process developed further with progress in the sugar industry, leading to stills with enlarged scales and improved performance.

In the early of 20th century, the desalination industry progressed markedly. In Egypt, a six-effect desalination plant was installed, with a freshwater production capacity of 75 m^3/day (El-Dessouky and Ettouney, 2002). Then, between 1929 and 1937, freshwater production capacities further increased, corresponding to the advent of the oil industry. Further improvements in and enlargements of desalination systems occurred from 1935 to 1960, a period that saw an annual freshwater production growth rate of 17% (El-Dessouky and Ettouney, 2002).

The historical development of thermal desalination systems in various regions is now described.

In 1957, a four-stage flash desalination plant (i.e., an MSF unit) was constructed by Westinghouse company in Kuwait (Al-Wazzan and Al-Modaf, 2001). At the same time, a patent by Silver led to a significant improvement in the performance of MSF systems, in which the number of stages was about three times the performance ratio (i.e., the ratio of freshwater produced to the steam used as a heat source to drive MSF system) (Al-Wazzan and Al-Modaf, 2001). In the Silver configuration, there was a considerably smaller specific heat transfer area for condensation compared to the Westinghouse configuration. This resulted in a significant reduction in the capital cost of plant, as the condenser tubes, which have a high cost, were substituted by inexpensive partitions in the MSF system.

In 1960, an MSF plant with 19 stages was installed in Shuwaikh, Kuwait. The freshwater production capacity was 4550 m^3/day and the performance ratio was 5.7 (Al-Wazzan and Al-Modaf, 2001).

In 1959, the first antiscaling material known as Shuwaikh Mix or SALVAP was tested in a thermal desalination plant in Kuwait (Al-Falah and Al-Shuaib, 2001). Utilization of this polyphosphate-based material increased the operation time from 200–600 h to 8000 h. The operation of a desalination plant using Shuwikh Mix was limited to a top brine temperature (*TBT*) of 95 °C. Although scale formation was successfully eliminated, the desalination process operation exhibited some problems with performance ratio and productivity.

During 1960–1969, an improved MSF plant with 40 stages was installed in the Channel Islands, with a freshwater production capacity of 6800 m^3/day (Al-Falah and Al-Shuaib, 2001). The performance ratio of this plant was enhanced from 10 to 12.5, representing the highest known performance ratio for an MSF desalination system. Furthermore, during this period of time, several developments were achieved such as acid cleaning, improved antiscaling chemical additives, feed deaeration, and cogeneration plant designs for seawater desalination.

In 1970–1979, advances were made in the construction and operation of desalination plants, in prevention and control of corrosion, and in chemical treatment. Furthermore, large desalination plants capacities were constructed in the 1970s in Japan, capable of freshwater production rates of 22,500–25,500 m^3/day (El-Dessouky and Ettouney, 2002; Maciver et al., 2005).

In 1980–1989, a polymer-based antiscaling material was introduced which had advantages over polyphosphate-based materials, such as higher TBT operation (around 110 °C), which resulted in a higher freshwater production rate (27,200 m^3/day) and performance ratio (about 8.65). Also, primary low-temperature single and multiple effect evaporation desalination processes were introduced in the 1980s. An MVC process is used in the single-effect desalination system, and a thermal vapor compression process in MED. The relatively low-temperature operation of single and MED systems allows for the use of inexpensive aluminum alloys and leads to reduced scaling phenomena during desalination.

Further enhancements and improvements in both MSF desalination and MED systems were made since 1995, in which a plant factor of 90% and continuous plant operation of 2–5 years were attained.

From 1990 to the present, there has been a tendency to shift from MSF desalination systems to MED systems. For instance, from 2001 to

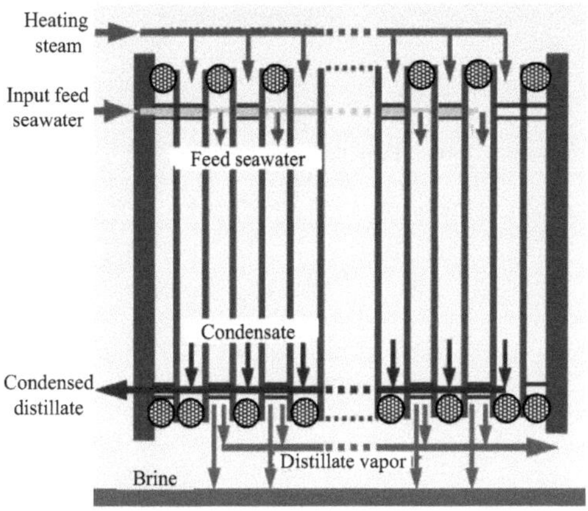

Fig. 1.5 Left: Pressed plate falling film configuration used in the effect of the MED system (modified from Maciver et al. (2005)). Right: Plate pack arrangement in the evaporator of the MED system installed in Japan *(from Maciver et al., (2005))*.

2003, two MED units with a thermal vapor compressor (TVC) replaced an aging MSF system in Japan (Maciver et al., 2005). Each MED unit was able to produce 100 m^3/day freshwater, and consisted of four effects, three tubular type preheaters and a tubular condenser within each evaporator. Titanium-pressed plates were used in the evaporators (Fig. 1.5). For each plate, falling-film evaporation occurs on one side and condensation on the other side. The steam enters from the top on the condensation side and condenses while moves down the plate and collected in the distillate headers on the bottom. The feed seawater flows down on the evaporation side of the pressed plate, absorbs the latent heat of condensation of the steam, and partially evaporates. The remaining part of the feed (brine stream) moves down and collects on the bottom. The generated steam is transferred into the next effect where it condenses and releases its latent heat. The plant was placed outdoors but, to prevent any effects due to weather changes between summer and winter, the plant was insulated completely. Fig. 1.6 shows a photograph of the MED plant installed in Japan.

Between 1960 and 1980, most of the large capacity thermal desalination systems installed in the Middle East were of the MSF kind. Although MSF desalination plants exhibited large freshwater production capacities, using a TVC in MED desalination plants led to energy efficient and competitive thermal desalination plants. The TVC module in MED plants increases freshwater production capacity, reduces brine temperature, and reduces the amount of steam heat source used to drive the MED unit. MED-TVC

Fig. 1.6 ***View of MED desalination plant built in 2003 in Japan.*** *Source: (Maciver et al., 2005).*

desalination plants are thermodynamically efficient owing to their lower operation temperatures and lower pressure drops at high vapor flow rates. In recent decades, issues surrounding fossil fuel resources, their costs, and environmental impacts have led to significant developments in cogeneration plants. But, the capability of MED-TVC plants to be driven by low-cost/low-grade energy or waste heat has often made them more favorable than MSF desalination systems.

1.2.2 History of solar still and HDH desalination systems

The idea of dehumidification of humid air comes from natural/artificial caves where the cool stones act as a dehumidifier to condense the moisture in humid air (Seifert et al., 2013). The solar still can be treated as a predecessor of the HDH desalination process. Solar stills are considered one of the first human-made desalination technologies, dating back to the 19th century. In a conventional solar still, a basin is filled with seawater or brackish water. The basin is commonly painted black to increase solar energy absorption. Fig. 1.7 shows a solar still desalination system. A transparent tilted cover (typically glass or plastic) is used to close the basin and form a greenhouse effect, creating a heat trap. Solar radiation passes through the glass and heats the seawater, vaporizing a part of it. The humid air carrying

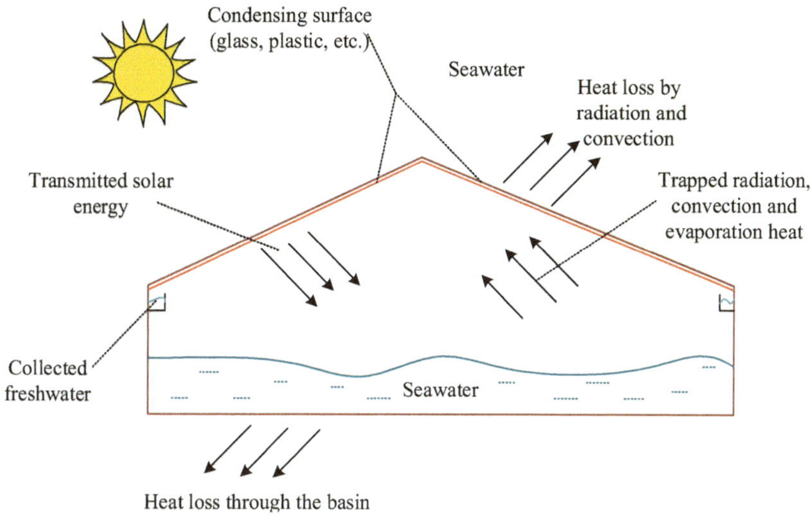

Fig. 1.7 ***Solar still desalination system.***

the produced vapor is condensed on the inner surface of the glass, which has a lower temperature than the seawater in the basin. On the surface of glass, the latent heat of condensation is transferred to the air. The condensed distillate drops and collected. In fact, in solar stills the area close to the surface of the seawater acts as a humidifier and the glass surface acts as a dehumidifier.

In 1872, a noteworthy solar still was built in Las Salinas, Chile (Delyannis, 2003). The solar still had a cover area of 4700 m^2 and a freshwater production capacity of 4.9 kg/m^2 of glass surface area. The plant operated for more than 40 years (Delyannis, 2003; Mahian et al., 2015). As noted earlier, HDH desalination processes originated from solar still desalination systems. The thermal inefficiencies in solar stills, such as those due to natural convection and infrared radiation from the warm brine to the condenser surface, reduce the brine temperature and result in the need for a large solar collector area for producing freshwater. These thermal inefficiencies also lead to a low gained output ratio (GOR), which is defined as the ratio of the latent heat of evaporation of the freshwater produced to the total heat input. As an alternative, using separate components for each thermal process leads to better thermal efficiencies. HDH desalination systems were designed to allow for better flexibility in thermodynamic design of the vaporization and condensation processes. Consequently, main advantages of HDH over solar still desalination are a relatively higher GOR and a lower required area for freshwater production.

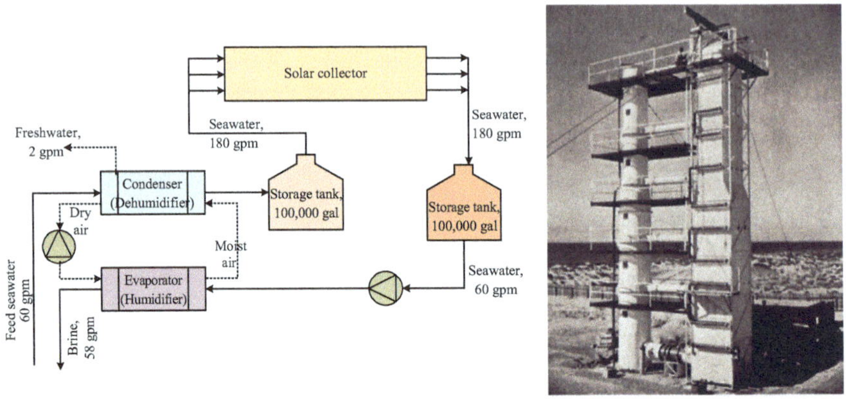

Fig. 1.8 ***Early solar driven humidification-dehumidification desalination plant in Puerto Peñasco, Mexico.*** *From Hodges et al. (1964).*

During 1962–1965, University of Arizona investigated the technical feasibility of a small pilot plant for a solar multieffect HDH desalination, which was subsequently built in Puerto Peñasco, Mexico (Hodges and Kassander Jr, 1962; Hodges and Thompson, 1966). The solar collectors used in the HDH desalination system were of the double-glazed type, with two plastic covers and a small air gap between them to reduce convection heat loss. The collectors have a surface area of 292 m^2 and include a basin slightly filled with seawater. Fig. 1.8 shows a flow diagram and view of the constructed solar HDH desalination system in Puerto Peñasco, Mexico. Seawater is heated in the collectors to 65 °C and stored in the storage tank, which enables 24-h operation. Then, the seawater is pumped from the storage to the evaporator of the evaporation/condensation tower located at a height of 13 m. The evaporation/condensation tower comprises two compartments, one for evaporation of seawater (the humidifier) and another containing finned aluminum condensing tubes (the dehumidifier). The heated seawater enters the top of the evaporator and heats and humidifies air rising from below. The saturated air moves from the evaporator to the condenser compartment and the latent heat of condensation preheats the feed seawater entering at 25 °C. The plant was designed to yield 18 m^3/day of freshwater in the summer.

1.2.3 History of membrane-based desalination systems

The initial idea of using membranes in desalination systems was drawn from biological membranes such as skin, lungs, and cell membranes. The elementary separation process is based on the difference in particle size (e.g., separation of fine grains from coarse grains). Furthermore, the membrane-based separation process was developed based on differences in diffusion rates of diverse constituents through the membranes. The concepts of osmosis (solvent passage through a membrane from low to high concentration) and dialysis (solvent passage through a membrane from high to low concentration) were explained by Dutrochet (1995), Graham (1858), and Fick (1995,1855). The concept of using seawater as a source for freshwater was introduced by Hassler (1949,1950), and expanded on subsequently (Hassler and McCutchan, 1960), but his idea was not successful in practice. Nonetheless, other scientists carried on with research following Hassler's concept, leading to the development of present-day membranes.

Reid and Brenton (1959) introduced early cellulose acetate RO membranes in 1959 at University of Florida. A permeator was developed to measure the semipermeability of RO membranes. The permeator is a container divided into two by a semipermeable membrane (Fig. 1.9). In a permeator, pressure is applied via compressed air to the feed solution. The pressurized feed passes through the membrane and is collected at the

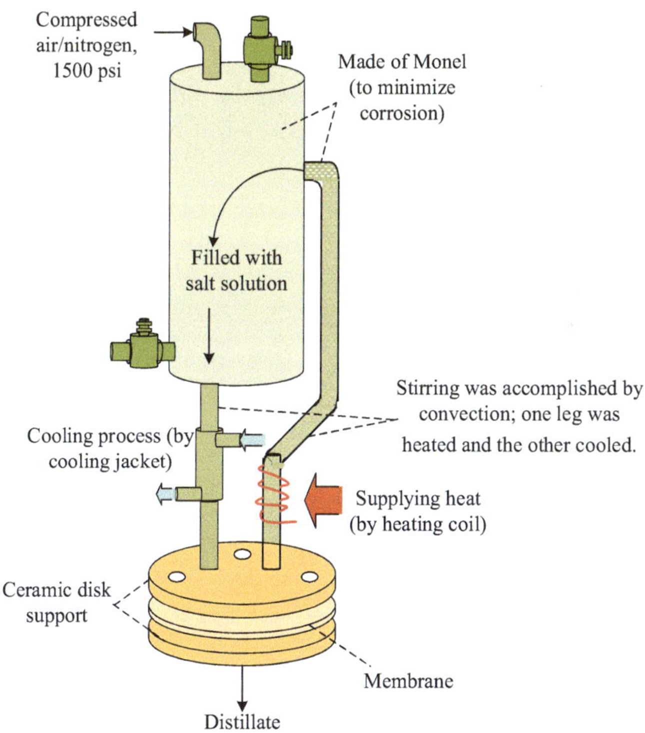

Fig. 1.9 ***Apparatus for measuring the semipermeability of imperfect osmotic membranes.*** *Modified from (Reid and Breton, 1959).*

bottom of the permeator, and then its constituents are determined. The salt rejection percentage was used as a measure of semipermeability. Although a salt rejection of 98% with 0.10M sodium chloride was obtained at pressures up to 5860.5 kPa, the permeate flux value was very low (around 0.5–4 μL/ cm^2h) (Reid and Breton, 1959). To address this issue, the cellulose acetate RO membrane was subsequently modified by Loeb and Sourirajan (1963, 1960) to produce practical amounts of permeate while maintaining high-levels of salt rejection.

A major breakthrough in membrane technology was achieved when Loeb and Sourirajan (1960,1963) made a cellulose acetate polymer membrane that was homogeneous but physically asymmetric. Fig. 1.10 shows an early RO desalination system developed at the University of California, Los Angeles by Loeb and Sourirajan (1960,1963). In preparation of the casting solution, aqueous magnesium perchlorate was used, which acted as a pore former (or swelling agent). The presence of a swelling agent was found to be necessary, as water molecules could be absorbed into the polymer to form sufficiently large pores for a practical RO desalination process. The first practical RO membranes, which exhibited high salt rejection and permeate flux rates due to the porous structure in a thin surface, resulted from the efforts of Loeb and Sourirajan. In the 1980s, cellulose-based RO MD systems were used for industrial processes. Nonetheless, there were some drawbacks associated with these types of membranes, which fostered research into new types of membranes.

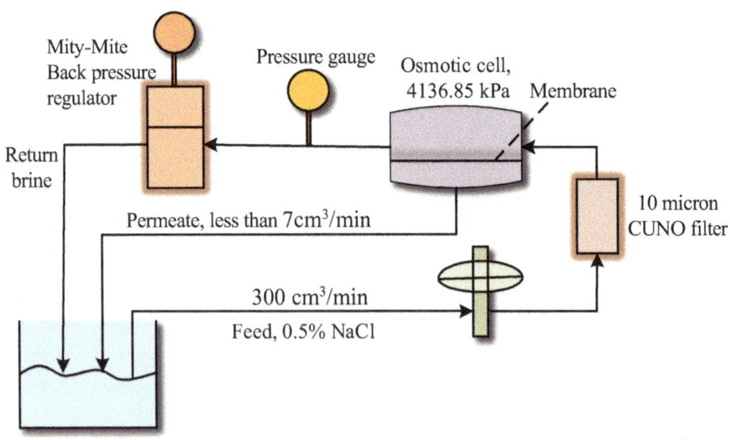

Fig. 1.10 ***Early reverse osmosis desalination system developed at University of California, Los Angeles by Loeb and Sourirajan (1960,1963).***

In the mid-1960s, two chemical companies, Dow and Dupont, focused on the development of hollow fiber MD modules. Dow focused on cellulose acetate fibers (Mahon, 1966), while Dupont was involved with polyamides. In 1985, Dupont developed a hollow fiber membrane from aromatic polyamide. Note that early studies on aliphatic compounds (Richter et al., 1971) in part led to the development of aromatic polyamide hollow fibers. In1971–1973, aromatic polyaramid hollow fine fibers known as Permasep B-9 and B-10 permeators were introduced by Dupont for brackish water and seawater desalination, respectively (Eckman et al., 1994; Richter et al., 1971). After 20 years, these two permeators were commercialized and became standards for freshwater production with appropriate quality. In1994, a new double bundle permeator known as TWIN was developed and commercialized by Dupont based on optimized bundle design and flow patterns (Eckman et al., 1994). Fig. 1.11 shows the flow pattern of the TWIN permeator. The permeate flow is increased by increasing the length of fiber, while reducing the length of fiber lowers costs and reduces salt passage (Eckman et al., 1994). The high performance of these polyamide membranes encouraged their adoption by industries over cellulose-based membranes.

In 1961, a tubular membrane configuration was developed for casting membranes inside cylindrical fiberglass tubes (Glater, 1998). The tubular

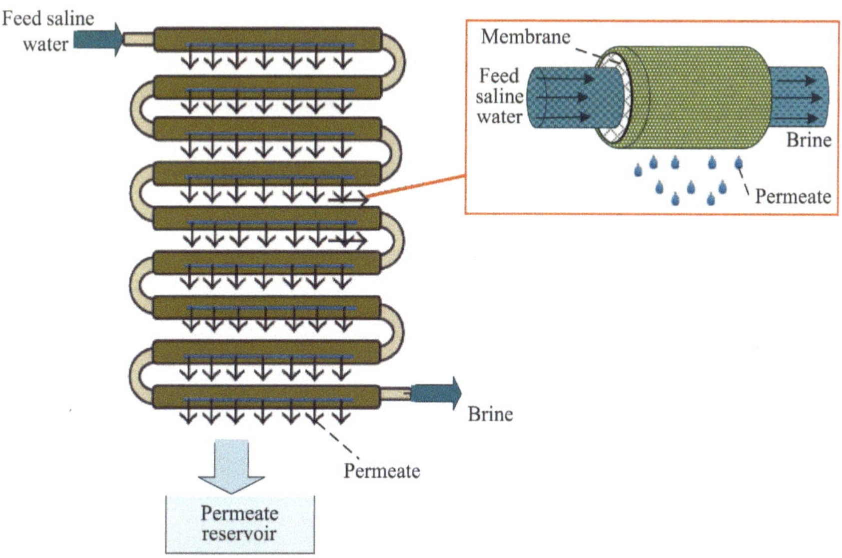

Fig. 1.11 ***Tubular membrane filtration module.***

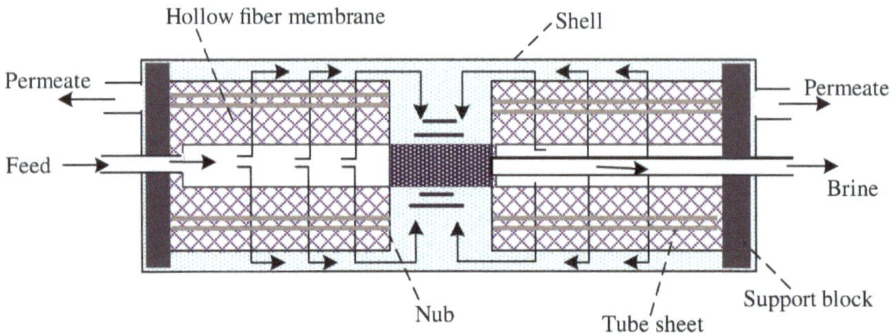

Fig. 1.12 ***Flow pattern of hollow fiber TWIN permeator developed and commercialized by Dupont in 1994.*** *Based on Eckman et al. (1994).*

configuration served two functions: a tube surface to support the membrane and a tube wall to serve as a pressure vessel. The membrane is located at the inner side of the tube and the saline water flows inside the tube. Permeate passes across the membrane and reaches the surface of the tube. The tubes are porous to allow permeate transfer from the inside to the outside of the tube. Fig. 1.12 shows a tubular membrane used in RO desalination. Tubular membranes were modified later by Loeb (1966), who used techniques such as solution casting, membrane tube casting, wrapping, applying seals and end fittings and heating of the assembly to fabricate a composite tubular assembly for RO desalination. The methods used by Loeb were previously developed at the University of California, Los Angeles for flat sheet acetate cellulose membranes.

Around 1975, a spiral wound polyamide thin film composite membrane was introduced (Channabasappa, 1975). Enhancements of this membrane continued until 1989 for higher salt rejection and higher permeate production. Fig. 1.13 shows a simplified schematic of a spiral wound reverse

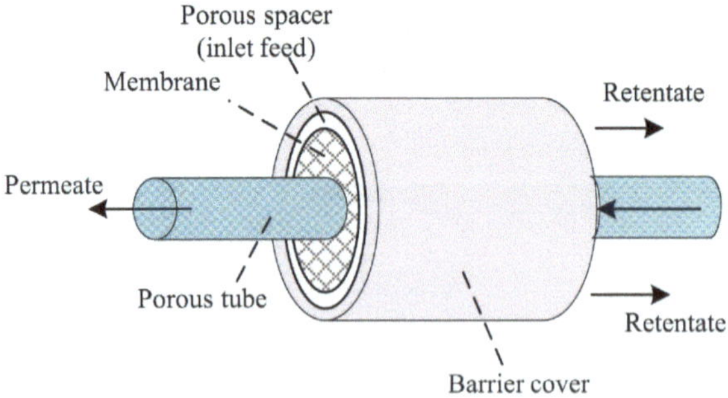

Fig. 1.13 ***Simplified schematic of a spiral wound reverse osmosis membrane.***

membrane. It consists of number of membrane envelopes attached with a porous spacer wound around permeate collection tube. By 2000, both spiral wound and hollow fiber RO membrane configurations were commercialized and used for large-scale seawater desalination, permitting the production of more than two billion gallons of freshwater per day.

Due to the high capital cost of tubular configuration of RO membrane systems, they are used only for specific applications such as industrial waste treatment. Nowadays, most RO desalination plants employ polyamide membranes rather than tubular and spiral wound membranes. This is mainly due to the lower specific energy consumption (i.e., energy use per unit freshwater production) of polyamide membranes compared to other types of membranes. For current RO membranes, the energy consumption typically is 2–5 kWh/m^3. Nonetheless, research to develop a better performing, chlorine- and fouling-resistant membrane carries on.

1.3 Review of sustainable energy technologies

Renewable energy can be defined as naturally persistent and repetitive flows of energy that occur in the environment. Renewable energy technologies are devices for harvesting energy from such energy flows. In contrast, nonrenewable energy resources (sometimes referred to as brown energy resources) are finite in extent, usually in static underground stores that can be released by human interactions.

Sustainable development is a concept suggesting that human society should live in a way that permits it to meet present needs without impeding the ability of future generations to meet their needs. Many factors relating to energy affect sustainable development, and energy-actions such as using sustainable energy resources and improving energy efficiency can contribute to achieving it (Rezaie and Rosen, 2020; Rosen, 2021a). Reliable energy supplies are necessary for all countries for tasks such as heating, lighting, industrial processing, transportation and communications. In the United States, 6.2% of gross national product is associated with purchases of energy (CSS, 2020). In developing countries, however, the cost of importing fossil fuels can amount to a significant portion of the total value of exports, up to half in some cases. The development of such developing countries often is unsustainable. In such cases, supportive policies are required for establishing or shifting toward sustainable development (Alvarez-Risco et al., 2021).

The Intergovernmental Panel on Climate Change (IPCC) is a United Nation body established in 1988 to provide regular scientific based assessments of climate change, its impacts and future risks, and of approaches for adaptation and mitigation. The IPCC does not carry out its own research, but instead presents consensus and agreements in scientific communities so as to provide a credible information source for policymakers. The IPCC is currently preparing its sixth assessment report (IPCC, 2020). The following is an IPCC statement on the sustainability of renewable energy resources (IPCC, 2011):

> *"For a renewable energy resource to be sustainable, it must be inexhaustible and not damage the delivery of environmental goods and services including the climate system. For example, to be sustainable, biofuel production should not increase net CO_2 emissions, should not adversely affect food security, nor require excessive use of water and chemicals, nor threaten biodiversity. To be sustainable, energy must also be economically affordable over the long term; it must meet societal needs and be compatible with social norms now and in the future. Indeed, as use of RE technologies accelerates, a balance will have to be struck among the several dimensions of sustainable development. It is important to assess the entire lifecycle of each energy source to ensure that all of the dimensions of sustainability are met."*

Given the expectation of continued increases in energy needs in the future, limited energy resources such as fossil fuels are not sustainable over the long term. Based on the reserve lifetime of a resource, which is defined as the ratio of known accessible amount of the resource to its current rate of use, the lifetime of coal (the most abundant fossil fuel) reserves is a few centuries and of oil and gas are a few decades (EIA, 2020). It is expected that, as the reserve lifetime of a fossil fuel shortens, the corresponding price of the fossil fuel will increase, in many cases significantly. Therefore, alternatives to fossil fuels need to be introduced and developed to respond to the need for energy, and efforts are required make fossil fuel reserves last longer.

In addition, emissions from the combustion of fossil fuels (specifically CO_2) to the atmosphere are agreed by most of the technical community to be causing the greenhouse effect and leading to global climate change. These emissions can have major effects on water supplies, food production, and society. Greenhouse gas (GHG) emissions contributed to two-thirds of all emissions over the last half century (IAEA, 2020). To meet climate goals under the Paris agreement, emissions from transportations, electricity production, building, and industry are required to be eliminated by 2050, but they have almost double over last 50 years (IAEA, 2020).

In contrast, renewable energy resources add little or no GHGs to the environment when used for supplying energy. The manufacture of renewable energy facilities emits some GHGs, but the magnitude is very small compared to emissions from fossil fueled plants. According to the International Atomic Energy Agency (IAEA), the increasing use of nuclear and renewable energy resources leads to "deep decarbonization" and helps to meet worldwide climate goals (IAEA, 2020). Expansion of energy production from nuclear and renewable energies can reduce GHG emissions in all sectors of the economy.

Fig. 1.14 shows variations in power sector CO_2 emissions, broken down by energy form, for 2010 to 2020 (IEA, 2020a). It is seen that CO_2 emission declined by 3.3% in 2020 which is the largest decrease on record. Although the COVID-19 pandemic reduced electrical power demand notably in 2020, the rapid increase in the share of renewable energy in the electrical power-generation sector is the main reason for lower CO_2 emissions. In fact, the share of renewable energy resources in global electrical power production increased from 27% in 2019 to 29% in 2020, which is the highest annual rise on record (IEA, 2020a).

The sustainable development goals established by International Energy Agency (IEA, 2020b) can be achieved in large part by a "sustainable development scenario" in which a major transformation of the global energy system occurs that provides universal access to energy, controls climate change, and reduces the health impacts of air pollution, simultaneously. Fig. 1.15 shows worldwide electricity generation by power sector type from 1980 to the present and predicted out to 2050. It is seen that hydropower, solar photovoltaic (PV), and onshore wind are expected to be the greatest contributors to electricity generation by 2050. Note that this figure does not account for the effect of the COVID-19 pandemic, which has slowed or cancelled some renewable energy projects in 2020. The global capacity of solar PV is anticipated to grow by about 107 GW in 2020, indicating stable growth (about 18%) from 2019 (IEA, 2020c). Solar PV utilization is on track to meet the sustainable development scenario level by 2030, which requires a 15% increase annually in electricity generation with that technology, corresponding to an increase from 720 TWh in 2019 to nearly 3300 TWh in 2030 (IEA, 2020c).

Electricity generation from onshore wind plants increased by 12% in 2019, but more extensive efforts are required through 2030 to get on track with the sustainable development scenario (IEA, 2020d). For the sustainable development scenario, the yearly net capacity additions for onshore wind plants need to increase from 55 GW in 2019 to 108 GW in 2030 (IEA, 2020e).

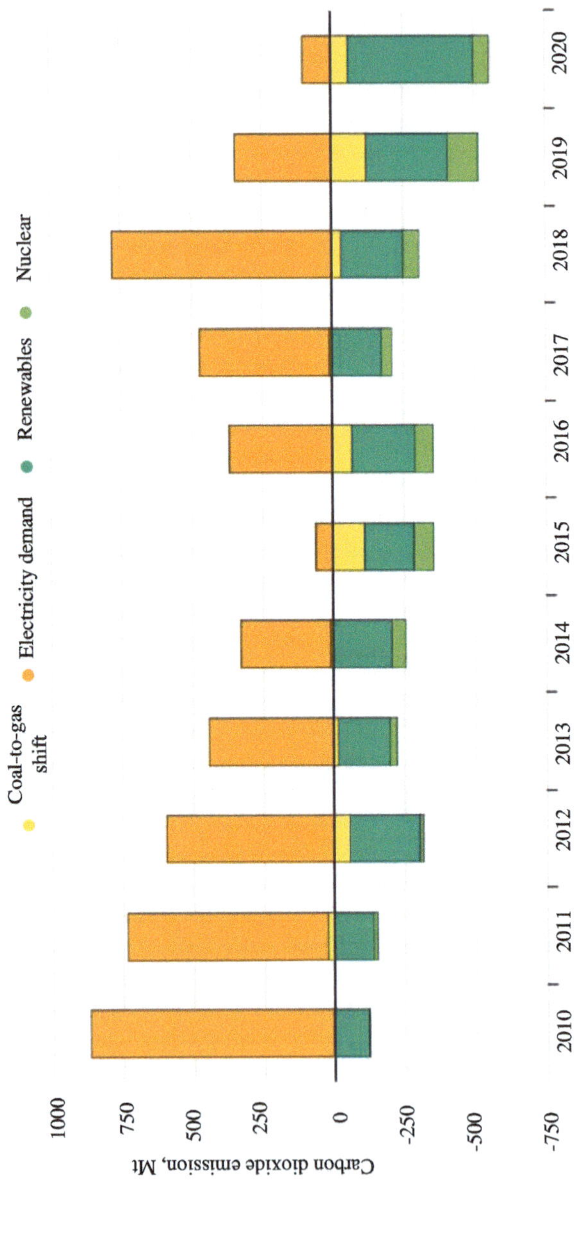

Fig. 1.14 *Variations in electrical power sector emissions by energy type from 2010 to 2020*. *Source: IEA(2020a), Change in power sector emissions by effect.*

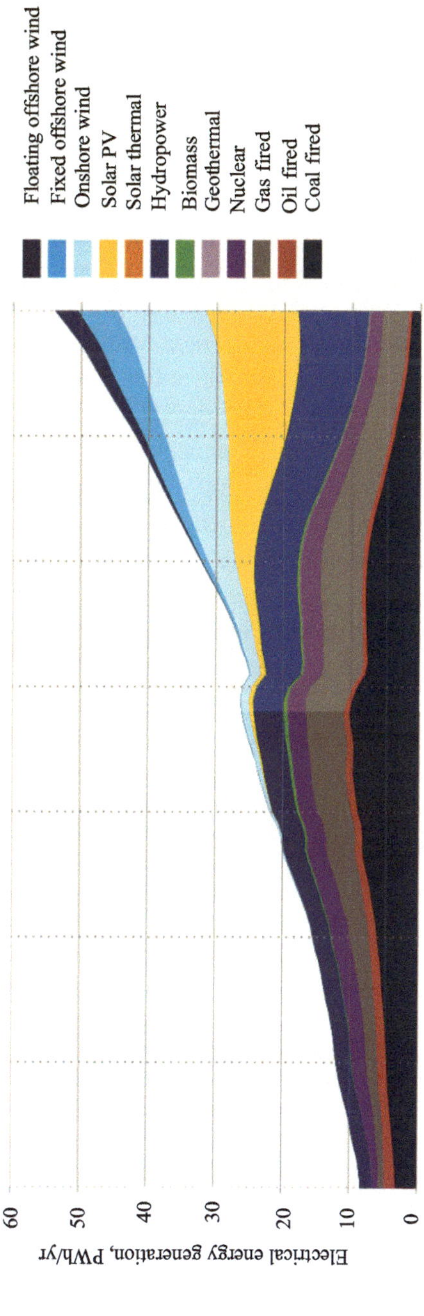

Fig. 1.15 Global electricity generation, by electrical power generation technology, from 1980 to 2050. *Data source (IEA, 2020g; IRENA, 2018).*

The capacity for hydropower was 12.7 GW in 2019, 45% lower than in 2018 due to the decreases in China and Brazil (the largest hydropower using countries). Hydropower nonetheless remains the biggest contributor to electricity generation among renewable energy technologies (IEA, 2020f). The current hydropower-generation capacity needs to increase to meet the sustainable development scenario, which requires hydropower generation to rise 3% per year to 2030 (IEA, 2020f).

Sustainable energy engineering methods are increasingly being used in engineering areas to design, develop, and promote sustainable energy-based systems. A key facet of that activity is the use of energy sources in a way that is sustainable, irrespective of the nature of the resource (renewable or nonrenewable). Advanced and practical engineering thermodynamic methods such as exergy analysis can be employed as the basis of or to support sustainable energy engineering. Other disciplines, such as environmental engineering, chemical and physical sciences, policy and economics, also need to contribute to successfully develop sustainable energy systems. Some approaches such as paying an energy penalty for consuming fossil fuels as well as CO_2 capture and sequestration are currently being employed for mitigating the effects of global climate change.

The ultimate primary sources of renewable energies are solar radiation, geothermal heat, the motion and gravitational potential of the moon, sun, and earth (Koohi-Fayegh and Rosen, 2020a). Other renewable energies stem from these ultimate primary sources. For instance, tidal energy is generated from planetary motion and the gravitational forces associated with the earth–moon and earth–sun systems. Many renewable energy forms, such as wind, biomass, and hydro, are derived from solar energy.

Fig. 1.16 shows the flows of the ultimate primary sources of renewable energy on the earth. The dominance is clearly seen of solar radiation compared to the two other primary renewable sources (geothermal and spinning and gravitational forces). Of the 120,000 TW energy rate from solar radiation which enters the atmosphere of the earth, $1.6 \times 10^{-8}\%$ of that is sufficient to supply the electrical power requirement for about 50,000 people. In the following sections, primary sustainable energy sources are described in greater detail.

1.3.1 Solar energy

The sun is the source of solar energy and is mainly composed of hydrogen gas. The average sun-earth distance is 150 Gm (i.e., 150 million km) (NASA, 2007) and the average temperature of the sun is 5500 °C (Sharp, 2017).

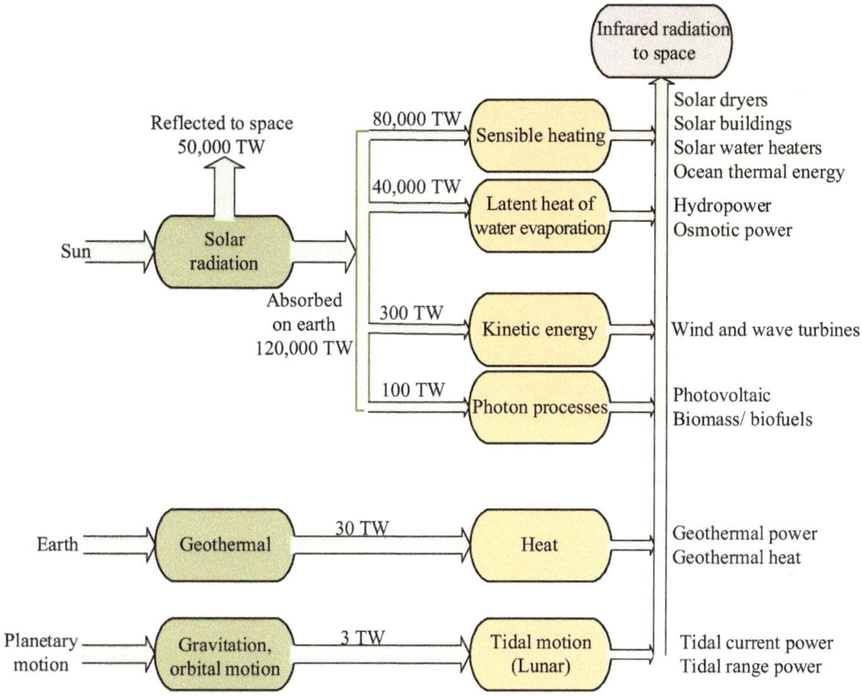

Fig. 1.16 ***Flows of ultimate primary energy sources (solar radiation, geothermal heat, and the planet's spinning and gravitational forces) into and from the earth, and sample applications.*** *Data extracted from Dincer and Zamfirescu (2011).*

A broad spectrum of electromagnetic radiation forms the sunlight that reaches the terrestrial atmosphere, after a transit time of 8 min and 20 s. Almost all of the life on the earth, as well as the climate and weather, are driven by the energy of solar radiation.

To understand solar energy better, the earth can be considered as a heat engine operating between $T_{Earth,H}$ and $T_{Earth,L}$ (Fig. 1.17). In this system, the earth can be considered as a closed thermodynamic system with two boundaries. One boundary is heated by the sun and has a surface area of A_H and temperature of $T_{Earth,H}$. The other boundary is a cold region of area A_L covering two polar zones with a temperature of $T_{Earth,L}$; those regions are cooled through radiation heat transfer with the universe. In simple terms, solar radiation is delivered from the sun to the earth and rejected to the sink.

If the solar radiation temperature is T_{Sun} = 5500 K and the average temperature of the earth is $T_{Earth,H}$ = 300 K, the radiation heat rate input $\dot{Q}_H$ to this system is proportional to $A_H({T_{Sun}}^4 - {T_{Earth,H}}^4)$ or

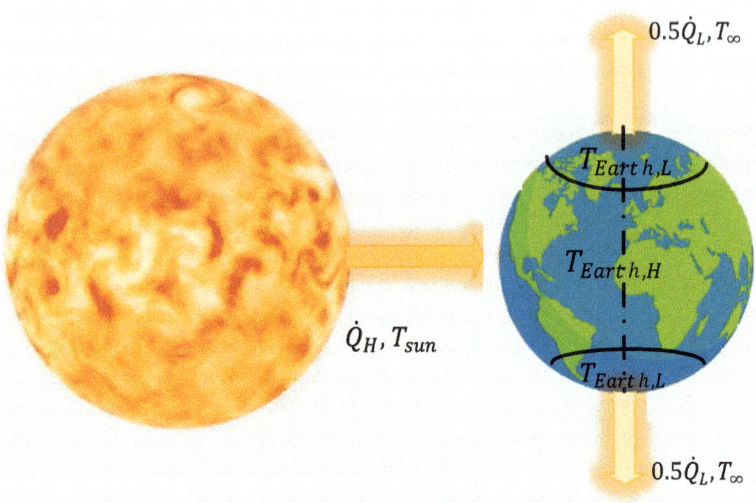

Fig. 1.17 ***Sun-earth heat engine for modeling solar energy conversion.***

$A_H((5500\ \text{K})^4 - (300\ \text{K})^4)$. As $T_{Earth,H} \ll T_{Sun}$, the radiation heat rate can be approximated as $\dot{Q}_H \sim A_H T_{Sun}^4$ or $\dot{Q}_H \sim A_H (5500\ \text{K})^4$. The heat rate output $\dot{Q}_L$ from the earth to the universe can be approximated in an analogous way, using the universe background temperature (i.e., actual temperature of the cosmic microwave background, $T_\infty = 2.725$ K) and $T_{Earth,L}$. Neglecting T_∞ compared to T_L gives $\dot{Q}_L \sim A_L T_{Earth,L}^4$.

If the energy efficiency of a heat engine is defined as $\eta = (\dot{Q}_H - \dot{Q}_L)/\dot{Q}_H$, a value of almost unity would be obtained for the efficiency (i.e., $\eta = 0.99$), mainly owing to the high temperature of the sun compared to the earth surface temperature. The energy efficiency for this heat engine of almost unity indicates a high efficiency of solar radiation conversion to work. The atmosphere and hydrosphere flows such as winds in the atmosphere and oceanic currents in the hydrosphere consume a major part of the energy received by the sun.

The distribution of incident solar energy entering the earth and its atmosphere is shown in Fig. 1.18. The percentages in Fig. 1.18 are based on the incident shortwave solar radiation. It is seen that about 47% of the incoming shortwave solar radiation is absorbed by land and oceans. The rest of the incoming shortwave solar radiation is reflected by clouds (17%), the surface of the earth (6%), and absorbed by dust, water vapor, and ozone molecules in the atmosphere (19%) and by clouds (4%).

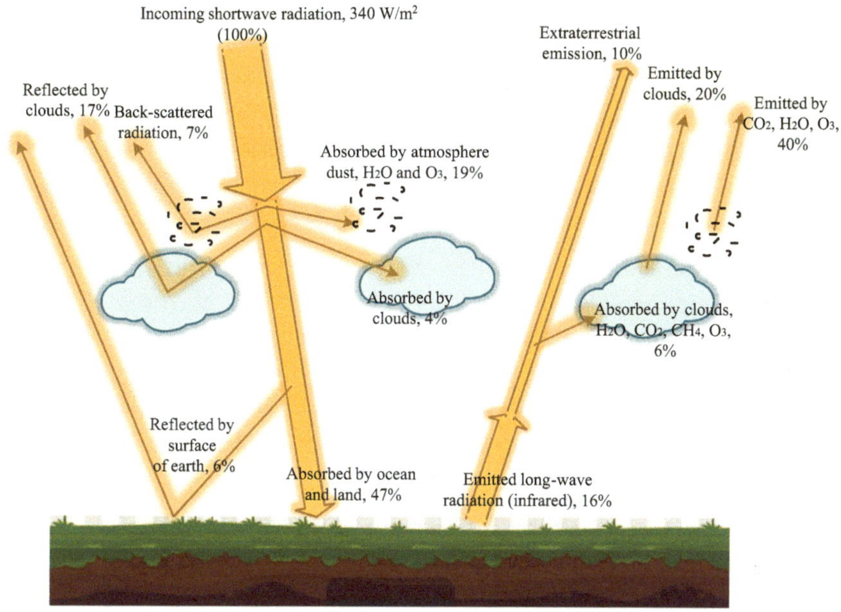

Fig. 1.18 ***The earth-atmosphere energy balance, showing incoming energy flux from the sun balanced with outgoing energy flux from the earth.*** The percentage values are relative to the incident shortwave solar radiation flux, which is 340 W/m^2 *(Tiwari and Ghosal, 2007; IPCC, 2013).*

A part of the solar radiation entering the earth's atmosphere is emitted at the average surface temperature of the earth (about 300 K and 10 μm wavelength in the infrared spectrum). From Fig. 1.18 the emitted solar radiation from the land and ocean surface is seen to be 16% of the incident solar radiation. About 10% of this amount is transferred to extraterrestrial space and the remainder is absorbed by various atmospheric gases and clouds. The infrared radiation emitted by the earth heats atmosphere through conduction, radiation, and water evaporation. About 20% of incident solar radiation is emitted to extraterrestrial space by clouds and about 40% is emitted by water vapor, CO_2, and ozone molecules.

From Fig. 1.18, it can be observed that the summation of reflected radiation percentages, including those for both long- and short-wave types, is equal to 100%. A comparison with the solar heat engine in Fig. 1.17 suggests that the work generated by the solar heat engine is completely converted back to heat. In fact, the generated work is completely destroyed as it transferred beyond the terrestrial atmosphere. This balance indicates that no net energy is gained by the earth, but this incident solar radiation and emission to the terrestrial atmosphere is necessary to maintain a constant temperature of the earth.

Distribution of solar energy received by the earth: Many processes can be driven by the solar energy incident on the earth. About 43% of the incident solar energy, consisting of the direct *beam* radiation, heats the surface of lands and oceans. Harvesting all of this solar beam radiation flux (146.2 W/m^2) can potentially generate 1000 times more electrical power than the global power consumption, demonstrating the abundance of solar energy on the earth. Currently, solar beam radiation is used by several technologies for various purposes:

- PV systems that convert incident solar energy to electricity.
- Concentrating and nonconcentrating solar collectors that collect solar radiation in the form of thermal energy for use in space/water or process heating, desalination, electrical power generation, etc.
- Solar ponds in which the temperature difference at the top and bottom of pool is the result of thermal stratification. The surface of pool is exposed to the solar radiation and is heated to 90 °C. The temperature difference between top and bottom of pool can be used for space heating, electricity generation, or desalination.
- Ocean thermal energy conversion in which the temperature difference between the surface and deep waters in the ocean is used to drive a heat engine for production of electricity or synthetic fuels.

In addition, about 22% of incident solar radiation flux (i.e., 74.8 W/m^2) is used by the hydrological cycle, in which the solar energy heats water in the oceans and seas, evaporating a portion of it. The generated vapor rises in the atmosphere and forms clouds, and subsequently condenses through rain and snow and falls to the earth. By this natural cycle, a significant amount of water is transferred from the low-elevation regions to relatively high-elevation regions, creating the *hydro energy*. The energy of water flows over water falls or running in rivers can be harvested for useful purposes such as electricity production. In ancient times, hydraulic energy from watermills had been used for irrigation. Then, in the late 19th century, hydropower began to be used for electricity production. The first commercial hydroelectric power plant was constructed at Niagara Falls.

About 0.21% of the incident solar radiation flux produces *wind* (around 0.71 W/m^2). Thus, wind energy is an indirect form of solar energy. Wind is caused by nonuniform heating of atmosphere by the sun and the rotation of the earth. The kinetic energy of wind can be converted to electrical energy as it flows through a wind turbine. The electricity produced from distributed wind energy can be used for homes, farms, etc. There are two main kinds of wind power plants: land-based (or onshore) and offshore. Details on these follow:

- *Land-based (onshore) wind energy*: Wind turbines can be connected to a utility power grid or used in stand-alone applications. A large number of wind turbines can be constructed close together so as to form a wind plant or wind farm. The electricity produced is transferred to power grids by power system operators or electric utilities. Utility-scale wind turbine plants are capable of electrical generation of more than 100 kW, while distributed wind turbine plants are typically relatively smaller, with electrical generation capacities under 100 kW. The attributes of distributed wind turbine plants include their small and modular power-generation systems. The latter attribute provides the option for them to be integrated to improve operation of the electricity delivery system.
- *Off-shore wind energy*: Off-shore wind energy systems involve the use of wind farms installed in bodies of water (usually oceans) to harvest wind energy to generate electricity. There usually is a higher wind speed available offshore than on land which results in a higher electricity production in offshore wind turbine plants, for a given installed capacity. Currently, the electrical power production capacity from wind using offshore wind turbines is up to 20 TW worldwide, and higher capacities, exceeding 75 TW, are predicted by 2030 (IEA, 2019a). The cost of electricity produced from offshore wind power plants is higher than that of land-based. A report from Wind Vision stated that by 2050 offshore wind can be made available in all coastal regions of the world (Wind Vision, 2020).

Finally, around 0.02% of the incident solar energy on the earth is used for *photosynthesis* (around 0.04 PW). Through photosynthesis, solar energy is converted to chemical energy and stored in plants, which can subsequently be recovered in the form of *biomass energy*. Biomass energy generally refers to the energy stored in materials such as woods and crops that can be combusted or converted to a synthetic fuel. The energy content per unit mass of different kinds of biomass varies, for example 15 GJ/kg for dry wood and 30 GJ/kg for charcoal. Note that the biomass that is fossilized over great spans of time in the underground reserves can be converted to *fossil fuels*.

1.3.2 Geothermal energy

Geothermal energy is a form of thermal energy derived from the origin and formation of the earth and from the core of the earth where radioactive decay occurs (Dinçer and Zamfirescu, 2011; Ziagos et al., 2013). Geothermal energy is stored in the earth's crust. The core temperature of the earth is estimated at about 5200 °C and the pressure is nearly 3.6 million atm (National Geographic, 2017). The underground temperature at about 4 km below the surface of earth is about 90 °C (Dinçer and Zamfirescu, 2011). Geo-fluid sources (contained mostly water with different amounts of salts and minerals dissolved) may vary significantly from one place to another, with respect to temperature, depth, and structure of rock. The intensity of geothermal energy is about 0.1 W/m^2 which is low compared to solar radiation (about 249 W/m^2) (Blackwell et al., 1991). However in hot springs, geysers, volcanoes, and hot rock regions, there is a much greater thermal energy potential.

Geothermal energy is relatively reliable, sustainable, environmentally friendly, and cost effective. Some important advantages of geothermal energy are its independence from weather conditions and very high capacity factors. Water and/or steam can be used to carry the geothermal energy to the surface of the earth. The main stages in developing geothermal power systems are resource exploration, drilling, reservoir/plant development, and power generation. Recent advances in geothermal-based technologies have significantly expanded the scale and range of geothermal-based systems for the industrial, domestic, and agricultural sectors for heating, cooling, and electricity production. Although some trapped GHGs are released from geothermal wells to the atmosphere, theses emissions are much lower per energy unit than those from fossil fuel plant operations.

There are three main types of geothermal power plants are as follows:

- *Dry steam power plants*: In these systems, dry steam is drawn from an underground source and directly transferred to a turbine/generator unit. The steam exiting the turbine is normally injected back to the underground, so as to make this a sustainable resource. The Geysers in northern California and Yellowstone National Park in Wyoming are two known underground resources of steam in the United States.
- *Flash steam power plant*: In such systems, high temperature water at a temperature more than 182 °C flows up through the wells. As it enters a flash chamber, its pressure drops and a portion of the water is vaporized. The resulting steam is then separated from the water and used to drive a turbine/generator. The remaining liquid water and condensed steam are usually injected back into the reservoir.
- *Binary cycle power plants*: These systems use water at relatively low temperature (between 105 °C and 182 °C) in an electricity-generation

process. An intermediate heat exchanger is used to transfer heat from underground hot water to the working fluid (often an organic compound with a low boiling point) in the power cycle. The working fluid is vaporized in the heat exchanger and input to the turbine in the power plant cycle, then injected back into the earth to be reheated. In binary cycle power plants, the underground water and the working fluid are kept separate in all steps.

Some geothermal technologies are mature, while others are in demonstration stages. Many technologies for direct use of geothermal heat, such as greenhouses, district heating systems, and geothermal heat pumps, are widely used, reliable, and mature. Technologies for electrical power production from geothermal energy resources such as dry steam, flash steam, and binary systems also are mature (Jalili et al., 2021). Some technologies, such as enhanced geothermal systems (EGSs) (IPCC, 2012), described below, are relatively new and are undergoing research and development.

Regional tectonic and geological phenomena have an important role in the sustainability of naturally occurring geothermal resources. These factors limit conventional geothermal energy systems to regions rich in natural geothermal reservoirs. EGSs, however, do not rely on the occurrence of natural geothermal reservoirs. With the EGS approach, geothermal energy reservoirs are enhanced or engineered, which makes EGSs adaptable in more regions, specifically those lacking natural reservoirs. Using the EGS technique extends lifetimes, expands resources, increases productivity, enhances flexibility, and provides environmental advantages relative to conventional geothermal energy facilities. Fig. 1.19 depicts the five main steps in the development of EGS reservoirs.

1.3.3 Nuclear energy

Nuclear energy makes a significant contribution to electricity production and can also meet nonelectrical demands (Rosen, 2021b), and often is treated as a clean energy technology. Like renewable energy resources, nuclear can potentially make a significant contribution to sustainable development and enhance energy security. In some jurisdictions, nuclear power projects form a key part of efforts to achieve sustainable energy systems.

Nuclear energy is obtained through controlled nuclear fission which yields heat; the heat can be used directly or for electricity generation. Based on the IAEA, nuclear power supplies around 10% of the global electricity generation, which is around one-third of the overall low carbon electricity and second to hydro (IAEA, 2019). In 2018, more than 400 nuclear reactors

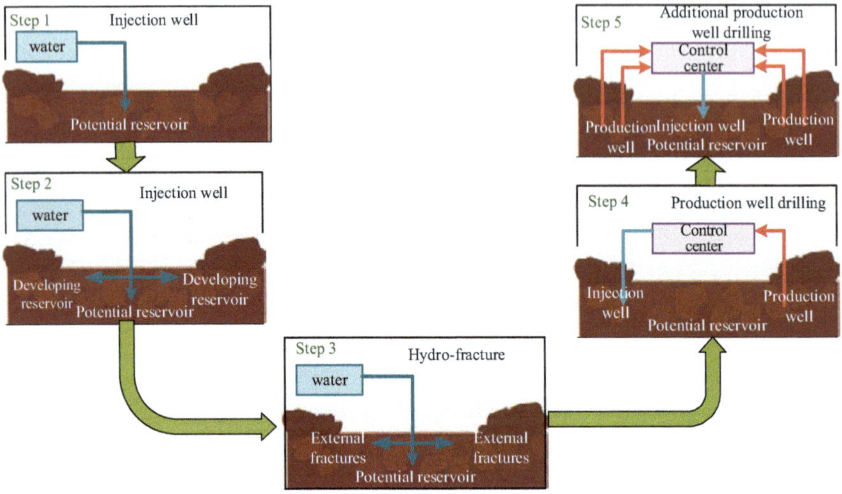

Fig. 1.19 ***Steps in the development of an enhanced geothermal system reservoir.*** *Based on Hussain et al., (2017).*

were operational worldwide, with an overall electricity production capacity of over 375 GW. Annual increases in nuclear power capacity of 15 GW are needed between 2020 and 2040 to achieve the sustainable development scenario (IEA, 2020h). However in 2019, only 5.5 GW of new nuclear generation capacity was connected to the grid (IEA, 2020h). Also in 2019, around 13 nuclear reactors (with a total power-generation capacity of 9.4 GW) were retired and shut down in the United States, Japan, Germany, Russia, Korea, Switzerland, Sweden, and Taiwan (IEA, 2020h).

Uranium-235 (U-235) is a conventional source of nuclear energy. A mass of 5.9 kg of U-235 can potentially generate about 180 MeV of fission energy, which can be converted to 40 GWh of electrical energy (Dinçer and Zamfirescu, 2011). The residues from nuclear fission are radioactive with high energy densities. The energy density of uranium is much higher than that of fossil fuels. For instance, 1 ton of coal can generate about 2.8 MWh of electrical energy, leading to an energy density that is 14,000 times that of nuclear fission (Dinçer and Zamfirescu, 2011).

Nuclear reactor designs are typically classified based on their generation, into generation I, II, III, III+, and IV. The main features of the nuclear reactors are advanced through each generation. These key differentiating features are cost effectiveness, safety, security and nonproliferation, capacity of the electric grid, commercialization roadmap, and nuclear fuel cycle. A detail explanation of each of these attributes, in the context

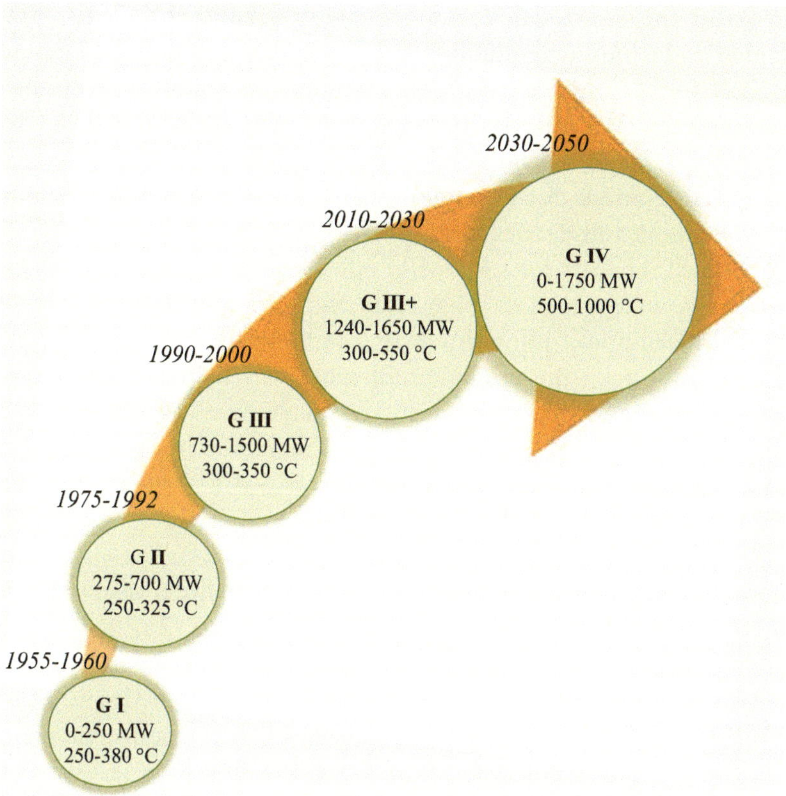

Fig. 1.20 ***Evolution of nuclear reactors from generation I to IV and their respective power production capacity and outlet temperature ranges.*** *Data from (Naterer et al., 2013).*

of the generation of the reactor, can be found elsewhere (Goldberg and Rosner, 2011).

The temperature of the core and the capacity of power generation are two primary parameters defining a nuclear reactor system. In Fig. 1.20, the trends in these parameters are shown, broken down by reactor generation (I, I, III, III^+, and IV). From 1954 to the present and on to 2030, nuclear reactors are likely to be mainly utilized for electricity generation. For safety reasons, the outlet temperature of the peak working fluid in these nuclear reactors is maintained at 300–350 °C. As the outlet temperature of the working fluid is limited, further enhancements in power production efficiency can be achieved by enlarging the production capacity of nuclear plants (because large plants have higher efficiency compared to smaller plants). For Generation IV nuclear technology, the allowable peak working fluid temperature is much higher, leading to an increase in the associated Carnot efficiency (about 60%–75%), which is higher than the corresponding value for Generation I–III$^+$ reactors (45-52%). In fact, the relatively higher temperatures of the heat produced in Generation IV reactors expands the power production scale from 20 MWe to 1700 MWe and enables nuclear facilities to be used in a wide range of applications, for example, desalination, thermochemical hydrogen production, and district heating (Naterer et al., 2013).

The small modular reactor (SMR) offers flexible electrical power generation for a wider range of applications and users and can replace ageing fossil fuel-based power plants. An SMR can be utilized as a single or multimodule plant, with the possibility for synergistic hybrid energy systems that combine nuclear and alternate energy sources (such as renewable energies). SMRs, with electricity production capabilities of up to 300 MW per module, are often considered advanced reactors. SMR plants are purported to have enhanced safety and economic benefits compared to larger existing nuclear reactor designs (Khan et al., 2017; ONE, 2021). There are several designs for SMRs ranging from small versions of existing nuclear reactor designs to the new-generation IV designs (Berniolles, 2019). SMRs are suitable for nonelectric applications and can be used in remote areas with less developed infrastructures. Currently, there are four SMRs, in Russia, Argentina, and China, which are in advanced stages of construction.

Fig. 1.21 represents a view of an SMR assembly. One feature of the SMR is that all of the main components, including steam generator, pressurizer, and reactor coolant pumps, are placed inside of the reactor pressure vessel. The integral design features of the SMR enhance the safety characteristics of the nuclear reactor, reducing the possibility of severe mishaps such as a loss of coolant accident (LOCA) (Zeliang et al., 2020). During a LOCA, which can occur due to a break in the reactor coolant pressure boundary or other reasons, the reactor core temperature increases because of radioactive decay in the fuel without coolant to remove that heat.

Despite advances in nuclear power plants, investments often do not proceed because of various obstacles, such as concerns about future changes in policy, risks in construction, long lead times, likelihood of delays and cost overruns, and large scale of investment. These challenges raise the possibility that investments in lifetime extensions or new nuclear projects will decline from 2018 to 2040, although there are also predictions that the use of nuclear energy will increase (IEA, 2019b). Simply as an example, a possible scenario of a declining trend in the use of nuclear energy for several countries and regions is shown in Fig. 1.22. It is seen

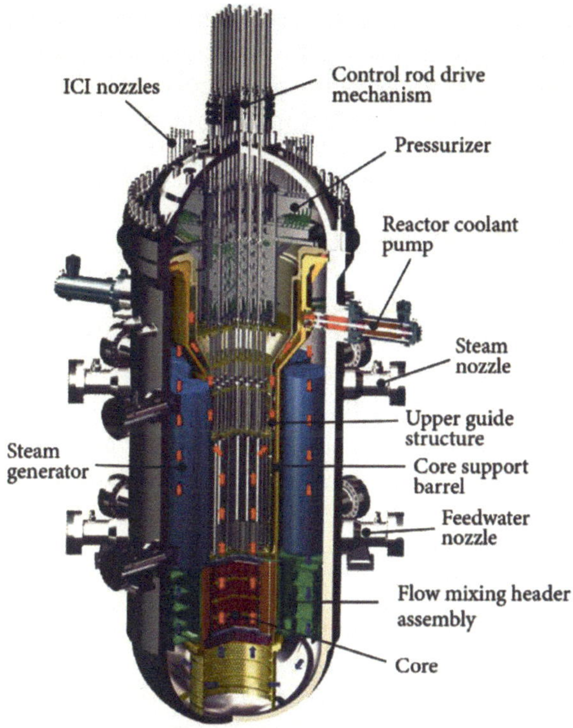

Fig. 1.21 ***Assembly of a small modular reactor (SMR).*** *From (Chung et al., 2014).*

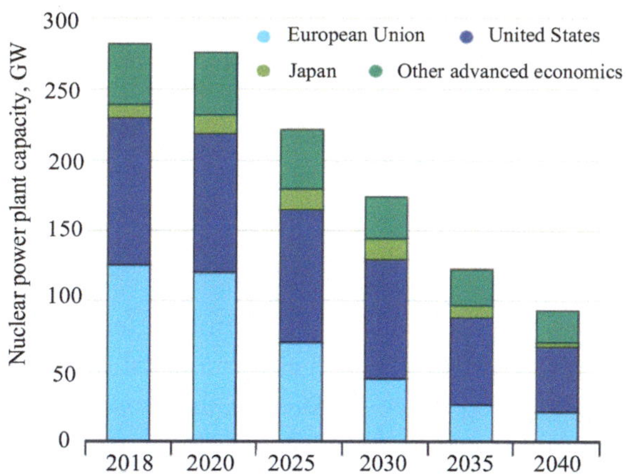

Fig. 1.22 ***Possible scenario in which nuclear power production capacity in advanced economies declines from 2018 to 2040.*** *Source: IEA (2019b), Nuclear Power in a Clean Energy System. All rights reserved.*

for the scenario in that figure that the European Union could experience the greatest decline in the share of nuclear capacity by 2040. Under this scenario, nuclear capacity in advanced economies will reduce by two-thirds by 2040, meaning that it declines roughly from 280 GW in 2018 to 90 GW in 2040 (IEA, 2019b).

1.4 Sustainable freshwater production

To have a sustainable water supply for a given need, water of an adequate quantity and quality is required to be available now and in the future. The sustainability of a water supply can be assessed in terms of social responsibility, economic feasibility, and environmental integrity. These key features are coupled with water users, such as those in the domestic, industry, and agriculture sectors. The sustainability of a water system also depends on its efficiency. In practice, when the efficiencies of both demand and supply increase, the sustainability of a water supply system tends to rise. On the demand side, efficient technologies can be developed to reduce water consumption, as can less water-intensive industrial and domestic processes. On the supply side, improving operation and maintenance of water facilities, careful economic planning regarding investment costs and cost recovery through a fair tariff system can help make a water system more sustainable.

Desalination technologies to produce freshwater have existed for over 50 years. Freshwater production from desalination plants requires energy, which is most often supplied by fossil fuel resources. The extraction and refining of fossil fuels to drive freshwater production facilities not only adds to the demand for water in long term but also contributes to the pollution of air and water resources (Gude and Nirmalakhandan, 2010). Therefore, the increasing need for freshwater supplies in the future may accelerate the depletion of fossil fuel resources and increase environmental contaminants and GHG emissions. Thus, sustainable desalination approaches that employ renewable energy sources are an important option for freshwater production.

As noted earlier, the sustainability of desalination technologies depends on social, economic, technological, and environmental factors. As desalination is a cost- and energy-intensive process that contributes notably to environmental impact, the technology used in such systems has a significant impact on their sustainability (e.g., some technologies can greatly reduce emissions). Environmental concerns of desalination processes also include chemical and brine discharges, GHG emissions, and ecological impacts.

Many factors are included in the economic impacts of desalination systems such as type of energy source used, availability and quality of water sources, geographical location, and operation and maintenance costs. Finally, social factors affecting the sustainability of desalination processes include social acceptance of desalination technologies, trust in the water providers, and confidence in the quantity and quality of water supplies.

As pointed out previously, most desalination technologies are energy and cost intensive and are operated with fossil fuels. The energy needs for selected desalination technologies and the corresponding CO_2 emissions if fossil fuels used to supply energy are presented in Table 1.1. The environmental impacts associated with fossil fuel consumption and the increasing cost and limited sources of fossil fuels tend to make fossil fuel–based desalination systems unreliable, unaffordable, and unsustainable for meeting future freshwater demands, especially in areas with low incomes and small fossil fuel reserves. Thus, sustainable approaches for freshwater production can include utilizing renewable energy resources, developing low-cost and energy-efficient technologies, applying cogeneration processes, and reusing water. These options are discussed further subsequently.

Presently, more than 75% of worldwide energy consumption is in the form of fossil fuels (Table 1.2, CAPP, 2020). The remaining sources are nuclear, biofuel, hydro, and other renewable energy sources. To enhance sustainability, renewable energy contributions likely need to increase significantly. The renewable energy contribution to global electricity generation is projected by the year 2040 to be about 45% (C2ES 2021; CAPP, 2020). Fig. 1.23 shows various renewable energies (such as solar, wind, tidal, hydro, and geothermal energy) that can be used to supply heat and electricity for

Table 1.2 Specific energy (electrical/thermal) consumed by various desalination technologies and the corresponding specific carbon dioxide emissions, when fossil fuels drive the desalination process.

	Desalination technology				
Parameter	**RO**	**ED**	**MVC**	**MSF**	**MED**
Specific energy consumption					
• Electrical (kWh/m^3H_2O)	5–9	2.6–5.5	11–12	3.5–5	1.5–2.5
• Thermal (kJ/kgH_2O)	0	0	0	250–300	150–220
• Total electric equivalent (kWh/m^3H_2O)	5–9	2.6–5.5	11–12	15–25	8–201
Average CO_2 emission ($kg\ CO_2/kgH_2O$)	0.032	0.38	0.051	0.09	0.04

Data from (Gude et al., 2010).

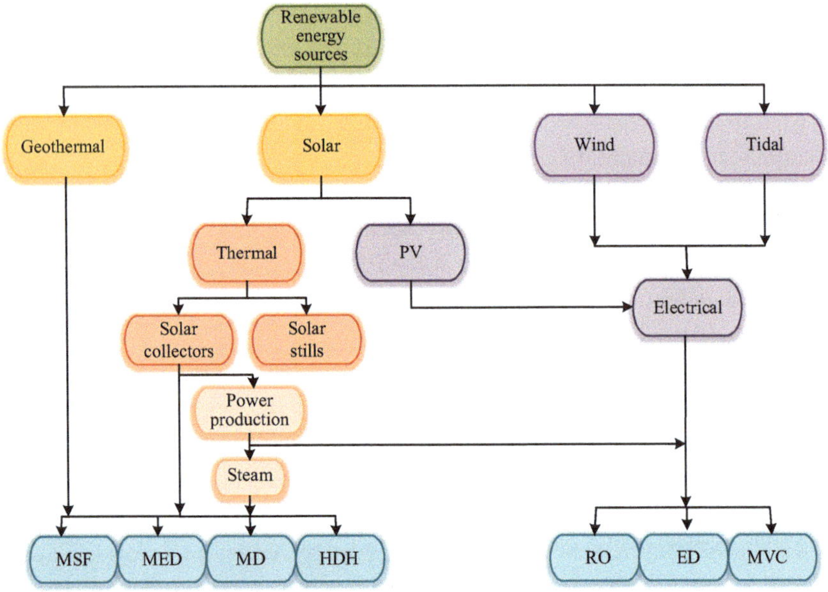

Fig. 1.23 ***Potential pathways for using renewable energy resources in desalination for sustainable freshwater production.***

desalination processes, and pathways for each. The pathways in Fig. 1.23 involve specific technologies with their own efficiencies and yields. Note that, regardless of whether nuclear energy is considered renewable or not, as nuclear energy meets numerous sustainable criteria (i.e., it can provide energy for long time in a way that is environmentally friendly, safe and economically viable, without depriving future generations), it often is considered as a sustainable energy source for freshwater production (Brook et al., 2014). Possible combinations of renewable energies with desalination processes are described, modeled, and assessed in Chapter 8 of this book.

Furthermore, an efficient combination of renewable energy resources and desalination technologies depends on many factors such as type and potential of renewable source, feed water salinity, plant capacity, electrical grid availability, remoteness, and technical infrastructure. Among possible combinations of renewable sources with desalination processes, solar and wind energies are found to be more economical and technological feasible in the areas where there are sufficient sources of these two renewable energies (Tzen et al., 2003). Geothermal energy also is a promising option due to its nonintermittent nature (compared to solar and wind) for freshwater production with a reasonable cost.

The cost of freshwater produced from desalination facilities has decreased significantly over the past few decades, due to improvements in materials (such as membranes), process schemes, durability, and energy recovery possibilities. In many countries, The cost of freshwater production from desalination systems is affected by external factors, such as population, environmental limitations related to climate change and pollution, economic growth/downturns, and energy costs.

The cost of energy required for desalination comprises a considerable percentage of the product freshwater cost. Cogeneration of freshwater and electricity is an effective practice to generate freshwater and electricity, often allowing good financial returns and lower impacts with respect to the atmospheric emissions compared to separate processes. Also, using low-grade heat sources from renewable or nonrenewable power systems in desalination processes can reduce the overall energy consumption and improve the economics of freshwater systems that combine renewable energy–based electricity generation with desalination. Developing desalination technologies such as MED systems compatible with low-grade waste heat sources (up to 60 °C) not only improves economics, but also reduces corrosion, scaling, and energy costs and raises thermodynamic efficiency.

An efficient and effective combination of a desalination system and renewable energy source(s) should be identified and optimized to achieve the best use of renewable energy resources to drive desalination (Dincer et al., 2017; Keshavarzzadeh et al., 2020). Some desalination technologies require an on-site energy source while others do not, allowing for an off-site energy scheme. For instance, solar-MSF/MED and geothermal-MSF/MED are on-site renewable energy–based desalination schemes. Examples of off-site schemes include wind-RO/MVC, PV-RO/MVC, and geothermal-RO/MVC systems.

Reusing water can help reduce pressures on conventional freshwater supplies and existing natural water sources (such as rivers, lakes, aquifers). Potential wastewater sources for water reuse include municipal, agricultural, industrial, and domestic wastewaters. Recycled water can be used for many nonpotable purposes such as agriculture, green spaces, and nonpotable domestic uses such as toilet water.

A range of renewable energy sources is considered in this chapter, with a focus on some of the more common ones. Due to the intermittent nature of some renewable energy sources energy storage devices play key roles in ensuring reliable continuous energy supplies (Koohi-Fayegh and Rosen,

2020b). Energy storage devices can be used to store electrical energy (such as batteries) or thermal energy (such as latent or sensible thermal energy storages) and then to recover it when renewable energy supplies are insufficient to operate desalination plants. Also, back-up generators, for example, units using natural gas or diesel fuel, can be used to provide energy when the generated and stored energy is insufficient. In subsequent chapters, comprehensive descriptions, models, and assessments of renewable energy–based desalination systems are presented.

Reducing the energy use in desalination technologies can help mitigate their environmental impacts and enhance their sustainability. Applying thermodynamic analyses provides useful information on how seawater properties and environmental conditions can influence the actual and minimum energy required for desalination. In addition, entropy generation and exergy methods can quantify the irreversibilities and inefficiencies in desalination system operations. These and other topics are covered in the remainder of this book, including design concepts for desalination and sustainable energy technologies accounting for thermodynamics, heat transfer, mass transfer, and economics.

1.5 Closing remarks

Water and energy are essential entities for humanity and civilization. The development in the past of noteworthy civilizations near large water sources demonstrates that water and civilization are both necessary commodities. Population growth and industrialization are significantly increasing water and energy consumption, leading to major challenges for meeting the needs for those two entities now and especially in the future. These challenges make desalination technologies promising options for supplying global freshwater needs, even though they are usually energy and cost intensive. The use of fossil fuels to drive desalination plants mainly occurs in wealthy countries, and increases energy-related emissions and concerns about climate change. Consequently, sustainable and more efficient desalination technologies for freshwater production are increasingly sought, exploiting technologies such as energy recovery, waste heat utilization, cogeneration and water reuse. The most beneficial combination of renewable and nuclear energy resources with desalination technologies depends on many factors, especially type and availability of renewable energy resource(s), feed water salinity, and plant size. Also, combining renewable energy resources with technologies for recovery and reuse of both water and energy is another potentially beneficial approach to meet global freshwater needs sustainably. The relatively high-cost freshwater produced from conventional (fossil-fueled) desalination facilities can sometimes be offset by increasing the contribution of renewable energy resources, by reducing energy, and operating and maintenance costs of desalination facilities. The optimization of renewable energy–driven desalination plants is very important for attaining the most advantageous design and operation of such systems. These and other topics are covered in the remainder of this book.

Nomenclature

A	area (m^2)
$\dot{Q}$	heat rate (W)
T	temperature (°C, K)
TBT	top brine temperature (°C, K)

Greek symbols

ϵ	emissivity
σ	Stephan–Boltzmann constant ($W/m^2.K^4$)

Subscripts

H	high-temperature heat source
L	low-temperature heat source
s	solar

Acronyms

EGS	Enhanced geothermal system
IAEA	International Atomic Energy Agency
IPCC	Intergovernmental Panel on Climate Change
GHG	Greenhouse gas
GOR	Gained output ratio
HDH	Humidification–dehumidification
LOCA	Loss of coolant accident
MED	Multieffect desalination
MENA	Middle East and North Africa
MD	Membrane desalination
MSF	Multistage flash
MVC	Mechanical vapor compression
RO	Reverse osmosis
SMR	Small modular reactor
TVC	Thermal vapor compressor

References

Al-Falah, E., Al-Shuaib, A.A., 2001. Impact of on-site training on performance of MSF plants: a case study. Desalination 141 (2), 129–143.

Alvarez-Risco, A., Del-Aguila-Arcentales, S., Rosen, M.A., García-Ibarra, V., Maycotte-Felkel, S., Martínez-Toro, G.M., 2021. Expectations and interests of university students in COVID-19 times about sustainable development goals: evidence from Colombia, Ecuador, Mexico, and Peru. Sustainability 13 (6), 3306.

Al-Wazzan, Y., Al-Modaf, F., 2001. Seawater desalination in Kuwait using multistage flash evaporation technology—historical overview. Desalination 134 (1-3), 257–267.

Berniolles. J.M, 2019. De-mystifying small modular reactors. https://www.sustainability-times.com/low-carbon-energy/de-mystifying-small-modular-reactors/ (Accessed 02 June 2021).

Blackwell, D.D., Steele, J.L., Carter, L.C., 1991. Heat flow patterns of the North American continent: a discussion of the DNAG geothermal map of North America. In: Slemmons, D.B., Engdahl, E.R., Zoback, M.D., Blackwell, D. (Eds.), Neotectonics of North America. Geological Society of America, United States.

Brook, B.W., Alonso, A., Meneley, D.A., Misak, J., Blees, T., van Erp, J.B., 2014. Why nuclear energy is sustainable and has to be part of the energy mix. Sustain. Mater. Technol. 1, 8–16.

C2ES, 2021. Center for climate and energy center. https://www.c2es.org/content/renewable-energy/ (Accessed 02 June 2021).

CAPP, 2020. World energy needs. https://www.capp.ca/energy/world-energy-needs/ (Accessed 02 June 2021).

CSS, 2020. U.S. Energy System Factsheet. Center for Sustainable Systems. University of Michigan. Pub. No. CSS03-11. http://css.umich.edu/factsheets/us-energy-system-factsheet (Accessed 02 June 2021).

Channabasappa, K.C., 1975. Status of reverse osmosis desalination technology. Desalination 17 (1), 69–73.

Chung, Y.J., Lim, S.W., Bae, K.H., 2014. Investigation of TASS/SMR capability to predict a natural circulation in the test facility for an integral reactor. Sci. Technol. Nucl. 2014, 181802.

Delyannis, E., 2003. Historic background of desalination and renewable energies. Sol. Energy 75 (5), 357–366.

Dincer, I., Rosen, M.A., Ahmadi, P., 2017. Optimization of Energy Systems. Wiley, London.

Dincer, I., Zamfirescu, C., 2011. Sustainable Energy Systems and Applications. Springer, US.

Dutrochet, R.H., 1995. New observations on endosmosis and exosmosis, and on the cause of this dual phenomenon. J. Membr. Sci. 100 (1), 5–7.

Eckman, T.J., Moch Jr, I., Shields, C.P., 1994. The TWIN™ permeator—development and commercialization of a new RO device. Desalination 96 (1-3), 11–22.

EIA, 2020. Coal explained, How much coal is left. Energy Information Administration, U.S. https://www.eia.gov/energyexplained/coal/how-much-coal-is-left.php (Accessed 02 June 2021).

El-Dessouky, H.T., Ettouney, H.M., 2002. Fundamentals of Salt Water Desalination. Elsevier, AmsterdamThe Netherlands.

Falkenmark, M., 1989. The massive water scarcity now threatening Africa: why isn't it being addressed? Ambio 18 (2), 112–118.

Fick, A., 1995. On liquid diffusion. J. Membr. Sci. 100 (1), 33–38.

Fick, A., 1855. Ueber diffusion. Ann. Phys. Lpz. 170 (1), 59–86.

Glater, J., 1998. The early history of reverse osmosis membrane development. Desalination 117 (1-3), 297–309.

Goldberg, S., Rosner, R., 2011. Nuclear reactors: Generation to generation March. American Academy of Arts and Sciences, Cambridge, MA.

Government of Canada, 2018. Water: frequently asked questions. https://www.canada.ca/en/environment-climate-change/services/water-overview/frequently-asked-questions.html (Accessed 02 June 2021).

Graham, T., 1858. Elements of Inorganic Chemistry: Including the Applications of the Science in the Arts. Blanchard and Lea, Philadelphia.

GRID-Arendal, 2009. Increased global water stress. https://www.grida.no/resources/5625 (Accessed 02 June 2021).

Gude, V.G., Nirmalakhandan, N., 2010. Sustainable desalination using solar energy. Energy Convers. Manage. 51 (11), 2245–2251.

Gude, V.G., Nirmalakhandan, N., Deng, S., 2010. Renewable and sustainable approaches for desalination. Renew. Sustain. Energy Rev. 14 (9), 2641–2654.

Hussain, A., Arif, S.M., Aslam, M., 2017. Emerging renewable and sustainable energy technologies: state of the art. Renew. Sustain. Energy Rev. 71, 12–28.

Hassler, G.L., McCutchan, J.W., 1960. Saline water conversion. Adv. Chem. (27), 192.

Hassler, G.L., 1949. The Sea as a source of fresh water. Department of Engineering Research Summary, University of California, Los Angeles.

Hassler, G.L., 1950. The use of molecular oil films in a pressure method for obtaining irrigation water from the sea. The University of California, Los Angeles.

Hodges, C.N., Kassander Jr, A.R., 1962. Distillation of Saline Water Utilizing Solar Energy in a Multiple-Effect System Consisting of Separate Collector, Evaporator, and Condenser. Solar Energy Laboratory, Institute of Atmospheric Physics, University of Arizona, US.

Hodges, C.N., Thompson, T.L., Groh, J.E., Sellers, W.D., 1964. The utilization of solar energy in a multiple-effect desalinization system. J. Appl. Meteorol. Climatol. 3 (5), 505–512.

Hodges, C.N., Thompson, L., 1966. Solar distillation utilizing multiple-effect humidification. University of Arizona, US. Research and development progress report. Report No. PB-206068; OSW-PR-194.

IAEA, 2020. The Game-Changing Power of Nuclear Energy to Reduce Greenhouse Gas Emissions. International Atomic Energy Agency. https://www.iaea.org/newscenter/news/iaea-data-animation-the-game-changing-power-of-nuclear-energy-to-reduce-greenhouse-gas-emissions (Accessed 02 June 2021).

IAEA, 2019. IAEA-Preliminary Nuclear Power Facts and Figures for 2019. International Atomic Energy Agency. https://www.iaea.org/newscenter/news/preliminary-nuclear-power-facts-and-figures-for-2019 (Accessed 02 June 2021).

IEA, 2020a. Change in power sector emissions by effect. International Energy Agency, Paris. https://www.iea.org/data-and-statistics/charts/change-in-power-sector-emissions-by-effect-2010-2020 (Accessed 02 June 2021).

IEA, 2020b. World energy model. International Energy Agency, Paris. https://www.iea.org/reports/world-energy-mode (Accessed 02 June 2021).

IEA, 2020c. Solar PV. International Energy Agency, Paris. https://www.iea.org/reports/solar-pv (Accessed 02 June 2021).

IEA, 2020d. Renewables. International Energy Agency, Paris. https://www.iea.org/reports/renewables-2020 (Accessed 02 June 2021).

IEA, 2020e. Onshore Wind. International Energy Agency, Paris. https://www.iea.org/reports/onshore-wind (Accessed 02 June 2021).

IEA, 2020f. Hydropower. International Energy Agency, Paris. https://www.iea.org/reports/hydropower (Accessed 02 June 2021).

IEA, 2020g. Renewables. International Energy Agency, Paris. https://www.iea.org/reports/renewables-2020 (Accessed 02 June 2021).

IEA, 2020h. Nuclear Power. International Energy Agency, Paris. https://www.iea.org/reports/global-energy-review-2020/nuclear (Accessed 02 June 2021).

IEA, 2019a. Offshore Wind Outlook. International Energy Agency, Paris. https://www.iea.org/reports/offshore-wind-outlook-2019 (Accessed 02 June 2021).

IEA, 2019b. Nuclear Power in a Clean Energy System. International Energy Agency, Paris. https://www.iea.org/reports/nuclear-power-in-a-clean-energy-system (Accessed 02 June 2021).
IPCC, 2020. The IPCC and the Sixth Assessment cycle. Intergovernmental Panel on Climate Change, Geneva, Switzerland. https://www.ipcc.ch/site/assets/uploads/2020/05/2020-AC6_en.pdf (Accessed 02 June 2021).
IPCC, 2013. Climate Change 2013: The Physical Science BasisContribution of Working Group I to the Fifth Assessment Report of the Intergovernmental Panel on Climate Change. Intergovernmental Panel on Climate Change. Cambridge University Press, Cambridge, United Kingdom and New York, NY, USA, p. 1535. https://www.ipcc.ch/report/ar5/wg1/ (Accessed 02 June 2021).
IPCC, 2012. Renewable energy sources and climate change mitigationSpecial Report of the Intergovernmental Panel on Climate Change. Intergovernmental Panel on Climate Change. Cambridge University Press, Cambridge, United Kingdom and New York, NY, USA, p. 1075. https://archive.ipcc.ch/pdf/special-reports/srren/SRREN_FD_SPM_final.pdf (Accessed 02 June 2021).
IPCC, 2011. Summary for Policy makers eds). In: Edenhofer, O., Madruga, R.Pichs, Sokona, Y., Seyboth, K., Matschoss, P., Kadner, S., Zwickel, T., Eickemeier, P., Hansen, G., Schlömer, S., Stechow, C.von (Eds.), IPCC Special Report on Renewable Energy Sources and Climate Change Mitigation. Cambridge University Press, Cambridge, United Kingdom and New York, NY, USA. https://www.uncclearn.org/wp-content/uploads/library/ipcc15.pdf (Accessed 02 June 2021).
IRENA, 2018. Global Energy Transformation: A roadmap to 2050. International Renewable Energy Agency, Abu Dhabi. https://www.irena.org/-/media/Files/IRENA/Agency/Publication/2018/Apr/IRENA_Report_GET_2018.pdf (Accessed 02 June 2021).
Jalili, M., Chitsaz, A., Hashemian, M., Rosen, M.A., 2021. Economic and environmental assessment using energy of a geothermal power plant. Energy Convers. Manage. 228, 113666.
Jones, E., Qadir, M., vanVliet, M.T., Smakhtin, V., Kang, S.M., 2019. The state of desalination and brine production: a global outlook. Sci. Total Environ. 657, 1343–1356.
Keshavarzzadeh, A.H., Ahmadi, P., Rosen, M.A., 2020. Technoeconomic and environmental optimization of a solar tower integrated energy system for freshwater production. J. Clean. Prod. 270, 121760.
Khan, S.U.D., Khan, S.U.D., Haider, S., El-Leathy, A., Rana, U.A., Danish, S.N., Ullah, R., 2017. Development and techno-economic analysis of small modular nuclear reactor and desalination system across Middle East and North Africa region. Desalination 406, 51–59.
Koohi-Fayegh, S., Rosen, M.A., 2020a. A review of renewable energy options, applications, facilitating technologies and recent developments. Eur. J. Sustain. Dev. 4 (4), em0138.
Koohi-Fayegh, S., Rosen, M.A., 2020b. A review of energy storage types, applications and recent developments. J. Energy Storage 27, 101047.
Kucera, J., 2019. Desalination: Water from Water, Second ed. John Wiley & Sons, NJ.
Loeb, S., Sourirajan, S., 1963. Saline Water Conversion II. Advances in Chemistry Series 38. American Chemical Society, Washington, DC, p. 117.
Loeb, S., Sourirajan, S., 1960. UCLA Dept. of Engineering Report 60-60. UCLA, Los Angeles, CA, USA.
Loeb, S., 1966. A composite tubular assembly for reverse osmosis desalination. Desalination 1 (1), 35–49.
Maciver, A., Hinge, S., Andersen, B.J., Nielsen, J.B., 2005. New trend in desalination for Japanese nuclear power plants, based on multiple effect distillation, with vertical titanium plate falling film heat transfer configuration. Desalination 182 (1-3), 221–228.
Mahian, O., Kianifar, A., Jumpholkul, C., Thiangtham, P., Wongwises, S., Srisomba, R., 2015. Solar distillation practice for water desalination systems. J. Therm. Eng. 1 (4), 287–288.

Mahon, H.I., Dow Chemical Co, 1966. Permeability separatory apparatus and process utilizing hollow fibers. U.S. Patent 3, 228, 877.
MWD, 2021. Merriam webster dictionary. https://www.merriam-webster.com/ (Accessed 02 June 2021).
NASA, 2007. Solar system math – NASA explorer schools pre-algebra unit, Lesson 1 Teacher guide. https://www.nasa.gov/pdf/622130main_SSML1Tchr.pdf (Accessed 02 June 2021).
Naterer, G.F., Dincer, I., Zamfirescu, C., 2013. Hydrogen Production from Nuclear Energy. Springer, London.
National Geographic, 2017. Earth's core. https://www.nationalgeographic.org/encyclopedia/core/#:~:text=Temperature in the inner core, not liquid or even molten (Accessed 02 June 2021).
National Research Council, 2004. Review of the Desalination and Water Purification Technology Roadmap. The National Academies Press, Washington, DC. https://doi.org/10.17226/10912 (Accessed 02 June 2021).
ONE, 2021. Benefits of Small Modular Reactors (SMRs). Office of Nuclear Energy, US Department of Energy, United States https://www.energy.gov/ne/benefits-small-modular-reactors-smrs (Accessed 02 June 2021).
Reid, C.E., Breton, E.J., 1959. Water and ion flow across cellulosic membranes. J. Appl. Polym. Sci. 1 (2), 133–143.
Rezaie, B., Rosen, M.A., 2020. The energy-water-food nexus: a framework for sustainable development modeling. Energy Equip. Syst. 8 (2), 179–201.
Richter, J.W. and Hoehn, H.H., 1971. Permselective, aromatic, nitrogen-containing polymeric membranes. EI Du Pont de Nemours and Co., U.S. Patent 3567632.
Rosen, M.A., 2021a. Energy sustainability with a focus on environmental perspectives. Earth Syst. Environ. 5, 217–230.
Rosen, M.A., 2021b. Nuclear energy: non-electric applications. Eur. J. Sustain. Dev. 5 (1), em0147.
Seifert, B., Kroiss, A., Spinnler, M., Sattelmayer, T., 2013. About the history of humidification-dehumidification desalination systems, Proc. The International Desalination Association World Congress on Desalination and Water Reuse 2013. Tianjin, China.
Sharp, T., 2017. How Hot Is the Sun? https://www.space.com/17137-how-hot-is-the-sun.html (Accessed 02 June 2021).
The World Bank, 2017. Annual freshwater withdrawals. https://data.worldbank.org/indicator/ER.H2O.FWAG.ZS (Accessed 02 June 2021).
Tiwari, G.N., Ghosal, M.K., 2007. Fundamentals of renewable energy sources. Alpha Science International Limited, Oxford, United Kingdom.
Tzen, E., Morris, R., 2003. Renewable energy sources for desalination. Sol. Energy 75 (5), 375–379.
USGS, 2021a. Science for changing world, Where is earth's Water? https://www.usgs.gov/special-topic/water-science-school/science/where-earths-water?qt-science_center_objects=0#qt-science_center_objects (Accessed 02 June 2021).
USGS, 2021b. Saline water and salinity. https://www.usgs.gov/special-topic/water-science-school/science/saline-water-and-salinity?qt-science_center_objects=0#qt-science_center_objects (Accessed 02 June 2021).
WHO, 2017. Guidelines for Drinking-Water Quality, Fourth ed. World Health Organization, Geneva, Switzerland, incorporating the first addendum. License: CC BY-NC-SA 3.0 IGO.
Wind vision: a new era for wind power in the United States, 2020. US Department of Energy. https://www.energy.gov/sites/prod/files/wind_vision_highlights.pdf (Accessed 02 June 2021).

White, C., 2012. Understanding water scarcity: Definitions and measurements. Australian National University, Australia. https://globalwaterforum.org/2012/05/07/understanding-water-scarcity-definitions-and-measurements/. (Accessed 02 June 2021).

WHO, 2008Guidelines for Drinking-Water Quality, Third ed, 1. WHO Press, Geneva, Switzerland.

Worldometer, 2021. Canada Population. https://www.worldometers.info/world-population/canada-population/. (Accessed 02 June 2021).

WRI Aqueduct, 2019. World Resources Institute. aqueduct.wri.org. (Accessed 24 August 2021).

Zeliang, C., Mi, Y., Tokuhiro, A., Lu, L., Rezvoi, A., 2020. Integral PWR-type small modular reactor developmental status, design characteristics and passive features: a review. Energies 13 (11), 2898.

Ziagos, J., Phillips, B.R., Boyd, L., Jelacic, A., Stillman, G., Hass, E., 2013. A technology roadmap for strategic development of enhanced geothermal systems, Proc. 38th Workshop on Geothermal Reservoir Engineering. Stanford, CA, 11–13.

CHAPTER TWO

Desalination technologies and their working principles

2.1 Introduction

Desalination has been extensively used for supplying freshwater to help meet the growing demand and reducing the environmental impacts of industrial wastewater. There are numerous desalination techniques for removing salts and impurities, for a wide range of solute concentrations in water. Large-scale seawater/brackish water desalination plants are important for sustainable freshwater production for large areas of the world. Energy use in desalination technologies is an important performance metric. The energy use in a desalination process varies based on the degree of salt removal, intake feed salinity, water recovery, and productivity of the desalting process (see Fig. 2.1).

Many desalination methods are shown in Fig. 2.2, categorized by driving energy form. Some of these technologies (marked in red) are commercialized and widely used at various scales. The main commercial desalination technologies are work driven processes like reverse osmosis (RO) and mechanical vapor compression (MVC) desalination, and thermal driven processes like multi-stage flash (MSF), humidification–dehumidification (HDH), and membrane distillation (MD) desalination, as well as multi-effect desalination (MED).

Desalination plants generally include several steps that, taken together, yield freshwater:

- Saline water intake: This process can be carried out using pumps and pipes to transfer saline water from sources (sea, ocean, etc.) to the desalination plant.
- Pretreatment: This step includes filtration processes to remove solid particles from the intake water and the addition of chemicals to avoid salt precipitation and corrosion within the desalination equipment.
- Desalination: This process constitutes the core of a desalination plant, and is where freshwater is extracted from saline water.
- Post-treatment: This step is carried out to control the pH of the water through the desalination process, so as to meet the requirements of final users.

Sustainable Energy Technologies for Seawater Desalination
DOI: https://doi.org/10.1016/B978-0-323-99872-7.00009-7

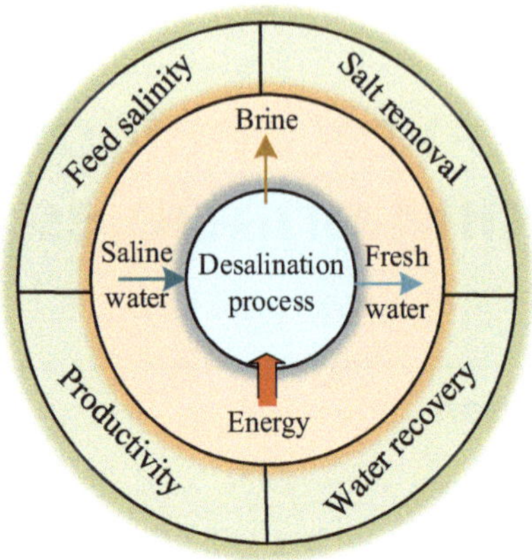

Fig. 2.1 ***Parameters related to freshwater production via desalination.***

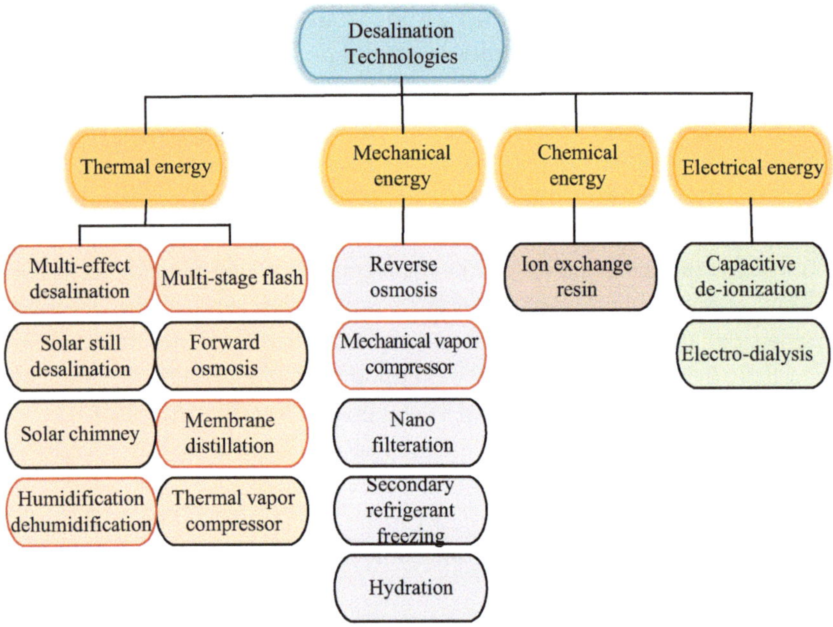

Fig. 2.2 ***Desalination technologies broken down by the primary energy inputs.***

In RO desalination, the applied pressure should exceed the osmotic pressure difference between the feed water and permeate. RO technology is widely used for brackish water and seawater, and usually achieves more than 90% salt rejection. Thermal desalination technologies involve the vaporization of saline water and the subsequent condensation of the generated vapor to yield distilled water. MSF desalination and MED systems differ in configuration details, but both are designed so as to enhance the recovery of condensation latent heat, as doing so is important in achieving high energy efficiencies. The MVC desalination technique uses an MVC process driven by electricity to increase the temperature and pressure of vapor. Then, the latent heat of the compressed vapor is reused for vaporizing the sprayed feed water. An emerging innovative technology, MD has been actively investigated. This thermal-driven technique operates based on the partial pressure difference that forces vapor to transfer across a hydrophobic membrane. MD desalination systems appear to have very good commercial potential, if the operational flux can be improved.

2.2 Definition and classification of industrial seawater desalination processes

Desalination technologies can be divided into two main categories: phase change (or thermal processes) and single-phase (or membrane processes). Some widely used desalination technologies like MSF, MED, HDH, and solar stills fall into the phase change desalination category. Other technologies like RO are categorized as membrane-based desalination. MD is a desalination technique that requires both thermal energy and a membrane to produce freshwater. All these techniques require pretreatment of the saline water to avoid fouling/scaling, foaming, and corrosion. Analyses of desalination technologies can be made by simplified or detailed mathematical models. With detailed models, a large number of equations are solved and greater information is attained on variables and properties of the system, while simplified models provide adequately accurate values for the main design parameters of the desalination system in an efficient and expedited manner.

In the following subsections, the working principles and equipment design details of each of the aforementioned desalination technologies are described.

2.2.1 Multi-stage flash desalination

Industrial-scale MSF desalination systems include once-through MSF (MSF-OT) and brine circulation MSF processes. In this subsection, the operating principles for both MSF-OT and brine circulation MSF are presented.

2.2.1.1 Once-through multi-stage flash desalination system

In the MSF-OT desalination process, seawater is desalted via flashing a portion of the water to a vapor through multiple stages, which have heat exchanger and condensate collectors. The stages have different pressures corresponding to the boiling temperature of seawater at each stage. Feed seawater is pumped from the cold end to the hot end and is preheated before entering the brine heater (see Fig. 2.3). The hot seawater receives additional heat in the brine heater and returns to the stages at lower temperatures and pressures. Since hot seawater has a temperature above the boiling temperature of the first stage, a small part of the seawater flashes on entering the first stage and its temperature reduces to the first stage temperature. The generated vapor from the flashing process preheats the seawater moving toward the brine heater. The feed seawater absorbs the latent heat of condensation of the vapor and the condensed vapor is collected in a tray. In fact, most of the thermal energy carried by the vapor is transferred to the colder feed seawater being pumped toward the brine heater, allowing the energy to be recovered. The pressure of each stage remains constant through the flashing and condensation of vapor in the heat exchanger. If more vapor is generated at some points, the pressure of the stage increases and the flashing process eases, but the condensation process increases to maintain a stable equilibrium.

In the final stage, the condensate (freshwater) and brine streams take on a temperature close to the inlet feed seawater temperature, and they are pumped out of the plant at the ambient pressure. The practical operating temperature of the last effect is between 30°C and 40°C. Reducing the last stage temperature below the aforementioned values results in a considerable increase in the specific volume of the generated vapor, which leads to an increase in the stage volume and dimensions. It is also important to maintain a constant velocity for the generated vapor in the stage to control the rate of entrainment of brine droplets in the generated vapor.

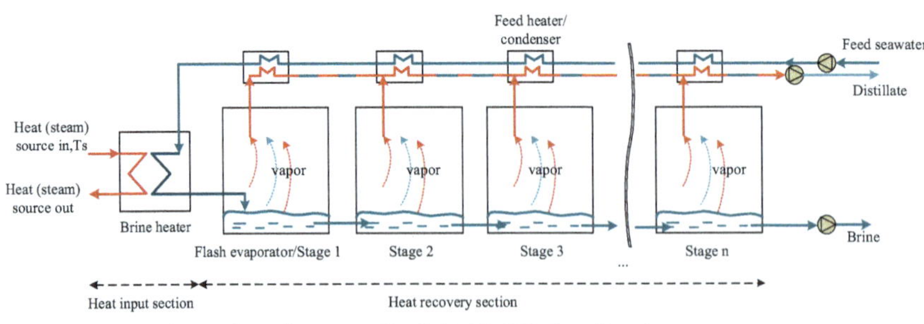

Fig. 2.3 ***Once-through multi-stage flash (MSF-OT) desalination system.***

The heat provided in the brine heater is typically in the form of steam from industrial plants (often power plants in a cogeneration scheme) near the desalination plant. The latent heat of steam is released through the condensation process adjacent to the tubes carrying the preheated feed seawater. The range of specific energy consumption (SEC) for MSF plants is typically 20 kWh–25 kWh per cubic meter of freshwater produced (Gude, 2018; El-Dessouky and Ettouney 2002; Kucera, 2019).

The total evaporation rate for an MSF system depends on the operating temperature. Increasing the operating temperature can lead to problems like scale formation and corrosion. For instance, an operating temperature range between 110°C and 120°C appears to be around the maximum, while operating temperatures below 70°C are required to avoid scale formation. Despite the relatively high operating temperature of an MSF desalination system, the use of freely moving brine into each effect, rather than using submerged heating tubes, to evaporate water constitutes one of the advantages of MSF over other thermal evaporation desalination systems. This approach avoids the buildup of salt scales over the surfaces of the evaporating tubes. Since most of the stages operate below 100°C, the pressures of these stages are at vacuum conditions which may lead to a leakage of air into the system. The presence of air as a non-condensable gas (NCG) reduces the heat transfer rate in each stage and flashing process and increases the rate of corrosion. In addition, dissolved gases that are not removed in the deaeration pretreatment can lead to an accumulation of NCGs. To avoid the adverse effects of NCG on desalination process, venting of stages is necessary, as doing so improves the flashing process and the MSF system efficiency.

The main reason for using multiple stages instead of a single stage at the lowest temperature and pressure, as is done in a single stage desalting process, is that the feed seawater temperature cannot reach a relatively high temperature. In fact, the feed seawater is heated to an intermediate temperature, between the heater and inlet feed temperatures. Most of the steam cannot be condensed in a single stage and consequently, the pressure and temperature of a single-stage cannot be maintained at low values.

2.2.1.2 Brine circulation multi-stage flash desalination system

In the brine circulation MSF process (see Fig. 2.4), the stages are divided into two sections: heat recovery and heat rejection. In this configuration, brine is recycled from the last stage in the heat rejection section and is introduced to the condensing tubes of the last stage of the heat recovery

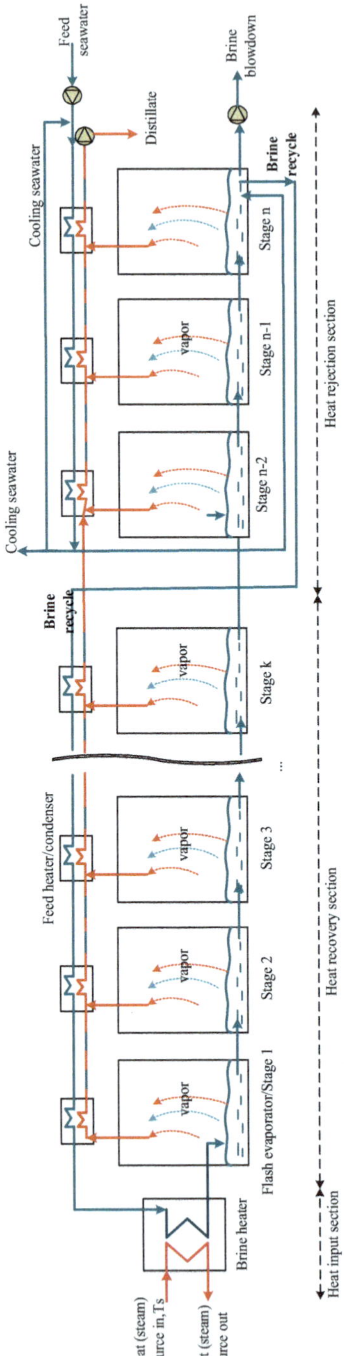

Fig. 2.4 ***Brine circulation multi-stage flash desalination system.***

section, where it enters the brine heater unit. In the brine heater, the recycled brine stream absorbs the latent heat of condensation of the external steam and its temperature rises to the maximum design value, known as top brine temperature (*TBT*). The value of *TBT* depends on the type and concentration of chemicals used to control scaling phenomena. The brine enters the heat recovery section, containing a number of stages, where a small amount of brine flashes to steam due to the decrease in the temperature and pressure of the sequential stages. The generated vapor is condensed on the outlet surface of the condenser tubes containing brine recycle stream flows from the cold side to the hot side of the system. This heat recovery increases the desalination process efficiency, since it increases the feed water temperature.

In the heat rejection section, the intake seawater is split into cooling seawater which is discharged to the sea and feed seawater which is chemically treated and deaerated and then combined with the brine stream from the last stage. In winter conditions, when the intake seawater temperature drops to 5–15°C, a portion of the warm cooling water can be recycled and mixed with the intake feed water. The intake seawater can be treated through deaeration and the addition of some chemicals to control foaming, scaling, and corrosion.

In recent years, there has been a tendency to shift from MSF desalination to MED systems, although the MSF plant still is the dominant thermal desalination process. The brine circulation MSF process offers a control on the temperature of feed seawater, which is a notable advantage that allows it to dominate the thermal desalination market, while the once-through MSF is being utilized on a limited scale due to the lack of this feature. There are several major differences between OT-MSF and brine circulation MSF configurations:

- The temperature of feed seawater cannot be controlled in a OT-MSF system, while in the brine circulation MSF system it take place at the heat rejection section shown in Fig. 2.3. This can be treated as the main factor that limits use of OT-MSF systems. In other words, when the intake feed seawater temperature drops to 5–15°C in the winter season, the operation of a OT-MSF system faces a problem because the volume of the last stage needs be increased drastically to respond to the increase in the specific volume of the vapor generated at a relatively lower temperature. Consequently, the OT-MSF system should is more suited for equatorial regions, where the seawater temperature is almost constant throughout the year (around 25°C). An alternative approach to this

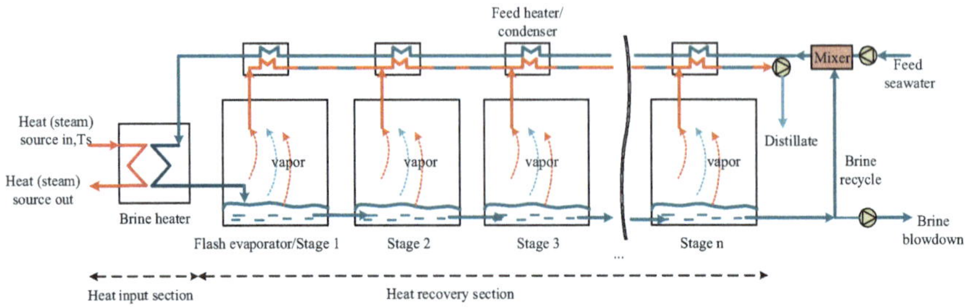

Fig. 2.5 ***Brine mixing multi-stage desalination system.***

problem is to use a brine mixing method to control the temperature of feed seawater, as shown in Fig. 2.5.

- A OT-MSF desalination system requires larger equipment for chemical treatment of materials because the intake seawater flow rate is around four times higher than that for the brine circulation MSF desalination configuration.
- The number of pumps for a brine circulation MSF system is more than for a OT-MSF configuration. Pumps for feed, brine, freshwater, and chemicals are common in both configurations, but in the brine circulation MSF system additional pumps are needed for brine recycling and intake seawater and cooling water flows. It is reported that the additional pumps in a brine circulation MSF system lead to 10%–30% higher electricity consumption compared to a MSF-OT system (Al-Falah et al., 2001; Ettouney et al., 2002).
- Since the concentration of salt in the brine recycle stream is higher than that of input feed seawater, the risk of scale formation in a brine circulation is greater for a MSF system than a OT-MSF system. As a result, the antiscaling dosing rate and frequency of acid cleaning of the condenser tubes in a brine circulation MSF desalination system are much greater than in a OT-MSF system.

In addition to the above differences between once-through MSF and brine circulation MSF desalination systems, the performance ratio (ratio of freshwater mass flow rate to the external steam supply mass flow rate) is almost same for these two configurations. Also, the freshwater production capacity and specific heat transfer area in those two configurations are similar.

The selection of preheater/condenser tube material in a MSF system depends on the operating temperatures of the stages. For instance,

Cu/Ni 70/30 is suitable for temperatures above 80°C, while Cu/Ni 90-10, titanium, high steel alloys, and aluminum brass can be used for relatively lower temperatures (El-Dessouky and Ettouney, 2002). Each of the aforementioned materials has particular characteristics of relevance. For instance, Cu/Ni 90/10 has the highest thermal conductivity and titanium the highest erosion resistance. Aluminum bronze is the least expensive material for use in the stages, although its copper content may dissolve in the water and cause some problems (El-Dessouky and Ettouney, 2002).

2.2.1.3 Tube arrangements in multi-stage flash desalination system

The tubes containing feed seawater in a MSF desalination system can be arranged in cross or long configurations (see Fig. 2.6). The cross tube arrangement attains a lower freshwater production rate compared to the long configuration. In cross tube arrangements, tubes are placed along the width of stages (flashing chambers) and are connected with water boxes, as shown in Fig. 2.6. In the long tube configuration, a single bundle of tubes is placed along the entire length of the stages. This configuration eliminates the need for the external water boxes found in the cross configuration and also permits for higher flow rates of feed seawater per width of stage and consequently yields a higher freshwater production rate. Other features of the long tube arrangement include lower pressure drops (usually by 25%–30%) compared to the cross tube arrangement, and more difficult cleaning and maintenance of tubes, as specific equipment is required (El-Dessouky and Ettouney, 2002).

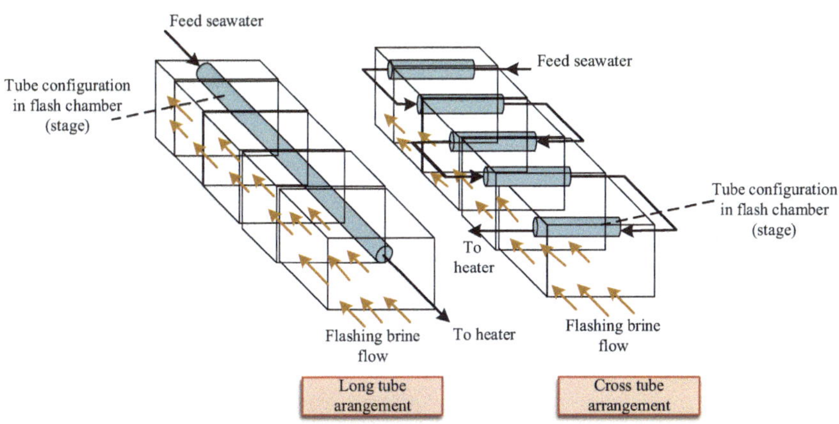

Fig. 2.6 ***Types of tube arrangements in multi-stage flash desalination systems.***

2.2.2 Multi-effect desalination

MED is a thermal desalination process in which heat is primary energy form used, mainly to drive successive evaporation/condensation processes. Pumps are also used to transport water streams (feed, distillate, and brine) within the system. The main feature of MED desalination is the relatively lower *TBT* (between 60°C and 70°C) compared to MSF desalination. The freshwater production capacity is smaller for MED than MSF desalination. Another feature of MED is that it can be adapted to a wide range of heat sources including waste heat from electrical generation plants, hot water systems and solar energy facilities. MED also produces a consistently high freshwater purity.

There are several types of MED feed configurations, including forward (co-current), backward (counter-current), parallel, and combinations thereof. A MED system consists of sequential evaporator/condenser heat exchangers called effects, in which the tubes can be arranged horizontally or vertically. All arrangements involve the same thermodynamic process, with the difference being the configuration of the heat transfer surfaces. Fig. 2.7 shows a flow diagram for a simplified MED desalination system. The external steam is fed to the first effect, vaporizing a part of it as it is sprayed over the heat exchanger tubes. The generated vapor enters the second effect, vaporizing a part of the feed stream by releasing its latent heat of condensation. This process is repeated in subsequent effects until the temperature of the vapor rises to near the temperature of the feed seawater. In the last effect, the generated vapor is condensed at the external condenser unit and preheats the feed seawater input to the system. Based on the first law of thermodynamics, when the pumping power is negligible, the heat rate removed from the last stage is almost equal to the heat rate injected to the first effect. Cooling water (or cooling seawater) is used in the condenser to adjust the design feed seawater temperature. When the cooling water is exhausted by the condenser, it is returned to the sea. The distillate and brine from the last effect are pumped to ambient pressure and exit the plant.

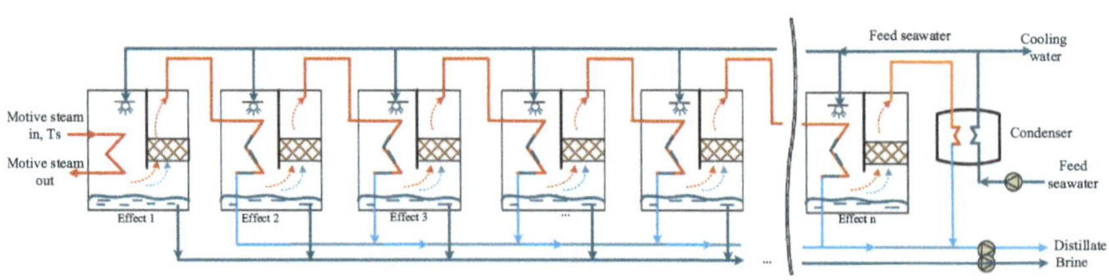

Fig. 2.7 ***Simplified flow diagram of a multi-effect desalination (MED) system.***

There is an important trade-off in the design of MED desalination systems. Reducing the thickness of tubes in each effect results in thinner layers of liquid on either side of tube surface, which leads to more efficient heat transfer. Increasing the number of effects reduces the temperature difference between the subsequent effects and consequently lowers the heat transfer per unit area of tube. However, the supplied heat is reused more times to vaporize more seawater, and the desalting process takes more time. The freshwater production rate in each effect is a function of the rate of energy transport. When the energy transport rate is slowed, the heat transfer area in each effect rises, resulting in an increase in the capital cost of the heat exchangers and consequently the desalination system.

2.2.2.1 Forward feed MED configuration

Fig. 2.8 shows a forward feed MED configuration, in which freshwater vapor and feed seawater flows are in the same direction, moving from higher to lower temperature effects. Feed seawater enters the first effect and is partly evaporated. The remaining part as brine is transferred to the second effect. In the second effect, part of input brine is evaporated and the unevaporated brine is transferred to the next effect, and so on. In the forward feed configuration, the relatively lower feed concentration combines with high temperature at the first effect, which decreases the risk of scaling. However, the forward feed (co-current) configuration benefits from the ability to operate at high *TBT*s. This configuration is not often used for desalination as it has a more complex structure compared to the parallel feed configuration, described subsequently.

Increasing the temperature of the feed seawater to the saturation temperature of each effect would significantly improve the performance of a desalination process because it maintains a high heat transfer coefficient and removes the sensible heat transfer process used to raise the feed stream temperature to the saturation temperature. This raising of temperature can be accomplished via preheaters that preheat the feed water to the saturation temperature corresponding to the temperature of each effect. Furthermore, in MED desalination systems, flash chambers can be used for each effect to produce more distillate; a small amount of vapor is generated and collected from flashing the brine stream. The generated distillate vapor can be transferred from the flash box to the preheaters for preheating the feed seawater.

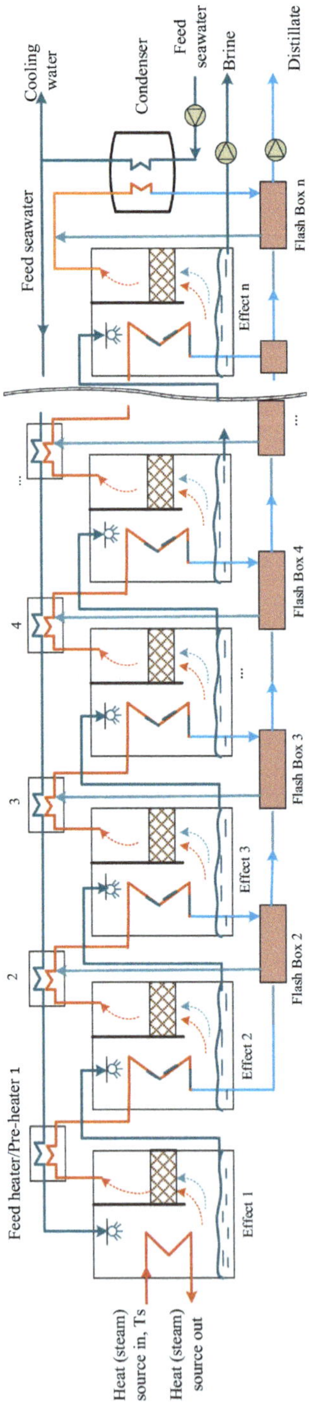

Fig. 2.8 *Forward feed MED desalination system with preheaters and flash chambers.*

2.2.2.2 Backward feed MED configuration

Fig. 2.9 shows a backward feed MED configuration, in which vapor and seawater move in opposite directions. First, feed seawater after passing from the end condenser enters the last effect, which is the effect with the lowest temperature and pressure. Then, brine pumps transport the brine of the first effect to the next effect where its temperature, pressure, and salinity increase and this carries on sequentially in each effect. The brine stream leaving the first effect is discharged to the sea as rejected brine. The backward (counter-current) feed configuration can attain a high thermal efficiency. However, it requires more electrical power to pump the brine stream through the sequential effects, which reduces the overall efficiency of the system. In addition, in the backward feed configuration, there is a relatively higher risk of scaling because the brine stream with the highest salinity is in contact with the first effect, which has the highest temperature. These drawbacks make the backward configuration not commonly chosen for desalination.

2.2.2.3 Parallel feed MED configuration

Fig. 2.10 shows a parallel feed MED configuration, which is the simplest and most commonly used arrangement in the desalination industry. In this configuration, feed seawater is distributed equally between the effects using a single pump. All effects are fed by seawater at the same temperature and salt concentration. In this configuration, the risk of scaling is higher in the first effect compared to the other effects, as it operates at a relatively higher temperature. In parallel feed MED, the brine in each effect is transferred to the next effect. This design is known as a parallel/cross configuration and it permits the energy content of brine to be recovered for heating the feed stream and a small part flashes into vapor. In an alternative design, brine is collected from each effect and then rejected to the sea. An important feature of parallel/cross feed configuration is that only a single pump is required to distribute the feed stream to the effects. In fact, no pump is needed for transporting the brine stream from one effect to another, because the brine moves spontaneously from higher to lower pressure effects.

Al-Mutaz (2014) compared forward, backward, and parallel feed MED desalination systems operating at identical conditions. For instance, last effect of MED and cooling water temperatures were considered as 42°C and 25°C, respectively. The *TBT* ranged from 58 to 70°C and the heating steam flow rate from 8 to 20 kg/s with the temperature of 62–75°C.

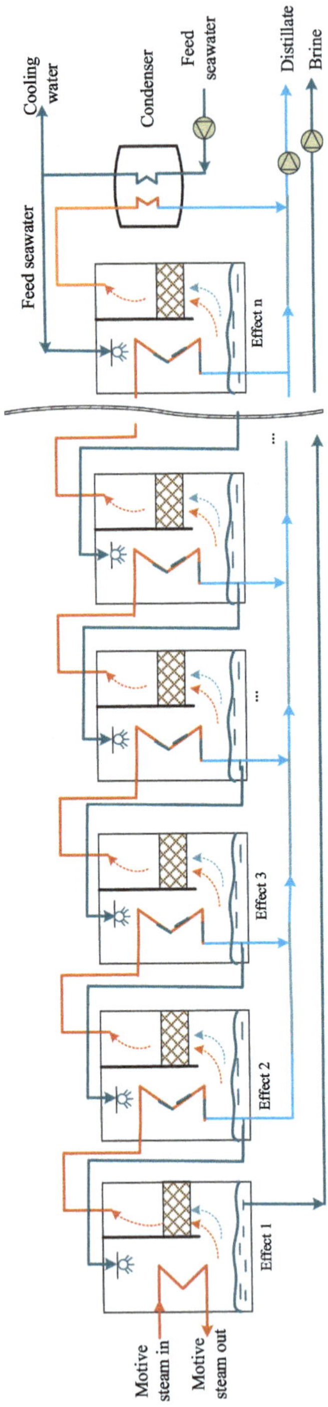

Fig. 2.9 ***Backward feed MED desalination system. MED***, multi-effect desalination.

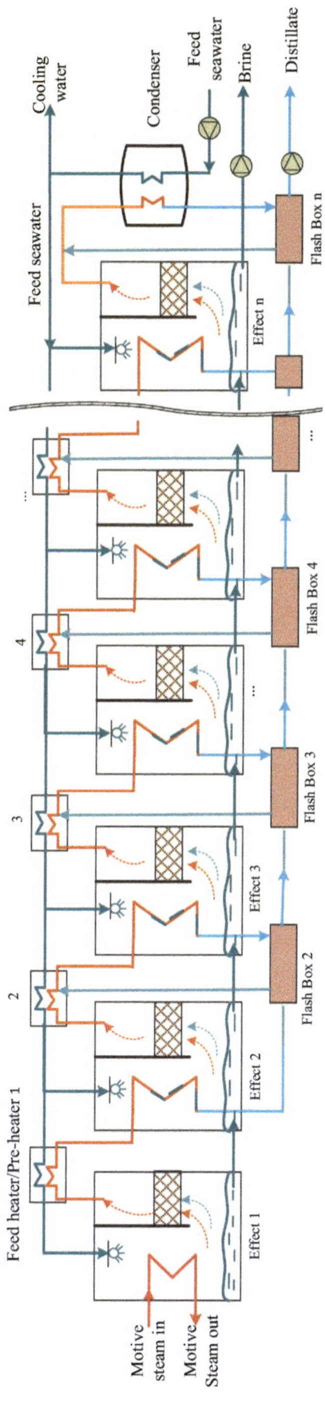

Fig. 2.10 ***Parallel/cross feed MED desalination system with preheaters and flash chambers.*** MED, multi-effect desalination.

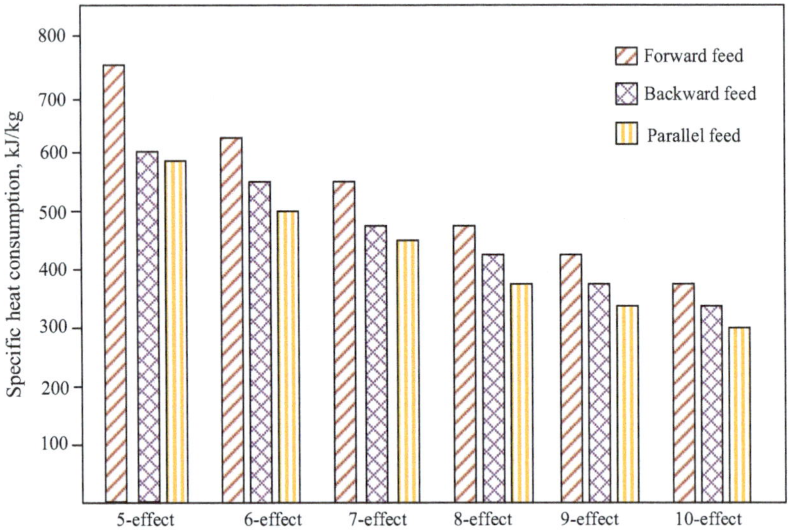

Fig. 2.11 ***Variation of specific heat transfer consumption with number of effects for various feed configurations in a MED system.*** *Data from Al-Mutaz (2014). MED, multi-effect desalination.*

Fig. 2.11 shows the variation of specific heat consumption with number of effects for various MED configurations. It is seen that the specific heat consumption is highest for the forward feed MED system and lowest for the parallel-cross feed configuration. Also, as the number of effects increases, the specific heat consumption decreases for all three configurations because, by increasing the surface area, the temperature difference across the effect decreases.

2.2.2.4 Types of evaporators in MED systems

Basically, two types of evaporators are used in MED desalination systems: vertical tube and horizontal tube. In some MED plants, the sequential effects are stacked vertically, placed on top of one another with the first effect at the top, and in other MED plants the effects are arranged sideways. For both types of evaporators, heat is transferred through a thin liquid film falling down on the inside or outside surface of tube. Each type of evaporator is now described.

Horizontal tube evaporator (HTE): The horizontal tube evaporator configuration is the most common evaporator type used in MED desalination plants due to its efficient heat transfer process. In this type of evaporator, tubes are placed horizontally in several rows and steam flows in the tubes.

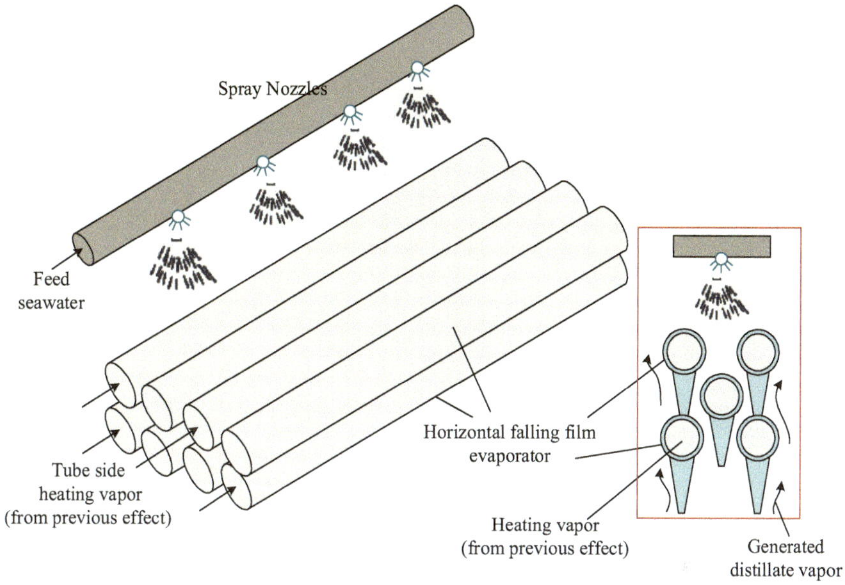

Fig. 2.12 ***Horizontal tube falling film evaporator.***

Fig. 2.12 shows a horizontal tube evaporator arrangement. As the steam is condensed inside the tubes, the latent heat of condensation is utilized to evaporate a falling film of feed water over the outside surface of the tubes. A thin film is in fact formed as the feed is sprayed over the tubes. The spray feed stream in an evaporator heat exchanger (effect) provides a more efficient heat transfer process than the flooded evaporator design because the motion of sprayed seawater across the heat exchanger surface allows the generated vapor bubbles to be carried away from the tube surface. Some more complex evaporator designs improve the heat transfer process. The tubes are often arranged in a square pitch to simplify the cleaning process.

In addition to the efficient heat transfer associated with the thin falling film in an HTE, the risk of mixing of generated vapor and NCG is kept low in this tube configuration, because positive venting results when vapor enters the tubes on one side and the condensed distillated exits on the other. If there is ineffective water distribution in the HTE, leaving some unwetted areas on the outer surface of the tubes, scale formation and fouling may occur on those surfaces.

Vertical tube evaporator (VTE): The tubes in a VTE are placed vertically, and the feed seawater flows inside of tubes downward from the top (by gravity) creating a falling film. Fig. 2.13 shows a MED effect with a

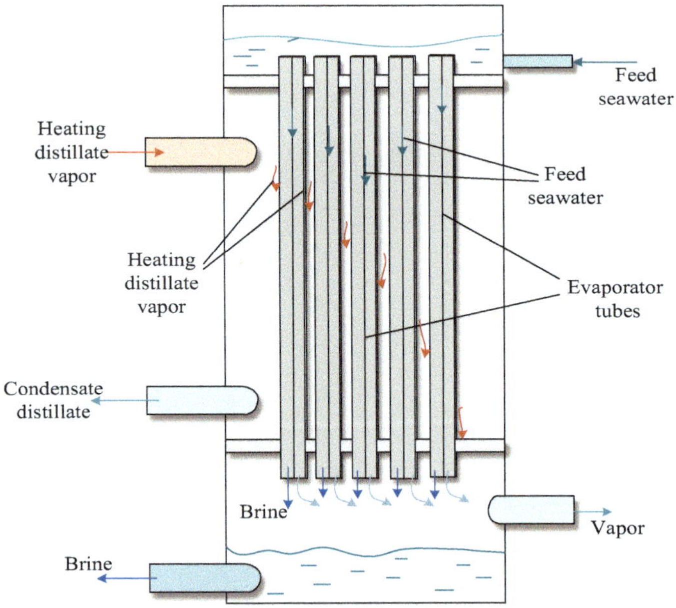

Fig. 2.13 ***MED effect of a vertical tube evaporator.*** MED, multi-effect desalination.

vertical tube evaporator. The generated vapor from the previous effect flows over the outside surface of the tubes, providing heat for the feed stream flowing inside the tubes. Thus, there are two thin films on the tube, one on the inside and the other on the outside, resulting in a high heat transfer rate. However, the film may collapse at a specific heat flux and feed flow rate, potentially yielding a poor average heat transfer rate. To avoid this, a uniform distribution of feed among the tubes is necessary, to match the heat transfer rates and tube dimensions. If the liquid feed inside the tube is completely evaporated, salt and other chemicals may precipitate on the surface of tubes, which leads to an inefficient desalting process.

2.2.2.5 Steam jet ejector

In a MED desalination system, a steam jet ejector can be used to compress the primary steam and a portion of the vapor generated in the last effect. In fact, a thermal vapor compression process can be achieved via a steam jet ejector to improve the performance of the MED system. Such a system is known as a multi-effect desalination-thermal vapor compressor (MED-TVC) system, and is shown in Fig. 2.14. The compressed vapor is transferred to the evaporating tubes of the first effect and provides heat in the first effect on release of the latent heat of condensation. Part of the

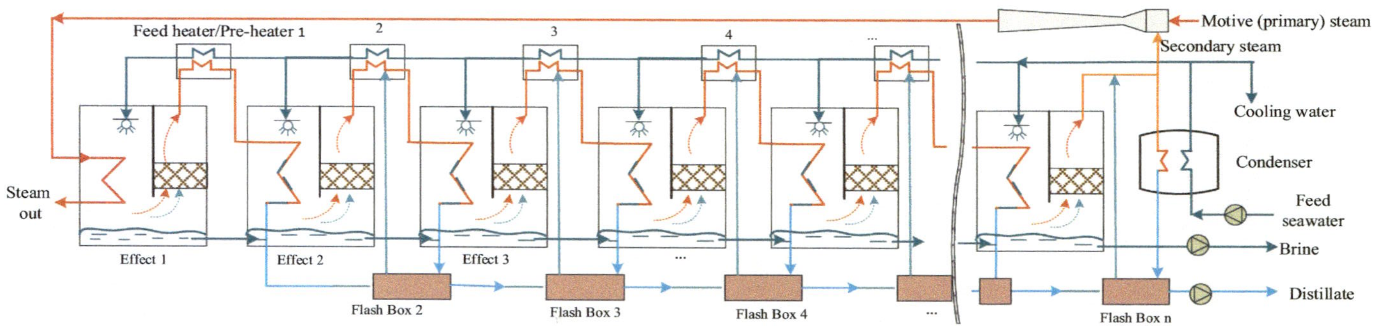

Fig. 2.14 ***Parallel-cross feed MED-TVC desalination system.*** MED-TVC, multi-effect desalination-thermal vapor compressor.

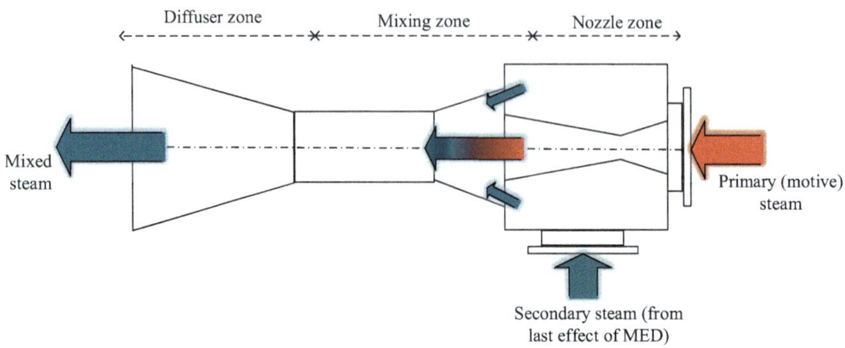

Fig. 2.15 ***Steam jet ejector.***

condensate is returned to the boiler and the remainder is collected as the distillate stream of the first effect. Through using a steam jet ejector, the latent heat of the low-pressure steam in the last effect of a MED system is recovered, which reduces the boiler load that supplies primary steam to drive the system. In fact, the energy contained in the primary steam can be transferred to the lower pressure steam (secondary steam) to produce a steam mix at an intermediate pressure.

Fig. 2.15 shows a steam jet ejector that can be used in a MED-TVC system. The ejector is mainly comprised of a nozzle zone, a mixing zone and a diffuser zone. First, the high-pressure primary steam enters the nozzle zone (a converging-diverging nozzle), where pressure is converted to kinetic energy. At the end of nozzle zone (i.e., the beginning of mixing zone), the artificially created low-pressure region draws secondary steam into the ejector. This phenomenon, called the entrainment effect, can be considered adiabatic because it occurs at a high speed leaving insufficient time for heat transfer. The primary and secondary steam flows are unified in the mixing zone. In the diffuser zone, the kinetic energy of the mixture is reconverted to pressure. The intermediate discharge pressure (i.e., the condensation pressure in the first effect of MED) is between the primary and secondary steam pressures and can be specified by the compression ratio. The compression ratio is defined as the pressure of the primary steam divided by the pressure of the secondary steam.

Another important ratio in the design and modeling of a TVC system is the entrainment ratio, which can be defined as the ratio of the mass flow rate of the primary steam to that of the secondary steam. An optimum value of entrainment ratio can be obtained by lowering mass flow rate of primary steam (Ji, 2010). An empirical correlation that can be used to determine the entrainment ratio is proposed by Power (2005) using a graphical data method. This diagram is in principle similar to the performance diagram of a centrifugal pump. Note that the function of an ejector can be explained by the Bernoulli equation, but this equation is not adequately accurate for writing the energy balance of the steam ejector, which uses compressible steam flow. Instead, the first law of thermodynamics along with momentum transport correlations can be used for a steam ejector.

The capacity of a TVC depends on the amount of primary steam, the primary and secondary steam pressures, and the intermediate steam pressure. An optimum compression ratio and, more generally, the optimum performance of a TVC improves significantly MED system efficiency by reducing the amount of motive steam supplied from an external boiler.

2.2.2.6 Non-condensable gases

In a desalination process, NCGs can flow within the condenser/evaporator and cause a local reduction in performance and efficiency and an increase in the cost of the desalination unit. NCGs primarily contain air (mainly O_2 and N_2) and CO_2. The main reasons for the presence of NCGs follow: penetration of air through leakage from small pores and holes in the connection pipes and containers; liberation of dissolved atmospheric gases in the feed seawater input (the solubility of CO_2 decreases as the pressure drops and CO_2 is released into the distillate vapor); and release of CO_2 from the carbonic acid decomposition chemical reaction. Methods for determining the amount of CO_2 released via carbonic acid decomposition can be found elsewhere (Glade, 2008; Millero, 1996).

The temperature difference between sequential effects in MED is small (about 1–4°C). The concentration of NCGs increases through the sequential effects. The presence of NCGs may cause a local reduction in the effect temperature (due to the condensation of vapor at a reduced pressure) to a level that results in a non-effective heat transfer process. Also, as the vapor condenses, the concentration of NCGs increases, forming a NCG film between the vapor and condensed distillate. This film prevents the vapor from coming into direct contact with the cool tube condensing surface, resulting in a diminished heat transfer rate. For instance, in a HTE, the presence of 0.11 mole % of nitrogen results in a 65% reduction in the overall heat transfer coefficient (Khan, 1973). To compensate for this reduction in heat transfer rate, the heat transfer area is often increased to 8%–10% more than its original value.

There are various prevention methods to remove NCGs from thermal desalination systems. Pumping out the low-concentration NCGs can be performed via suction from the distillate vapor, which prevents NCGs from participating in the condensation process. This requires a proper analysis to find the optimization characteristics required for a design. Also, using a deaerator in each effect can remove dissolved atmospheric gases. Adding certain chemicals to the deaerator can pretreat and decarbonate the feed seawater input to each effect. A proper venting system is also needed to eliminate the carbon dioxide released from the carbonic acid decomposition reaction in each effect.

2.2.2.7 Scaling and corrosion

Scale formation is common operational problem in desalination plants. Scale can be alkaline or sulfate and it is formed by suspended solids and impurities being precipitated from seawater. Scale often forms on heat transfer surfaces. In other words, scale is caused by constituents (like salts) that have limited solubility in the seawater as they precipitate out when solubility limits are exceeded.

In addition, corrosion problems can occur in desalination plants because of the presence of dissolved oxygen in water at high operating temperatures. Corrosion is particularly problematic as it reduces or limits the lifetime of a desalination system. Thus, it is usually necessary to utilize corrosion-resistant materials in desalination facilities.

Scale formation adversely effects the heat transfer performance of desalting processes as it raises the tube wall thermal resistance. In fact, the scale formed has a low thermal conductivity that acts as an insulating layer between the boiling water and heated steam which significantly reduces the evaporation rate within each effect. In a vertical tube evaporator, scale formation on the inside the surface of tubes increases the pressure drop, and thus increases the pumping power. In a horizontal tube evaporator, scale is formed on the outside surface of tube. Based on the experimental studies, the heat transfer rate is seen to decline by about 25% when the thickness of scale on the surface of a tube grows to 5 mm (Khan, 1973).

The heat transfer area of a condenser/evaporator is often increased (by up to 40%) to offset the adverse thermal effects of scaling. This increase in heat transfer area leads to the increase in the capital cost of desalination system as well as pumping power. Some approaches that can be employed to reduce the effects of scaling follow: reducing the plant operational temperature, cleaning scales by brushing the pipes, and spraying high-pressure water on the outside of the evaporators, pretreatment of feed seawater with chemicals (e.g., antiscaling compounds, acid) and controlling brine concentration.

2.2.3 Membrane distillation desalination

MD is a promising technology for desalination of high salinity waters like brackish water and seawater. MD is driven by thermal energy using microfiltration (MF) techniques that permit only water vapor molecules to pass across a porous hydrophobic (non-wetting) membrane (see Fig. 2.16). Specifically, the surface tension forces of a hydrophobic membrane do not allow liquid molecules to pass through the membrane pores, while the vapor pressure difference, resulting from the temperature difference between both sides of the hydrophobic membrane, drives vapor to pass through the membrane pores.

It is interesting and informative to compare MD and other desalination systems. A MD system is capable of operating at relatively low-temperature ranges (60–90°C) compared to other thermal desalination technologies. Note in particular that it is not required with MD to heat saline water

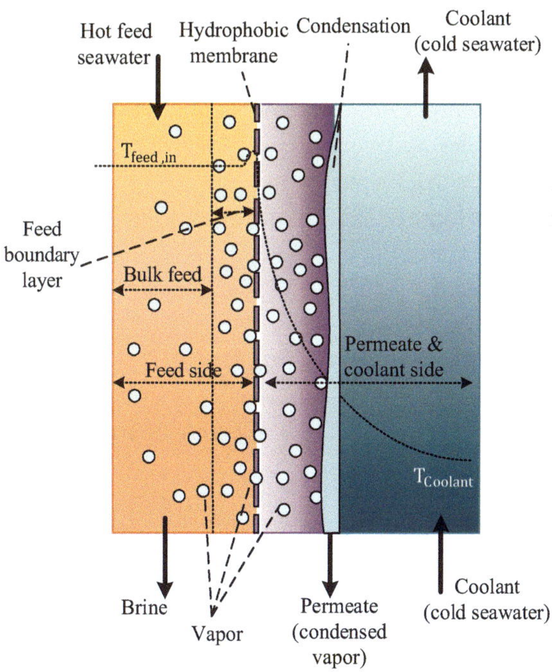

Fig. 2.16 ***Membrane distillation using a hydrophobic membrane.***

to its boiling temperature. In addition, MD systems operate at relatively low pressures (zero to a few hundred kPa) compared to membrane-based pressure-driven desalination techniques such as RO. This is because the driving force in the MD distillation process is not based on a hydrostatic pressure difference. As a consequence, MD desalination systems are not energy and cost-intensive. Also, the size of pores in MD systems is larger than those for RO desalination devices. This leads to less fouling phenomena in MD systems. Fouling results from the accumulation of suspended solids, microbes, or organics on the surface of the membrane, usually on the feed/concentrate side.

Another advantage of MD systems is their high rejection factor, which theoretically allows permeate to be produced with no nonvolatile species such as ions, colloids, and macromolecules. The presence of surface-active agents in the saline water can cause membrane pore wetting because these substances reduce the hydrophobic characteristics and surface tension of the membrane, leading to a decrease in the performance of the MD process. Thus, MD systems are more appropriate for applications in which water is the main constituent in the saline solution. In addition, MD also provides an effective method for removing organic and heavy metals, radioactive wastes, and brine wastes. It is also noted that MD systems can be combined with other desalination systems using conventional or sustainable energy sources.

Studies on MD processes approximately stem from 1980. Although the working principles and benefits are understood and have been demonstrated successfully, MD systems are still undergoing investigations to permit utilization at large scales. Disadvantages like inefficient membranes and conductive heat losses need to be addressed before MD can be applied significantly in industry.

2.2.3.1 Membrane distillation configurations

Various configurations of MD systems are shown in Fig. 2.17 and described in this subsection. They differ mainly in terms of vapor pressure difference across the membrane.

Direct contact membrane distillation (DCMD): The DCMD system is the simplest MD system configuration that is widely used for desalination. The relatively high-temperature feed stream (hot solution) is in direct contact with the hot side of the membrane. The feed stream evaporates at the feed membrane interface. The pressure difference across the membrane moves the generated vapor to the permeate side and then the vapor is

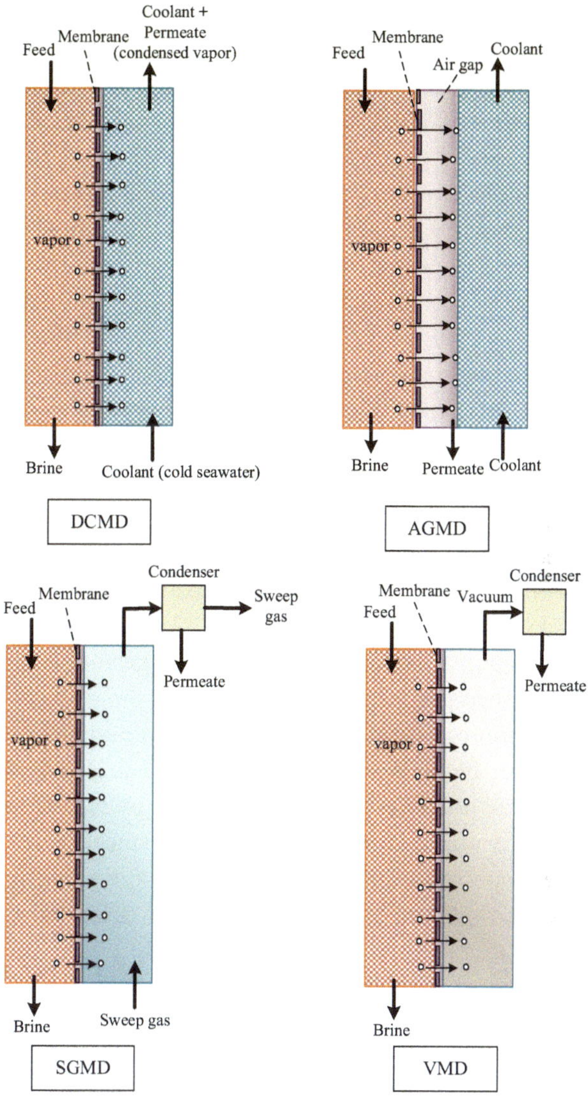

Fig. 2.17 ***Configurations of membrane distillation.***

condensed on the other side of the membrane. As mentioned before, the hydrophobic characteristic of the membrane allows only vapor to pass through the membrane pores and blocks liquid feed water. The condensation of vapor on the cooler side of the membrane significantly increases the sensible heat loss (i.e., conduction heat loss). Note that this effect can be diminished by using the air-gap MD configuration, which is discussed next.

Air-gap membrane distillation (AGMD): In the AGMD configuration in Fig. 2.17, the hot feed solution is in direct contact with the hot side of the membrane, while a stagnant air gap exists between the membrane and the condensation surface (cold side). The vapor passes through the air gap and condenses over the cold side of the membrane. However, the conductive heat transfer resistance increases considerably by introducing this air gap, and an additional mass transfer resistance is formed that lowers the permeate flux. Increasing the temperature difference between the two sides of the membrane can compensate for this disadvantage.

Sweeping gas membrane distillation (SGMD): To mitigate the mass transfer resistance deficiency of the AGMD configuration, a cold inert gas can be applied to sweep the vapor at the permeate side of the membrane, carrying away the vapor molecules and condensing the vapor outside of the membrane unit (see Fig. 2.17). In fact, there is a moving gas barrier in the SGMD configuration (unlike in AGMD where a stationary gas barrier is used). This moving gas barrier improves the mass transfer coefficient. One of the drawbacks of this configuration is that a large condenser is required to condense the vapor because a small volume of permeate is diffused in a large volume of sweep gas.

Vacuum membrane distillation (VMD): The conductive heat loss in the membrane can be reduced if a vacuum pressure (a pressure lower than the saturation pressure of pure water) is applied. To achieve this, a pump is used to create a vacuum on the permeate membrane side. Condensation occurs outside of the membrane unit. Fig. 2.17 shows a vacuum MD. Although the conductive heat loss is low in a VMD configuration, the membrane can be wetted if the hydrostatic pressure across the membrane exceeds the liquid entry pressure of the pores (usually between 1 and 4 bar). The liquid entry pressure of pores depends on the physical properties of the membrane (pore size, material, etc.) and the surface tension of the feed solution. Consequently, the pore size in VMD membranes (i.e., a diameter less than 0.45 μm) is smaller than for other types of MD systems.

The MD configurations in Fig. 2.17 can be used for different applications. For instance, DCMD is often applied in desalination operations, the food industry, and acid manufacturing. The AGMD configuration can be employed in desalination processes and also in separating volatile components from solutions. The SGMD and VMD configurations can be used for separation of aqueous volatile solutions.

2.2.3.2 *Mass transfer resistance in membrane distillation*

In the MD system, a liquid–vapor equilibrium state corresponding to the pore pressure and membrane temperature can be assumed. The mass transport process may occur in three regions (see Fig. 2.16): from bulk feed to membrane surface at the feed side in the liquid phase; through the pores of membrane in the gaseous phase; and at the permeate side to the bulk permeate (in liquid or gas phase depending on the type of MD). Mass transfer via MD is limited by mass diffusion across the boundary layers. Analyses of the mass transfer processes in an MD distillation system are important. Basically, the mass transfer process is dominated by Knudsen diffusion, Poiseuille flow (viscous flow), and molecular diffusion mechanisms.

These three mechanisms result in various mass transfer resistances, such as viscous resistance due to momentum transfer to the membrane, molecular resistance due to the collisions of molecules with each other and Knudsen resistance associated with the membrane itself. The surface resistance can be considered insignificant since the surface area of the MD membrane is small compared to the pore area. Fig. 2.18 shows an electrical analogy of the mass transfer resistances within a MD system.

During the operation of a MD system, a concentration polarization layer may be created due to the concentration of solutes in the membrane exceeding the corresponding concentration of the bulk feed. The concentration polarization raises the mass transfer resistance and in some severe conditions may cause scaling on the surface of the membrane. Scaling is a result of precipitation of saturated salts on the surface of the membrane. The mass transfer resistance within the pores of the membrane can be attributed to the collisions of molecules diffusing into the membrane, Knudsen diffusion (mass transfer resistance within the membrane itself), and Poiseuille flow (i.e., viscous flow) resistances.

In MD systems, mass and heat transfer through the hydrophobic membrane take place simultaneously. Mass transfer occurs through the

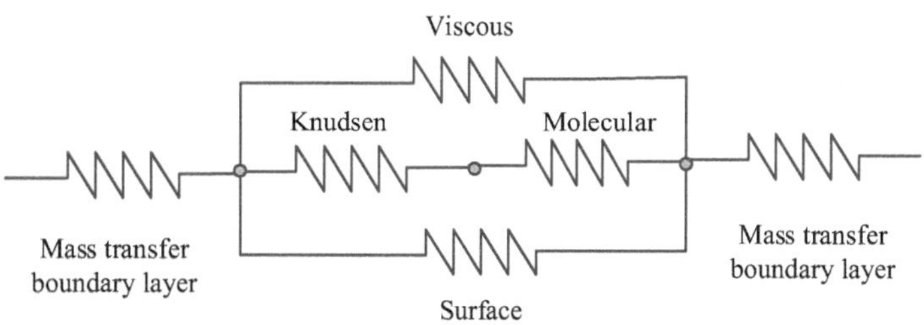

Fig. 2.18 ***Mass transfer resistances in a membrane distillation process.***

membrane pores while heat is transferred through both the pores and the membrane material. The heat transfer process from feed to the surface of the membrane provides the strongest resistance to the mass transfer within the membrane. This is due to a phenomenon called temperature polarization. Temperature polarization takes place intrinsically due to the water evaporation process that leads to a decline in the solution temperature at the surface of membrane and results in a loss of driving force. In fact, the formation of thermal boundary layers on the feed and permeate sides decreases the permeate flux because of the resistance imposed on the heat transfer process from the bulk feed to the permeate. Basically, there are two main heat transfer mechanisms across the membrane: (1) conduction across the membrane and the pores filled with gas, which is considered as the main source of heat loss and needs to be kept low or minimized; and (2) latent heat transfer via evaporation, which is unavoidable because it is related to the mass transfer mechanism.

Despite the advantages of MD systems, there are some disadvantages that limit their applications and sizes. First, the permeate (freshwater) flux is typically lower in MD systems than RO systems. Also, production rate of permeate is highly dependent on the temperature and concentration of feed water input. Furthermore, air can be trapped within the membrane, which increases the mass transfer resistance and thus affects the permeate flux. Finally, heat loss by conduction is quite high in MD systems, so it needs to be controlled.

Both temperature and concentration polarizations adversely affect the performance of MD systems. Conduction heat loss through the membrane and mass transfer resistance in the pores are the results of temperature polarization and concentration polarization, respectively. Some technological methods have been investigated to diminish the negative effects of temperature polarization. Improving hydrodynamic conditions by increasing Reynolds number or applying turbulence promoters can potentially improve the mass and heat transfer mechanisms at the same time (Ali et al., 2013; Martínez-Díez and Vázquez-González, 1999). In addition, the latent heat of evaporation can be recovered by the coolant flow so as to preheat it for use as a feed stream. This method can potentially improve the thermal efficiency of a MD system. This method is applicable to DCMD and AGMD configurations and can potentially improve their thermal efficiencies. For the other two MD configurations, SGMD and VMD, heat recovery is not straightforward as the condensation process occurs at the external condenser.

2.2.3.3 Membrane characteristics

The general desirable characteristics of membrane modules in MD systems include low thermal conductivity (to prevent heat loss in the membrane), high liquid entry pressure, and low resistance to mass transfer. Other characteristics required of the membrane module are sufficient thermal stability at high temperature and high resistance to chemicals (like bases and acids). Membranes should be hydrophobic and microporous. They are typically made from polypropylene, polyvinylidene fluoride, or polytetrafluoroethylene.

Increasing the porosity and size of pores in a MD membrane results in an increase in the permeate flux, while the membrane tortuosity and thickness conversely effects the permeate flux. The tortuosity of a membrane can be described as a relation between the membrane thickness and average length of the pores. The common size of pores usually varies between 0.01 and 1 μm depending on hydrostatic pressure needed across the membrane, which should be below the liquid entry pressure of the pores (Lienhard et al., 2012). The smaller is the pore size, the higher is the mass transfer resistance. The porosity of membranes typically ranges from 30% to 80% (Lienhard et al., 2012). An increase in the porosity of a membrane causes the conductive heat loss to decline, leading to a large evaporation surface area. Increasing the thickness of the membrane raises the mass transfer resistance, while reducing the conductive heat loss. Thus, an optimum membrane thickness exists; a value between 30 and 60 μm is suggested (Alkhudhiri et al., 2012). The thicker is the membrane, the lower is the permeate flux. This is generally true except for the air-gap MD configuration, in which the resistance from the air gap greatly exceeds that from the thickness of the membrane.

2.2.3.4 Membrane module designs

A membrane module should be designed so as to provide high heat and mass transfer rates within the membrane. Heat and mass transfer resistances in boundary layers and fouling are the limiting factors in a well-operated of MD process. The membrane modules can be plate and frame (flat sheet), hollow fiber, tubular (capillary), or spiral wound.

In a flat sheet configuration, the membrane and spacers are layered together between two porous support plates. Fig. 2.19 shows a flat sheet membrane module. The retentate (or concentrate or brine) is the part of the feed that does not pass through the membrane. A number of plates are packed together in a module, and each plate includes frames with two

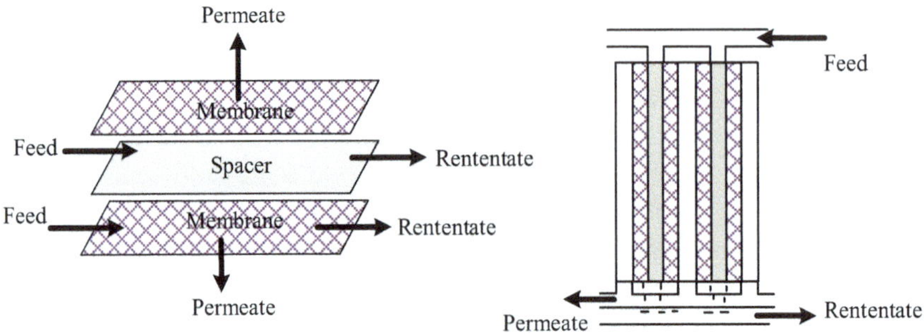

Fig. 2.19 ***Plate and frame membrane module.***

membranes, condensing walls, and an intermediate feed channel. Packing density, which is defined as the ratio of membrane area to the packing volume, is relatively low in this type of module and it depends on the number of membranes. Flat sheet membranes are easy to clean and replace and are often used in MD applications like desalination and water treatment.

In hollow fiber modules, thousands of hollow fibers are sealed in a shell tube (see Fig. 2.20). There are two flow passages, for the feed and the permeate. The feed flows from outside of the hollow fibers and the permeate is collected from inside of the hollow fiber, or the feed flows inside the hollow fibers and the permeate is collected from outside of the hollow fibers. In the hollow fiber module, the package density is high and the energy consumption is relatively low, two key advantages of hollow fiber module. But, there is a high risk of fouling and damage in this type of module which is not easy to clean and maintain.

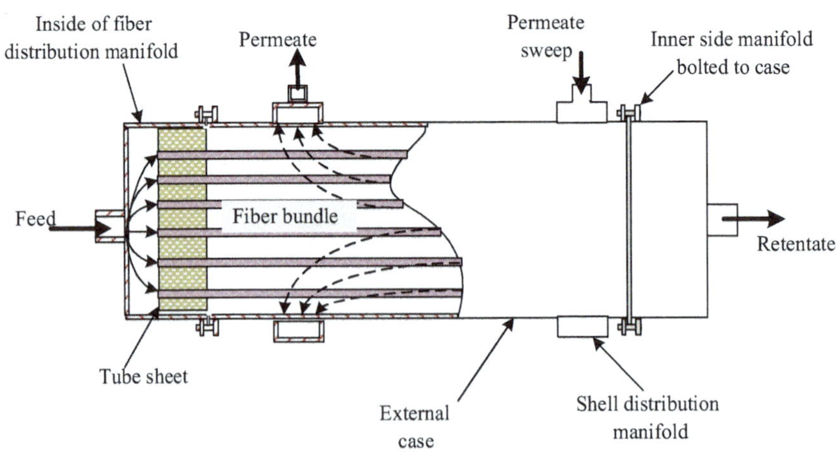

Fig. 2.20 ***Cut-away view of a typical hollow fiber membrane module.***

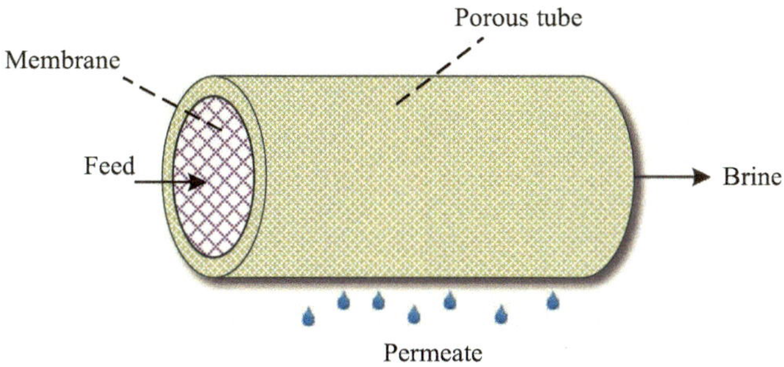

Fig. 2.21 ***Tubular membrane module.***

In the tubular (capillary) membrane module (see Fig. 2.21), tube-shaped membranes are placed between two cylindrical chambers containing hot and cold fluids. The membranes resemble a shell and tube heat exchanger in which radial flow occurs. The flow velocities need to be high to reduce or minimize the effect of polarization. Some advantageous features of this type of membrane module are low risk of fouling, high effective area, and easy maintenance and cleaning. These features make the tubular module suitable for MD and other commercial applications. There are some drawbacks to this type of module, the main ones being a high operating cost and a low packing density.

In a spiral wound membrane module, shown in Fig. 2.22, flat sheet membrane and condenser foils are rolled around a perforated collection tube. The feed solution flows across the surface of the membrane in an axial direction and the permeate flows toward the perforated collection tube in a radial direction. The packing density is in the moderate range and varies based on the channel height. Thermal energy can be recovered from the

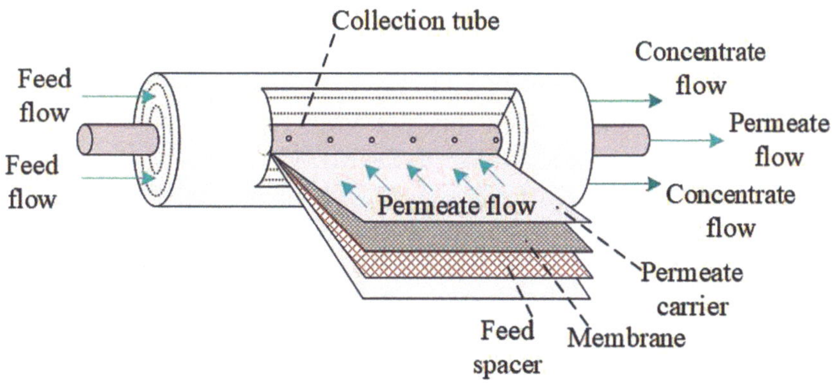

Fig. 2.22 ***Spiral wound membrane module.***

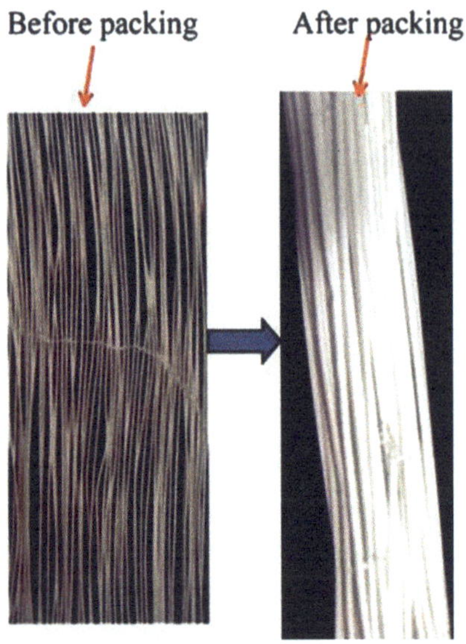

Fig. 2.23 ***Structured array membrane module.*** *From Yang (2011).*

surface of the membrane, but in a controlled manner to avoid a large drop in the transmembrane temperature difference. The levels of fouling and energy consumption in this type of membrane module tend to be in the acceptable range.

A well-operated MD system is achieved by designing membrane modules that lower the effects of temperature and concentration polarizations and fouling within the membrane. A recent design has been fabricated using hollow fiber module to yield a more uniform flow distribution by weaving all fibers into a fabric sheet that is rolled up and packed (see Fig. 2.23). However, an adequate design for scaling-up of MD membrane modules has not yet been developed. Most of the pilot-scale membranes deployed (like spiral wound or hollow fiber types) were originally designed for other membrane processes (like RO and membrane contactors) that can be scaled up by upsizing the module capacity.

The concept and theory of cascade DCMD was introduced by He et al. (2013). With this approach, heat recovery is accomplished by heat recovery stages. Fig. 2.24 shows a schematic of a cascade module design with heat recovery stages. In the heat recovery process, the permeate from the previous stage is used to heat the cold feed. Based on an analysis of the cascade configuration of DCMD with heat recovery (He, 2013), water recovery of up to 60% and a significant improvement in the energy efficiency of the system were achieved.

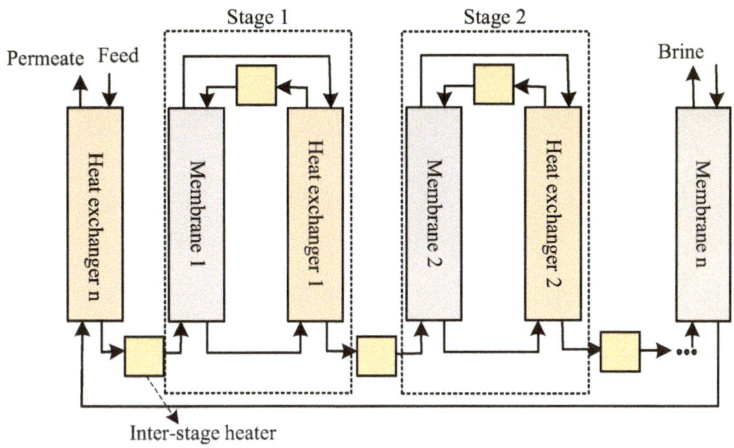

Fig. 2.24 ***Cascaded MD module design with inter-stage heating.*** *Modified from He (2013). MD, membrane distillation.*

Overall, MD systems have significant potential for scale-up for industrial desalination and water treatment applications, due to their attractive features and characteristics. With further technological improvements, the capacities MD systems can likely be increased while permeate production costs and energy requirements are decreased. Developments in efficient modules, highly permeable membranes, innovative hybrid integrations, cascade modular configurations, and energy recovery approaches are necessary to establish economically feasible and large-scale desalination processes using MD units.

2.2.4 Reverse osmosis desalination system

The phenomenon of osmosis has been recognized for many years. This phenomenon is illustrated in Fig. 2.25 and can be defined as a spontaneous movement of water molecules through a semipermeable membrane from

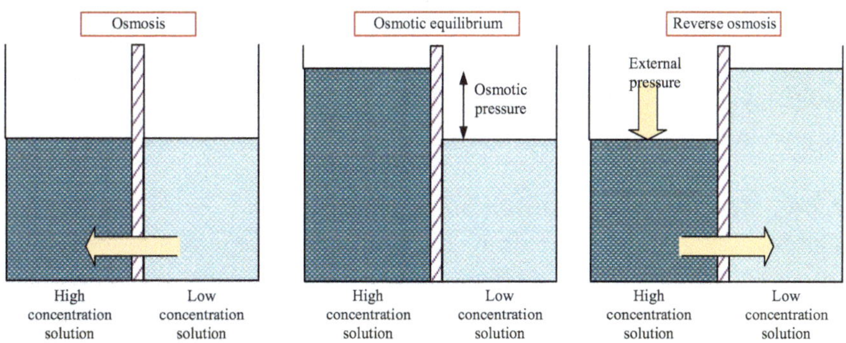

Fig. 2.25 ***Concepts of osmosis, osmotic equilibrium, and reverse osmosis.***

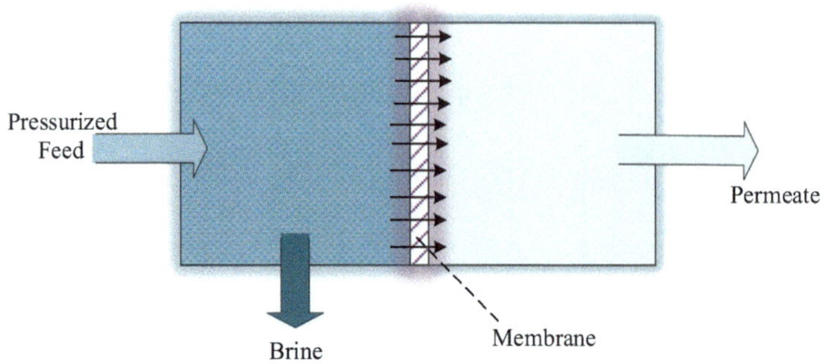

Fig. 2.26 ***Continuous reverse osmosis desalination process.***

a region of low concentration of solute, which has low osmotic pressure, to a region of high concentration of solute with high osmotic pressure. The solutes are rejected by the semipermeable membrane and only water is allowed to pass through it. The osmosis process continues until the equilibrium state is attained, at which equal chemical potentials exist across the membrane (see Fig. 2.25). Osmotic pressure is related to the colligative properties of a solution like boiling point elevation and freezing point depression. For an ideal dilute solution, the osmotic pressure is a function of molar concentration and absolute temperature of the solution. For non-ideal solutions, in addition to the temperature and molar concentration, properties like molar volume and activity of solvent also contribute to the osmotic pressure.

An external pressure can be applied to the high concentration solute to drive molecules of water in a reverse direction to that of the natural spontaneous movement that occurs with osmosis. The applied pressure needs to be greater than the osmotic pressure difference across the membrane to force water molecules to move in the opposite direction of osmosis. This process is known as RO and is illustrated in Fig. 2.25. RO can be used for desalination and water treatment. In a continuous RO process, the feed water input is split into permeate (low solute concentration) and brine (high solute concentration) (see Fig. 2.26).

2.2.4.1 Specific energy consumption of RO process

The SEC of a RO process can be defined as the energy consumption per volume of permeate. SEC mathematically is a function of the pressure at the feed and permeate sides and also the recovery ratio (ratio of permeate flow rate divided by feed flow rate). The recovery ratio depends on the feed salinity, temperature, efficiencies of devices (e.g., pumps, energy recovery devices (ERDs)), design configuration (single-stage, two-stage, etc.), pre-treatment and brine recovery considerations. The higher is the feed salinity, the higher is the osmotic pressure, which results in a higher SEC for RO desalination. In addition, temperature affects membrane physiochemical properties. As the temperature of the feed solution increases, both water and salt permeability coefficients and osmotic pressure increase. Thus, increasing the feed temperature throughout the range of interest (e.g., 15–40°C) results in inconsistent effects on the SEC. For instance, for a low-salinity fluid-like brackish water, increasing the feed temperature from 15°C to 40°C results in a decrease in the SEC of the RO process, while for higher feed salinities, increasing the feed temperature up to only 25°C might be beneficial in terms of the SEC of the RO process (Koutsou, 2020).

The SEC of industrial seawater RO plants is typically within the range of 2.5–4 kWh/m^3 (Voutchkov, 2018). The recovery ratio varies from 50% to 85% for the brackish water RO and 35% to 50% for seawater RO (Greenlee, 2009). The feed pump pressure is typically 6–30 bar for brackish water and 55–80 bar for seawater (Okamoto, 2019). The salt rejection in both brackish water and seawater systems is higher than 95%. The recovery design parameter should be selected in a balanced fashion, to balance the purity of the permeate and the concentration of the brine stream. A higher recovery ratio indicates a lower brine volume that needs to be discharged at the expense of a lower purity of permeate.

2.2.4.2 Pretreatment

In a RO desalination process, pretreatment of saline water is important before it enters the RO membrane module. There are several pretreatment methods; these are often custom designed based on the types of solutes and their concentrations in saline water. In a general water treatment process, several types of filtration technologies may be used, such as filtration, MF, ultrafiltration (UF), RO, and nanofiltration (NF). Fig. 2.27 shows the required pressure difference applied to the membrane and the range of membrane pore sizes in these filtration technologies. It is seen that filtration and MF membranes have the largest pore sizes, which normally prevents large particles and microorganisms from passing through the membrane. These membranes are commonly used in pretreatment processes to remove particulates from intake feed water. In this method, a two-stage filters are usually used to remove particles with diameters larger than 5 μm. The size of pores is smaller in UF than MF, preventing the passing of bacteria and

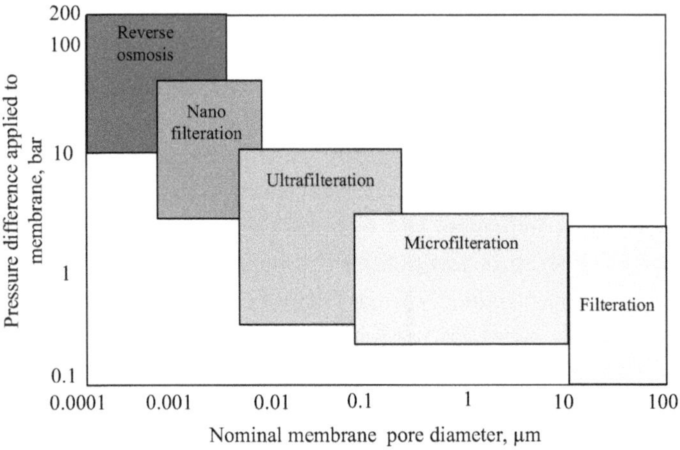

Fig. 2.27 ***Range of membrane pore size and pressure difference applied to the membrane for selected filtration technologies.*** *Based on Fritzmann et al. (2007).*

soluble macromolecules (in addition to large particles and microorganisms). RO membranes are effectively non-porous and prevent many low molar mass constituents like salt ions and organics. NF porous membranes are relatively new and are extensively used for pretreatment of saline feed water in RO desalination systems.

Furthermore, carbon filters can be used to remove free chlorine from the feed water, and chlorination or ozonation can be utilized for removing high levels of bacteria. Also, antiscalants can be used to prevent mineral scaling, especially for brackish water RO desalination processes. Antiscalants can help extend the lifetimes of RO membranes.

2.2.4.3 Concentration polarization in reverse osmosis desalination

In RO desalination, a convective flow of solutes from the bulk feed toward the membranes increases the solute concentration on the surface of the membrane, which forms a boundary layer. In this boundary layer, the solute concentration exceeds the bulk solute concentration. The higher solute concentration in the boundary layer leads to the back diffusion of solutes away from the membrane. This phenomenon, the accumulation of solutes near the membrane of RO, is called concentration polarization.

Concentration polarization adversely impacts the performance of RO desalination by increasing the osmotic pressure across the membrane surface. This leads to a decrease in the permeate flux, an increase in the salt flux across the membrane, and an increase in the risks of scaling and fouling.

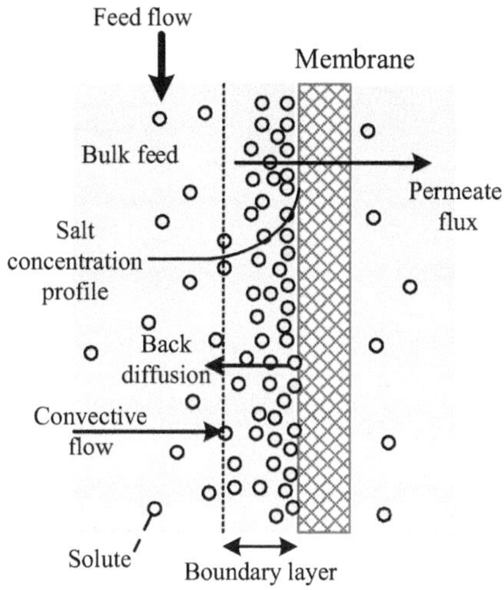

Fig. 2.28 ***Concept of concentration polarization in a reverse osmosis process.***

Membrane lifetime and permeate flux are primarily affected by concentration polarization and fouling/scaling at the membrane surface. A representation of concentration polarization and fouling phenomena on the surface of a membrane is provided in Fig. 2.28. The concentration polarization decreases the permeate flow rate because it increases the hydraulic resistance of the water flow. As a consequence, the net pressure across the membrane decreases due to the increase in osmotic pressure in the boundary layer. Measures that help to reduce concentration polarization include increasing the turbulence (velocity) of the brine flow, reactive micro-mixing, pulsation, the use of feed spacers and vibrating modules.

2.2.4.4 Membrane module designs

The membrane module designs for RO are the same as for the membranes in MD, as described in Section 2.2.1.3. Fig. 2.29 shows membrane module designs for RO processes. A well-performing industrial RO desalination process requires membrane modules with high packing density (which indicates the compactness of packing the membrane area), easy cleaning and replacement and low capital cost. The characteristics of various membrane module designs are presented in Table 2.1. Due to the low packing density of plate and frame and tubular configurations, these two designs are gradually being replaced by hollow fiber and spiral wound modules.

The plate and frame module design for a RO membrane is almost the same as that for a MD membrane (shown in Fig. 2.19). A flat sheet

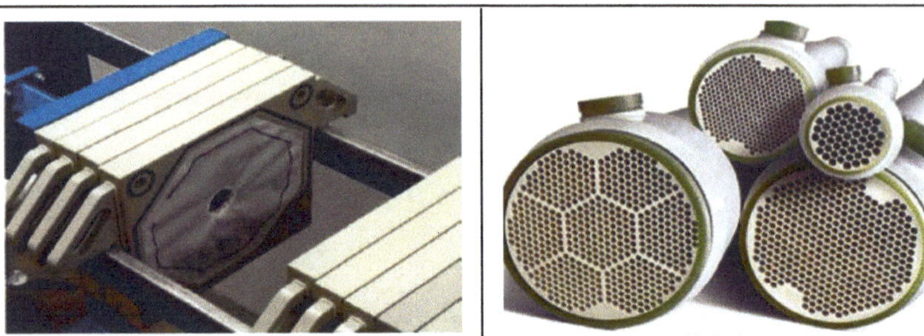

Fig. 2.29 ***Plate and frame (ECOGRACE, 2020) and tubular (Berghof, 2021) membrane modules.***

Table 2.1 Comparison of various membrane design modules in reverse osmosis process.

Module type	Packing density (ft^2/ft^3)	Manufacturing cost	Fouling propensity	Ease of cleaning
Plate and frame	45–150	High	Moderate	Good
Tubular	6–120	Very high	Low	Excellent
Hollow fiber	150–1500	Low	Very high	Poor
Spiral wound	150–380	Moderate	High	Poor

Data extracted from (Accepta, 2021) and (Kucera, 2010).

membrane is sealed to a plate which is made from porous fiber glass, solid plastic, or reinforced porous paper. The plates are stacked together by a pressurized support frame. There are grooved channels on the surfaces of the plates to direct the permeate flow. The feed flow stream is introduced at one end of module and the molecules of water pass through the membrane and then collect in a permeate collection manifold. The brine stream exits the membrane at the other end. For the plate and frame module design, cleaning of the plate and frame modules is easy, the packing density is low and there is a high tendency for fouling, due to the existence of waste spaces in the design. Also, due to the difficulties associated with the design and construction of these modules, they are quite expensive.

The tubular membrane design includes a tube-shaped membrane, which is typically made from fiberglass, carbon, paper, plastic, or ceramic. The feed flow is introduced at one end and water molecules pass through the membrane, while the remaining concentrated saline water (brine) leaves at the other end of the module. The tube-shaped membrane modules can be arranged in parallel or series to achieve higher permeate production capacities. Like plate and frame modules, tubular designs are easy to clean, but they are not economic. In addition, this module type has a low packing density, which has limited their applications mainly to UF and MF processes, rather than RO desalination.

Hollow fiber membrane modules are utilized in some of the largest scale seawater and brackish water RO desalination applications due to their high packing density and reasonable cost. In this type of system, feed water is pumped through a hollow fiber membrane water filter, where a hundreds of fibers (resembling tiny straws) are bundled together to form a filter matrix. The fiber walls are filled with microscopic pores or holes. The end of the bundle is sealed to prevent bypassing of feed solution to the brine outlet. The pressurized feed flows through a core porous tube, in which water molecules can pass through the fibers, and exits the module from the open end of the fiber bundle. There is a high tendency of fouling in this module due to the small fiber spacing, which makes the cleaning process difficult.

Spiral wound membrane modules are the other widely used design for RO desalination, primarily because they have a high packing density and are economic. In such a module, a plate and frame module is wrapped around a central permeate collection tube (pipe) to create a spiral configuration. The permeable membrane is sealed along three edges to build a membrane envelope and a spacer is used in between to form a leaf. The leaves are held together with a brine/feed mesh spacer to induce turbulence and reduce the risk of concentration polarization. The fourth edge is left open and connected to a central perforated permeate collector tube. A pressurized feed solution enters from one end and flows across the surface of the membrane in an axial direction, and a backpressure drives permeate toward the perforated collection tube in a radial direction. In spite of improved designs of spiral wound modules like multi-envelope designs in which the pressure drop due to the permeate movement is minimized, they have a high tendency for fouling and the cleaning of the membrane modules is not simple. Thus, a pre-filtration is important to prevent such problems.

The RO membrane modules are available in several sizes, ranging from small to large. For large industrial-scale seawater RO desalination plants, usually five to eight modules, each having dimensions of 10.16 cm by 20.32 cm by 40.64 cm (4 in by 8 in by 16 in), are connected in series in an RO pressure vessel (Haidari, 2018; Johnson, et al., 2010).

2.2.4.5 Membrane characteristics

A well-performing RO membrane exhibits high permeability and selectivity with little fouling to avoid a reduction in the permeate flux. Also, the membranes in a RO system need to be sufficiently tough so that they can be sustained at high operating pressures, and at the same time sufficiently thin so as to have a large membrane surface area per unit volume. Most of the currently used RO membranes are thin-film composite polyimide types, which have high salt rejection, water permeability, and tolerance to wide range of PH and temperature. Thin-film composite membranes are usually made from three layers of polyester, a polysulfonic porous support and cross-linked aromatic polyamide active layers (Kurihara et al., 2017) (see Fig. 2.30). The water permeability from these RO membranes is between 1 and 2 LMH/bar for seawater desalination, and between 2 and 8 LMH/bar for brackish water RO desalination. Note that LMH/bar is the unit for mean permeability, which is an average value of permeate flux divided by hydraulic pressure during one cycle. Also, newer RO membranes are capable of 99.7% salt rejection from seawater.

As noted earlier, membrane water permeability affects RO performance. For instance, many studies have demonstrated how increasing the water permeability decreases the SEC. Fig. 2.31 combines the results of three studies to show the relationship between membrane water permeability and SEC of the RO process for various operating conditions (feed salinity and pump efficiency). It is seen that SEC is more sensitive to variations in membrane permeability at higher feed salinities. For example, Cohen et al. (2014) demonstrate that, at a feed salinity of 42,000 ppm and a pump efficiency of 75%, increasing the membrane water permeability from 0.1 to 1.0 LMH/bar reduces the SEC by 64%. It is also observed that,

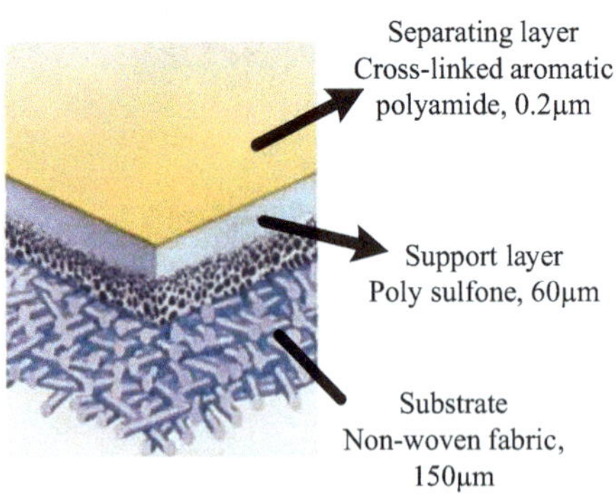

Fig. 2.30 ***Structure of composite reverse osmosis membrane.*** *From Kurihara et al. (2017).*

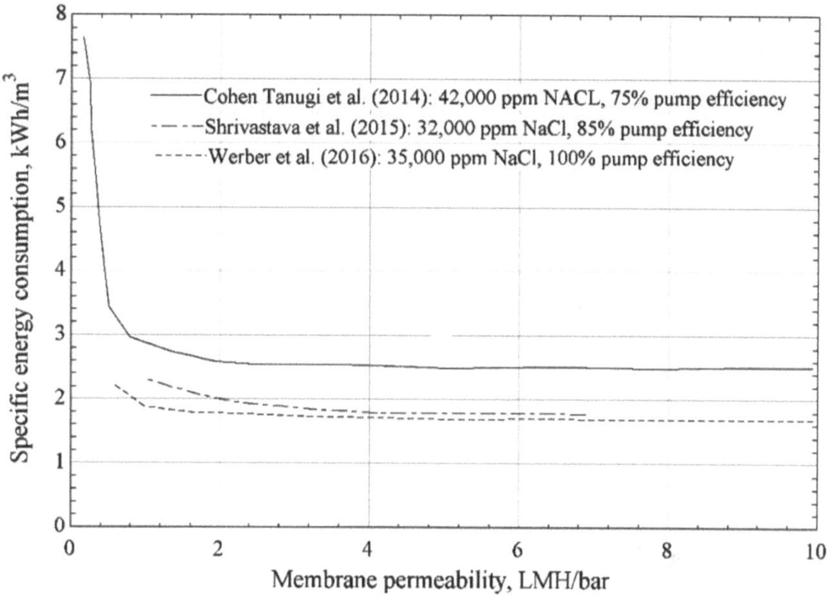

Fig. 2.31 ***Specific energy consumption in a RO module as a function of membrane water permeability at various values of salt concentration and pump efficiency.*** *Data extracted from Cohen-Tanugi et al. (2014), Shrivastava et al. (2015), and Werber et al. (2016).*

for a membrane water permeability exceeding 3 LMH/bar, the effect of increasing membrane permeability on the SEC of the RO process is almost negligible.

2.2.4.6 Reverse osmosis configurations for enhanced energy efficiency

RO desalination conventionally consists of a single-stage RO process. In a single-stage RO desalination system, approximately 45% of the feed input is converted to permeate (i.e., the system has a 45% recovery ratio). Hence, more than half of the feed stream is discharged as brine. To increase the recovery ratio, a second-stage RO component can be added to the single-stage RO. This configuration is known as two-stage RO and permits a higher recovery ratio (60%) to be achieved (Kim et al., 2020). Recently, a two-stage RO systems exhibit better performance in terms of SEC, especially for seawater desalting. In a two-stage RO desalination system, shown in Fig. 2.32A, the brine stream from the first-stage is pressurized further to overcome the osmotic pressure of the brine concentrated solution, and then fed to the second stage. In another configuration, the permeate stream is further desalted by passing it through a two-stage RO system

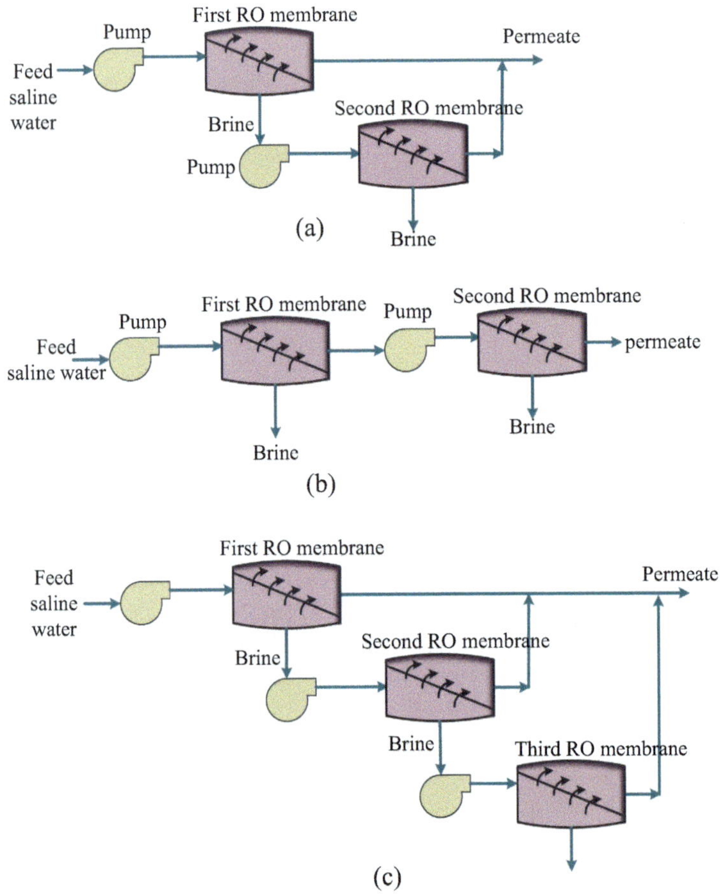

Fig. 2.32 (A) Two-stage RO in which brine is desalted in the second stage, (B) two-stage RO in which permeate is further desalted in the second stage, (C) three-stage RO desalination process. RO, reverse osmosis.

(see Fig. 2.32B). As mentioned previously, a higher recovery ratio can be obtained by two-stage seawater RO desalination at the expense of higher energy consumption. However, for a same recovery for both single- and two-stage seawater RO desalination systems, the two-stage configuration consumes less energy than the single-stage configuration.

For a multi-stage configuration of RO desalination, a serial arrangement of RO membrane modules is more energy efficient than a parallel arrangement. In a serial arrangement, there is an increase in concentration through the process toward the last stage. If the theoretical minimum energy required for desalting is defined as the energy required to pressurize

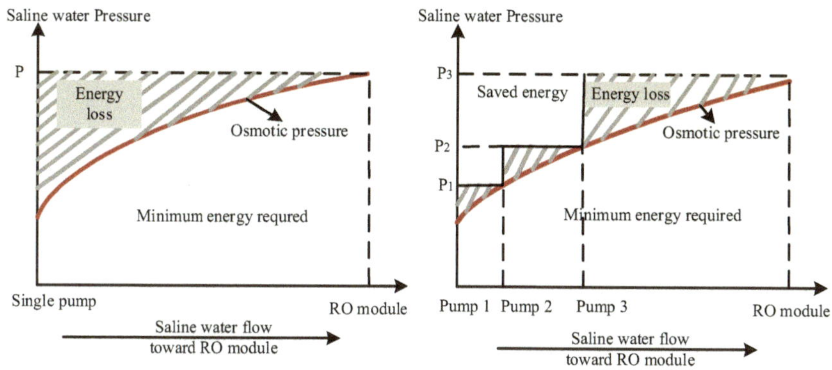

Fig. 2.33 ***Illustration of energy saving attained by three-stage reverse osmosis desalination compared to single-stage reverse osmosis desalination.***

saline feed water to the maximum osmotic pressure in the corresponding module, for a single-stage RO process the applied pressure must be at least equal to the osmotic pressure of the brine at the outlet of RO membrane. A three-stage RO desalination process is shown in Fig. 2.32C. Since the first stage in such a process operates at lower salt concentration, the corresponding applied pressure (P_1) is smaller than the pressure of the second stage which operates at a relatively higher water salinity and consequently higher pressure (P_2). Similarly, the applied pressure of the third stage (P_3) is higher than that of the second stage due to the higher water salinity. Fig. 2.33 provides an illustration of the energy saved by a three-stage RO desalination system compared to a single-stage RO desalination system. It is seen from Fig. 2.33 that in three-stage RO configuration the energy loss is reduced.

2.2.4.7 Energy recovery device

The concentrated brine stream leaves the RO membrane module at high pressure. The energy of this high-pressure brine stream can be harnessed to produce work via an ERD. The energy efficiency of seawater RO desalination systems can be significantly improved by using an ERD, with some showing that the SEC of seawater RO plants can be reduced by up to 25% (Farooque et al., 2008). There are various types of ERDs used in the seawater desalination plants.

Francis and Pelton turbines were the first ERDs used in RO plants. These turbines allow the pressure of brine to be converted to mechanical energy (see Fig. 2.34). The turbines are coupled to a shaft to supply a portion of the electrical power used by the feed pumps. These types of ERDs are not very efficient, in part because two energy conversion processes are

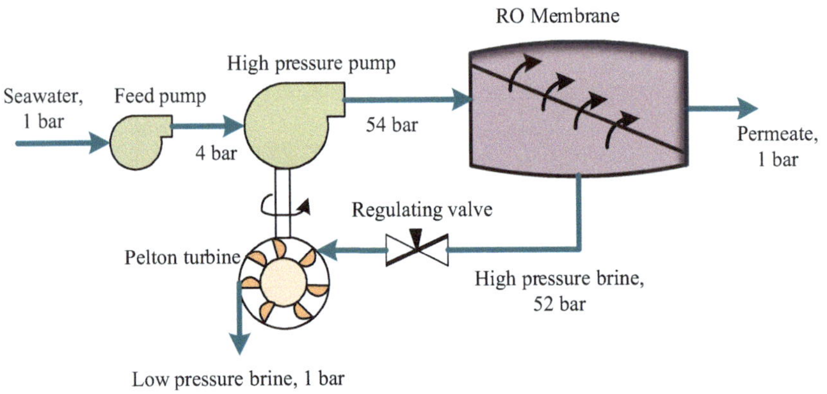

Fig. 2.34 ***Pelton turbine used as an energy recovery device in a RO desalination system.*** RO, reverse osmosis.

involved: conversion of the hydraulic energy in the brine to mechanical energy (to rotate the shaft) and the conversion of mechanical energy to hydraulic energy in the feed stream.

Turbochargers are a widely adopted ERDs that recover hydraulic energy from the brine stream and transfer it to the feed stream. Turbochargers consist of a pump section and turbine section. Fig. 2.35 shows a typical turbocharger installation for a seawater RO system. First, a brine stream is introduced to the turbine section, where the hydraulic energy of brine is converted to mechanical energy. Then, in the pump section, the mechanical energy produced by the turbine is converted to pressure in the feed flow. This type of ERD is designed for reliable performance and high availability. Turbochargers have relatively high energy efficiencies compared to Pelton turbines (Gude, 2018).

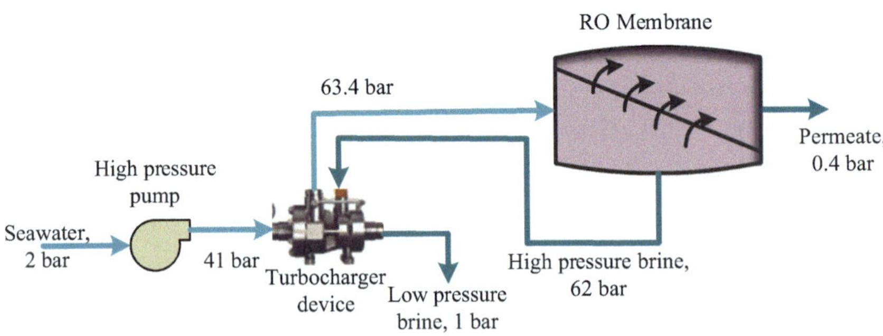

Fig. 2.35 ***Typical turbocharger used in a seawater RO desalination system.*** RO, reverse osmosis.

Alternatively, isobaric ERDs like dual work exchanger energy recovery devices and pressure exchangers (PXs) can be utilized in RO desalination process to yield higher energy efficiencies. However, a slight increase in the feed salinity may occur in the pressure exchange process. In isobaric ERDs, the hydraulic energy of brine is directly transferred to the feed stream without an intermediate conversion to mechanical energy. Fig. 2.36 shows a typical pressure exchanger used in a seawater RO process. In this unit, a ceramic matrix is caused to rotate by the brine flow. The channels are alternatively connected to the high-pressure brine and high-pressure feed pipes or to the low-pressure feed and low-pressure brine pipes. In the first case, channels that were previously filled with feed water are now refilled with brine, forcing the feed stream to flow into the high-pressure feed pipe. In the second case, the feed water forces saline water to refill the channel with feed at low pressure.

These types of ERD units usually have high energy efficiencies (above 95%), due to the direct energy transfer process they entail. Pressure exchanger devices require lower operating pressures and they are often

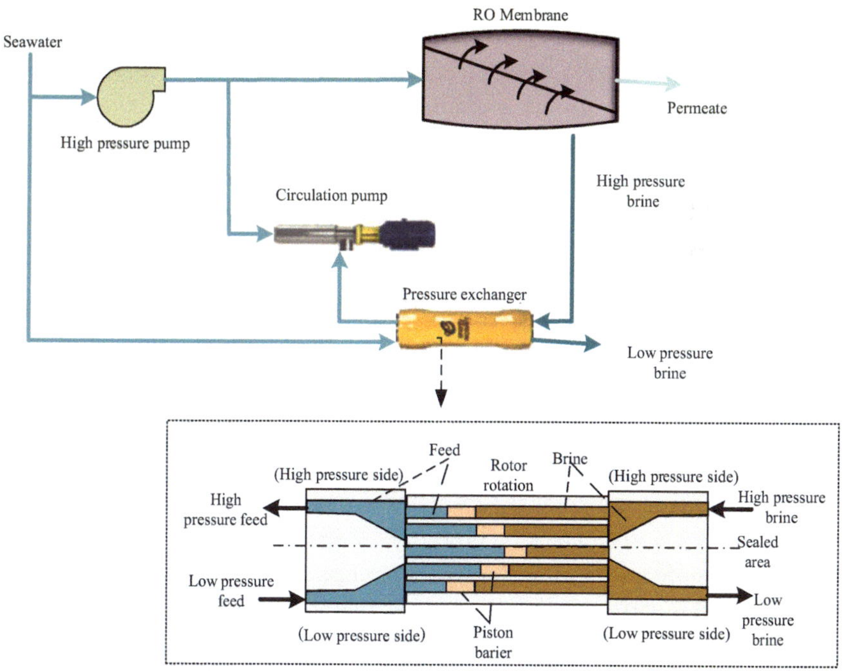

Fig. 2.36 ***Use of a pressure exchanger in the form of an isobaric energy recovery device in a RO desalination system.*** RO, reverse osmosis.

used for large-scale desalination plants with energy efficiencies exceeding 90%. They are noisy while working. Pelton turbines and hydraulic pressure boosters require higher operating pressures and they are commonly used for small-scale desalination systems with lower energy efficiencies. More details on designs and working principles of ERD units can be found elsewhere (Gude, 2018).

Note that to achieve relatively higher water recoveries at lower SEC values, ERD systems can be used in multi-stage (e.g., two-stage and three-stage) RO configurations. The use of optimal designs for such systems is very important to yield the desired permeate quality, with high energy efficiency and low operating and capital costs.

2.2.5 Humidification–dehumidification desalination process

The humidification–dehumidification (HDH) desalination method basically originated from the natural water cycle in which solar energy vaporizes water from the oceans. The generated vapor is carried by air, then condensed as rain when the temperature of moist air drops.

The solar still constitutes the most basic configuration of a HDH desalination system and is based on the same process as the natural water cycle. Although the design and operation of a solar still is relatively simple, the energy efficiency of solar stills is very low, mainly due to the loss of latent heat of condensation to the environment. By separating the evaporation and condensation process and introducing regenerative heating, a more energy-efficient form of desalination process evolved, which is now known as HDH desalination.

Although HDH desalination has been improved significantly over the last several decades, it still in general is not competitive with RO, MED, and MSF desalination technologies for medium and large-scale freshwater production applications. The main drawback of the HDH method is its relatively higher thermal energy requirement compared to the other technologies. Nevertheless, HDH desalination systems are applicable and advantageous for off-grid and decentralized desalination applications, where freshwater demand typically does not exceed 7000 m^3 per day. Some important advantages of HDH desalination are the fact that is does not need membranes and its ability to operate over a wide range of water salinities without the need for complex maintenance and expensive corrosion resistance materials. HDH technology has been commercialized successfully in certain applications, particularly in producing water from gas wells.

2.2.5.1 Classification of HDH desalination systems

The basic components of a HDH desalination system are a humidifier (evaporator), a dehumidifier (condenser) and a heater. HDH systems can be classified based on whether water or air is heated and based on whether the water or air circuit is open or closed.

One type of HDH desalination system, a water-heated, closed-air open-water HDH system, is shown in Fig. 2.37. In this configuration, cold dry air enters the humidifier where the high-temperature feed saline water is sprayed over a packed bed. The temperature and water content of the air increases in the humidifier. Then the high temperature moist are is directed to the dehumidifier, where the water content of moist air is condensed through condensing coils containing feed water. The feed stream is preheated as it absorbs the latent heat of the condensation vapor and then it is further heated by a heater in which solar energy, natural gas or another energy source is used to supply heat for the feed water. The condensed water vapor exits the dehumidifier as freshwater and the remaining concentrated saline water in the humidifier is discharged as brine. One method to improve the energy efficiency of this HDH system is to make the dehumidifier design more effective, so it preheats the feed water as much as possible before it passes to the heater.

Another type of HDH desalination system is the air-heated, closed-air, open-water system, which is shown in Fig. 2.38. In this configuration, air is heated instead of water in the heater. The air heater can be located either prior to the humidifier or prior to the dehumidifier. In the first case, hot unsaturated air which can hold much water vapor, enters the humidifier

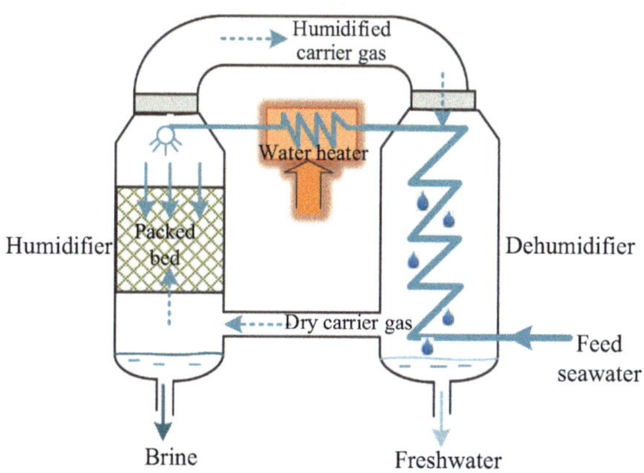

Fig. 2.37 ***A water-heated, closed-air open-water HDH desalination system.*** HDH, humidification–dehumidification.

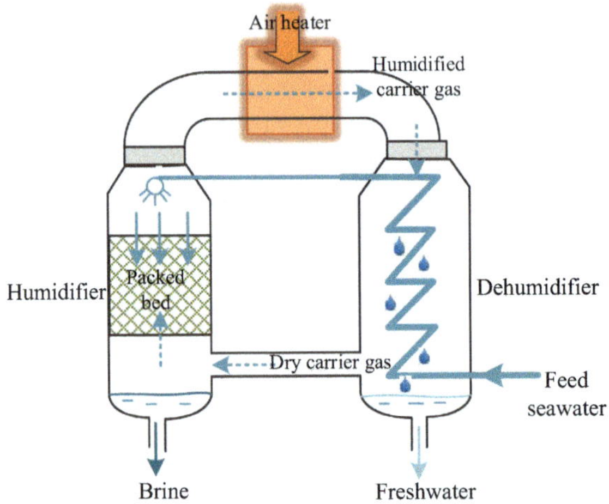

Fig. 2.38 ***An air-heated, closed-air open-water HDH desalination system (with the heater located before the dehumidifier).*** HDH, humidification–dehumidification.

resulting in a more efficient humidification process of air. In the second case, where the air heater is located before the dehumidifier, hot unsaturated moist air enters the dehumidifier, which then preheats the feed saline water before it enters the humidifier.

The first configuration of air-heated, closed-air open-water HDH desalination system is less efficient than the second. This is due to the fact that, when the air heater is placed prior to the humidifier, the heat and mass transfer processes are in opposite directions. Since heat and mass transfer compete, a less efficient process results. But when the heater is placed before the dehumidifier, most of the heat from the dehumidifier is transferred to the feed saline water (via heat recovery). Then, in the humidifier, both the heat and mass transfer processes are from the water stream to the air stream, due to the large temperature and concentration differences between the inlet air and sprayed saline water.

There also are other HDH configurations in which the air connection between humidifier and dehumidifier is broken. These configurations, open-air and open-water, can be either water-heated or air-heated. Fig. 2.39 shows an air-heated open-water, open-air HDH desalination system. The working principles of these configurations are very similar to those of the closed air counterparts. Note that, since open-air cycles take air directly from the environment, their operational performance is dependent on the environmental conditions.

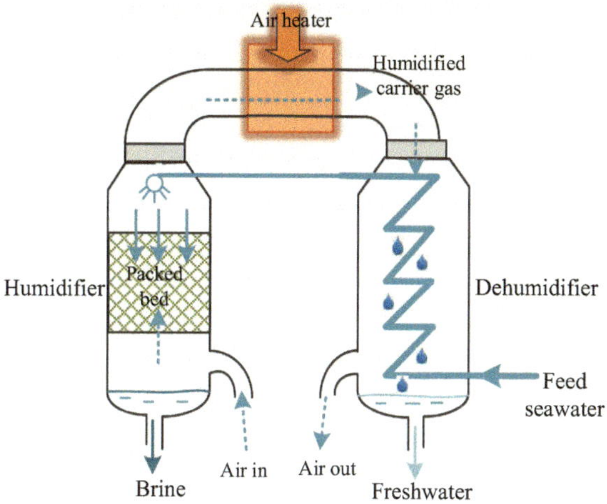

Fig. 2.39 ***An air-heated open-water, open air HDH desalination system.*** HDH, humidification–dehumidification.

The closed-water cycle is another HDH desalination system configuration. In this cycle, shown in Fig. 2.40, a brine stream from the humidifier is recirculated and a desirable recovery is achieved using makeup water. By brine recycling, relatively higher recovery ratios can be obtained. In some cases, where the brine stream needs to have a certain temperature before entering the humidifier, intake water is required to maintain a fixed temperature. In essence, a closed-water cycle operates like an open-water configuration, drawing intake water at a fixed temperature.

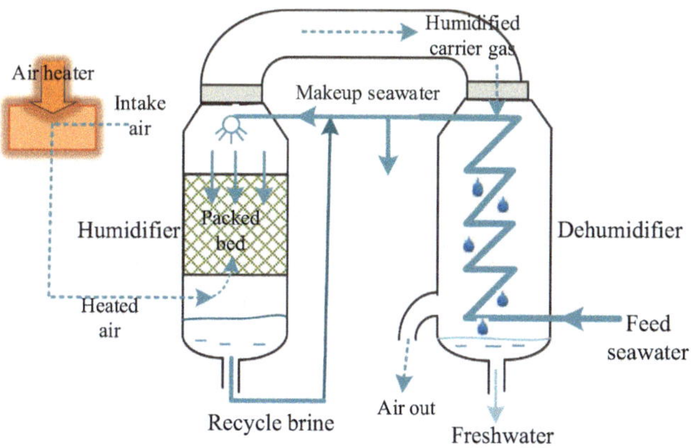

Fig. 2.40 ***An air-heated open-air, closed-water HDH desalination system.*** HDH, humidification–dehumidification.

The air flow in all HDH desalination configurations can be circulated by mechanical blowers (forced convection) or natural convection. Although the mechanical blower increases the energy consumption of a HDH system, a stable flow of air can be advantageous for attaining the optimum value of water to air mass flow ratio, which can result in higher energy efficiencies.

Selecting the best HDH configuration in terms of energy efficiency, cost of produced freshwater and other factors is dependent on an understanding of the relative technical, economic, and other characteristics of each configuration. The performance of a HDH system depends on such operating conditions as the humidity ratio of air at the inlet and outlet of the humidifier, the maximum temperature of the cycle, the humidifier/dehumidifier effectiveness, and the top and bottom temperatures in the humidifier/dehumidifier. The effectiveness of a humidifier and dehumidifier expresses the heat transfer performance within these heat exchangers and can be defined as the ratio of actual heat transfer to theoretical maximum heat transfer. Note that increasing the effectiveness of the humidifier and dehumidifier can improve the operating conditions since the heat and mass transfer processes improve the humidifier and dehumidifier performance through certain temperature and concentration differences. To achieve a higher humidifier/dehumidifier effectiveness, the surface area of the humidifier/dehumidifier should be increased, even though this raises the size and capital cost of the components.

2.2.5.2 Component design for humidification–dehumidification desalination systems

Additional information is useful regarding the design of components for humidification- dehumidification desalination systems. Some important points are described here.

A packed bed is typically used in the humidifier of a HDH desalination system. The saline water is sprayed over the packed bed and the air flows from the bottom. The packing bed is generally made of inexpensive polymeric materials. In addition to providing a large compact area of packing bed for evaporation, the packed bed should have enough open area to avoid excessive pressure drops. The packed bed structure materials are not necessarily corrosion-resistant metals, and they are not very expensive as the humidifier operates at a moderate temperature and atmospheric pressure. Moreover, in a packed bed humidifier, fouling and scaling phenomena at the packing material surfaces do not adversely affect the heat and mass transfer processes in the water–air interface in a significant manner.

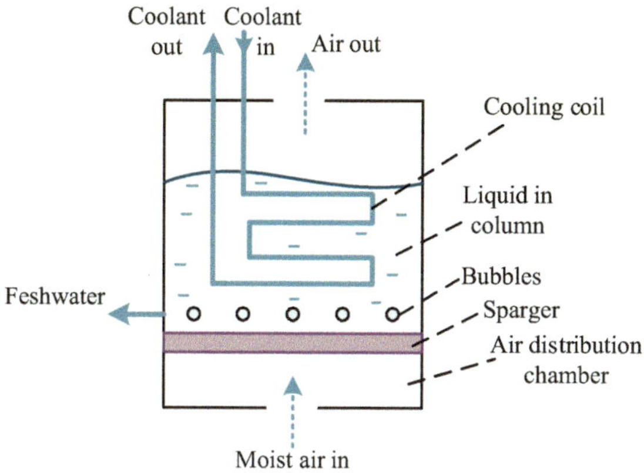

Fig. 2.41 ***Bubble column dehumidifier.***

In contrast to the humidifier, the operation of the dehumidifier requires more stringent level of thermal design due to the accumulation of NCGs on the surface of the condenser. The presence of NCGs on the surface of condenser adversely affects the heat and mass transfer processes within the dehumidifier. The condenser fins or plates can be enlarged to reduce the transport resistance of NCGs which increases the capital cost associated with the surface area of the dehumidifier. One approach is to employ bubble columns, in which a moist air is sparged into cool freshwater to condense on the surface of bubbles. Spargers are important tools that deliver air into the column and ensure that bubbles are dispersed evenly throughout the column. The spargers are arranged in a countercurrent configuration (see Fig. 2.41) to yield an efficient condensation process in which the latent heat of condensation can be effectively recovered.

Multi-stage bubble column dehumidifiers can be also designed to enhance the energy effectiveness of the dehumidifier; these can provide an order-of-magnitude reduction in the area and volume of the dehumidifier (Narayan, 2012). Both single- and multi-stage bubble columns are compact and cost-effective and are used in commercial HDH desalination processes.

For cases in which the feed salt concentration is high, the circulation of the feed saline solution through the coil of a bubble column may cause fouling, which cannot be cleaned easily. To avoid this, free coil bubble columns have been designed in which the temperature of cool freshwater increases as it flows from upper trays to lower trays, absorbing the heat of vapor condensation in each tray (see Fig. 2.42). The warm freshwater is collected from the bottom and the cool dry air is removed at the top. It

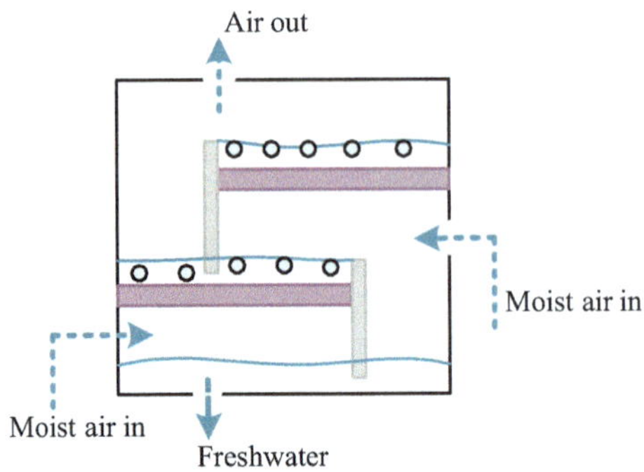

Fig. 2.42 ***Free coil bubble dehumidifier.***

is noted that, in the coil-free dehumidifier, a separate heat exchanger is required to recover the energy of warm freshwater and to transfer it to the saline feed water.

2.2.6 Mechanical vapor compression desalination system

The single-effect MVC desalination system is one of the most established desalination technologies for desalting seawater or higher salinity water, at various freshwater production scales. An MVC desalination system is driven by electricity and is suitable for remote areas that have access to grid electric power.

Fig. 2.43 shows an MVC desalination system. Its main components are as follows: evaporator/condenser, mechanical vapor compressor,

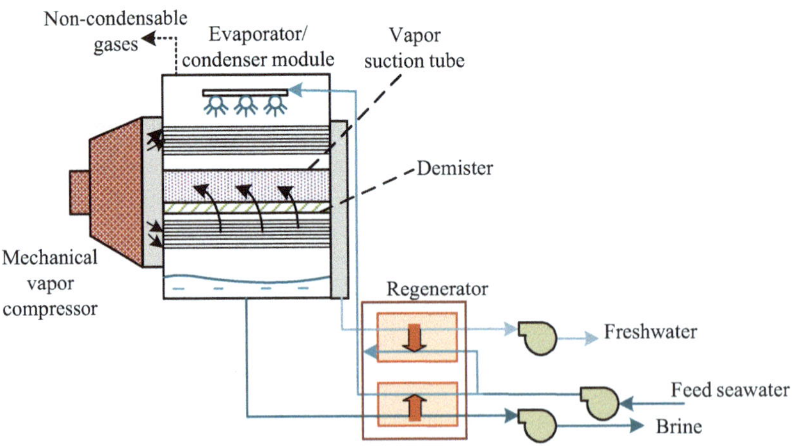

Fig. 2.43 ***Mechanical vapor compression (MVC) desalination system.***

regenerator, venting system, and freshwater/brine pumps. The evaporator/condenser module consists of a spray nozzle, falling film horizontal tubes, a wire-mesh mist eliminator (demister), and a suction vapor tube. The saline feed intake stream is preheated in the regenerator, which receives heat from the relatively high-temperature brine and freshwater streams. The preheated feed stream is sprayed over the hot horizontal tubes, forming falling films over the tube surfaces. This enhances the heat transfer process in the evaporator/condenser module. The feedwater temperature rises to the boiling temperature and then the evaporation process begins, generating distillate vapor. This vapor is transferred to the mechanical vapor compressor via a vapor suction tube. The vapor is compressed in the compressor, and then transferred back to the evaporator/condenser module to provide heat for vaporization of the sprayed feed stream. The brine and distillate (freshwater) streams exit the evaporator/condenser module at about the boiling and condensation temperatures, respectively. The thermal energy of these streams can be recovered in the regenerator (commonly a plate-type heat exchanger) to preheat the intake saline feed water. Note that a demister is used along with the vapor suction tube to prevent entrainment of brine droplets in the distillate vapor, which is important as they can damage the compressor blades.

The SEC of MVC systems typically varies within the range 4 to 17 kWh/m^3, depending on the compressor isentropic efficiency, *TBT,* and temperature difference between the superheated vapor and the saturation temperature of the feed stream (El-Dessouky and Ettouney, 2002).

To ensure the electricity required by the mechanical vapor compressor and the heat transfer area of the evaporator/condenser and regenerator units are practical, the temperature difference in the various streams or in one stream from inlet to outlet should be maintained within reasonable ranges. For instance, the temperature difference of the inlet and outlet temperatures of the compressor should be between 4°C and 10°C, while the temperature difference for the brine (or distillate) and feed should be between 1°C and 5°C, and for the regenerator input saline water and output brine (or distillate) should be between 1°C and 5°C (El-Dessouky and Ettouney, 2002).

MVC desalination systems are compact and do not require an external condenser and cooling water. MVC systems have some advantages over RO process, like higher product purity, longer life of the separating surface, and simpler pretreatment. This type of desalination technology typically operates at relatively low-temperatures (about 60°C), reducing the likelihood of

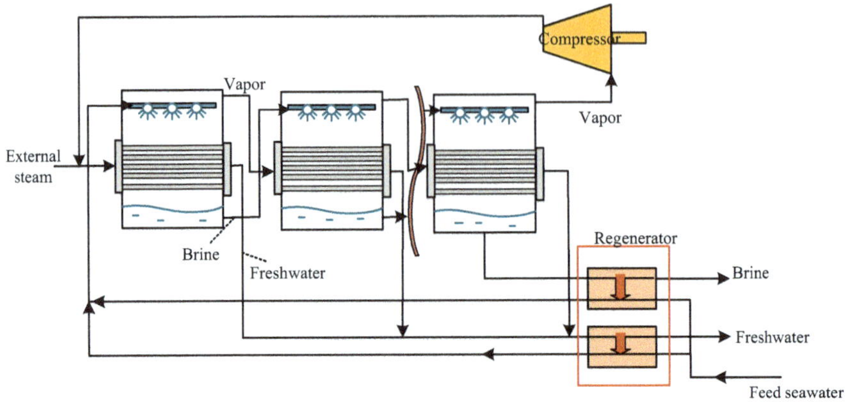

Fig. 2.44 ***Multi-effect forward-feed MVC desalination system.*** MVC, mechanical vapor compression.

fouling/scaling. In addition, an MVC system is a modular type of desalination system that can be enlarged simply by adding modules. In this way, the thermal energy of brine and vapor are recovered through the next effects, resulting in an increase in the efficiency of desalination process.

Fig. 2.44 shows a multi-effect forward-feed MVC desalination system. In this system, the brine stream in the first effect is transferred to the next evaporator/condenser module, sprayed over the horizontal falling film evaporating tube to generate vapor. This process is repeated in subsequent effects. It realizes noted that the working principle of this system is similar to that for forward-feed MED using TVC (see Section 2.2.2.5).

Various studies have been carried out to analyze and improve the performance of MVC systems. Some have investigated combinations of MVC systems with other desalination technologies, such a multi-stage flash–mechanical vapor compression (MSF-MVC) (Mabrouk et al., 2007) and mechanical vapor compression–membrane distillation (MVC-MD) (Swaminathan et al., 2016). In the latter case, a MD desalination system is used instead of the regenerator of an MVC desalination system to produce additional distillate, while preheating the feed water using the brine stream. This hybrid system, shown in Fig. 2.45, provides thermal energy for MD and reduces the capital cost of the MVC unit as the regenerator, typically an expensive unit, as it is replaced by a MD module. It is shown that the freshwater production cost can be decreased by about 9% for this combined configuration compared to a stand-alone MVC system (Swaminathan et al., 2016).

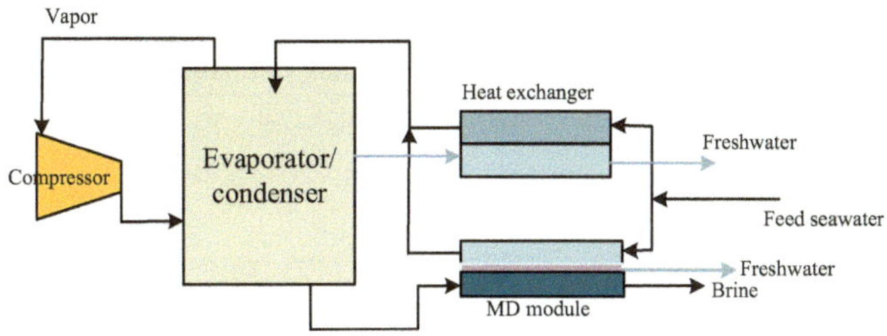

Fig. 2.45 ***Hybrid MVC-MD desalination system.*** Modified from Swaminathan et al. (2016). MVC-MD, mechanical vapor compression–membrane distillation.

2.3 Future expectations of water supply strategies

Freshwater demands for irrigation activities, energy production facilities, and commercial, industrial, and domestic applications have risen significantly, mainly due to population growth and economic development. Supplying freshwater demand is often accomplished through one of the following five water development strategies:

- Exploring for and developing new groundwater or local freshwater resources.
- Transferring freshwater from areas that are rich in freshwater sources to those that are not.
- Improving the efficiency of water use and decreasing water demand in water-intensive sectors like industrial, agricultural (mainly for irrigation), and domestic.
- Treatment of low-quality water to convert it to high-quality freshwater.
- Lowering the water quality requirements for cooling water as well as industrial and other processes that do not require high-quality freshwater.

The general energy requirements for various water supply strategies have been evaluated by Sandia National Laboratories (2009). The cost of energy use in many water supply development options is the largest cost element. Thus, the cost of energy use is a useful and meaningful criteria for comparing the costs of water supply strategies.

The energy required for various water supply strategies is presented in Table 2.2. Increasing water use efficiency and water conservation are the most cost-effective water supply strategies because almost no energy is required for them. According to Table 2.2, brackish water desalination and treatment and wastewater reuse and storage both require less energy than freshwater transportation from freshwater-rich areas to areas with little freshwater supplies. The energy demand for seawater desalination varies

Table 2.2 Energy consumption for various water supply strategies.

Water supply strategy	Energy consumption range (kWh/kg)
Freshwater importation (160–480 km)	3785–68,130
Seawater desalination reverse osmosis	45,420–75,700
Brackish groundwater desalination	
Reverse osmosis treatment	26,495–34,065
Pumping/concentrate management	3785–11,355
Waste water aquifer storage and recovery	
Pretreatment and post-treatment processes	11,355–15,140 (for each process)
Pumping	7570–11,355

Data adapted from Sandia National Laboratories (Kucera, 2019).

with location and can overlap with the freshwater transportation option. However, seawater desalination process can be a cost-effective option to meet the freshwater demand in many regions.

Water reuse and desalination constitute emerging trends in water supply strategies in which the technologies used are very important. For instance, in a RO desalination process, the type of membrane technology significantly affects the rate of growth in the utilization of such desalination systems. According to the high rates of growth of using RO desalination technology, it is emerging at present as the major new water supply approach. Additionally, advances in pretreatment, concentrate disposal, and scale/fouling inhibition all are expected to play important roles in the future growth of RO desalination systems.

RO desalination technologies typically requires 9–90 kJ of energy to produce one kilogram of freshwater from seawater. The cost of 3785 liters (about 1000 gallons) of freshwater produced from RO plants varies within the range of 2 to 4 US$ (Kucera, 2019). Fig. 2.46 shows a cost breakdown

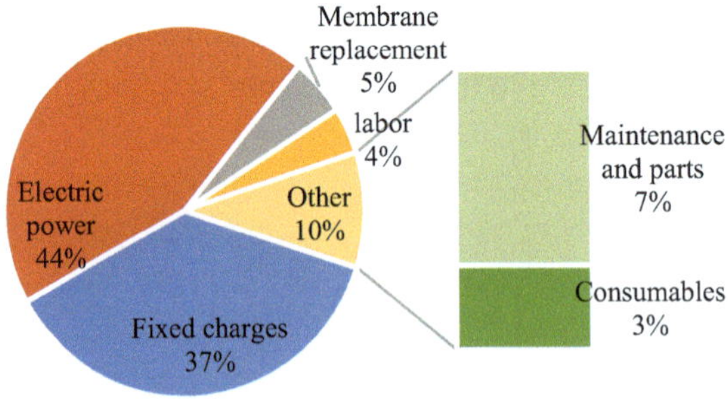

Fig. 2.46 ***Cost breakdown for a seawater reverse osmosis desalination system.*** *From Semiat (2000) and Taylor & Francis, www.tandfonline.com.*

for a seawater RO desalination process and indicates that almost half of the total cost of freshwater production is attributable to energy consumption. It is highly likely that lower energy desalination techniques will remain a central goal in future developments, among others. For instance, for membrane-based desalination technology, research is required for:

- Developing advanced pretreatment processes and optimizing pretreatment conditions.
- Removing contaminates via integrated treatment approaches.
- Developing alternatives to current membrane module configurations (e.g., spiral modules) and configuration optimization.
- Developing high-pressure membranes.
- Enhancing MD and nanostructured high flux membranes.

Also, additional hybrid technologies can be analyzed for desalination and wastewater treatment, like RO/ED (electrodialysis), or RO/FO (forward osmosis), RO/MD, and FO/MD (Subramani et al., 2015). For other desalination technologies research can be targeted at:

- Enhancing evaporation processes in thermal desalination techniques.
- Elimination of silica and minimization of antiscalants and biocides.
- Processing useful contaminants from brine for sale.
- Identifying energy recovery potentials within desalination and other plants in cogeneration schemes.

In addition to all aforementioned considerations for improving desalination technologies, nanotechnology and biotechnology for biomimetic desalination and nanostructured membranes also merit investigation, as potential means to improve desalination in the future. In fact, major research and development investments have already been made in applications of nanotechnology in desalination.

Much research related to desalination has been reported. For instance, many studies have examined the energy sources used to drive desalination, including nuclear energy (Orhan et al., 2010; Khalid et al., 2016b), renewable energy (Hamut et al., 2014; Akbari Kordlar et al., 2014; Maleki et al., 2016.; Keshavarzzadeh et al., 2020; Salehi, et al., 2018; Peng et al., 2018) and multiple energy sources (Maleki et al., 2016; Khalid et al., 2016a; Wu et al., 2018; Peng et al., 2018). Studies have also been reported on polygeneration systems including desalination as a component (Haghghi et al., 2019; Chitgar et al., 2019; Hamrang et al., 2020).

It can be informative to ponder the future by examining the past. Fig. 2.47 shows a timeline of developments in desalination technologies. One of the central goals in desalination research has been to develop

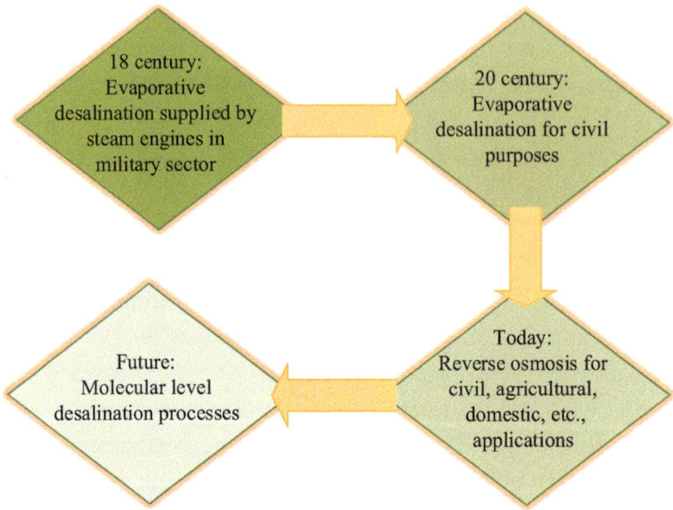

Fig. 2.47 ***Evolution timeline of developments in desalination.***

an understanding of desalination processes at the molecular level. Such knowledge in the past initiated the development of technologies to achieve higher performance and lower costs for desalination, and developments are expected to continue. For example, the utilization of nanoparticles and carbon or graphene-based nanotubes holds the promise of new desalination technologies with superior water permeability and salt rejection. Microbial desalination units, that are able to desalt high salinity water without the need for external energy sources, also have significant future potential, even though scale-up of this technique has not been methodically evaluated (Subramani et al., 2015).

2.4 Closing remarks

Freshwater scarcities in many regions have led to advances in desalination technology and its increasing adoption. Various technologies have been developed to desalt saline water for a wide range of salinities. In this chapter, current commercial desalination technologies are described, including MED as well as MSF, MVC, HDH, RO, and MD desalination. Classifications considering working principles, SEC, and the main energy input required for desalination are presented. Thermal desalination methods (MSF, MED, solar stills) have been historically the primary methods utilized, while today most desalination technologies are based on membranes. Among the membrane-based desalination techniques, reverse osmosis in recent years has been the most well-established and widely used technology, largely due to its good efficiency, low unit cost of freshwater, and relatively straightforward capability for coupling with renewable energy systems. In addition, MD exhibits the most promise for enhanced performance with low-temperature heat sources (e.g., waste heat). The main objectives of future research on desalination are to develop an understanding of desalination at the molecular and nano-scale levels, to achieve better efficiency and performance and to lower the unit cost of producing freshwater.

Nomenclature

Symbols

P	pressure (kPa, bar)
SEC	specific energy consumption (kWh/m^3)
TBT	top brine temperature (°C, K)

Acronyms

AGMD	air-gap membrane distillation
DCMD	direct contact membrane distillation
ERD	energy recovery device
HDH	humidification–dehumidification
HTE	horizontal tube evaporator
MD	membrane distillation
MED	multi-effect desalination
MSF	multi-stage flash
MVC	mechanical vapor compression
NCG	non-condensable gas
OT	once through
RO	reverse osmosis
SGMD	sweeping gas membrane distillation
TVC	thermal vapor compressor
VMD	vacuum membrane distillation
VTE	vertical tube evaporator

References

Accepta, 2021. Membrane separation technology: Minimising waste and effluent. The Water Treatment Products Company. https://accepta.com/environmental-water-wastewater-knowledge/reverse-osmosis-membrane-technologies-knowledge/294-membrane-separation-technology-minimising-waste-and-effluent, (Accessed 01 December, 2021).

Al-Falah, E., Al-Shuaib, A.A., Ettouney, H., El-Dessouky, H., 2001. On-site training programmes in desalination plants. Eur. J. Eng. Edu. 26 (4), 407–418.
Al-Mutaz, I.S., Wazeer, I., 2014. Comparative performance evaluation of conventional multi-effect evaporation desalination processes. Appl. Thermal Eng. 73 (1), 1194–1203.
Ali, A., Macedonio, F., Drioli, E., Aljlil, S., Alharbi, O.A., 2013. Experimental and theoretical evaluation of temperature polarization phenomenon in direct contact membrane distillation. Chem. Eng. Res. Des. 91 (10), 1966–1977.
Alkhudhiri, A., Darwish, N., Hilal, N., 2012. Membrane distillation: a comprehensive review. Desalination 287, 2–18.
Berghof, 2021. Tubular UF membrane modules. https://www.berghofmembranes.com/tubular-uf-membrane-modules/ (accessed July 27, 2021).
Cohen-Tanugi, D., McGovern, R.K., Dave, S.H., Lienhard, J.H., Grossman, J.C., 2014. Quantifying the potential of ultra-permeable membranes for water desalination. Energy Envir. Sci. 7 (3), 1134–1141.
Chitgar, N., Emadi, M.A., Chitsaz, A., Rosen, M.A., 2019. Investigation of a novel multigeneration system driven by a SOFC for electricity and fresh water production. Energy Convers. Manage. 196, 296–310.
ECOGRACE, 2020. Membrane filter press membrane filter Plate 1200x1200 Filter Press Plate. http://www.eco-grace.com/sale-9570986-membrane-filter-press-membrane-filter-plate-1200x1200-filter-press-plate.html (accessed June 7, 2021).
El-Dessouky, H.T., Ettouney, H.M., 2002. Fundamentals of Salt Water Desalination. Elsevier, Amsterdam.
Ettouney, H., El-Dessouky, H., Al-Juwayhel, F., 2002. Performance of the once-through multistage flash desalination process. Proc. Inst. Mech. Eng., Part A: J. Power Energy 216 (3), 229–241.
Farooque, A.M., Jamaluddin, A.T.M., Al-Reweli, A.R., Jalaluddin, P.A.M., Al-Marwani, S.M., Al-Mobayed, A.A., Qasim, A.H., 2008. Parametric analyses of energy consumption and losses in SWCC SWRO plants utilizing energy recovery devices. Desalination 219 (1-3), 137–159.
Fritzmann, C., Löwenberg, J., Wintgens, T., Melin, T., 2007. State-of-the-art of reverse osmosis desalination. Desalination 216 (1-3), 1–76.
Glade, H., Al-Rawajfeh, A.E., 2008. Modeling of CO2 release and the carbonate system in multiple-effect distillers. Desalination 222 (1-3), 605–625.
Greenlee, L.F., Lawler, D.F., Freeman, B.D., Marrot, B., Moulin, P., 2009. Reverse osmosis desalination: water sources, technology, and today's challenges. Water Res. 43 (9), 2317–2348.
Gude, V., 2018. Renewable Energy Powered Desalination Handbook, Application and Thermodynamics. Elsevier, Oxford.
Haidari, A.H., Heijman, S.G.J., Van Der Meer, W.G.J., 2018. Optimal design of spacers in reverse osmosis. Sep. Purif. Technol. 192, 441–456.
Haghghi, M.A., Shamsaiee, M., Holagh, S.G., Chitsaz, A., Rosen, M.A., 2019. Thermodynamic, exergoeconomic, and environmental evaluation of a new multi-generation system driven by a molten carbonate fuel cell for production of cooling, heating, electricity, and freshwater. Energy Convers. Manage. 199, 112040.
Hamrang, F., Shokri, A., Mahmoudi, S.M.S., Ehghaghi, B., Rosen, M.A., 2020. Performance analysis of a new electricity and freshwater production system based on an integrated gasification combined cycle and multi-effect desalination. Sustainability 12 (19), 7996.
Hamut, H.S., Dincer, I., Rosen, M.A., 2014. Assessment of desalination technologies integrated with renewable energy sources in Turkey. In: Dincer, I., Midilli, A., Kucuk, H. (Eds.), Progress in Exergy, Energy, and the Environment. Springer, New York, NY, pp. 279–292.
He, F., Gilron, J., Sirkar, K.K., 2013. High water recovery in direct contact membrane distillation using a series of cascades. Desalination 323, 48–54.

Ji, M., Utomo, T., Woo, J., Lee, Y., Jeong, H., Chung, H., 2010. CFD investigation on the flow structure inside thermo vapor compressor. Energy 35 (6), 2694–2702.
Johnson, J., Busch, M., 2010. Engineering aspects of reverse osmosis module design. Desalin. Water Treat. 15 (1-3), 236–248.
Keshavarzzadeh, A.H., Ahmadi, P., Rosen, M.A., 2020. Technoeconomic and environmental optimization of a solar tower integrated energy system for freshwater production. J. Clean. Prod. 270, 121760.
Khalid, F., Dincer, I., Rosen, M.A., 2016a. Analysis and assessment of a gas turbine-modular helium reactor for nuclear desalination. J. Nuc. Eng. Rad. Sci. 2 (3), 031014.
Khalid, F., Dincer, I., Rosen, M.A., 2016b. Comparative assessment of CANDU 6 and sodium-cooled fast reactors for nuclear desalination. Desalination 379, 182–192.
Khan, A.R., 1973. Final report horizontal spray film (HTEX) and vertical tube (VTEX) evaporator. Office of Saline Water (O.S.W). Burns and Roe Construction Corporation, Paramus, New Jersey.
Kim, J., Park, K., Hong, S., 2020. Optimization of two-stage seawater reverse osmosis membrane processes with practical design aspects for improving energy efficiency. J. Memb. Sci. 601, 117889.
Koutsou, C.P., Kritikos, E., Karabelas, A.J., Kostoglou, M., 2020. Analysis of temperature effects on the specific energy consumption in reverse osmosis desalination processes. Desalination 476, 114213.
Kucera, J., 2019. Desalination: Water from Water, 2nd ed. John Wiley & Sons, Hoboken, New Jersey, USA.
Kucera, J., 2010. Reverse Osmosis: Design, Processes, and Applications for Engineers, 2nd ed. Wiley, Massachusetts.
Kurihara, M., Sasaki, T., 2017. The pursuits of ultimate membrane technology including low pressure seawater reverse osmosis membrane developed by "Mega-ton Water System" Project. J. Memb. Sci. Res. 3 (3), 157–173.
Lienhard, J.H., Antar, M.A., Bilton, A., Blanco, J., Zaragoza, G., 2012. Solar desalination. Ann. Rev. Heat Transfer 15 (15) 1049-0787.
Mabrouk, A.A., Nafey, A.S., Fath, H.E.S., 2007. Analysis of a new design of a multi-stage flash–mechanical vapor compression desalination process. Desalination 204 (1-3), 482–500.
Maleki, A., Gholipour Khajeh, M., Rosen, M.A., 2016. Weather forecasting for optimization of a hybrid solar-wind-powered reverse osmosis water desalination system using a novel optimizer approach. Energy 114, 1120–1134.
Martínez-Díez, L., Vazquez-Gonzalez, M.I., 1999. Temperature and concentration polarization in membrane distillation of aqueous salt solutions. J. Memb. Sci. 156 (2), 265–273.
Millero, F.J., 1996. Chemical Oceanography, 2nd ed. CRC press, New York, NY.
Narayan, G.P., Lienhard, J.H., 2012. Thermal design of humidification–dehumidification systems for affordable small-scale desalination. IDA J. Desal. Water Reuse 4 (3), 24–34.
Okamoto, Y., Lienhard, J.H., 2019. How RO membrane permeability and other performance factors affect process cost and energy use: a review. Desalination 470, 114064.
Orhan, M.F., Dincer, I., Naterer, G.F., Rosen, M.A., 2010. Coupling of copper-chloride hybrid thermochemical water splitting cycle with a desalination plant for hydrogen production from nuclear energy. Int. J. Hydr. Energy 35, 1560–1574.
Peng, W., Maleki, A., Rosen, M.A., Azarikhah, P., 2018. Optimization of a hybrid system solar-wind-based water desalination by reverse osmosis: comparison of approaches. Desalination 442, 16–31.
Power, R.B., 2005. Steam Jet Ejectors for the Process Industries, second ed. McGraw-Hill, New York, NY.
Salehi, S., Mahmoudi, S.M.S., Yari, M., Rosen, M.A., 2018. Multi-objective optimization of two double-flash geothermal power plants integrated with absorption heat transformation and water desalination. J. Clean. Prod. 195, 796–809.

Semiat, R., 2000. Present and future. Water Int. 25 (1), 54–65.
Shrivastava, A., Rosenberg, S., Peery, M., 2015. Energy efficiency breakdown of reverse osmosis and its implications on future innovation roadmap for desalination. Desalination 368, 181–192.
Subramani, A., Jacangelo, J.G., 2015. Emerging desalination technologies for water treatment: a critical review. Water Res. 75, 164–187.
Swaminathan, J., Nayar, K.G., Lienhard, V.J.H., 2016. Mechanical vapor compression—membrane distillation hybrids for reduced specific energy consumption. Desal. Water Treat. 57 (55), 26507–26517.
Voutchkov, N., 2018. Energy use for membrane seawater desalination–current status and trends. Desalination 431, 2–14.
Werber, J.R., Deshmukh, A., Elimelech, M., 2016. The critical need for increased selectivity, not increased water permeability for desalination membranes. Environ. Sci. Technol. Lett. 3 (4), 112–120.
Wu, B., Maleki, A., Pourfayaz, F., Rosen, M.A., 2018. Optimal design of stand-alone reverse osmosis desalination driven by a photovoltaic and diesel generator hybrid system. Solar Energy 163, 91–103.
Yang, X., Wang, R., Fane, A.G., 2011. Novel designs for improving the performance of hollow fiber membrane distillation modules. J. Memb. Sci. 384 (1-2), 52–62.

CHAPTER THREE

Sustainability and sustainable energy

3.1 Introduction

Sustainability and sustainable energy are important topics, both in general and it terms of the focus of this book on sustainable desalination. This chapter is drawn extensively from a recent publication by one of the authors on renewable energy and energy sustainability (Rosen, 2021), due to its particular fit with the needs of this chapter.

Sustainable energy is a holistic concept that encompasses, but extends well beyond, sustainable energy resources. Sustainable energy necessitates the sustainable energy use in energy systems, which include processes for obtaining energy sources, converting them to utilizable forms of energy when necessary or desirable, transporting and storing energy, and using energy. Energy is typically used to provide energy services such as providing thermal comfort, transportation, lighting, and operating a phone. Thus, sustainable energy necessitates the existence and use of sustainable energy systems and processes.

Sustainable energy is a subset or component of overall sustainability. Most countries, regions, and cities are striving for sustainability and thus are reevaluating their use of energy, which at present is for the most part far from sustainable. In simple terms, sustainability can be viewed as having environmental, economic, and social dimensions. Energy interacts with each dimension, making sustainable energy a very important component of sustainability. For instance, energy wastes are discarded into the natural environment while energy resources are extracted or harvested from the environment. As energy is needed for most activities, the attainment of sustainable energy is important to achieving sustainability.

Significant environmental, social, and economic challenges are associated with energy, including climate change, increasing pollutant emissions, rapid resource consumption, affordability, and social inequities. These must be addressed adequately for sustainable energy to be achieved. But doing so can be difficult and complex. For instance, energy prices are in part

Sustainable Energy Technologies for Seawater Desalination
DOI: https://doi.org/10.1016/B978-0-323-99872-7.00007-3

artificial, being affected by taxes as well as incentives and rebates, while the economics and politics of energy change across countries and over time.

The extent to which sustainability can be attained by individual countries varies, in large part because countries differ in many ways. For instance, wealth, population, living standards, culture, and political systems often are notably different from one country to the next. Lack of advanced technology and wealth can make it difficult for nonindustrialized countries to achieve sustainability. The desires of people and the societies in which they live and work often require increasing energy use with corresponding increases in the problems associated with energy. Urban energy use is significant in most countries and this trend is growing as the world's population and economies become increasingly urbanized. So addressing the sustainability of urban energy use is increasingly important in many countries as part of their efforts to achieve sustainable energy.

This chapter aims to inform about sustainability and sustainable energy, and to help society move toward it. In this chapter, sustainability, energy, societal energy use, sustainable energy, and methods and technologies that can be used to enhance sustainable energy are all described. The necessary conditions for sustainable energy are examined in detail. Finally, a sustainable energy evaluation of a particular energy system, the net-zero energy building, is presented.

3.2 Sustainability and sustainable development

There are many definitions and interpretations of sustainability, but none are universally accepted and applicable. The theoretical concept of sustainability can be taken to mean "lasting" or "persisting" or "enduring."

However, the time scale to be considered can be contentious. A period reflecting perhaps two to four generations (e.g., 50–100 years) is often viewed as a practical and tractable timeframe for sustainability considerations (Graedel and Allenby, 2010). This period reflects a trade-off between the facts that little (or nothing) is sustainable forever, while many (or most) activities are sustainable over short timescales such as months or years.

From an environmental viewpoint, sustainability can be considered as carrying capacity, that is, the maximum population supportable in a given region, given the ability of the environment to accept waste emissions and resource availability. The carrying capacity of a region is dependent on the demand and supply of natural resources. But sustainability goes beyond environmental factors, as it embodies other perspectives (Rosen, 2018; Baleta et al., 2019; Hengst et al., 2020; Pauliuk, 2020; Dragicevic, 2020; Chen et al., 2020).

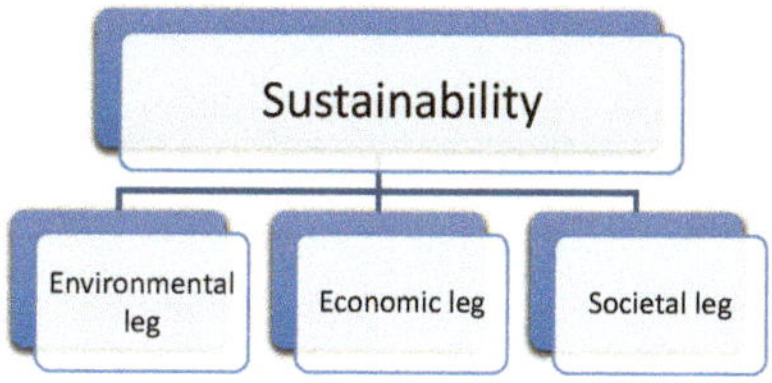

Fig. 3.1 ***Three primary legs of sustainability.***

Sustainability is often viewed as multidimensional with economic, social, and environmental legs. This viewpoint extends sustainability beyond environmental aspects, to incorporate economic and social factors. Achieving a sustainable balance is challenging as these three legs are often in tension. For instance, economic and environmental sustainability may be attained at the expense of social sustainability. Fig. 3.1 shows the main legs of sustainability.

Making sustainability practical and implementable is a significant challenge, given the multidisciplinary nature of the subject as well as the numerous definitions and interpretations of sustainability. Significant efforts have recently been taken in this regard (Rosen, 2017c) with the adoption of the UN Sustainable Development Goals for 2015–2030 at the 70th Session of the United Nations General Assembly in 2015, as part of the 2030 Agenda for Sustainable Development (United Nations, 2015). The UN Sustainable Development Goals encompass the 17 broad goals illustrated in Fig. 3.2.

Fig. 3.2 ***United Nations Sustainable Development Goals (SDGs) (public domain material provided by United Nations at http://www.un.org/sustainabledevelopment/news/communications-material/).***

Table 3.1 Energy resources broken down by renewable and nonrenewable.

Renewable	Non-renewable
Biomass	Biomass (if inadequately regenerated)
Geothermal	Coal
Hydraulic	Fossil fuel alternatives (oil sands, peat, wood)
Ocean current	Natural gas
Solar radiation	Oil
Tidal	Uranium
Wave	Wastes
Wind	

3.3 Sustainable energy

3.3.1 Energy

Energy resources are found in the environment. Some such as solar and wind energy are renewable while others such as fossil fuels are nonrenewable as there are finite quantities of them. Energy resources are listed in Table 3.1, broken down into renewable and nonrenewable types.

Energy carriers are the energy forms that are used. The range of energy carriers is diverse and includes fuels, electricity, and thermal energy (Rosen, 2015a). Energy carriers can be separated into chemical fuels and nonchemical forms (Table 3.2). Chemical fuels can include fossil fuel products in addition to nonconventional chemical fuels such as hydrogen and methanol. Some energy carriers exist naturally while others do not (e.g., electricity and work).

From Tables 3.1 and 3.2, it is clear that energy exists in various forms, for example, petroleum and refined products such as diesel fuel, electricity,

Table 3.2 Energy carriers broken down by chemical and nonchemical.

Nonchemical	Chemical
Direct thermal energy (heat or cold)	Ammonia
Electricity	Coke and other coal products
Indirect thermal energy (heated or cooled media)	Fossil fuels (coal, petroleum, natural gas)
Work	Hydrogen
	Methanol
	Petroleum products (e.g., diesel fuel, gasoline)
	Substitute natural gas

work, heat, uranium, and solar radiation. Energy-conversion technologies permit us to convert energy from one form to another. The energy interactions that occur in energy-conversion processes can be explained with thermodynamics. The first law embodies the principle of conservation of energy, and the second law embodies the principle of nonconservation of entropy, and relates to energy quality and exergy.

The services provided by energy provide good living standards and lifestyles, as well as social and cultural development. Energy is needed for most activities, in all sectors of the economy. For instance, energy is needed for space and water heating as well as cooling in the residential and commercial sectors, for the provision of motive force in the transportation sector, for an array of activities in the industrial sector such as process heating, manufacturing and chemical production. Present-day energy systems in most countries are predominantly driven by fossil fuels, but use of renewable energy is steadily increasing (Hansen et al., 2019; Kumar J. and Majid, 2020).

3.3.2 Sustainable energy and energy use

Societal energy use is very broad and varied, and thus is important to consider in efforts to shift toward or achieve sustainable energy. Changes in lifestyle and societal structures or norms can lead to reductions in energy use. Many measures can be undertaken to help efforts to achieve sustainable energy, and some of these are considered here. Raising energy efficiencies often can enhance key societal metrics, such as quality of life, standard of living, and satisfaction of people. For instance, Sweden utilizes 40% less energy (per capita) than the United States and exceeds it in most social indicators, yet both countries have similar a gross domestic products (per capita) (Rosen, 2002). Some reasons for this are better public transit, smaller automobiles, higher gasoline taxes, and less energy wastes in buildings in Sweden, as well as the compactness of its geography.

Energy measures can be voluntary or mandatory, and both methods can be successful. Governments can also apply incentives and enforcement activities to support energy measures. This is especially evident for societal measures. For example, North America has a preference for travel by automobile rather than public mass transit, and this is in large part due to moderate environmental constraints and relatively inexpensive energy. Changing such a preference is challenging. Options for energy issues to improve living standards can be technically focused and/or societally directed (e.g., increasing public awareness, improving education, and altering lifestyles to lower energy use). The historical transition in energy patterns from traditional fuels such as wood to fossil fuels and electricity is likely to continue toward renewable energy, as sustainability pressures increase.

The energy needs of cities are large and increasing, and urbanization in many developing countries is accelerating. In general, urbanization causes changes the ways societies use energy as well as shifts in land-use patterns

As noted above, nations in which energy is expensive typically use of mass transit pervasively and have efficient transportation systems, whereas countries with abundant and cheap sources of energy often prefer automotive travel and have less efficient transportation systems. In essence, inexpensive energy typically breeds a culture of individual travel in one's own vehicle, while high energy rates do not do so. Of course, situations can change quickly as a result of sudden energy disruptions or shocks, such as rapid escalations in energy rates. For example, oil shortages resulted and prices increased markedly in a short time period during the "oil crises" of the 1970s, which occurred due to oil embargos by numerous Middle East exporting nations. The resulting marked shift toward efficient vehicles and mass transit followed in response to the oil crises.

An informative recent example of sustainable energy challenges can be seen by examining China and its development. Energy use climbed greatly in China in 2000–2005, as did wealth and living standards, leading to increased demand for consumer goods and resources. Such a behavioral change can reinforce the trend, raising further resource demands. Although there are many benefits to the people of China associated with these changes, there are also difficulties. Increased environmental impact has been one undesired consequence, as has scarcities of material and energy resources. Other challenges in achieving sustainable energy in China were also observed. For instance, lifestyle, culture, and behavior are important determinants of success of energy efficiency technologies in buildings in that country (Zhang and Wang, 2013). The reasons are that most people in China are used to present lifestyles, do not wish to change to highly energy-efficient buildings, and believe government should be responsible for sustainable energy (rather than modifying their own behavior).

3.3.3 Interpretations and definitions of sustainable energy

If efforts to enhance sustainable energy are to succeed, it is important to interpret and define the term appropriately. However, there are no clean and universal definitions of sustainable energy. It is possible that a definition of sustainable energy can be developed by extending general sustainability definitions to the discipline of energy. For instance, sustainable energy can perhaps be defined by extending the description of the term sustainable development, coined by the World Commission on Environment and Development (1987) of the United Nations in its 1987 report "Our Common Future." Sustainable development was defined there as "development that meets the needs of the present without compromising the ability of future generations to meet their own needs." But extending such a general definition for sustainable energy is neither simple nor straightforward, as sustainable energy is complex and multifaceted.

But, energy is utilized in all countries in virtually all aspects of living, and is the basis of civilizations as well as life and ecosystems. Despite the fact that different countries and societies adapt to their circumstances and the energy resources available to them in deciding how to supply and use energy, the living standards attained in countries are often dependent on, or at least related to, energy-related factors. So, given the pervasiveness of energy systems, it is important to interpret and define sustainable energy appropriately.

As pointed out above, a universally accepted interpretation or definition of sustainable energy has not yet been developed. Yet, various definitions and descriptions have been proposed (Rosen, 2002, 2017a; Zvolinschi et al., 2007; Hennicke and Fischedick, 2006; Lior, 2008; Chen et al., 2020; Razmjoo et al., 2020; Suganthi, 2020; Kumar J. and Majid, 2020). Sustainable energy is defined by the present author as the provision of energy services at present and in the future, for all people, in a manner that is sustainable, that is, sufficient to provide basic necessities, not overly detrimental to the environment, affordable by all people, and acceptable to people and their communities. This definition leads to several necessary conditions for sustainable energy, which are now discussed.

3.4 Necessary conditions for sustainable energy

Defining sustainable energy leads to numerous conditions necessary for it. These are examined in this section.

Various factors impact how energy resource in society can be sustainable. Through these, key necessary conditions for sustainable energy can be developed, as illustrated in Fig. 3.3 and discussed in this section.

3.4.1 Necessary condition 1: obtain sustainable energy resources

Renewable energy sources are energy resources that can be sustained extremely long periods of time, or even indefinitely. Following Table 3.1, the primary types of renewable energy can be identified as follows: hydraulic energy, biomass energy (unless its utilization rate exceeds its replenishment

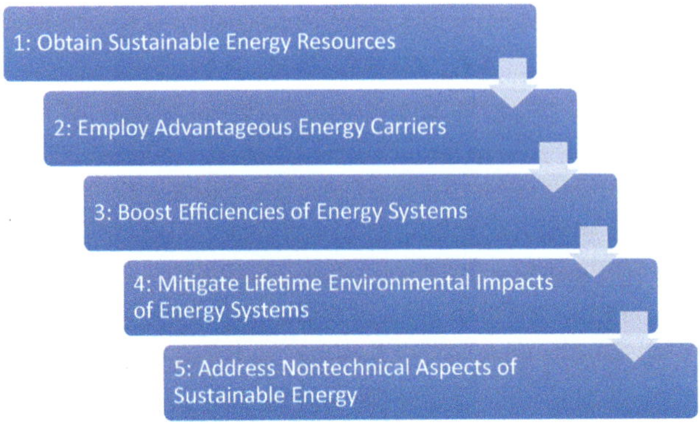

Fig. 3.3 ***Principal necessary conditions for achieving or shifting toward sustainable energy.***

rate), solar energy (including both solar thermal energy and solar photovoltaics), wind energy, and geothermal energy (whether it is ground energy at the natural background temperature or regions heated to high temperatures by the internal heat of the earth). Many of these types of renewable energy are derived from solar energy, including hydraulic energy, biomass energy and wind energy, geothermal energy in the form of ground energy at the natural background temperature. Among other benefits, renewable energy resources avoid or at least mitigate greenhouse gas emissions, facilitating energy options that are sustainable.

Fossil fuels, particularly coal, oil, and natural gas, are by far the most common nonrenewable energy resources. Wastes can be treated as a kind of renewable energy, as they can drive waste-to-energy incinerators to generate heat and/or electricity. But this notion is not accepted universally, as wastes are not a natural resource that can be sustained for extremely long periods of time if people choose to reduce wastes as much as possible. Others nonrenewable energy resources are uranium (nuclear energy fuel) and fusion material (e.g., deuterium). But the lifetimes of nuclear fuel using advanced breeder reactors and other nuclear technologies are thought to exceed 1000 years. Furthermore, although nuclear energy is not renewable, it does not contribute significantly to climate change as it has very little greenhouse gas emissions. Nuclear energy has received much attention (Al-Zareer et al., 2020a).

In general, finite energy sources are nonrenewable. Renewable energy options are sustainable over very long time periods and reduce greatly greenhouse gas emissions and many other environmental impacts, and thus have the potential to support and play a key role in sustainable energy options.

3.4.2 Necessary condition 2: employ advantageous energy carriers

Energy carriers, which include work, electricity, heat, and secondary chemical fuels, play a significant, although somewhat a hidden role, in sustainable energy. This is due to the fact that both nonrenewable and renewable energy resources are not necessarily directly exploitable in their natural forms. For instance, solar photovoltaic panels are needed to produce electricity from solar energy, while energy conversion systems using upgraders and refineries are often needed to make petroleum resources directly utilizable in user technologies.

Much attention has been paid to a future hydrogen economy, in which hydrogen plays an important role as a chemical energy carrier (Hennicke and Fischedick, 2006; Marban and Valdes-Solís, 2007; Sigfusson, 2007; Scott, 2007; Rosen, 2017b; Gnanapragasam and Rosen, 2017; Moharamian et al., 2019; Abe et al., 2019; Endo et al., 2019; Fonseca et al., 2019; Chapman et al., 2020; Al-Zareer et al., 2020a). Hydrogen energy can be used to upgrade heavy hydrocarbons and to support the use of nonfossil fuels. It achieves the latter by allowing renewable energy forms to be converted to either

- hydrogen or
- electricity

which are two primary energy carriers. Sustainable energy is supported well by this combination of energy carriers as, when employed strategically and advantageously, hydrogen (and select hydrogen-derived fuels) can be used for most chemical energy needs, whereas electricity can play its conventional role in meeting nonchemical energy demands.

3.4.3 Necessary condition 3: boost efficiencies of energy systems

Boosting efficiency is taken here to means much more than raising device efficiencies. Rather, efficiency improvement measures are considered here holistically as encompassing device efficiency improvement, enhanced energy management, expanded energy conservation, beneficial fuel substitution, strategic utilization of energy quality in addition to energy quantity, and enhanced matching of energy demands and supplies.

Advanced methods are available to support efforts to boost efficiencies. For instance, exergy analysis, a method based on the second law of thermodynamics, typically provides insights not found with conventional energy methods (Rosen, 2012; Dincer and Rosen, 2013, 2015). This has been demonstrated through many applications (Morosuk and Tsatsaronis, 2019; Sciubba, 2019; Veiga and Romanelli, 2020; Kumar et al., 2020).

Boosting efficiency assists efforts to achieve sustainable energy by stretching the benefits of any energy technologies. This is true whether for renewable or nonrenewable energy, although the benefits are much more pronounced for nonrenewable energy forms such as coal and other fossil fuels. This is mainly because boosting efficiency stretches the lives of finite energy resources. In addition, it lowers the needed capacities for energy devices compared to what is needed for less efficient devices.

3.4.4 Necessary condition 4: mitigate lifetime environmental impacts of energy systems

Achieving sustainable energy necessitates that these energy-related environmental impacts be adequately addressed. Renewable energy options are needed to help with climate change mitigation, among other benefits. This is largely because the substitution of renewable energy resources for fossil fuels can greatly lower emissions of the primary greenhouse gas, carbon dioxide. That gas is a direct and unavoidable consequence of any fossil fuel combustion, whether in electricity-generation devices such as power plants, transportation devices such as vehicles, and heating devices such as furnaces. More generally, mitigating lifetime environmental impacts of energy systems would foster sustainable energy by helping in providing sustainable energy services (Rosen et al., 2012; Sciubba, 2019; Veiga and Romanelli, 2020).

Various environmental impacts are associated with energy systems over their lifetimes. The effects of these impacts range from local and regional up to national and global. Some important environmental impacts related to energy systems are as follows:

- *Global warming and climate change.* Global warming is caused mainly by greenhouse gas emissions (especially carbon dioxide emissions). These disrupt the earth-sun-space energy balance, and cause increases in mean global temperatures as well as changes in climates. These issues are considered the most urgent environmental concerns facing humanity by many experts.
- *Stratospheric ozone depletion.* This involves the reduction in the concentration of ozone in the ozone layer in the region of the atmosphere containing the stratosphere. Such reduced ozone concentrations allow increased levels of ultraviolet radiation to reach the surface of the earth, leading to effects such as increased incidences of cancers.

- *Acid precipitation and acidification.* This impact includes (1) acid precipitation (rain or snow) due to emissions of substances such as sulfur dioxide and nitrogen oxides into the atmosphere, where they form acids such as sulfuric acid and nitric acids, and (2) the impacts that acidic emissions have on anthropogenic infrastructure such as building materials as well as on natural environmental resources such as soil and water.
- *Abiotic resource depletion.* This impact concerns the depletion of nonbiological and nonrenewable raw materials due to their extraction and utilization.
- *Ecotoxicity and radiological exposures.* Ecotoxicity encompasses the exposure of living entities to toxic substances and the consequent health problems that can arise in those entities due to the toxicity, while radiological releases cause radiation exposures that can raise rates of radiogenic cancer morbidity and mortality.

It is very important that the overall life cycle of an energy system or activity be assessed if a comprehensive and true assessment of its environmental impact is desired. This necessitates considering all life cycle steps, from the acquisition of energy and other resources, to their processing, use, and eventual disposal. Life cycle assessment (LCA) is an effective tool that does this by developing an inventory of releases and other environmental effects such as resource depletion, waste generation, and energy consumption, and then evaluating the resulting environmental impacts and developing recommended actions (Graedel and Allenby, 2010). LCA guidelines have been drawn up by the International Organization for Standardization (e.g., ISO 14040 Life Cycle Assessment—Principles and Guidelines), the Society for Environmental Toxicology and Chemistry, and others. Analyses using LCA have been performed for various activities such as the conversion of energy sources (e.g., natural gas, wind and solar energy) to useful energy carriers (Granovskii et al., 2007; Utgikar and Thiesen, 2006; Jianu et al., 2016; Mendecka et al., 2020), as well as other applications (Ben-Alon et al., 2019; Lodato et al., 2020; Lu and Halog, 2020).

3.4.5 Necessary condition 5: address nontechnical aspects of sustainable energy

Although we have focused up to now on technical issues, numerous nontechnical factors affect the sustainability of energy processes. Despite being different in nature, these factors also need to be addressed in striving

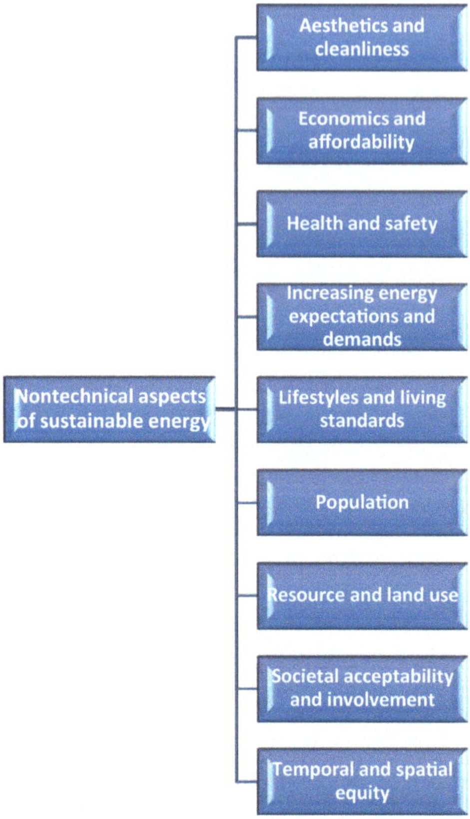

Fig. 3.4 ***Selected nontechnical aspects of sustainable energy.***

for sustainable energy. Several of factors are illustrated in Fig. 3.4 and discussed next:

- *Population.* Increasing global population stresses the carrying capacity of the planet and the environment. For sustainable energy, energy options must account for population growth and the increased demands it encompasses.
- *Increasing energy expectations and demands.* The increasing demand for energy resources with rising wealth and the desire for energy-intensive devices, especially as populations grow, poses a threat to attaining sustainable energy. This issue is particularly notable as developing countries attain higher living standards through becoming more industrialized, and expectations for improved services become more prevalent.
- *Lifestyles and living standards.* Sustainable energy can be supported by measures involving transformations in lifestyles and reductions in energy-related desires, although having such measures adopted is

often difficult because people's aspirations tend to increase rather than decrease as time passes. Asking people to sacrifice living standards to assist sustainable energy objectives is also extremely difficult as it runs counter to the nature of most people. Changing behaviors requires recognition and acceptance that present development trends are unsustainable, especially over the long term. Making future energy-related threats present-day priorities is almost always a difficult task for governments and policy makers.

- *Societal acceptability and involvement.* To attain sustainable energy, society, including communities and the people who live and work in them, must be involved in energy-related decisions, and informed and supportive of them. The support of communities is very important, and usually requires careful consultation. This can be particularly important when special or unique communities are involved, such as indigenous communities, especially when they are disadvantaged.
- *Economics and affordability.* The economics of sustainable energy measures need to be reasonably competitive with conventional approaches, accounting for the benefits provided by the sustainability measures if they are to have a chance at being accepted by most people. Furthermore, the energy services required for basic needs must be economically affordable by most if not all societies and people for sustainable energy to be attained (Rosen, 2011). Measures for boosting efficiency and mitigating environmental impact in many cases are revenue neutral and sometimes can save money over time. In such cases affordability can be achieved through loans that are repaid through energy savings. Government incentives can also enhance affordability
- *Resource and land use.* It is sometimes necessary to preserve resources that serve energy and nonenergy purposes for the latter, even if it hinders sustainable energy efforts. Similarly, land uses for energy (e.g., growing biofuel feedstocks or installing ground-mount solar collectors) need to be appropriately balanced with other uses such as agriculture, recreation and ecosystem preservation. Contentious arguments in this area are not uncommon. For instance, debates often arise in discussing (1) the need to offset or accept the damage caused by flooding large tracts of land when such land is employed for hydroelectric generation, (2) the need to balance land use for electricity transmission with the needs to preserve the ecosystems they traverse when such ecosystems are sensitive, and (3) deciding whether fertile lands should be used agricultural purposes or for growing energy plants.

- *Aesthetics and cleanliness.* Sustainable energy measures should not degrade significantly the aesthetic appeal and cleanliness of the environment, as doing so affects many people and their well-being. The avoidance of damage to the aesthetic appeal and cleanliness of the environment can be difficult for energy technologies, as they almost always have some environmental impact. This is not only true for nonrenewable energy resources, but also for renewable energy resources. Although the latter are often thought of as clean, renewable energy resource use can degrade the landscape, for example, the installation of large wind farms and photovoltaic panels (rooftop or ground-mount).
- *Health and safety.* In striving for sustainable energy, the preferred energy options must be both healthy and safe. In particular, energy options, and any factors associated with them, must not cause undue health or safety risks or effects (short or long term). This often needs to be assessed by considering the situations and views of the people who are affected by an energy options.
- *Temporal and spatial equity.* For sustainable energy, equity is needed among developed and developing countries in terms of energy access and opportunity. More generally, all societies need to have access to necessary energy resources, regardless of geographic location and circumstance. This applies not just for present generations, but also for future generations, as they too must be able to access the energy resources they will need.

  Note the following:
- The nontechnical aspects of sustainable energy are often addressed if the other necessary conditions (1–4) are addressed with a sustainability focus. This is not surprising as the other necessary conditions are indirectly related to the factors listed in this section. For instance, to obtain sustainable energy resources, one must account for such factors as economics, public acceptability, and geographic and intergenerational equity for sustainability.
- The nontechnical factors are at times related and occasionally overlap, as they are often interconnected.

3.5 Selected measures for enhancing sustainable energy

Numerous methods and technologies can be used to enhance sustainable energy. Some energy technologies that can be used to enhance sustainable energy are illustrated in Fig. 3.5 and described next.

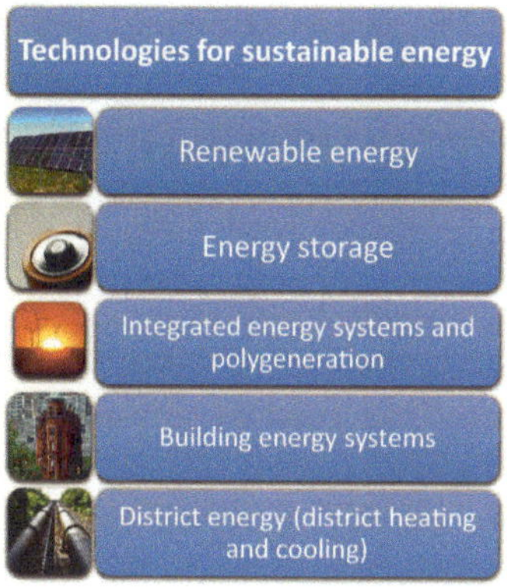

Fig. 3.5 ***Selected energy technologies for sustainable energy.***

- *Renewable energy.* Expanding the use of renewable energy sources can contribute greatly to achieving sustainable energy, particularly because they can be sustained for very long time periods and have very low environmental emissions and impacts. The main types of renewable energy, as noted earlier, include hydraulic energy, solar energy, wind energy, geothermal energy, biomass energy, wave energy, tidal energy, and ocean thermal energy.
- *Energy storage.* Energy storage can be used to mitigate the challenges of intermittent renewable energy forms such as solar and wind energy, by allowing them to be harvested when available and have the energy they produce stored until needed. Energy storage can also store energy until the economics justify using it, or to hold it between times when it is available and when it is needed in instances when they are not coincident. A simple example of this is the collection of solar thermal energy when the sun is shining during the day for use at night through the provision of space heating. Regardless of application and purpose, energy storage can help enhance the use of renewable energy, boost system efficiency, and improve energy management (Dincer and Rosen. 2011, 2015; Sliwa et al., 2019; Krishan and Suhag, 2019; Koohi-Fayegh and Rosen, 2020; Al-Zareer et al., 2020b; Rekioua, 2020; Al-Ghussain et al., 2020).
- *District energy (district heating and cooling).* As pointed out earlier, thermal energy is an energy carrier and it can be beneficially and usefully employed. Thermal energy can be generated in heating facilities, using renewable energy or conventional resources, and transported to users in district energy systems (also known as district heating and cooling systems). This can be done over short or long distances. District energy

systems are used in many cities, university campuses, and industrial parks (Rosen and Koohi-Fayegh, 2016). District heating is the use to produce a heated medium which is transported to multiple user buildings via a thermal network, while district cooling is the central production of a cold medium, which is transported to users for cooling. Buildings in the cores of many cities are often connected by pipes through which hot water or steam flows to provide space and water heating. Renewable energy–based district heating systems can utilize solar thermal energy, biomass, and geothermal energy.

- *Integrated energy systems and polygeneration.* Efficiency can be increased integrating energy systems. Polygeneration systems, including cogeneration and trigeneration systems, are one category of integrated energy system (Rosen and Koohi-Fayegh, 2016; Calise et al., 2019; Rokni, 2020; Mendecka et al., 2020; Kasaeian et al., 2020). For instance, solar PV/T (photovoltaic/thermal) systems convert sunlight into usable thermal energy and electricity. Other types of integrated energy systems link separate systems beneficially (Campana et al., 2019; Liu et al., 2020). For instance, systems can be combined advantageously by using wastes from some processes as inputs to others, where it is feasible to do so (Rokni, 2020). Many integrated energy systems are based on renewable energy technologies. Others integrate fossil fuel–based energy systems with renewable energy systems to exploit the benefits while mitigating the negative attributes of each.
- *Building energy systems.* Building energy systems can use renewable energy resources such as solar energy (thermal and/or photovoltaic) and geothermal energy, so as to enhance their sustainability. These are active systems. Passive methods and technologies can also be used to enhance sustainable energy. From the perspective of solar energy, for instance, buildings can be heated using passive systems such as Trombe walls, windows with selective coatings can be used to permit sunlight to enter a building through a window only when desired, daylight harvesting can replace some artificial lighting and window shades such as awnings and trees can be advantageously utilized to keep buildings cool during hot periods in summer when the sun is high in the sky. Furthermore, the energy efficiency of a building can be improved via insulation to reduce heat gains in summer and losses in winter, weather stripping, and caulking, to reduce air leakages, high-efficiency windows with multiple glazings, and low-emissivity window coatings, and window shades or blinds equipped with sensors that adjust to keep out excessive sunlight.

Fig. 3.6 ***Selected methods for sustainable energy.***

Some methods relating to renewable energy that can be used to enhance sustainable energy are illustrated in Fig. 3.6 and described next.

- *Efficiency, loss prevention, and waste recovery.* Sustainable energy can be facilitated by the use of appropriately high-efficiency systems. Lighting provides an informative case, as new lighting systems have significantly higher efficiencies than older equipment through the use of reduced lighting intensities, task lighting, timers, occupancy sensors, and more efficient bulbs (e.g., the efficiencies in lumens of light delivered per watt of electricity are 8–18 for incandescent, 16–24 for halogen, 20–55 for mercury, 20–104 for fluorescent, 30–170 for LED, 50–130 for high-pressure sodium, and 80–155 for low-pressure sodium lights, based on numerous sources). Sustainable energy is also supported through the utilization of high-efficiency user devices, heaters, chillers and air conditioners, pumps and compressors, and motors and fans. By corollary, efficiency can be improved by preventing losses, for example, using technologies to avoid losses such as building insulation. Efficiency can also be supported by recovering energy wastes, for example, waste heat recovery.
- *Exergy analysis and other tools.* Thermodynamic performance can be assessed and enhanced with exergy analysis, as a replacement or complement to energy analysis. Doing so can enhance sustainable energy, for any type of energy system. Exergy is a measure of the usefulness or quality or value of energy (Dincer and Rosen, 2013). Exergy analysis provides more meaningful efficiencies energy analysis, and specifies process inefficiencies (types, causes, and locations) better. Many exergy analyses and optimizations have been reported (Dincer et al., 2017; Moharamian et al., 2019).

- *Quality matching of energy supply and demand.* As pointed out above, energy storage can be used to hold energy between times when it is available and when it is needed. That addresses matching of energy quantity supply and demand. Another type of matching relates to energy quality. Rather than supplying energy of a quality much higher than necessary for a given demand, it can be more efficient to supply energy with a quality that is better matched to the demand. For instance, to provide space heating to maintain a building at 21 °C, using solar thermal energy produced by a concentrating solar collector at 400 °C is not as well matched as providing heat with a simple solar flat plate collector at 45 °C.

3.6 Illustration example: net-zero energy buildings

A net-zero energy building over an average year generates as much energy from renewable energy sources as it uses. Such a building is illustrated in Fig. 3.7. The building uses energy for a variety of tasks and generates energy from various renewable energy resources. Net-zero energy buildings can contribute to sustainable energy (Rosen, 2015b; Endo et al., 2019; Delavar and Sahebi, 2020; Tumminia et al., 2020; Singh and Das, 2020), and this is important as building energy use in many countries comprises a noteworthy portion of the national energy use, as well as national environmental impacts. To attain net-zero energy status in buildings, careful design is necessary for the building components and envelope. This involves the utilization of extensive automation, advanced controls, and careful integration of such building components as renewable energy systems, energy storage, lighting systems, and heating, ventilation and air conditioning (HVAC) equipment.

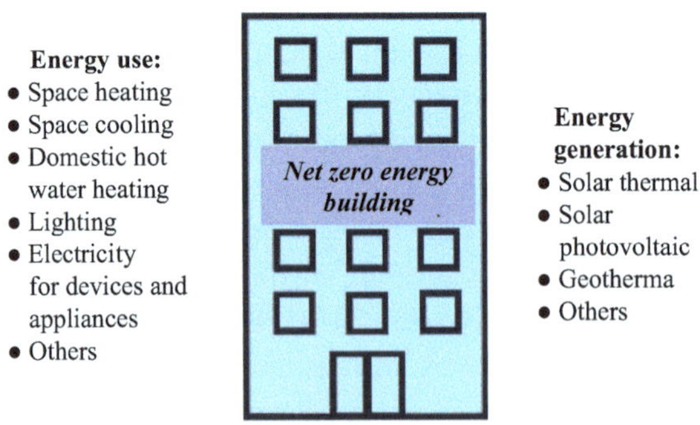

Fig. 3.7 ***A net-zero energy building.*** Energy use is satisfied over a typical year by energy generation from renewable energy resources.

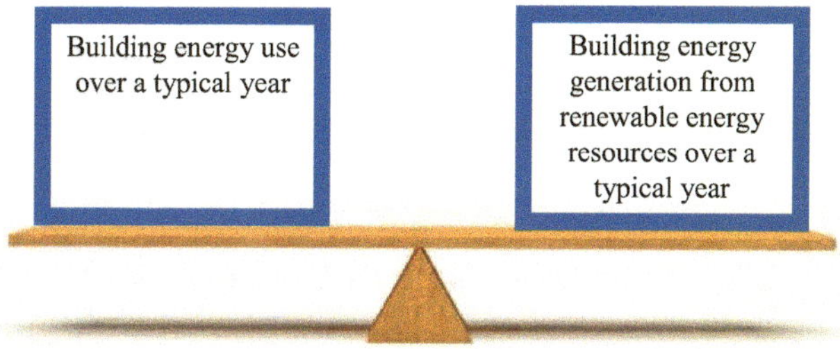

Fig. 3.8 ***Concept of a net-zero energy building in which energy use is in balance over a typical year with energy generation from renewable energy resources.***

The concept of a net-zero energy building is illustrated in Fig. 3.8. There it is seen that a net-zero energy building generates as much electrical plus thermal energy from renewable energy sources as it consumes, over an average year. In essence, this means that buildings can be designed and operated to be energy generators, which is a much different scenario that what exists for most buildings today. But there are barriers, some challenging, such as higher initial costs. These can likely be overcome, but today they limit the use of net-zero energy buildings and their contributions to sustainable energy. Once net-zero energy buildings become more prevalent; they will provide benefits such as lower utility electrical demands, electrical load management, localized generation strategies, and reduced environmental impacts.

Significant research has been undertaken into net-zero energy buildings, especially their potential in the residential and commercial sectors. The linkage between urban density of solar buildings and net energy use has been examined (O'Brien et al., 2010). Electrical grid interaction issues and load matching problems stemming from the use of net-zero energy buildings have been investigated (Gaiser and Stroeve, 2014). Issues associated with the integration into the electrical grid of net-zero energy buildings have been studied (Stylianou, 2011). Modeling, design, and optimization aspects of net-zero energy buildings have been investigated (Bucking et al., 2013; Berggren et al., 2013; Cellura et al., 2014; Mohamed et al., 2014). The benefits of considering near net-zero energy buildings for various scenarios have been examined (O'Brien et al., 2011; Doiron et al., 2011). On a broader scale, the International Energy Agency included an annex on "Towards Net-zero Energy Solar Building," while Canada launched in 2011 the Smart Net-zero Energy Buildings Strategic Research Network (http://www.solarbuildings.ca).

Important and noteworthy contributions to sustainable energy can be made by net-zero energy buildings. These are explained next, following the five necessary conditions for sustainable energy listed and described earlier:

- *Necessary condition 1: obtain sustainable energy resources.* The main advantage of net-zero energy buildings is that they utilize renewable energy resources and limit greatly or totally the use of nonrenewable energy resources, averaged over a typical year. The renewable energy is often in the form of solar energy but it also can be geothermal energy (Rosen and Koohi-Fayegh, 2017).
- *Necessary condition 2: employ advantageous energy carriers.* Two benign energy carriers are mainly used in net-zero energy buildings. These are electricity and heat (i.e., thermal energy). The latter is sometimes accommodated in a district energy system. Using these energy carriers facilitates the use of solar energy, geothermal energy, and other renewable energy resources.
- *Necessary condition 3: boost efficiencies of energy systems.* The efficiency of net-zero energy buildings is typically high, because great effort is put into the design of such buildings to reduce thermodynamic losses as much as is reasonably possible—so as to reduce the need for energy generation in the building system. This benefit is in addition to the fact that all the energy-derived services required in net-zero energy buildings are delivered with no net use of finite energy resources.
- *Necessary condition 4: mitigate lifetime environmental impacts of energy systems.* Low emissions and environmental impacts are associated with the operation of net-zero energy buildings. This is particularly important from a life cycle perspective as the operating phase of the life cycle is the dominant phase for buildings in that they usually have operating lifetimes of decades or more. There are environmental emissions and resource needs associated with other phases of the life cycles of the buildings, especially the construction and decommissioning phases, but the operating environmental benefits tend to greatly exceed the environmental impacts during the nonoperating phases of the technologies, which usually occur only once during the lifetime.
- *Necessary condition 5: address nontechnical aspects of sustainable energy.* Numerous nontechnical aspects of sustainable energy are impacted by the use of net-zero energy buildings. Some examples follow. First, net-zero energy buildings are expected to help mitigate resource demands on societies, which is particularly significant as populations rise and

developing countries become more industrialized. Second, net-zero energy buildings are anticipated to contribute to economic affordability of energy, especially when energy prices increase. Third, net-zero energy buildings are likely to be implemented only where they are viewed as socially acceptable by the communities in which they are located.

The concept of a net-zero energy building can be extended in various ways:

- The net-zero energy building concept can be expanded to include the transportation devices that are part of or housed in the building energy. Benefits can accrue from linking building-integrated renewable energy systems that are part of net-zero energy buildings with electric or hybrid cars that use electricity (Garmsiri et al., 2016). This can make the net-zero energy concept more holistic for buildings and expand their contribution to sustainable energy.
- The concept of a net-zero energy building can be broadened to a net-zero energy community. The basic concept remains similar, but the benefits are potentially greater dealing with entire communities. A net-zero energy community over an average year generates as much energy from renewable energy sources as it uses. A key distinction with net-zero energy buildings is that net-zero energy communities permit some buildings within them to be net energy users where that makes sense (e.g., a building is old and/or retrofit costs are not justifiable economically and for other reasons), while having enough buildings that are net energy producers to make up for any energy shortfalls. Some research has been undertaken on net-zero energy communities. For example, infrastructure-related interactions affecting sustainable community design have been investigated, as has the design of solar-optimized communities (Hachem et al., 2011). Also, the advantages of utilizing seasonal thermal energy storage for solar energy that is to be used for space heating has been examined for the Drake Landing Solar Community in Canada (Rad et al., 2017).

3.6.1 Example

As an example, the Falconridge net-zero energy house in Guelph, Ontario, Canada is considered. This example draws extensively on information from NRC (2019). The house was built in a new suburban, development, by Sloot Construction Ltd., with construction completed in 2016. The house is a two-story single detached residence, with a floor area (including basement) of 327 m^2 and with 4 bedrooms and 4 bathrooms.

Table 3.3 Annual energy performance of Falconridge net-zero energy house in Guelph, Ontario, Canada.

	Energy consumption	
Device	**Energy units (GJ)**	**Percentage of total energy consumption**
Production equipment		
Solar PV system	-43.82	-108.04
Total energy production	-43.82	-108.04
Consumption equipment		
Space heating	9.70	23.92
Domestic water heating	6.87	16.94
Space cooling	1.50	3.70
Mechanical ventilation	0.25	0.62
Lighting, appliances and other plug loads	22.24	54.83
Total energy consumption	40.56	100.00
Net annual energy consumption	-3.26	-8.04

The house design focused on the building envelope, aiming at very low heat transmission, through high levels of insulation, and of air leakage (achieving 0.93 air changes per hour at 50 Pa). The house has low-E, argon-filled, triple pane windows. Space heating and cooling is provided by an air source heat pump and a natural gas furnace and a heat recovery ventilator with 67% efficiency. Water heating is carried out with a natural gas–driven, tank-less heating system including a drain water heat recovery unit having 59% efficiency. Energy production by the house is accomplished through electricity generation by a solar PV system of capacity 10 kW (41 panels at 235 W each).

The performance of the building is described in Table 3.3. It is seen that, on an annual basis, the house produces 8% more energy than it consumes, making it a net-zero energy building (actually the building is a slight positive energy building). The solar PV system produces 43.8 GJ yearly, while the building consumes 40.6 GJ yearly. For comparison, it is noted that a reference house of similar size but without energy-efficiency features, consumes 109 GJ annually, more than double the consumption of the Falconridge house.

3.7 Closing remarks

This chapter describes sustainability, energy and its use, sustainable energy, and methods and technologies that can be used to enhance sustainable energy. Sustainable energy is important for or a prerequisite to achieving broader sustainability. Numerous necessary conditions must be addressed appropriately to attain or shift toward sustainable energy, including obtaining sustainable energy resources, employing advantageous energy carriers, boosting efficiencies of energy systems, mitigating lifetime environmental impacts of energy systems, and addressing nontechnical aspects of sustainable energy, such as living standards and lifestyles, economics and affordability, and societal acceptability and equity. These necessary conditions are examined in detail in this chapter. This chapter is intended to inform readers about sustainable energy, to provide a stronger focus on sustainable energy, and to help society shift toward it.

References

Abe, J.O., Popoola, A.P.I., Ajenifuja, E., Popoola, O.M., 2019. Hydrogen energy, economy and storage: review and recommendation. Int. J. Hydrogen Energy 44 (29), 15072–15086.

Al-Ghussain, L., Samu, R., Taylan, O., Fahrioglu, M., 2020. Sizing renewable energy systems with energy storage systems in microgrids for maximum cost-efficient utilization of renewable energy resources. Sustain. Cities Soc. 55, 102059.

Al-Zareer, M., Dincer, I., Rosen, M.A., 2020a. Assessment and analysis of the integrated generation IV gas-cooled fast nuclear reactor and copper-chlorine cycle for hydrogen and electricity production. Energy Convers. Manage. 205, 112387.

Al-Zareer, M., Dincer, I., Rosen, M.A., 2020. A thermal performance management system for lithium-ion battery packs. Appl. Therm. Eng. 165, 114378.

Baleta, J., Mikulčić, H., Klemeš, J.J., Urbaniec, K., Duić, N., 2019. Integration of energy, water and environmental systems for a sustainable development. J. Clean. Prod. 215, 1424–1436.

Ben-Alon, L., Loftness, V., Harries, K.A., DiPietro, G., Hameen, E.C., 2019. Cradle to site life cycle assessment (LCA) of natural vs conventional building materials: a case study on cob earthen material. Build. Environ. 160, 106150.

Berggren, B., Hall, M., Wall, M., 2013. LCE analysis of buildings: taking the step towards net zero energy buildings. Energy Build. 62, 381–391.

Bucking, S., Athienitis, A., Zmeureanu, R., 2013. An information driven hybrid evolutionary algorithm for optimal design of a net zero energy house. Sol. Energy 96, 128–139.

Calise, F., Dentice d'Accadia, M., Vicidomini, M., 2019. Optimization and dynamic analysis of a novel polygeneration system producing heat, cool and fresh water. Renew. Energy 143, 1331–1347.

Campana, P.E., Wästhage, L., Nookuea, W., Tan, Y., Yan, J., 2019. Optimization and assessment of floating and floating-tracking PV systems integrated in on- and off-grid hybrid energy systems. Sol. Energy 177, 782–795.

Cellura, M., Guarino, F., Longo, S., Mistretta, M., 2014. Energy life-cycle approach in net zero energy buildings balance: operation and embodied energy of an Italian case study. Energy Build. 72, 371–381.

Chapman, A., Itaoka, K., Farabi-Asl, H., Fujii, Y., Nakahara, M., 2020. Societal penetration of hydrogen into the future energy system: impacts of policy, technology and carbon targets. Int. J. Hydrogen Energy 45 (7), 3883–3898.

Chen, S., Kharrazi, A., Liang, S., Fath, B.D., Lenzen, M., Yan, J., 2020. Advanced approaches and applications of energy footprints toward the promotion of global sustainability. Appl. Energy 261, 114415.

Delavar, H., Sahebi, H., 2020. A sustainable mathematical model for design of net zero energy buildings. Heliyon 6 (1), 03190.

Dincer, I., Rosen, M.A., 2011. Thermal Energy Storage: Systems and Applications, Second ed. Wiley, London.

Dincer, I., Rosen, M.A., 2013. Exergy: Energy, Environment and Sustainable Development, Second ed. Elsevier, Oxford, UK.

Dincer, I., Rosen, M.A., 2015. Exergy Analysis of Heating, Refrigerating and Air Conditioning: Methods and Applications. Elsevier, Oxford, UK.

Dincer, I., Rosen, M.A., Ahmadi, P., 2017. Optimization of Energy Systems. Wiley, London.

Doiron, M., O'Brien, W., Athienitis, A.K, 2011. Energy performance, comfort and lessons learned from a near net-zero energy solar house. ASHRAE Trans. 117 (2), 585–596.

Dragicevic, A.Z., 2020. Concentric framework for sustainability assessment. J. Clean. Prod. 248, 119268.

Endo, N., Shimoda, E., Goshome, K., Yamane, T., Nozu, T., Maeda, T., 2019. Construction and operation of hydrogen energy utilization system for a zero emission building. Int. J. Hydrogen Energy 44 (29), 14596–14604.

Fonseca, J.D., Camargo, M., Commenge, J.-M., Falk, L., Gil, I.D., 2019. Trends in design of distributed energy systems using hydrogen as energy vector: a systematic literature review. Int. J. Hydrogen Energy 44 (19), 9486–9504.

Gaiser, K., Stroeve, P., 2014. The impact of scheduling appliances and rate structure on bill savings for net-zero energy communities: application to West Village. Appl. Energy 113, 1586–1595.

Garmsiri, S., Kouhi-Fayegh, S., Rosen, M.A., Smith, G.R., 2016. Integration of transportation energy processes with a net zero energy community using captured waste hydrogen from electrochemical plants. Int. J. Hydrogen Energy 41 (19), 8337–8346.

Gnanapragasam, N.V., Rosen, M.A., 2017. A review of hydrogen production using coal, biomass and other solid fuels. Biofuels 8 (6), 725–745.

Graedel, T.E., Allenby, B.R., 2010. Industrial Ecology and Sustainable Engineering. Prentice Hall, Upper Saddle River, NJ.

Granovskii, M., Dincer, I., Rosen, M.A., 2007. Exergetic life cycle assessment of hydrogen production from renewables. J. Power Sources 167, 461–471.

Hachem, C., Athienitis, A., Fazio, P., 2011. Design of solar-optimized neighborhoods, Proc. ASHRAE Annual Conference. Montreal.

Hansen, K., Breyer, C., Lund, H., 2019. Status and perspectives on 100% renewable energy systems. Energy 175, 471–480.

Hengst, I.-A., Jarzabkowski, P., Hoegl, M., Muethel, M., 2020. Toward a process theory of making sustainability strategies legitimate in action. Acad. Manag. J. 63, 246–271.

Hennicke, P., Fischedick, M., 2006. Towards sustainable energy systems: the related role of hydrogen. Energy Policy 34, 1260–1270.

Jianu, O., Pandya, D., Rosen, M.A., Naterer, G., 2016. Life cycle assessment for thermolysis and electrolysis integration in the copper-chlorine cycle of hydrogen production. WSEAS Trans. Environ. Dev. 12 (26), 261–267 Art. #.

Kasaeian, A., Bellos, E., Shamaeizadeh, A., Tzivanidis, C., 2020. Solar-driven polygeneration systems: recent progress and outlook. Appl. Energy 264, 114764.

Koohi-Fayegh, S., Rosen, M.A., 2020. A review of energy storage types, applications and recent developments. J. Energy Storage 27, 101047.
Krishan, O., Suhag, S., 2019. An updated review of energy storage systems: classification and applications in distributed generation power systems incorporating renewable energy resources. Int. J. Energy Res. 43, 6171–6210.
Kumar, A., Sharma, M., Thakur, P., Kumar Thakur, V., Rahatekar, S.S., Kumar, R., 2020. A review on exergy analysis of solar parabolic collectors. Sol. Energy 197, 411–432.
Kumar. J, C.R., Majid, M.A., 2020. Renewable energy for sustainable development in India: current status, future prospects, challenges, employment, and investment opportunities. Energy Sustain. Soc. 10, 2.
Lior, N., 2008. Energy resources and use: the present situation and possible paths to the future. Energy 33, 842–857.
Liu, Z., Yu, H., Liu, R., Wang, M., Li, C., 2020. Configuration optimization model for data-center-park-integrated energy systems under economic, reliability, and environmental considerations. Energies 13, 448.
Lodato, C., Tonini, D., Damgaard, A., Astrup, T.F., 2020. A process-oriented life-cycle assessment (LCA) model for environmental and resource-related technologies (EASETECH). Int. J. Life Cycle Assess 25, 73–88.
Lu, T., Halog, A., 2020. Towards better life cycle assessment and circular economy: on recent studies on interrelationships among environmental sustainability, food systems and diet. Int. J. Sustain. Dev. World Ecol. 27 (6), 515–523.
Marban, G., Valdes-Solis, T., 2007. Towards the hydrogen economy? Int. J. Hydrogen Energy 32, 1625–1637.
Mendecka, B., Tribioli, L., Cozzolino, R., 2020. Life cycle assessment of a stand-alone solar-based polygeneration power plant for a commercial building in different climate zones. Renew. Energy 154, 1132–1143.
Mohamed, A., Hasan, A., Sirén, K., 2014. Fulfillment of net-zero energy building (NZEB) with four metrics in a single family house with different heating alternatives. Appl. Energy 114, 385–399.
Moharramian, A., Soltani, S., Rosen, M.A., Mahmoudi, S.M.S., Bhattacharya, T., 2019. Modified exergy and modified exergoeconomic analyses of a solar based biomass co-fired cycle with hydrogen production. Energy 167, 715–729.
Morosuk, T., Tsatsaronis, G., 2019. Advanced exergy-based methods used to understand and improve energy-conversion systems. Energy 169, 238–246.
NRC, 2019. Pilot Case Study: Falconridge - Sloot Construction Ltd. Cat. No. M134-57/3-2018E-PDF. Natural Resources Canada, Ottawa, Canada.
O'Brien, W.T., Kennedy, C.A., Athienitis, A.K., Kesik, T.J., 2010. The relationship between net energy use and the urban density of solar buildings. Environ. Plann. B Plann. Des. 37, 1002–1021.
O'Brien, W., Athienitis, A.K., Kesik, T., 2011. Parametric analysis to support the integrated design and performance modeling of net-zero energy houses. ASHRAE Trans. 117 (1).
Pauliuk, S., 2020. Making sustainability science a cumulative effort. Nat. Sustain. 3 (1), 2–4.
Rad, F.M., Fung, A.S., Rosen, M.A., 2017. An integrated model for designing a solar community heating system with borehole thermal storage. Energy Sustain. Dev. 36, 6–15.
Razmjoo, A.A., Sumper, A., Davarpanah, A., 2020. Energy sustainability analysis based on SDGs for developing countries. Energy Sources Part A 42 (9), 1041–1056.
Rekioua, D., 2020. Storage in hybrid renewable energy systems. In: Rekioua, D. (Ed.), Hybrid Renewable Energy Systems: Optimization and Power Management Control. Springer, Nature, pp. 139–172.
Rokni, M.M., 2020. Waste to energy and polygeneration systems based on solid oxide fuel cells. J. Electrochem. Energy Convers. Storage 17 (3), 031011.
Rosen, M.A., 2002. Energy efficiency and sustainable development. Int. J. Global Energy Issues 17 (1-2), 23–34.

Rosen, M.A., 2011. Economics and Exergy: An Enhanced Approach to Energy Economics. Nova Science Publishers, Hauppauge, NY.
Rosen, M.A., 2012. Environment, Ecology and Exergy: Enhanced Approaches to Environmental and Ecological Management. Nova Science Publishers, Hauppauge, NY.
Rosen, M.A., 2015a. Natural Energy versus Additional Energy. Article in Encyclopedia of Energy Engineering and Technology, 1273–1280.
Rosen, M.A., 2015b. Net-zero energy buildings and communities: potential and the role of energy storage. J. Power Energy Eng. 3 (4), 470–474.
Rosen, M.A., 2017a. Sustainable development: a vital quest. Eur. J. Sustain. Dev. Res. 1 (1), 2.
Rosen, M.A., 2017b. Enhancing renewable energy prospects via hydrogen energy systems. J. Fundam. Renew. Energy Appl. 7 (5), 22.
Rosen, M.A., 2017c. How can we achieve the UN sustainable development goals? Eur. J. Sustain. Dev. Res. 1 (2), 06.
Rosen, M.A., 2018. Issues, concepts and applications for sustainability. Glocalism: Journal of Culture, Politics and Innovation 3, 1–21.
Rosen, M.A., Koohi-Fayegh, S., 2016. Cogeneration and District Energy Systems: Modelling, Analysis and Optimization. Institution of Engineering and Technology, London.
Rosen, M.A., Koohi-Fayegh, S., 2017. Geothermal Energy: Sustainable Heating and Cooling Using the Ground. Wiley, London.
Rosen, M.A., 2021. Renewable energy and energy sustainability. In: El Haj Assad, M., Rosen, M.A. (Eds.), Design and Performance Optimization of Renewable Energy Systems. Academic Press/Elsevier, New York, pp. 17–31.
Sciubba, E., 2019. Exergy-based ecological indicators: from thermo-economics to cumulative exergy consumption to thermo-ecological cost and extended exergy accounting. Energy 168, 462–476.
Scott, D.S., 2007. Smelling Land: The Hydrogen Defense Against Climate Catastrophe. Canadian Hydrogen Association, *Ottawa*.
Sigfusson, T.I., 2007. Pathways to hydrogen as an energy carrier. Philos. Trans. R. Soc., A 365 (1853), 1025–1042.
Singh, G., Das, R., 2020. Comparative assessment of different air-conditioning systems for nearly/net zero-energy buildings. Int. J. Energy Res. 44, 3526–3546.
Sliwa, T., Sojczyńska, A., Rosen, M.A., Kowalski, T., 2019. Evaluation of temperature profiling quality in determining energy efficiencies of borehole heat exchangers. Geothermics 78, 129–137.
Stylianou, M., 2011. Smart net zero energy buildings and their integration in the electrical grid, Proc. ASHRAE Winter Meeting. Las Vegas, Nevada.
Suganthi, L., 2020. Sustainability indices for energy utilization using a multi-criteria decision model. Energy Sustain. Soc. 10, 16.
Tumminia, G., Guarino, F., Longo, S., Aloisio, D., Cellura, S., Sergi, F., Brunaccini, G., Antonucci, V., Ferraro, M., 2020. Grid interaction and environmental impact of a net zero energy building. Energy Convers. Manage. 203, 112228.
United Nations. 2015. Resolution adopted by the General Assembly on 25 September 2015. A/RES/70/1. Seventieth session. United Nations 1, https://undocs.org/en/A/70/49 (Vol.I).
Utgikar, V., Thiesen, T., 2006. Life cycle assessment of high temperature electrolysis for hydrogen production via nuclear energy. Int. J. Hydrogen 31, 939–944.
Veiga, J.P.S., Romanelli, T.L, 2020. Mitigation of greenhouse gas emissions using exergy. J. Clean. Prod. 260, 121092.
World Commission on Environment and Development, 1987. Our Common Future. Oxford University Press, New York.
Zhang, Y., Wang, Y., 2013. Barriers' and policies' analysis of China's building energy efficiency. Energy Policy 62, 768–773.
Zvolinschi, A., Kjelstrup, S., Bolland, O., van der Kooi, H.J., 2007. Exergy sustainability indicators as a tool in industrial ecology. J. Ind. Ecol. 11, 85–98.

CHAPTER FOUR

Energy and exergy methods

4.1 Introduction

Energy and exergy methods are important in general for analyzing, improving, and designing processes and systems, and in particular for sustainable desalination, the theme of this book. This chapter is based heavily on a recent publication by one of the authors on exergy analysis (Rosen, 2021), which meets the coverage required of this chapter. Exergy is also detailed by Dincer and Rosen (2021), Moran (1989), Sato (2005), Kotas (2012), and others, and the present chapter also draws on these presentations.

Thermodynamics describes the characteristics, performance, and efficiency of energy conversion systems. Conventional thermodynamic analysis predominantly is based on the principle of conservation of energy, as embodied in the First Law of Thermodynamics. Exergy analysis, a thermodynamic analysis technique based primarily on the Second Law of Thermodynamics, overcomes many of the weaknesses of energy analysis. The exergy associated with a quantity provides a quantitative assessment of its quality or usefulness or value. Energy cannot be created or destroyed, but exergy analysis quantifies how it can be degraded in quality. In the limit, energy is completely degraded in quality, that is, a state in complete equilibrium with the surroundings and hence of no further use for performing tasks, and then its exergy is zero.

Efficiencies, usually ratios of quantities, are commonly used to assess and compare processes and systems. Two key types of efficiencies are considered here:

- *Energy efficiency*: An energy analysis of a process or system accounts for the energy entering, accumulating, and exiting (as products or byproducts or wastes), and efficiencies based on ratios of energy quantities are common. Energy efficiencies are often misleading, however, as they do not always provide a measure of how nearly performance approaches ideality. Further, the thermodynamic losses that occur within a process or system (i.e., factors causing performance to deviate from ideality) often are not properly identified and assessed with energy analysis. This can lead to confusion and inappropriate actions, for example, energy

Sustainable Energy Technologies for Seawater Desalination
DOI: https://doi.org/10.1016/B978-0-323-99872-7.00004-8

analysis results can identify the wrong sections of the system as responsible for the main inefficiencies, and an erroneous technological state of efficiency.

- *Exergy efficiency*: Exergy analysis provides a rational, illuminating, and meaningful means of assessing and comparing processes and systems. Exergy efficiencies measure how nearly actual performance approaches ideality, and indicate more clearly than energy analysis the causes, types, and locations of thermodynamic losses. Consequently, exergy analysis can assist in improving and optimizing designs.

In recent decades, increasing recognition of the power and utility of exergy analysis by those in industry, government, and academe has been observed, and the application of exergy analysis has grown in many countries. Devices that have been evaluated and compared with energy-based measures of merit such as energy efficiencies, as well as with exergy efficiencies are widely varying and include electricity-generating stations, chemical producers, heaters, coolers, energy storages, and many others.

In this chapter, a step-by-step procedure for energy and exergy analyses is provided, and energy and exergy fundamentals and methods for applying exergy analysis are described. Theoretical and practical aspects of thermodynamics relevant to energy and exergy methods are discussed, including energy and exergy balances, the reference environment, efficiency definitions, and material properties. Although there is at present no generally agreed upon terminology and nomenclature for exergy analysis, the exergy analysis nomenclature used here follows that proposed by Kotas (2012) as a standard exergy nomenclature. Applications and implications of exergy analyses are examined, and contrasted with those of energy analyses. Finally, two case studies are assessed involving the application of energy and exergy methods: thermal storage and electric heating.

4.2 Exergy and procedure for energy and exergy analyses

The exergy of a commodity or flow is the maximum work that can be obtained from it as it passes reversibly to the environmental state, exchanging materials and heat only with the surroundings. Thus, exergy is a measure of the ability to perform work. Exergy analysis exploits this notion and indicates the theoretical limitations governing a system. The analysis also shows that exergy is conserved only in an ideal (reversible) system, and that a real system cannot conserve exergy. Exergy analysis thereby specifies

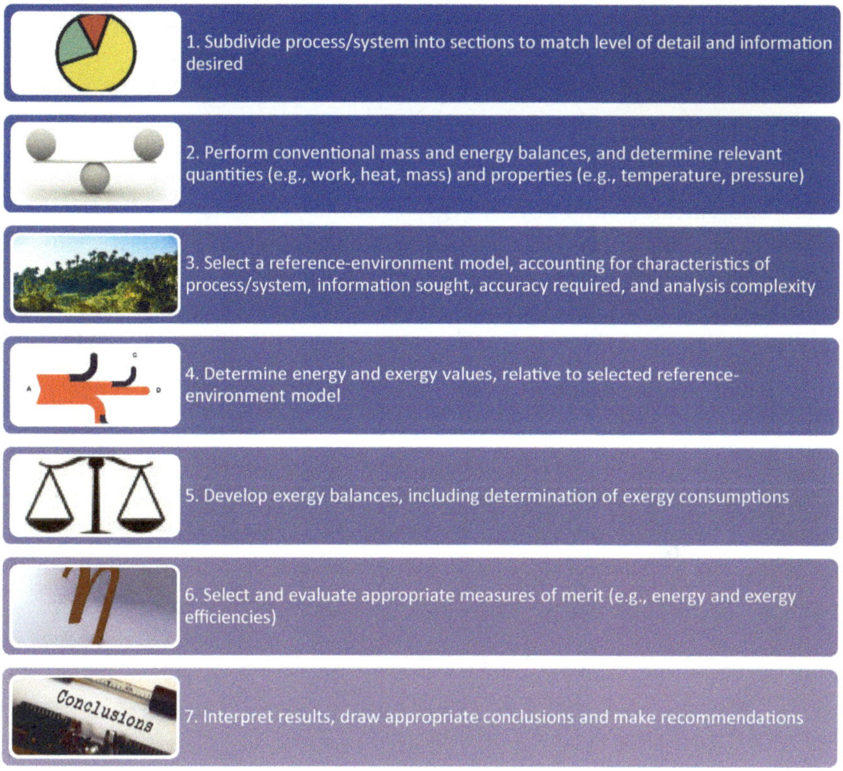

Fig. 4.1 ***Procedure for energy and exergy analyses of a process or system.***

quantitatively limitations by expressing losses in terms of exergy. Only a portion of the input exergy to a system or process can be recovered.

A straightforward procedure for performing energy and exergy analyses of a process or system involves several steps, which are listed in Fig. 4.1. The sections that follow in this chapter describe and explain many of the above steps as well as the quantities discussed in them.

4.3 Conventional balances for conserved and nonconserved quantities

Consider a nonsteady flow process in open (flow) system over a time interval t_1 to t_2. Taking mass flow into, heat transfer into and work transfer out of a system as positive, balances of mass, energy, and entropy, respectively, can be written for a control volume as follows:

$$\sum_i m_i - \sum_e m_e = m_2 - m_1 \tag{4.1}$$

$$\sum_i (e+Pv)_i m_i - \sum_e (e+Pv)_e m_e + \sum_r (Q_r)_{1,2} - (W')_{1,2} = E_2 - E_1 \quad (4.2)$$

$$\sum_i s_i m_i - \sum_e s_e m_e + \sum_r (Q_r/T_r)_{1,2} + \Pi_{1,2} = S_2 - S_1 \quad (4.3)$$

Here, m_i and m_e denote, respectively, the amounts of mass input across port i and exiting across port e; $(Q_r)_{1,2}$ denotes the amount of heat transferred into the control volume across region r on the control surface; $(W')_{1,2}$ denotes the amount of work transferred out of the control volume; $\Pi_{1,2}$ denotes the amount of entropy created in the control volume; m_1, E_1, and S_1 denote, respectively, the amounts of mass, energy, and entropy in the control volume at time t_1 and m_2, E_2 and S_2 denote, respectively, the same quantities at time t_2; and e, s, P, T, and v denote specific energy, specific entropy, absolute pressure, absolute temperature, and specific volume, respectively. Also, T_r is the temperature at the point on the control surface where the heat flow occurs, and heat transfers are assumed to occur at discrete regions on the control surface with constant temperatures. The total work W' done by a system excludes flow work, and can be written as

$$W' = W + W_x \quad (4.4)$$

where W is the work done by a system due to change in its volume and W_x is the shaft work done by the system. The term "shaft work" includes all forms of work that can be used to raise a weight but excludes work done by a system due to change in its volume. The specific energy e is given by

$$e = u + ke + pe \quad (4.5)$$

where u, ke, and pe denote, respectively, specific internal, kinetic, and potential (due to conservative force fields) energies.

An energy balance is illustrated in Fig. 4.2, where the principle of conservation of energy is evident.

For a closed system, $m_i = m_e = 0$ and Eqs. (4.1) through (4.3) become

$$0 = m_2 - m_1 \quad (4.6)$$

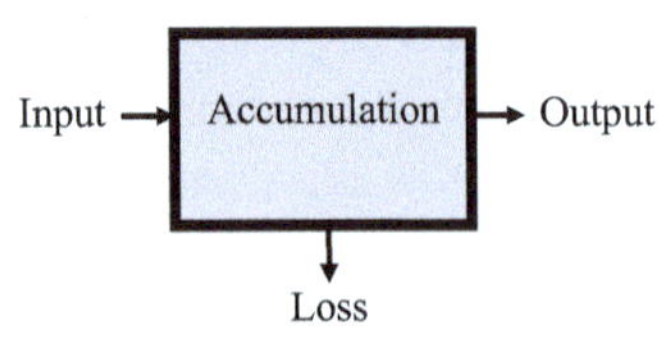

Fig. 4.2 ***Energy balance for a process in a system.***

$$\sum_r (Q_r)_{1,2} - (W')_{1,2} = E_2 - E_1 \tag{4.7}$$

$$\sum_r (Q_r/T_r)_{1,2} + \Pi_{1,2} = S_2 - S_1 \tag{4.8}$$

For irreversible processes $\Pi_{1,2} > 0$, and for reversible processes $\Pi_{1,2} = 0$.

4.4 Exergy balance and exergy consumption

An analogous balance to those given in Eqs. (4.1) through (4.3) can be written for exergy. For a nonsteady flow process during time interval t_1 to t_2,

$$\sum_i ex_i m_i - \sum_e ex_e m_e + \sum_r \left(Ex_{Q_r}\right)_{1,2} - \left(Ex_W\right)_{1,2} - \left(W_{NET}\right)_{1,2} - I_{1,2} = Ex_2 - Ex_1 \tag{4.9}$$

where exergy consumption is proportional to entropy creation (both due to irreversibilities) is as follows:

$$I_{1,2} = T_o S_{gen,1,2} \tag{4.10}$$

Here, I and S_{gen}, respectively, denote exergy consumption and entropy creation, and ex denotes specific exergy. The first two terms on the left side of Eq. (4.9) represent the net input of exergy associated with matter, the third term the net input of exergy associated with heat, the fourth and fifth terms the net input of exergy associated with work, and the sixth term the exergy consumption. The right side of Eq. (4.9) represents the accumulation of exergy.

An exergy balance is illustrated in Fig. 4.3. The exergy balance differs from the energy balance in Fig. 4.2, as energy is conserved while exergy is not, as it can be consumed or destroyed due to irreversibilities.

For a closed system, Eq. (4.9) simplifies to

$$\sum_r \left(Ex_{Qr}\right)_{1,2} - \left(Ex_W\right)_{1,2} - \left(W_{NET}\right)_{1,2} - I_{1,2} = Ex_2 - Ex_1 \tag{4.11}$$

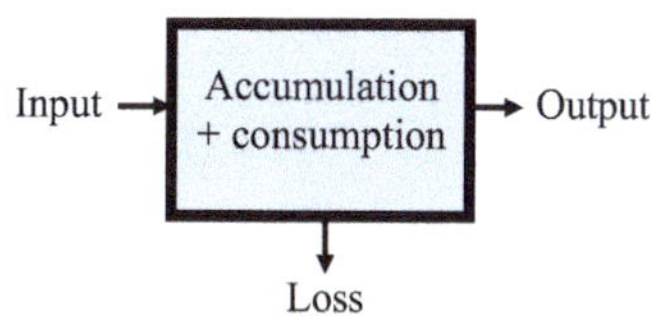

Fig. 4.3 ***Exergy balance for a process in a system.***

When volume is fixed, $(W_{NET})_{1,2} = 0$ in Eqs. (4.9) and (4.11). Also, when the initial and final states are identical as in a complete cycle, the right sides of Eqs. (4.9) and (4.11) are zero. The exergy balances in Eqs. (4.9) and (4.11) can be obtained by combining the conservation law for energy and nonconservation law for entropy.

For a process occurring in a system, the difference between the total exergy flows into and out of the system, less the exergy accumulation in the system, is the exergy consumption I, expressible as

$$I = T_o S_{gen} \tag{4.12}$$

which points out that exergy consumption is proportional to entropy creation.

4.5 Exergy of heat, work, and electricity interactions

4.5.1 Exergy of heat

Consider a control mass, initially at the dead state, being heated or cooled at constant volume in an interaction with some other system. The heat transfer experienced by the control mass is Q. The flow of exergy associated with the heat transfer Q is denoted by Ex_Q, and can be expressed as

$$Ex_Q = \int_i^f \left(1 - T_o/T\right) \delta Q \tag{4.13}$$

where δQ is an incremental heat transfer, and the integral is from the initial state (i) to the final state (f). This "thermal exergy" is the minimum work required by the combined system of the control mass and the environment in bringing the control mass to the final state from the dead state.

Often the dimensionless quantity in parentheses in this expression is called the "exergetic temperature factor" and denoted by τ. That is,

$$\tau = 1 - T_o/T \tag{4.14}$$

The relation between the factor τ and the temperature ratio T/T_o is interesting. Note that τ is equal to zero when $T = T_o$. For heat transfer at above-environment temperatures (i.e., $T > T_o$), $0 < \tau \leq 1$. For heat transfer at subenvironment temperatures (i.e., $T < T_o$), $\tau < 0$, implying that exergy and energy flow are in opposite directions in such cases. Note that the magnitude of exergy flow exceeds that of the energy flow when $\tau < -1$, which corresponds to $T < T_o/2$.

If the temperature T of the control mass is constant, the thermal exergy transfer associated with a heat transfer is

$$Ex_Q = \left(1 - T_o/T\right)Q = \tau Q. \tag{4.15}$$

4.5.2 Exergy of work and electricity

Eq. (4.4) separates total work W' into two components: W_x and W. The exergy associated with shaft work Ex_W is by definition W_x.

The exergy transfer associated with work done by a system due to volume change is the net usable work due to the volume change, and is denoted by W_{NET}. Thus, for a process in time interval t_1 to t_2,

$$\left(W_{NET}\right)_{1,2} = W_{1,2} - P_o\left(V_2 - V_1\right) \tag{4.16}$$

where $W_{1,2}$ is the work done by the system due to volume change ($V_2 - V_1$). The term $P_o(V_2 - V_1)$ is the displacement work necessary to change the volume against the constant pressure P_o exerted by the environment.

As for shaft work, the exergy associated with electricity is equal to the energy.

4.6 Exergy of matter

The exergy of matter is described here first for a closed system containing matter, and then for a flow of matter.

4.6.1 Matter in a closed system

The exergy $Ex_{nonflow}$ of a closed system of mass m, or the nonflow exergy, can be expressed as

$$Ex_{nonflow} = Ex_{nonflow,ph} + Ex_o + Ex_{kin} + Ex_{pot} \tag{4.17}$$

where

$$Ex_{pot} = PE \tag{4.18}$$

$$Ex_{kin} = KE \tag{4.19}$$

$$Ex_o = \sum_i \left(\mu_{io} - \mu_{ioo}\right)N_i \tag{4.20}$$

$$Ex_{nonflow,ph} = \left(U - U_o\right) + P_o\left(V - V_o\right) - T_o\left(S - S_o\right) \tag{4.21}$$

Here, the system has a temperature T, pressure P, chemical potential μ_i for species i, entropy S, energy E, volume V, and number of moles N_i of species i. The system is within a conceptual environment in an equilibrium state with intensive properties T_o, P_o, and μ_{ioo}. The quantity μ_{io} denotes the value of μ at the environmental state (i.e., at T_o and P_o). The terms on the right side of Eq. (4.17) represent, respectively, physical, chemical, kinetic, and potential components of the nonflow exergy of the system.

Exergy is a property of the system and conceptual environment, combining the extensive properties of the system with the intensive properties of the environment.

Physical nonflow exergy is the maximum work obtainable from a system as it is brought to the environmental state (i.e., to thermal and mechanical equilibrium with the environment), and chemical nonflow exergy is the maximum work obtainable from a system as it is brought from the environmental state to the dead state (i.e., to complete equilibrium with the environment).

4.6.2 Matter in flow

The exergy of a flowing stream of matter Ex_{flow} is the sum of nonflow exergy and the exergy associated with the flow work of the stream (with reference to P_o), that is,

$$Ex_{flow} = Ex_{nonflow} + (P - P_o)V \tag{4.22}$$

Alternatively, Ex_{flow} can be expressed following Eq. (4.17) in terms of physical, chemical, kinetic, and potential components

$$Ex_{flow} = Ex_{flow,ph} + Ex_o + Ex_{kin} + Ex_{pot} \tag{4.23}$$

where

$$Ex_{flow,ph} = (H - H_o) - T_o(S - S_o) \tag{4.24}$$

and the other terms are as defined previously.

4.6.3 Properties of materials for energy and exergy analyses

Material properties are needed for energy and exergy analyses. For most materials, conventional property data are widely available.

Energy values of material flows are relative. Enthalpies and internal energies are evaluated relative to a reference level. As energy analyses are typically concerned only with energy differences, the reference level used for enthalpy and internal energy calculations can be arbitrary. But the enthalpies and internal energies must be evaluated relative to specific reference levels to determine most energy efficiencies. The reference level is often selected so that the enthalpy of a material equals its higher heating value (*HHV*) for energy conversion processes.

For energy and exergy analyses to be comparable, the reference levels for enthalpy calculations should be determined relative to the stable components of the reference environment. Enthalpies calculated with respect to such conditions are referred to as "base enthalpies" (Rodriguez, 1980). The base enthalpy, which is the enthalpy of a compound (at T_o and P_o) relative to the elements (at T_o and P_o) from which it would be formed, is similar to the enthalpy of formation, which is the enthalpy of a component (at T_o and P_o) relative to the stable components of the environment (at T_o and P_o). Base enthalpies for many substances are given by Gaggioli and Petit (1977) and Rodriguez (1980), and some of these are listed in Table 4.1; these values are relative to the reference-environment model in the next section (Table 4.2), A compound that is a stable component of the reference environment has an enthalpy of zero at T_o and P_o.

Chemical exergy values are needed for most exergy analyses. Numerous methods have been developed for evaluating chemical exergies (e.g., Rodriguez, 1980; Szargut, 1967; Sussman, 1980), including approximation methods for complex materials such as tar and ash. Chemical exergy can be determined for gaseous streams using component chemical exergy values (i.e., values of $\mu_{io} - \mu_{ioo}$ in Table 4.1) by considering environmental air and gaseous process streams as ideal gas mixtures.

Table 4.1 Specific values of base enthalpy and chemical exergy for selected chemical species.

Chemical species		Specific property value (kJ/g-mol)	
Symbol	**Name**	**Base enthalpy**	**Chemical exergy**[a]
C	Carbon (graphite)	393.505	410.535
CH_4	Methane	890.359	2.478907 ln y + 830.212
CO	Carbon monoxide	282.964	2.478907 ln y + 275.224
CO_2	Carbon dioxide	0.000	2.478907 ln y + 20.108
H_2	Hydrogen	285.851	2.478907 ln y + 235.153
H_2O	Water	44.001	2.478907 ln y + 8.595
N_2	Nitrogen	0.000	2.478907 ln y + 0.693
O_2	Oxygen	0.000	2.478907 ln y + 3.948

Source: Compiled from data in Gaggioli and Petit (1977) and Rodriguez (1980).
[a] y denotes molar fraction for corresponding species.

4.7 Reference environment

Exergy is evaluated with respect to a reference environment. The reference environment is a sink and source for heat and materials, and acts as an infinite system. The reference environment is in stable equilibrium, with all parts at rest relative to one another. It experiences only internally reversible processes in which its intensive state remains unaltered (i.e., its temperature T_o, pressure P_o, and the chemical potentials μ_{ioo} for each of the i components present remain constant). No chemical reactions can occur between the environmental components. The intensive properties of the reference environment partly determine the exergy of a flow or system. The exergy of a flow or system is zero when it is in equilibrium with the reference environment, and the exergy of the reference environment is zero.

The natural environment does not have the theoretical characteristics of a reference environment, as the natural environment is not in equilibrium, its intensive properties exhibit spatial and temporal variations, and many chemical reactions in the natural environment are blocked because the transport mechanisms necessary to reach equilibrium are too slow at ambient conditions. Thus, the exergy of the natural environment is not zero.

As a result, models for the reference environment seek a compromise between the theoretical requirements of the reference environment and the actual behavior of the natural environment. Numerous classes of reference-environment models exist, selected ones of which follow:

- *Reference-substance models*: Here, a "reference substance" is selected and assigned zero exergy for every chemical element. A model in which reference substances were selected as the most valueless abundant substances in the natural environment was proposed by Szargut (1967). Another model was proposed by Sussman (1980, 1981), in which reference substances are selected arbitrarily.
- *Process-dependent models*: A model that contains only components that participate in the process being examined in a stable equilibrium composition at the temperature and total pressure of the natural environment was proposed by Bosnjakovic (1963). This model is process dependent, and therefore not general, making comparisons impractical.
- *Natural-environment-subsystem models*: They seek to simulate realistically subsystems of the natural environment. Such a model, consisting of saturated moist air and liquid water in phase equilibrium, was

Table 4.2 Reference-environment model that simulates natural environment.

		Composition		
Pressure, P_o (atm)	**Temperature, T_o (K)**	**Gaseous: atmospheric air saturated with H_2O at T_o and P_o**		**Condensed phases at T_o and P_o**
1	298.15	Air constituent	Mole fraction	$CaCO_3$ (limestone)
		Ar (argon)	0.0091	$CaSO_4 \cdot 2H_2O$ (gypsum)
		CO_2 (carbon dioxide)	0.0003	H_2O (water)
		H_2 (hydrogen)	0.0001	
		H_2O (water)	0.0303	
		N_2 (nitrogen)	0.7567	
		O_2 (oxygen)	0.2035	

Source: Adapted from Gaggioli and Petit (1977).

proposed by Baehr and Schmidt (1963). An extension of this model (Table 4.2), allowing for consideration of sulfur-containing materials, was proposed by Gaggioli and Petit (1977) and Rodriguez (1980). This reference environment temperature and pressure are normally set to be 25 °C and 1 atm, respectively, and the chemical composition is taken to consist of air saturated with water vapor, and the following condensed phases at 25 °C and 1 atm: water (H_2O), gypsum ($CaSO_4{\cdot}2H_2O$), and limestone ($CaCO_3$). The stable configurations of C, O, and N, respectively, are taken to be those of CO_2, O_2, and N_2 as they exist in air saturated with liquid water at T_o and P_o; of hydrogen is taken to be in the liquid phase of water saturated with air at T_o and P_o; and of S and Ca, respectively, are taken to be those of $CaSO_4{\cdot}2H_2O$ and $CaCO_3$ at T_o and P_o.

Researchers have examined the characteristics of and models for reference environments (e.g., Wepfer and Gaggioli, 1980; Sussman, 1981; Ahrendts, 1980), and the sensitivities of exergy values to different reference-environment models (Rosen and Dincer, 2004).

4.8 Efficiencies and other measures of merit

4.8.1 Efficiency conceptually

Efficiency is defined as "the ability to produce a desired effect without waste of, or with minimum use of, energy, time, resources, etc.," and is used conceptually to mean the effectiveness with which something is used to produce something else, or the degree to which the ideal is approached in performing a task. In general, nondimensional ratios of quantities are typically used to determine efficiencies.

4.8.2 Energy efficiencies and their deficiencies

Ratios of energy are conventionally used to determine efficiencies of systems whose primary purpose is energy conversion. These efficiencies are based on the First Law of Thermodynamics. A process has maximum efficiency according to the First Law if energy input equals recoverable energy output (i.e., if no "energy losses" occur). However, efficiencies determined using energy are misleading because in general they are not measures of "an approach to an ideal."

To determine more meaningful efficiencies, a quantity is required for which ratios can be established that provide a measure of approach to ideality. Thus, the Second Law must be involved, as this law states that maximum efficiency is attained (i.e., ideality is achieved) for a reversible process. The "increase of entropy principle," which states that entropy is created due to irreversibilities, quantifies the Second Law. Entropy is created for nonideal processes, while maximum efficiency is attained when entropy is conserved. In general, however, ratios of entropy do not provide a measure of approach to ideality.

4.8.3 Exergy and exergy-based efficiencies

For a steady-state process in a system, the energy efficiency η is often written as

$$\eta = \frac{\text{Energy in product outputs}}{\text{Energy in inputs}} = 1 - \frac{\text{Energy loss}}{\text{Energy in inputs}} \tag{4.25}$$

and the exergy efficiency ψ as

$$\psi = \frac{\text{Exergy in product outputs}}{\text{Exergy in inputs}} = \frac{\text{Exergy loss plus consumption}}{\text{Exergy in inputs}} \tag{4.26}$$

Efficiencies determined using ratios of exergy provide a measure of approach to ideality as, from an exergy perspective, maximum efficiency is attained for a process in which exergy is conserved. Exergy efficiencies are often more intuitively rational than energy efficiencies because exergy efficiencies always fall between 0% and 100%, unlike some energy efficiencies. Many refer to exergy efficiencies "real" or "true" efficiencies and energy efficiencies "approximations to real" efficiencies.

Two other common exergy-based efficiencies exist for steady-state devices. The first is the task efficiency, which is expressed as the ratio of the theoretical minimum exergy input required for a process, which applies to a reversible or ideal case, to the actual exergy input. That is,

$$\text{Task efficiency} = \frac{\text{Theoretical minimum exergy input required}}{\text{Actual exergy input}} \tag{4.27}$$

The rational efficiency is akin to the exergy efficiency in Eq. (4.26), but it only treats exergy consumptions as losses. It is defined as follows:

$$\begin{aligned} \text{Rational efficiency} &= \frac{\text{Total exergy output}}{\text{Total exergy input}} \\ &= 1 - \frac{\text{Exergy consumption}}{\text{Total exergy input}} \end{aligned} \tag{4.28}$$

Exergy efficiencies often provide more illuminating insights into process performance than energy efficiencies because (1) they weigh energy flows according to their exergy contents, and (2) they separate inefficiencies into those associated with effluent losses and those due to irreversibilities. In general, exergy efficiencies provide a measure of potential for improvement.

4.9 Applications and implications of exergy analysis

4.9.1 Thermodynamic applications

Thermodynamic applications of exergy have occurred in a diverse range of fields, including electricity generation and cogeneration, fuel processing, energy storage, transportation, industrial energy use, building energy systems, and many others.

For example, one of the present authors has applied exergy methods to:

- desalination (Hamrang et al., 2020; Keshavarzzadeh et al., 2020; Wu et al., 2018; Salehi et al., 2018; Peng et al., 2018; Haghghi et al., 2019; Chitgar et al., 2019),
- renewable energy systems (Rosen and Koohi-Fayegh, 2017; Rezaie, Reddy and Rosen, 2018; Kool et al., 2018; Compton, Rezaie and Rosen, 2018; Rahnama et al., 2019; Aghbashlo et al., 2020; Javanshir et al., 2020),
- hydrogen energy (Al-Zareer, Dincer and Rosen, 2018; Moharramian et al., 2019; Haghghi et al., 2019; Ehyaei and Rosen, 2019),
- electricity generation and multigeneration (Arsalis and Alexandrou, 2014; Khaljani et al., 2015; Karellas and Braimakis, 2016; Barberis et al.,

2016; Calise et al., 2016; Rosen and Koohi-Fayegh, 2016; Mehrpooya et al., 2018; Seyyedi, Hashemi-Tilehnoee and Rosen, 2018; Khoshgoftar Manesh and Rosen, 2018; Mahmoudi et al., 2019),
- energy storage (Rosen, 2012a), including thermal energy storage (Dincer and Rosen, 2021; Rosen et al., 2004),
- buildings and heating, refrigerating and air conditioning (Dincer and Rosen, 2015; Karimi et al., 2019; Ashrafi Goudarzi et al., 2019),
- countries and sectors of countries (Rosen, 2013a; 2013b; Rosen and Dincer, 1997b), and
- optimizing energy systems (Dincer, Rosen and Ahmadi, 2017).

Several books on exergy analysis, including numerous applications, have been published over the past three decades (Brodyanski et al., 1994; Edgerton, 1982; Kotas, 2012; Moran, 1989; Sato, 2005; Szargut, 2005; Szargut et al., 1988; Yantovskii, 1994; Dincer and Rosen, 2021; El-Sayed, 2004; Dincer, Rosen and Ahmadi, 2017), as have many research articles dealing with applications (Hammond, 2004; Sciubba, 2005, 2019). Also, exergy has been worked extensively into many books on thermodynamics (Moran et al., 2018).

4.9.2 Other applications

Exergy analysis has also been applied beyond thermodynamics, to such fields as biology and ecology, management of industrial systems and economics. Two particularly important areas are discussed below.

Although exergy is a measure of usefulness, it is also a measure of potential to cause a change. The latter point suggests that exergy may be, or provide the basis for, an effective indicator of the potential of a substance or energy form to impact the environment. Furthermore, it has been suggested that the environmental impact of energy utilization and the achievement of increased efficiency are best addressed using exergy.

Exergy has been applied in these environmental areas. For example, exergy has been applied to help improve environmental stewardship, ecology, and sustainability (Hammond, 2004; Edgerton, 1982; Szargut, 2005; Rosen, 2012b; Dincer and Rosen, 2021; Aghbashlo and Rosen, 2018), and exergy methods have been employed for environmental impact assessments (Crane et al., 1992; Rosen and Dincer, 1997a, 1999; Gunnewiek and Rosen, 1998).

In designing of energy systems, technical methods such as thermodynamics are combined with economics, with costs conventionally based on energy. But many suggest that costs are better distributed among outputs based on exergy. Based on this and other economic concepts, exergy has been applied extensively in conjunction with economics.

Numerous exergy-based economic methods have been developed. They are referred to be such names as thermoeconomics, second-law costing, and exergoeconomics. These methods recognize that exergy, not energy, is the commodity of value in a system, and assign costs and/or prices to exergy-related variables. These methods also help in determining the appropriate allocation of economic resources to optimize system design and operation and system economic feasibility and profitability.

Many applications combining exergy and economic concepts have been reported (Rosen, 2011; El-Sayed, 2004; Sciubba, 2005; Edgerton, 1982; Yantovskii, 1994; Mehrpooya et al., 2018).

4.9.3 Implications

The results of exergy analyses of processes and systems have direct implications on decision making regarding applications of technologies and directions for research and development. Of course, application and research and development allocation decisions generally are not be based exclusively on energy or exergy considerations, as factors such as economics, environmental impact, safety, and social and political implications need to be considered. Nonetheless, exergy considerations provide useful information to assist in such decision making.

In previous research, it has been shown that exergy analyses more than energy analyses provide insights into the "best" directions for research and development effort, that is, the most promising for significant efficiency gains (Rosen, 2013c). There are two main reasons for this statement:

- Exergy efficiencies always provide a measure of how nearly the operation of a system approaches the theoretical upper limit or ideality. In general, this is not true for energy efficiencies. By focusing on systems with low-energy efficiencies, one can expend research and development effort on topics for which little margins for improvement, even theoretically, exist. But by focusing on systems with the lowest exergy efficiencies, the effort is being directed to those areas that inherently have the largest margins for efficiency improvement.
- Exergy losses represent true losses of the potential that exists to generate the desired product from the given driving input. This is not true in general for energy losses. Thus, if the objective is to increase efficiency, focusing on exergy losses permits research and development to focus on reducing losses that will affect the objective.

Two other interesting points regarding the implications of exergy analyses merit consideration:

- Analysis results based on exergy usually suggest that research and development efforts should concentrate more on internal rather than external exergy losses, based on thermodynamic considerations. Of course, higher priority is appropriate for processes having larger exergy losses.
- Although focusing on areas having the greatest margins for improvement is sensible, research and development should also be devoted to those processes having low exergy losses in appropriate circumstances. This is particularly true when simple and cost-effective ways exist to increase efficiency by reducing small exergy losses.

4.10 Case studies

4.10.1 Case study 1: thermal energy storage

We consider as a case study of a thermal energy storage, a device that receives and holds thermal energy for subsequent use. Thermal energy storages come in many types (e.g., tanks, aquifers, ponds, caverns), and a storage in the form of a tank containing water as the storage fluid is considered here. This example serves to describe the information obtained from an exergy analysis, and to demonstrate that exergy efficiency provides an appropriate measure of performance of thermal energy storage while energy efficiency does not.

The case study is based on part on the long-term (seasonal) thermal energy storage used for a central solar heating plant that forms part of the Solarthermie 2000 Plus program. The system was built in 2007 in Munich, Germany (Bauer et al., 2010). The system integrates flat plate solar thermal collectors (of a total surface area of 2700 m^2) with thermal energy storage (of storage volume of 5700 m^3). The storage medium is hot water (about 5,700,000 kg when the storage is full).

The case study is simplified for convenience and to contrast exergy and energy analysis results. The ambient temperature is taken to be 15 °C, and the following hypothetical operating parameters are considered:

- The thermal energy storage is taken to be full of water and initially to be at 50 °C.
- The thermal energy storage is assumed to be perfectly insulated, so it is subject to no energy losses.

During the charging period, a quantity of 17,930 MJ of heat is input into the storage via a heat exchanger. 427,500 kg of water flows through

Table 4.3 Data for storage, charging, and recovery fluids (all water) in the thermal storage case study.

Fluid	Mass of fluid (kg)	Temperature range: Low temperature (°C)	State at low temperature	High temperature (°C)	State at high temperature
Storage tank fluid	5,700,000	50	Start of charging; end of discharging	50.75	End of charging; start of discharging
Charging transport fluid	427,500	80	Output from storage	90	Input to storage
Discharging (recovery) transport fluid	427,500	15	Input to storage	25	Output from storage

the external side of the heat exchanger. The flow cools from 90 °C to 80 °C as heat is transferred from it to the thermal storage. This heat addition raises the temperature of the thermal storage by 0.75 °C, to 50.75 °C. This is followed by a storage period, which in turn is followed by a discharging period.

During the discharging period, all of the input heat (i.e., 17,930 MJ) is recovered from the storage through the heat exchanger. Again, 427,500 kg of water flows through the external side of the heat exchanger. The flow is heated from 15 °C to 25 °C as heat is transferred to it from the thermal storage. The heat removal causes the thermal storage to return to 50 °C, its initial temperature.

The data for the operation of the storage are summarized in Table 4.3.

Efficiency data for the thermal storage operation for the case study are provided in Table 4.4. Some key observations follow:

- For the complete storage cycle for the thermal energy storage, the energy efficiency is seen to be 100%. The energy efficiency reflects the ratio of the heat reclaimed from the storage during charging to the heat

Table 4.4 Flows, losses and efficiencies based on energy and exergy for the thermal storage case study.

Commodity	Quantity (MJ): Charged (input)	Discharged (recovered)	Lost	Efficiency (%)
Energy	17,930	17,930	0	100
Exergy	3,659	299	3,360	8.2

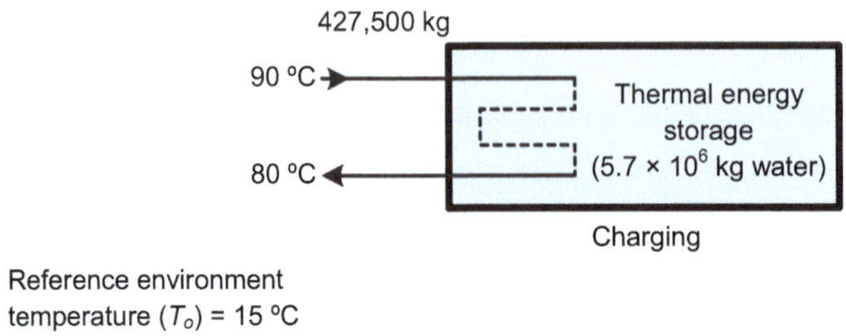

Fig. 4.4 ***Data and results for the operation of the thermal energy storage during charging.***

input during charging. However, for the complete cycle, the exergy recovered is 299 MJ, while the exergy supplied as 3659 MJ. The exergy efficiency is thus 8.2%, measured as the ratio of the thermal exergy recovered from storage to that input.

- The stark difference in the above point can be explained by noting that the recovered thermal energy is of little use, as it is at only 25 °C. The energy efficiency of the thermal storage is 100%, but energy analysis does not account for the degradation in the quality of the heat, as reflected by its temperature and reflected in its exergy. Exergy analysis therefore clearly accounts for the degradation in energy quality, and the exergy efficiency provides a more meaningful expression of the performance of the thermal energy storage.

Data and results for the operation of the thermal energy storage are illustrated in Fig. 4.4 for charging. It is seen that relatively high-quality energy is input to the storage, in terms of temperature.

Data and results for the operation of the thermal energy storage are illustrated in Fig. 4.5 for discharging. A key benefit of exergy analysis in thermal energy storage (TES) evaluation is seen, as the device appears ideal from an energy perspective even though it is far from ideal from an exergy

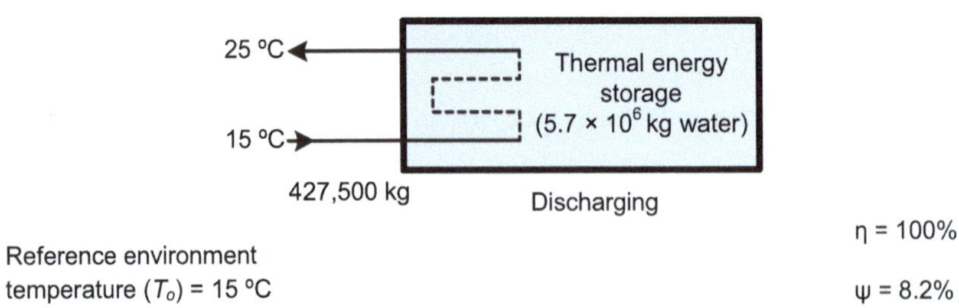

Fig. 4.5 ***Data and results for the operation of the thermal energy storage during discharging.***

perspective. It is seen that relatively low-quality energy is input to the storage, in terms of temperature.

This result is indeed practical and important, as the evaluation of thermal energy storage requires a rational, meaningful, and useful performance measure. The conventional energy efficiency of the thermal storage is not adequate for this purpose. The more perceptive basis provided by exergy analysis is needed for the true usefulness of thermal storage is to be assessed and optimized in terms of technical, economic, and other factors. The exergy efficiency measures the approach of the performance of a system to the ideal.

4.10.2 Case study 2: electrical resistance versus heat pump heating

In this case study, electrical heating is considered for the two main applications in buildings: domestic hot water production and space heating. The water heater converts electricity to heat at a temperature suitable for domestic hot water. Similarly, space heater converts electricity to heat at a temperature suitable for keeping a room at a comfortable temperature. Here, we consider a typical small commercial building, and investigate it a time when the external ambient temperature during the heating operations is taken to be 5 °C. This temperature is taken to be the reference-environment temperature in the analyses.

The heating needs of the building at the time considered are as follows:

- The space heater provides a thermal energy rate of 40 kW to maintain the building air temperature at 22 °C.
- The domestic hot water heater provides a heat supply rate of 8 kW to maintain the domestic hot water at a temperature of 55 °C.

The characteristics of the two technologies used to meet the heating needs of the building at the time considered are as follows:

- A coefficient of performance value of 3.5 is considered here for the electric heat pump, for space and both domestic hot water heating. This value is typical, as electric heat pumps have coefficients of performance for heating usually ranging from 2.5 to 5.5 for air-source units. The coefficient of performance is a measure of the heat pump energy efficiency, and is dependent on the temperatures of the product heating and the surroundings.
- An energy efficiency value of 98.5% is considered here for the electric resistance heaters. This value is typical, as electric resistance heaters have high energy efficiencies, normally over 98%.

Table 4.5 Energy and exergy flow rates and temperatures for an electrical resistance heater and a heat pump applied to space and domestic hot water heating.

	Electrical resistance heater		Heat pump	
Quantity	**Space heating**	**Domestic hot water heating**	**Space heating**	**Domestic hot water heating**
Input electricity rate (W)				
Energy	40,610	8,122	11,430	2,286
Exergy	40,610	8,122	11,430	2,286
Product heat rate (W)				
Energy	40,000	8,000	40,000	8,000
Exergy	2,304	1,219	2,304	1,219
Product heat temperature (°C)	22	55	22	55

Energy and exergy flow rates for the input electricity and product heat are listed in Table 4.5 for the electrical resistance heater and the heat pump, for the two applications considered: domestic hot water heating and space heating. Also listed is the temperature at which the product heat is delivered for each application.

Energy and exergy efficiencies are listed in Table 4.6 for the electrical resistance heater and the heat pump, for the two applications considered: domestic hot water heating and space heating. Note that the energy efficiency exceeds 100% as more product heat is delivered than the electrical energy input. The excess delivered product heat is heat extracted from the environment at the ambient temperature. To avoid the awkwardness of an efficiency exceeding 100%, the term coefficient of performance (COP) is used for a heat pump.

Several interesting and useful key points can be drawn from the information in Table 4.6, which demonstrate that useful and practical insights are obtained with exergy analysis:

Table 4.6 Energy and exergy efficiencies (in %) for an electrical resistance heater and a heat pump applied to space and domestic hot water heating.

	Electrical resistance heater		Heat pump	
Efficiency	**Space heating**	**Domestic hot water heating**	**Space heating**	**Domestic hot water heating**
Energy	98.5	98.5	350[a]	350[a]
Exergy	5.7	15.0	20.2	53.3

[a] A coefficient of performance (COP) for a heat pump of 3.5 corresponds to an energy efficiency of 350%.

- The exergy efficiencies in Table 4.5 for domestic hot water heating and for space heating using an electric heat pump are seen to be meaningful and the energy efficiencies misleading. For a heat pump with a COP of 3.5, domestic hot water heating and space heating are both seen to be achieved using only one-third of the electricity required for electric resistance heating. The heat pump exergy efficiencies, at 20.2% for space heating and 53.3% for domestic hot water heating, are seen to be much higher than the corresponding efficiencies for electrical resistance heating.
- The energy efficiency of electric resistance heating, at 98.5%, is very high and implies the maximum possible thermodynamic efficiency for electric heating is 100%. But this implication is incorrect, as an energy efficiency of 100% does not correspond to the most efficient heating device possible, as explained in the previous point.

The reason for the stark difference in energy and exergy perspectives is that energy analysis ignores the fact that, in electrical resistance heating, high-quality energy (electricity) is used to produce a relatively low-quality product (domestic hot water), or an even lower quality product (warm air). Exergy analysis recognizes this difference in energy qualities, and indicates the exergy of the thermal energy delivered by resistance heating to be 15% of the exergy entering the heater for domestic hot water heating, and to be 6% of the exergy entering the heater for space heating. Thus, the efficiency, based on exergy, of electric resistance heating is found to be about 15% for domestic hot water heating and 6% for space heating.

The processes considered in this example for providing heat from electricity are now illustrated and contrasted. Recall that 40 kW of heating is provided for space heating at a temperature of 22 °C and 8 kW for domestic hot water heating at a temperature of 55 °C.

The system using an electric resistance heater is shown in Fig. 4.6 for space heating and in Fig. 4.7 for domestic hot water heating. In both cases, electric resistance spacing heating is provided with an energy efficiency η of 98.5%.

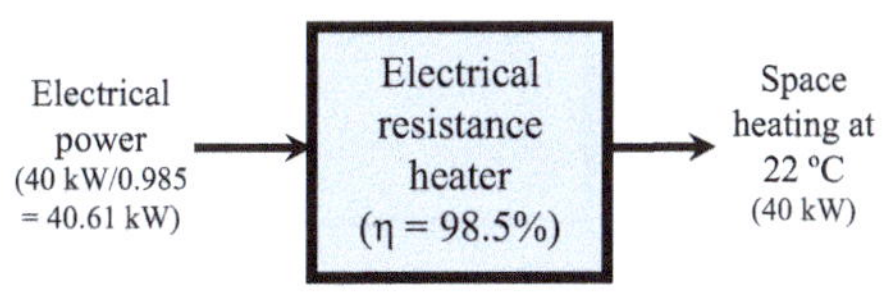

Fig. 4.6 ***Provision of space heating (40 kW) at a fixed temperature (22 °C) with electricity, using an electric resistance spacing heater with an energy efficiency η of 98.5%.***

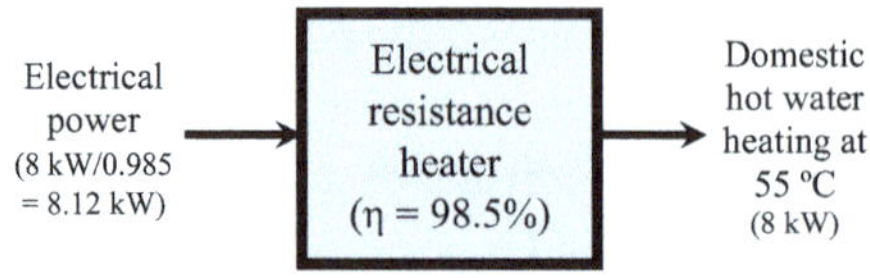

Fig. 4.7 ***Provision of domestic hot water heating (8 kW) at a fixed temperature (55 °C) with electricity, using an electric resistance spacing heater with an energy efficiency η of 98.5%.***

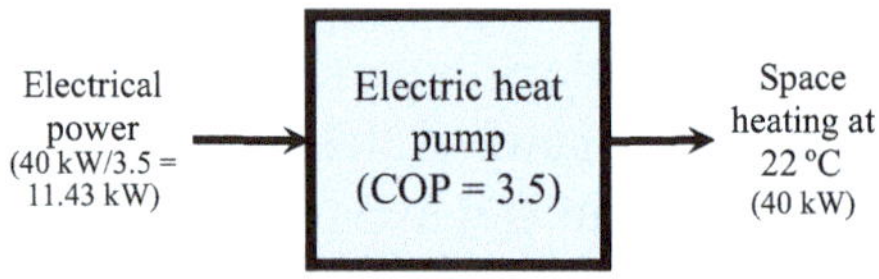

Fig. 4.8 ***Provision of space heating (40 kW) at a fixed temperature (22 °C) with electricity, using an electric heat pump with a coefficient of performance (COP) of 3.5.***

The system using an electric heat pump is shown in Fig. 4.8 for space heating and in Fig. 4.9 for domestic hot water heating. In both cases, heating with an electric heat pump is provided with a COP of 3.5.

Two interesting supplemental points are noted, as they provide further illustrations of the benefits of applying exergy analysis:

- The quality of the product's thermal energy for domestic hot water heating is higher than that for space heating because of the higher temperature involved, but both are nonetheless low quality compared to the high temperatures than can potentially be attained via electrical heating. Hence much higher exergy efficiencies can be attained for heating processes at higher temperatures, such as the refining of metal ores at temperatures exceeding 1000 °C.
- Thermodynamically ideal processes (purely theoretical considerations) for domestic hot water or space heating would have exergy efficiencies of 100%. Thus, the same space heating needed in this case study can in theory be achieved using much less electricity. That is, based on the present example, theoretically ideal electric heating operations

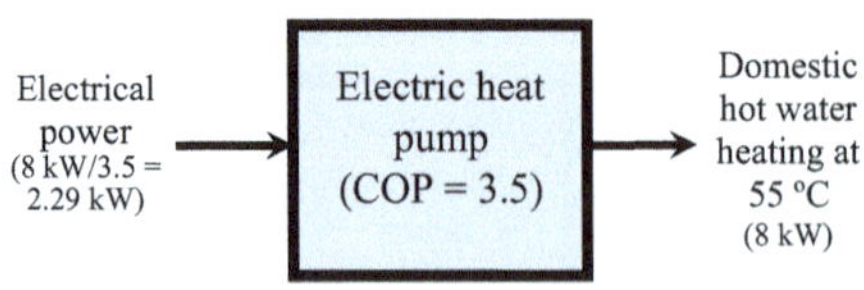

Fig. 4.9 ***Provision of domestic hot water heating (8 kW) at a fixed temperature (55 °C) with electricity, using an electric heat pump with a coefficient of performance (COP) of 3.5.***

would use 15% of the electricity used in conventional electric resistance domestic hot water heating, and as 6% of the electricity used in conventional electric resistance space heating.

4.11 Closing remarks

In this chapter, exergy is defined and a procedure is provided for energy and exergy analyses. Energy and exergy fundamentals and methods for applying exergy analysis are described, including thermodynamic balances, the reference environment, and efficiencies. Applications and implications of exergy analyses are examined, as are the relation of exergy to renewable energy and sustainability. The application of exergy analysis is assessed for two case studies, one involving thermal storage and the other different modes of electric heating. These examples demonstrate the advantages of exergy analysis as a substitute for or complement to energy analysis.

Nomenclature

COP	coefficient of performance
e	specific energy (kJ/kg)
E	energy (kJ)
ex	specific exergy (kJ/kg)
Ex	exergy (kJ)
Ex_Q	exergy associated with heat Q (kJ)
Ex_W	exergy associated with work W (kJ)
h	specific enthalpy (kJ/kg)
H	enthalpy (kJ)
HHV	higher heating value (kJ/kg)
I	irreversibility (kJ)
ke	specific kinetic energy (kJ/kg)
KE	kinetic energy (kJ)
m	mass (kg)
N	number of moles
P	pressure (atm)
pe	specific potential energy (kJ/kg)
PE	potential energy (kJ)
Q	heat (kJ)
Q_r	heat transfer into system across region r on system boundary (kJ)
s	specific entropy (kJ/kg K)
S	entropy (kJ/K)
S_{gen}	entropy generation (kJ/K)
t	time (s)
T	temperature (K)
u	specific internal energy (kJ/kg)

U internal energy (kJ)
v specific volume (m^3/kg)
V volume(m^3)
W work (kJ)
W_{NET} net usable work due to the volume change (kJ)
W_x shaft work (kJ)
W' total work done by a system (kJ)
y mole fraction

Greek Letters

η energy efficiency
μ chemical potential (kJ/mole)
Π entropy creation (kJ/K)
τ exergetic temperature factor
ψ exergy efficiency

Subscripts

e exit
f final
$flow$ flow
i inlet; initial
kin kinetic
$nonflow$ nonflow
o reference-environment state
oo dead state
ph physical
pot potential

References

Aghbashlo, M., Rosen, M.A, 2018. Exergoeconoenvironmental analysis as a new concept for developing thermodynamically, economically, and environmentally sound energy conversion systems. J. Clean. Prod. 187, 190–204.

Aghbashlo, M., Tabatabaei, M., Rahnama, E., Rosen, M.A., 2020. A new systematic decision support framework based on solar extended exergy accounting performance to prioritize photovoltaic sites. J. Clean. Prod. 256, 120356.

Ahrendts, J., 1980. Reference states. Energy Int. J. 5, 667–678.

Al-Zareer, M., Dincer, I., Rosen, M.A., 2018. Transient energy and exergy analyses of a multistage hydrogen compression and storage system. Chem. Eng. Technol. 41 (8), 1594.

Arsalis, A., Alexandrou, A., 2014. Thermoeconomic modeling and exergy analysis of a decentralized liquefied natural gas-fueled combined-cooling–heating-and-power plant. J. Nat. Gas Sci. Eng. 21, 209–220.

Ashrafi Goudarzi, S., Fazelpour, F., B. Gharehpetian, G., Rosen, M.A., 2019. Techno-economic assessment of hybrid renewable resources for a residential building in Tehran. Environ. Prog. Sustain. Energy 38 (5), 13209.

Baehr, H.D., Schmidt, E.F., 1963. Definition und berechnung von brennstoffexergien (Definition and calculation of fuel exergy). Brennst-Waerme-Kraft 15, 375–381.

Barberis, S., Rivarolo, M., Traverso, A., Massardo, A.F., 2016. Thermo-economic analysis of the energy storage role in a real polygenerative district. J. Energy Storage 5, 187–202.
Bauer, D., Marx, R., Nussbicker-Lux, J., Ochs, E., Heidemann, W., Muller-Steinhagen, H., 2010. German central solar heating plants with seasonal heat storage. Solar Energy 84, 612–623.
Bosnjakovic, F., 1963. Bezugszustand der exergie eines reagiernden systems (Reference states of the exergy in a reacting system). Forsch. Ingenieurw 20, 151–152.
Brodyanski, V.M., Sorin, M.V., Le Goff, P., 1994. The Efficiency of Industrial Processes: Exergy Analysis and Optimization. Elsevier, London.
Calise, F., Dentice d'Accadia, M., Macaluso, A., Piacentino, A., Vanoli, L., 2016. Exergetic and exergoeconomic analysis of a novel hybrid solar–geothermal polygeneration system producing energy and water. Energy Convers. Manage. 115, 200–220.
Chitgar, N., Emadi, M.A., Chitsaz, A., Rosen, M.A., 2019. Investigation of a novel multi-generation system driven by a SOFC for electricity and fresh water production. Energy Convers. Manage. 196, 296–310.
Compton, M., Rezaie, B., Rosen, M.A., 2018. Exergy approach for advancing sustainability of a biomass boiler. Int. J. Exergy 27 (1), 62–80.
Crane, P., Scott, D.S., Rosen, M.A., 1992. Comparison of exergy of emissions from two energy conversion technologies, considering potential for environmental impact. Int. J. Hydrogen Energy 17, 345–350.
Dincer, I., Rosen, M.A., 2021. Thermal Energy Storage: Systems and Applications, Third ed. Wiley: London.
Dincer, I., Rosen, M.A., 2021. Exergy: Energy, Environment and Sustainable Development, Third ed. Elsevier, Oxford, UK.
Dincer, I., Rosen, M.A, 2015. Exergy Analysis of Heating, Refrigerating and Air Conditioning: Methods and Applications. Elsevier, Oxford, UK.
Dincer, I., Rosen, M.A., Ahmadi, P., 2017. Optimization of Energy Systems. Wiley, London.
Edgerton, R.H., 1982. Available Energy and Environmental Economics. D.C. Heath, Toronto.
Ehyaei, M.A., Rosen, M.A., 2019. Optimization of a triple cycle based on a solid oxide fuel cell and gas and steam cycles with a multiobjective genetic algorithm and energy, exergy and economic analyses. Energy Convers. Manage. 180, 689–708.
El-Sayed, Y., 2004. The Thermoeconomics of Energy Conversion. Elsevier: Amsterdam.
Gaggioli, R.A., Petit, P.J, 1977. Use the second law first. ChemTech 7, 496–506.
Gunnewiek, L.H., Rosen, M.A., 1998. Relation between the exergy of waste emissions and measures of environmental impact. Int. J. Environ. Pollut. 10 (2), 261–272.
Haghghi, M.A., Shamsaiee, M., Holagh, S.G., Chitsaz, A., Rosen, M.A, 2019. Thermodynamic, exergoeconomic, and environmental evaluation of a new multi-generation system driven by a molten carbonate fuel cell for production of cooling, heating, electricity, and freshwater. Energy Convers. Manage. 199, 112040.
Hammond, G.P., 2004. Engineering sustainability: thermodynamics, energy systems, and the environment. Int. J. Energy Res. 28 (7), 613–639.
Hamrang, F., Shokri, A., Mahmoudi, S.M., Ehghaghi, B., Rosen, M.A., 2020. Performance analysis of a new electricity and freshwater production system based on an integrated gasification combined cycle and multi-effect desalination. Sustainability 12 (19), 7996.
Javanshir, N., Mahmoudi SM, S., Kordlar, M.A., A Rosen, M., 2020. Energy and cost analysis and optimization of a geothermal-based cogeneration cycle using an ammonia-water solution: thermodynamic and thermoeconomic viewpoints. Sustainability 12 (2), 484.
Karellas, S., Braimakis, K., 2016. Energy–exergy analysis and economic investigation of a cogeneration and trigeneration ORC–VCC hybrid system utilizing biomass fuel and solar power. Energy Convers. Manage. 107, 103–113.
Karimi, M.S., Fazelpour, F., Rosen, M.A., Shams, M., 2019. Techno-economic feasibility of building attached photovoltaic systems for the various climatic conditions of Iran. Environ. Prog. Sustain. Energy 38 (6), 13239.

Keshavarzzadeh, A.H., Ahmadi, P., Rosen, M.A., 2020. Technoeconomic and environmental optimization of a solar tower integrated energy system for freshwater production. J. Clean. Prod. 270, 121760.
Khaljani, M., Khoshbakhti Saray, R., Bahlouli, K., 2015. Comprehensive analysis of energy, exergy and exergo-economic of cogeneration of heat and power in a combined gas turbine and organic Rankine cycle. Energy Convers. Manage. 97, 154–165.
Khoshgoftar Manesh, M., Rosen, M.A., 2018. Combined cycle and steam gas-fired power plant analysis through exergoeconomic and extended combined pinch and exergy methods. J. Energy Eng. 144 (2), 04018010.
Kool, E.D., Cuomo, M.A., Reddy, B.V., Rosen, M.A., 2018. Multi-generation renewable energy system for dairy farms: exergy analysis. Eur. J. Sustain. Dev. Res. 2 (3), 37.
Kotas, T.J., 2012. The Exergy Method of Thermal Plant Analysis. Paragon Publishing, Malabar Reprint Ed.
Mehrpooya, M., Ansarinasab, H., Sharifzadeh, M.M.M., Rosen, M.A., 2018. Conventional and advanced exergoeconomic assessments of a new air separation unit integrated with a carbon dioxide electrical power cycle and a liquefied natural gas regasification unit. Energy Convers. Manage. 163, 151–168.
Moharramian, A., Soltani, S., Rosen, M.A., Mahmoudi, S.M.S., Jafari, M., 2019. Conventional and enhanced thermodynamic and exergoeconomic analyses of a photovoltaic combined cycle with biomass post firing and hydrogen production. Appl. Therm. Eng. 160, 113996.
Moran, M.J., 1989. Availability Analysis: A Guide to Efficient Energy Use. American Society of Mechanical Engineers, New York Revised ed.
Moran, M.J., Shapiro, H.N., Boettner, D.D., Bailey, M.B., 2018. Fundamentals of Engineering Thermodynamics, Ninth ed. Wiley, New York.
Peng, W., Maleki, A., Rosen, M.A., Azarikhah, P., 2018. Optimization of a hybrid system for solar-wind-based water desalination by reverse osmosis: comparison of approaches. Desalination 442, 16–31.
Rahnama, E., Aghbashlo, M., Tabatabaei, M., Khanali, M., Rosen, M.A., 2019. Spatio-temporal solar exergoeconomic and exergoenvironmental maps for photovoltaic systems. Energy Convers. Manage. 195, 701–711.
Rezaie, B., Reddy, B.V., Rosen, M.A., 2018. Exergy Assessment of a Solar-Assisted District Energy System. Open Fuels Energy Sci. J. 11 (1), 30–43.
Rodriguez, LSJ., 1980. Calculation of available-energy quantities. In: Gaggioli, R.A. (Ed.), Thermodynamics: Second Law Analysis, 122. ACS Symposium Series, American Chemical Society, Washington, pp. 39–60.
Rosen, M.A., 2011. Economics and Exergy: An Enhanced Approach to Energy Economics. Nova Science Publishers, Hauppauge, NY.
Rosen, M.A., 2012. Energy Storage. Nova Science Publishers, Hauppauge, NY.
Rosen, M.A., 2012. Environment, Ecology and Exergy: Enhanced Approaches to Environmental and Ecological Management. Nova Science Publishers, Hauppauge, NY.
Rosen, M.A., 2013a. Assessing global resource utilization efficiency in the industrial sector. Sci. Total Environ. 461-462, 804–807.
Rosen, M.A., 2013b. Using exergy to assess regional and national energy utilization: a comparative review. Arab. J. Sci. Eng. 38 (2), 251–261.
Rosen, M.A., 2013c. Using exergy to correlate energy research investments and efficiencies: concept and case studies. Entropy 15 (1), 262–286.
Rosen, M.A., 2021. Exergy analysis. In: El Haj Assad, M., Rosen, M.A. (Eds.), 2021. Design and Performance Optimization of Renewable Energy Systems. Academic Press, Elsevier, New York, pp. 43–60.
Rosen, M.A., Dincer, I., 1997a. On exergy and environmental impact. Int. J. Energy Res. 21, 643–654.
Rosen, M.A., Dincer, I., 1997b. Sectoral energy and exergy modeling of Turkey. ASME J. Energy Resour. Technol. 119, 200–204.

Rosen, M.A., Dincer, I., 1999. Exergy analysis of waste emissions. Int. J. Energy Res. 23 (13), 1153–1163.
Rosen, M.A., Dincer, I., 2004. Effect of varying dead-state properties on energy and exergy analyses of thermal systems. Int. J. Therm. Sci. 43 (2), 121–133.
Rosen, M.A., Koohi-Fayegh, S., 2016. Cogeneration and District Energy Systems: Modelling, Analysis and Optimization. Institution of Engineering and Technology, London.
Rosen, M.A., Koohi-Fayegh, S., 2017. Geothermal Energy: Sustainable Heating and Cooling Using the Ground. Wiley, London.
Rosen, M.A., Tang, R., Dincer, I., 2004. Effect of stratification on energy and exergy capacities in thermal storage systems. Int. J. Energy Res. 28, 177–193.
Salehi, S., Mahmoudi, S.M.S., Yari, M., Rosen, M.A., 2018. Multi-objective optimization of two double-flash geothermal power plants integrated with absorption heat transformation and water desalination. J. Clean. Prod. 195, 796–809.
Sato, N., 2005. Chemical Energy and Exergy: An Introduction to Chemical Thermodynamics for Engineers. Elsevier, Oxford, UK.
Sciubba, E., 2005. Exergo-economics: thermodynamic foundation for a more rational resource use. Int. J. Energy Res. 29, 613–636.
Sciubba, E., 2019. Exergy-based ecological indicators: from thermo-economics to cumulative exergy consumption to thermo-ecological cost and extended exergy accounting. Energy 168, 462–476.
Seyed Mahmoudi, S.M., Sarabchi, N., Yari, M., Rosen, M.A., 2019. Exergy and exergoeconomic analyses of a combined power producing system including a proton exchange membrane fuel cell and an organic Rankine cycle. Sustainability 11 (12), 3264.
Seyyedi, S.M., Hashemi-Tilehnoee, M., Rosen, M.A., 2018. Exergy and exergoeconomic analyses of a novel integration of a 1000 MW pressurized water reactor power plant and a gas turbine cycle through a superheater. Ann. Nucl. Energy 115, 161–172.
Sussman, MV., 1980. Steady-flow availability and the standard chemical availability. Energy Int. J. 5, 793–804.
Sussman, MV., 1981. Second law efficiencies and reference states for exergy analysis. Proc. Second World Congress Chem. Eng., Canadian Society of Chemical Engineers. Montreal, 420–421.
Szargut, J., 1967. Grezen fuer die anwendungsmoeglichkeiten des exergiebegriffs (limits of the applicability of the exergy concept). Brennst.-Waerme-Kraft 19, 309–313.
Szargut, J., 2005. Exergy Method: Technical and Ecological Applications. WIT Press, Southampton, UK.
Szargut, J., Morris, D.R., Steward, F.R, 1988. Exergy Analysis of Thermal, Chemical and Metallurgical Processes. Hemisphere, New York.
WJ, W., Gaggioli, RA., 1980. Reference datums for available energy. In: Gaggioli, R.A. (Ed.), Thermodynamics: Second Law Analysis. ACS Symposium Series, 122. American Chemical Society, Washington, pp. 77–92.
Wu, B., Maleki, A., Pourfayaz, F., Rosen, M.A., 2018. Optimal design of stand-alone reverse osmosis desalination driven by a photovoltaic and diesel generator hybrid system. Sol. Energy 163, 91–103.
Yantovskii, E.I., 1994. Energy and Exergy Currents: An Introduction to Exergonomics. Nova Science Publishers, New York.

CHAPTER FIVE

Seawater desalination process modeling

5.1 Introduction

In this chapter, various desalination technologies are modeled. We start by formulating the thermodynamic properties of pure water and seawater using property functions previously derived (Millero et al., 2008; Sharqawy et al., 2010; Wagner and Pruß, 2002). Then the modeling is described of various desalination technologies, including humidification–dehumidification (HDH) desalination, multieffect desalination (MED), multistage flash (MSF) desalination, direct contact membrane desalination (DCMD), mechanical vapor compression (MVC) desalination, and reverse osmosis (RO) desalination. Part of the modeling involves specifying the thermodynamic state of the streams interacting with each system. Through modeling, various characteristics and parameters of the desalination systems can be determined, including the heat transfer area required for the evaporation/condensation processes and the energy required for desalination.

5.2 Thermodynamic properties of seawater

The salinity or salt concentration of seawater refers to the total amount of solids dissolved in a unit mass of seawater. The seawater property package developed by Millero et al. (2008) is one of the most accurate for the absolute salinity of water. In that package, seawater is considered as a complex electrolyte solution. Details on fundamental property equations for water have been developed by the International Association for the Properties of Water and Steam (IAPWS) (Wagner and Pruß, 2002). Sharqawy et al. (2010) presented thermodynamic property equations for seawater, which are in a good agreement with property information provided by the IAPWS.

The fundamental correlation for Gibbs free energy as a function of salinity, pressure, and temperature is used to evaluate the thermodynamic properties of pure water and seawater. The thermodynamic property

Sustainable Energy Technologies for Seawater Desalination
DOI: https://doi.org/10.1016/B978-0-323-99872-7.00008-5

equations are in the form of polynomial expressions depending on salt concentration and temperature at the atmospheric pressure. The selection of an appropriate reference state is important for meaningfully assessing desalination systems.

We now present and discuss the following thermodynamic properties for seawater: specific enthalpy, specific entropy, specific volume, and chemical potential. The correlations presented for the thermodynamic properties are based on the seawater model utilized by the IAPWS (Wagner and Pruß, 2002), in which correlations for Gibbs free energy are used as a function of temperature, pressure, and salinity to determine the seawater properties.

Specific volume and density: Here we discuss the specific volume ν and density ρ of seawater. The dissolved salts in seawater increase its density compared to pure water. As the specific volume is the inverse of density (i.e., $\nu_{sw} = 1/\rho_{sw}$, where the subscript *sw* denotes seawater), the specific volume of seawater is less than that of pure water and it decreases with increasing salt concentration. Sharqawy et al. (2010) provide the following respective expressions for the density of pure water and seawater:

$$\rho_w = 999.9 + 2.034\times10^{-2}T - 6.162\times10^{-2}T^2 + 2.261\times10^{-5}T^3 - 4.657\times10^{-8}T^4 \tag{5.1}$$

$$\rho_{sw} = \rho_w + w_s\left(8.020\times10^2 - 2.001(T) + 1.677\times10^{-2}\left(T^2\right) - 3.060\times10^{-5}\left(T^3\right) - 1.613\times10^{-5}w_s\left(T^3\right)\right) \tag{5.2}$$

where ρ_w and ρ_{sw} are the density of pure water and seawater (kg/m^3), w_s is the salt concentration in kg$_s$/kg$_{sw}$ and T is the temperature (°C). These expressions are valid for $0 \le T \le 180$ °C and $0 \le w_s \le 0.16$ kg$_s$/kg$_{sw}$. The accuracy of these expressions is relatively high relative to the values provide by Millero (2008) and IAPWS (Wagner and Pruß, 2002). From the above polynomial expressions, it can be seen that the higher the seawater temperature, the higher is its specific volume. Note that the above expressions can be used for pressures below atmospheric pressure (up to the saturation pressure) or above atmospheric pressure (up to the critical pressure). This is mainly due to the independence of incompressible flow (such as seawater) with the specific volume.

In the thermodynamic literature, specific volume ν is sometimes used instead of density ρ the specific volume dealing with the flow work.

Specific enthalpy: The specific enthalpy of pure water slightly exceeds that of seawater as the heat capacity is greater for pure water than seawater. The following correlation for the specific enthalpy of pure water, fitted to data from the IAPWS (Wagner and Pruß, 2002), is valid for $5 \leq T \leq 200$ °C:

$$h_w = 141.355 + 4202.070\,T - 0.535T^2 + 4 \times 10^{-3}T^3 \tag{5.3}$$

where the temperature T is in units of °C. Alternatively, the specific enthalpy of pure water can be calculated using data for saturated liquid water and saturated vapor, as described in Farsi et al. (2017). The specific enthalpy of seawater can be expressed as follows, for $10 \leq T \leq 120$ °C and $0 \leq w_s \leq 0.12$ kg_s/kg_{sw}:

$$\begin{aligned} h_{sw} = h_w + \big(&2.348 \times 10^{-4} w_s - 3.152 \times 10^{-5} w_s^2 - 2.803 \times 10^6 w_s^3 \\ &+ 1.446 \times 10^7 w_s^4 - 7.826 \times 10^3 w_s T + 4.417 \times 10^1 w_s T^2 \\ &- 2.139 \times 10^{-1} w_s T^3 + 1.991 \times 10^4 w_s^2 T - 2.778 \times 10^4 w_s^3 T \\ &- 9.728 \times 10^1 w_s^2 T^2\big) \end{aligned} \tag{5.4}$$

where T is in °C. Again the expression is fitted to data from the IAPWS (Wagner and Pruß, 2002) and the correlation exhibits a good agreement with the data. The higher is the temperature and the lower is the salinity for seawater, the higher is the specific enthalpy. Unlike the specific volume, the specific enthalpy of seawater, treated as an incompressible fluid, varies with pressure as follows (Moran, 1989):

$$h_{sw}(T, P, w_s) = h_{sw}(T, P_0, w_s) + v_{sw}(P - P_0) \tag{5.5}$$

where h_{sw} (T, P_0, w_s) is the specific enthalpy of seawater evaluated from Eq. (5.4) and v_{sw} is the specific volume of seawater, obtainable as the inverse of the density expression Eq. (5.2).

Specific entropy: The specific entropy of pure water exceeds slightly that of seawater, and can be evaluated as follows:

$$\begin{aligned} s_w = &\,0.1543 + 15.383\,T - 2.996 \times 10^{-2}\,T^2 + 8.193 \times 10^{-5}T^3 \\ &- 1.370 \times 10^{-7}\,T^4 \end{aligned} \tag{5.6}$$

where s_w is the specific entropy of pure water (kJ/(kg.K)). The above equation is valid for $5 \leq T \leq 200$ °C and exhibits good agreement with data from the IAPWS (Wagner and Pruß, 2002). The specific entropy of seawater is given by Sharqawy et al. (2010):

$$\begin{aligned} s_{sw} &= s_w + 4.231\times10^{2} w_s - 1.463\times10^{4} w_s^{2} + 9.880\times10^{4} w_s^{3} \\ &- 3.905\times10^{5} w_s^{4} - 2.562\times10\, w_s T + 1.443\times10^{-1} w_s T^{2} \\ &- 5.879\times10^{-4} w_s T^{3} + 6.111\times10^{1} w_s^{2} T - 8.041\times10^{1} w_s^{3} T \\ &- 3.035\times10^{-1} w_s^{2} T^{2} \end{aligned} \tag{5.7}$$

which is valid for $10 \le T \le 120$ °C and $0 \le w_s \le 0.12$ kg_s/kg_{sw}. The specific entropy of seawater is more sensitive to temperature than salt concentration; seawater-specific entropy rises significantly with increasing temperature and declines slightly with increasing salinity. For an incompressible fluid such as seawater, the specific entropy is almost independent of pressure. Thus, the specific entropy expression for seawater (Eq. 5.6) can be employed at the atmospheric and nonatmospheric pressures (Moran, 1989).

Chemical potential: To determine the chemical potential of salt and water in seawater, the specific Gibbs free energy function can be differentiated in terms of composition (salt and water). Then,

$$\mu_s = \frac{\partial G_{sw}}{\partial m_s} = g_{sw} + (1 - w_s)\frac{\partial g_{sw}}{\partial w_s} \tag{5.8}$$

$$\mu_w = \frac{\partial G_{sw}}{\partial m_w} = g_{sw} + x_s \frac{\partial g_{sw}}{\partial w_s} \tag{5.9}$$

where μ_s and μ_w denote chemical potential of salt and pure water in (kJ/kg_s) and (kJ/kg_w), respectively. The specific Gibbs free energy of seawater (g_{sw}) can be evaluated with the specific enthalpy and specific entropy of seawater as

$$g_{sw} = h_{sw} - T s_{sw} \tag{5.10}$$

where T is the absolute temperature in K. The specific Gibbs free energy differentiated with respect to salt concentration $\left(\frac{\partial g_{sw}}{\partial w_s}\right)$ can be written as

$$\frac{\partial g_{sw}}{\partial w_s} = \frac{\partial h_{sw}}{\partial w_s} - T\frac{\partial s_{sw}}{\partial w_s} \tag{5.11}$$

where the differentials of specific enthalpy ($\partial h_{sw}/\partial w_s$) and specific entropy ($\partial s_{sw}/\partial w_s$) are determined through differentiating the expressions for specific enthalpy (Eq. 5.4) and specific entropy (Eq. 5.6) with respect to the salt concentration. It is clear that the differentiations of specific enthalpy and specific entropy for pure water with respect to salinity are zero (i.e., $\partial h_{sw}/\partial w_s = 0$ and $\partial s_{sw}/\partial w_s = 0$). The terms $\partial h_{sw}/\partial w_s$ and $\partial s_{sw}/\partial w_s$ can be written as

$$\begin{aligned}\frac{\partial h_{sw}}{\partial w_s} &= -2.348\times10^4 + 2\left(3.152\times10^5\right)w_s + 3\left(2.803\times10^6\right)w_s^2 \\ &+4\left(-1.446\times10^7\right)w_s^3 + 7.826\times10^3 T + \left(-4.417\times10^1\right)T^2 \\ &+2.134\times10^{-1}T^3 + 2\left(-1.991\times10^4\right)w_s T + 3\left(2.778\times10^4\right)w_s^2 \\ &-2\left(4.728\times10^1\right)w_s T^2\end{aligned} \tag{5.12}$$

$$\begin{aligned}-\frac{\partial s_{sw}}{\partial w_s} &= -4.231\times10^2 + 2\left(1.463\times10^4\right)w_s + 3\left(-9.880\times10^4\right)w_s^2 \\ &+4\left(3.095\times10^5\right)w_s^3 + 2.262\times10^1 T \pm 1.443\times10^{-1}T^2 \\ &+5.879\times10^{-4}T^3 + 2\left(-6.111\times10^1\right)w_s T + 3\left(8.041\times10^1\right)w_s^2 T \\ &+2\left(3.035\times10^1\right)w_s T^2\end{aligned} \tag{5.13}$$

where the temperature in Eqs. (5.12) and (5.13) are in K. Also, the temperature in Eq. (5.11) is an absolute temperature in K. Fig. 5.1 shows the variation of chemical potential of water and salt in seawater with temperature and salinity. It is seen that the chemical potential of pure water in seawater varies inversely with both temperature and salinity. But the chemical potential of salt in seawater increases with the increasing temperature and salt concentration.

As noted before, the correlations presented here for seawater and pure water are based on the seawater model. There are other models for specifying seawater properties such as the sodium chloride aqueous solution and ideal mixture models, which are frequently used. The predicted thermodynamic properties of seawater differ between these models. In this book, the most updated correlations are provided to establish an appropriate source for seawater property determination.

In the following sections, comprehensive models are presented for several desalination techniques, namely HDH desalination, MED, MSF desalination, DCMD, MVC desalination, and reverse osmosis desalination.

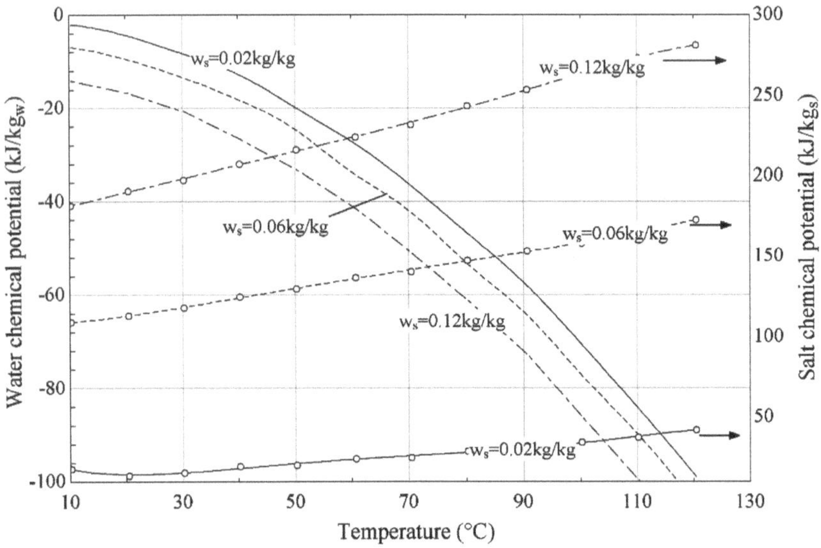

Fig. 5.1 ***Chemical potentials of water and salt in seawater for various values of salt concentration*** w_s.

5.3 HDH desalination model

The working principle for HDH desalination is similar to the natural water cycle. In the humidifier (Fig. 5.2), the air is moistened and heated by the hot water, after which its water content is reduced by condensing using the cold saline seawater within the dehumidifier. The preheated seawater in the dehumidifier is further heated to the top brine temperature by a heat source such as solar collector, before being sprayed in the humidifier. As the HDH system has low recovery in a single pass of the process, the brine stream leaving the humidifier can be recirculated through the heat rejection component where its temperature is reduced to the feed temperature, $T_{f,b}$.

As pointed out before, the gained output ratio (GOR) is the ratio of latent heat of produced water per unit of heat input, and expressible as

$$GOR_{HDH} = \frac{h_{fg}\dot{m}_{fw}}{\dot{Q}_{HDH}} \tag{5.14}$$

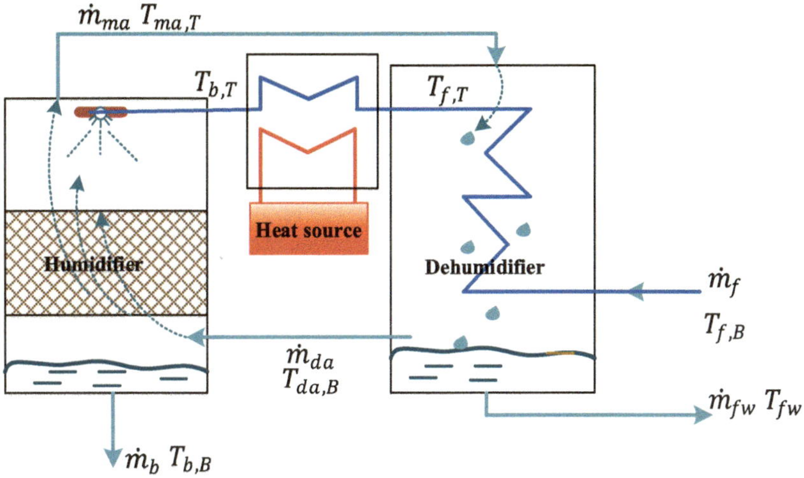

Fig. 5.2 ***Typical closed air, open water (CAOW) humidification–dehumidification (HDH) zero extraction system illustrating flow paths and their corresponding temperatures.***

where $\dot{Q}_{HDH}$ is the heat input to the HDH system, h_{fg} is the specific latent heat of vaporization at the inlet water temperature. Generally, a higher *GOR* means that more seawater is distillated per unit energy input. More specifically, *GOR* indicates how efficiently the heat input is utilized by a desalination system. A value of *GOR* of unity expresses that the exact energy input required for seawater evaporation is used. *GOR* values higher than unity indicate that the input thermal energy is regenerated within the cycle and that less heat input is provided than the exact heat input required when $GOR = 1$.Values of *GOR* less than one indicate that the heat input to the cycle exceeds that required for evaporation and a portion of the heat is degraded due to the losses. In solar-desalination facilities, for which *GOR* typically is equal to or less than one, a higher *GOR* permits less collector area to be utilized, while higher *GOR* values for fossil fuel–based desalination systems implies lower fuel costs.

In HDH desalination systems, many parameters contribute the gained output ratio of the system such as humidity ratio, mass flow rates of air and water streams, top and bottom temperatures at the humidifier/dehumidifier, and the effectiveness of the heat exchangers. Also, the performance of an HDH system depends on the mass flow rate ratio, which is defined as the mass flow rate of water at the inlet of the humidifier to the mass flow rate of air through the humidifier.

Common assumptions considered in developing the HDH model follow:

- Steady-state conditions and operation at atmospheric pressure are dominant.
- Potential and kinetic energy effects are negligible and pumping power is neglected compared to the heat input rate.
- The dehumidifier and humidifier are well insulated to avoid heat exchange with the environment.
- The moist air streams at the exits of the dehumidifier and humidifier are saturated.

Mass rate balances for the air and water flows are now described:

Air: In the closed air, open water HDH system, the mass flow rate of dry air ($\dot{m}_{da}$) remains constant through the humidifier (H) and dehumidifier (DH). Thus, a mass rate balance for the dry air can be expressed by

$$\dot{m}_{da} = \dot{m}_{da,DH,in} = \dot{m}_{da,DH,out} = \dot{m}_{da,H,in} = \dot{m}_{da,H,out}. \tag{5.15}$$

The humidity ratio (ω), which is the mass ratio of water vapor to dry air, varies throughout the cycle. But the mass flow rate of moist air ($\dot{m}_{ma}$) is not constant throughout the cycle. A mass rate balance for the moist air can be written as

$$\dot{m}_{ma} = \dot{m}_{da} + \dot{m}_{v} = \dot{m}_{da}\left(1 + \frac{\dot{m}_{v}}{\dot{m}_{da}}\right) = \dot{m}_{da}\left(1 + \omega\right) \tag{5.16}$$

where $\dot{m}_{v}$ is the mass flow rate of water vapor in the moist air, in kg/s.

Water: In the dehumidifier, the conservation of water indicates that the vapor mass rate condensed in the dehumidifier ($\dot{m}_{con}$), that is, the mass flow rate of freshwater product $\dot{m}_{fw}$, is

$$\dot{m}_{con} = \dot{m}_{fw} = \dot{m}_{da}\left(\omega_{ma,DH,in} - \omega_{ma,DH,out}\right) \tag{5.17}$$

In the humidifier, the brine (b) mass flow rate is obtained by subtracting produced freshwater (fw) from the feed seawater (f) sprayed into the humidifier (H) as follows:

$$\dot{m}_{b} = \dot{m}_{f} - \dot{m}_{fw} = \dot{m}_{f} - \dot{m}_{da}\left(\omega_{ma,H,out} - \omega_{ma,H,in}\right). \tag{5.18}$$

The governing equations for the humidifier, dehumidifier, and heater follow:

Dehumidifier: The dehumidifier is a two-phase heat exchanger in which the water vapor in moist air is condensed through the use of cold tubes containing feed water. An energy rate balance for the dehumidifier based on Fig. 5.2 can be written as follows:

$$\dot{m}_f\left(h_{f,B}-h_{f,T}\right)+\left(\dot{m}_{ma,T}h_{ma,T}-\dot{m}_{ma,B}h_{ma,B}\right)-\dot{m}_{fw}h_{fw}=0 \qquad (5.19)$$

where h denotes specific enthalpy (kJ/kg), and subscripts T and B indicate top and bottom of dehumidifier. On the left side of Eq. (5.19), the first term denotes the energy rate change associated with the feed stream (f), the second term denotes the energy rate change of moist air (ma), and the last term denotes the energy rate of produced freshwater (fw).

Mistry (2010) proposed the following expression for the condensate temperature in terms of the moist air inlet and outlet wet-bulb temperatures in the dehumidifier operating at atmospheric pressure:

$$\begin{aligned} T_{con} &= 0.0051918T_{wb,in}^2+0.0027692T_{wb,out}^2-0.007417T_{wb,in}T_{wb,out} \\ &\quad -0.41913T_{wb,in}+1.0511T_{wb,out}+61.6186 \end{aligned} \qquad (5.20)$$

where $T_{wb,in}$ and $T_{wb,out}$ are the inlet and outlet moist air wet-bulb temperatures in K. Note that, for the case in which the moist air stream is at the saturation state, the wet-bulb temperature is identical to the dry-bulb temperature.

Humidifier: The humidifier is a kind of direct-contact heat and mass exchanger and its working principle is the same as for the cooling tower. An energy rate balance for the humidifier in Fig. 5.2 can be written as

$$\dot{m}_{da}h_{da,B}-\dot{m}_{ma}h_{ma,T}+\dot{m}_f h_{b,T}-\dot{m}_b h_{b,B}=0 \qquad (5.21)$$

where the first two terms represent the energy rates associated with the dry air and moist air, respectively. The third term represents the energy rate of the feed seawater sprayed into the humidifier and the last term represents the energy rate of the brine stream rejected from the system.

Heater: The heat input rate to the HDH system ($\dot{Q}_{HDH}$) from the heat source shown in Fig. 5.2 can be expressed as

$$\dot{Q}_{HDH}=\dot{m}_f\left(h_{b,T}-h_{f,T}\right) \qquad (5.22)$$

where $\dot{m}_f$ is the feed seawater mass flow rate (kg/s), and $h_{b,T}$ and $h_{f,T}$ are the specific enthalpies of feed streams entering and leaving the heater in kJ/kg.

The saturated humidity ratio of moist air (ω^s) is defined as

$$\omega^s=\frac{M_v}{M_{da}}\left(\frac{P_s}{P-P_s}\right) \qquad (5.23)$$

where M is the molar mass in kg/mol, and P and P_s are the vapor pressure and saturation pressure, respectively. Using ω and the mass flow rate ratio ($m_r = \dot{m}_f / \dot{m}_{da}$), Eq. (5.19) can be normalized in terms of mass flow rate of dry air ($\dot{m}_{da}$) as follows:

$$m_r = \frac{\left(h_{ma,T} - h_{ma,B}\right) - \left(\omega^s_T - \omega^s_B\right)h_{fw}}{h_{f,B} - h_{f,T}} \approx \frac{\left(\omega^s_T - \omega^s_B\right)h_{fg}}{h_{f,B} - h_{f,T}} \tag{5.24}$$

Here the specific enthalpy of the product freshwater is equal to the latent heat of condensation ($h_{fw} = h_{fg}$). Assuming that the moist air passing through the dehumidifier experiences only a phase change process within the vapor to become freshwater, regardless of the sensible heat term, the first term in the numerator of Eq. (5.24) is eliminated. This assumption may lead to an error of about 10%. In a high-salinity desalination operation, the boiling point is higher for the saline mixture than the pure water solvent. In fact, the temperature difference between the pure and saline mixture with identical values of humidity ratio (i.e., the same vapor pressure) is defined as the boiling point elevation (*BPE*). In a detailed analysis of HDH desalination, when both the moist air mixture and the saline solution have the same specific enthalpies, their corresponding temperatures are not necessarily equal. This difference in the temperature is referred as the effective boiling point elevation (BPE_{eff}). Values of BPE_{eff} for NaCl (aq) are provided in Fig. 5.3. The specific enthalpies of the feed stream at the bottom and top of the dehumidifier, respectively, can be expressed in terms of the specific heat at constant pressure of the feed stream ($c_{p,f}$), the moist air temperature leaving the humidifier ($T_{ma,T}$), the pinch point temperature difference (ΔT_{pp}), and the BPE_{eff} as

$$h_{f,T} = c_{p,f}\left(T_{ma,T} - \Delta T_{pp} - BPE_{eff}\right) \tag{5.25}$$

$$h_{f,B} = c_{p,f} T_{f,B} \tag{5.26}$$

where ΔT_{pp} can be taken to be equal at both ends of the dehumidifier to satisfy the energy balance within this module.

Exchanger effectiveness: The effectiveness of the humidifier and dehumidifier heat exchangers can be defined as the ratio of actual heat transfer and theoretical maximum heat transfer, that is, $C_{min}(T_{hot,in} - T_{cold,in})$, where C_{min} denotes the minimum heat capacity rate of the two streams. The mass transfer between the water and air leads to the change in the enthalpy of the stream, which depends on the temperature and humidity.

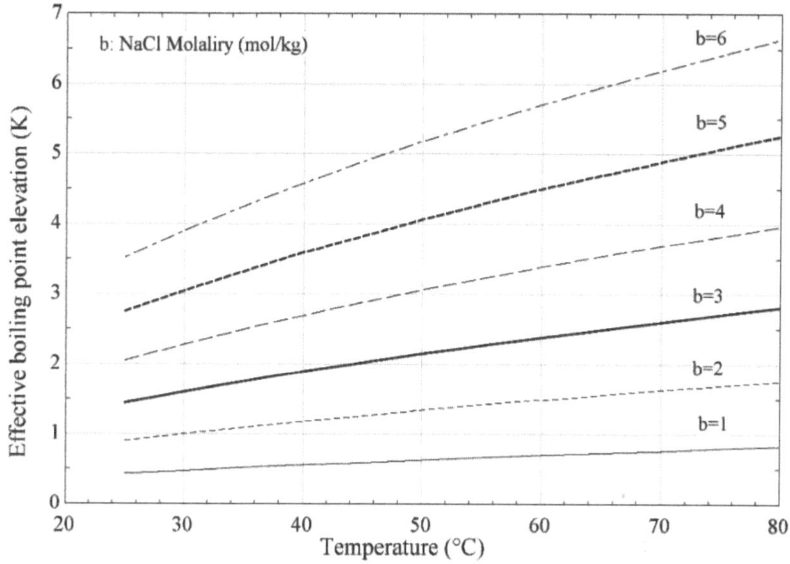

Fig. 5.3 ***Effective boiling point elevation BPE$_{eff}$ for various values of NaCl (aq) molality (mol/kg).*** *Data from Thiel et al. (2015).*

The heat exchanger effectiveness can be determined in two ways. The first assumes the air to be the lower maximum enthalpy rate change ε_a while the second assumes the water stream as the lower maximum enthalpy rate change (ε_w). That is,

$$\varepsilon_a = \frac{\Delta \dot{H}_{ma}}{\Delta \dot{H}_{ma}^{ideal}} \tag{5.27}$$

$$\varepsilon_w = \frac{\Delta \dot{H}_{w}}{\Delta \dot{H}_{w}^{ideal}} \tag{5.28}$$

where the ideal enthalpy rate change in the above two equations is obtained at the condition of zero terminal temperature difference at the bottom (or top) of the heat exchanger (i.e., $T_{w,out}^{ideal} = T_{ma,in}$ and $T_{ma,out}^{ideal} = T_{w,in}$). In addition, the saturated state ($\omega = 1$) is assumed at the exits for the moist air stream. Note that the smaller ideal enthalpy rate change is associated with the stream having the lower total capacity rate. Thus, the actual effectiveness is the greater of ε_a and ε_w, that is,

$$\varepsilon = \max\left(\varepsilon_a, \varepsilon_w\right) \tag{5.29}$$

For example, the effectiveness of the dehumidifier in an HDH desalination system is determined by

$$\varepsilon_{DH,a} = \frac{\left(h_{ma,DH,in} - h_{ma,DH,out}\right) - \left(\omega_{ma,DH,in} - \omega_{ma,DH,out}\right) h_{fw}}{\left(h_{ma,DH,in} - h_{ma,DH,out}^{ideal}\right) - \left(\omega_{ma,DH,in} - \omega_{ma,DH,out}^{ideal}\right) h_{fw}^{ideal}} \tag{5.30}$$

$$\varepsilon_{DH,w} = \frac{h_{w,DH,in} - h_{w,DH,out}}{h_{w,DH,in} - h_{w,DH,out}^{ideal}} \tag{5.31}$$

$$\varepsilon_{DH} = \max\left(\varepsilon_{DH,a}, \varepsilon_{DH,w}\right). \tag{5.32}$$

5.4 MED model

The reliability and resistance to fouling of the MED system have made it a practical and efficient method, especially for treating wastewaters from the industrial sector. MED is a well-established desalination technology in which a high *GOR* is achieved through successive evaporation/condensation mechanisms.

Fig. 5.4 presents a flow diagram of a forward-feed MED system where the heat source, steam, vaporizes a portion of sprayed feed in the first effect; the other portion, which is flashed once sprayed into the first effect, preheats the feed stream and then is condensed in the second effect. The heat released from condensation supplies thermal energy to vaporize a part of sprayed feed in the second effect. Also, the brine stream from the first effect is partially flashed within the flash box and then transferred into the second effect. The vapor of evaporation and flashing preheats the feed stream and then is condensed in the third effect, from which the required vaporization heat is provided for the next effect. This process is repeated through the sequential effects that depend on the scale of system (the larger the scale of the MED system, the more the number of effects).

Numerous models for MED desalination system have been reported (Darwish et al., 2006; El-Sayed and Silver, 1980; El-Dessouky and Ettouney, 2002). An accurate model of a system is necessary for better understanding and identifying potential improvements.

The model developed here for MED can be solved through a simultaneous equation solver rather than an iterative sequential solver. A modular method is used in the MED model in which each component is modeled

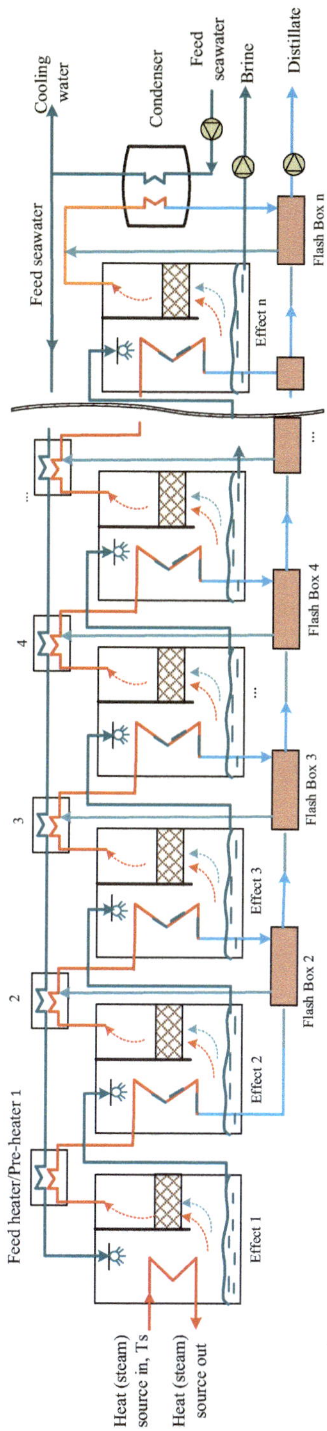

Fig. 5.4 *Typical forward-feed multieffect desalination system.*

individually; then by connecting the components and utilizing additional correlations, the overall system model can be developed. This method allows for studying different configurations of MED such as backward, forward, and parallel feed with the minimum duplicated code. Thus, thermodynamic models for the effects of MED, flash boxes, feed heaters, and the condenser are developed, and now described.

MED effect: In the MED effect, salt is removed from a part of the sprayed feed stream. The freshwater vapor generated through the evaporation process is transferred to the preheater and the remaining high-saline brine stream moves to the next effect. We start by defining the subscripts *bo* for boiling, *b* for brine, *f* for feed, *fw* for freshwater (vapor or liquid), and *fl* for flashing. Prior to vapor generation through boiling, a part of feed stream inside the effect is flashed, producing vapor and brine as the effect is maintained at a pressure slightly below the saturation pressure of the feed water. The remaining brine mass flow rate from the flashing process ($\dot{m}_{b,fl}$) is boiled through the evaporator tubes, yielding freshwater vapor ($\dot{m}_{fw,bo}$) and hot brine ($\dot{m}_{b,bo}$ or $\dot{m}_b$), as seen in Fig. 5.5. So, the mass rate balances in the *i*th effect of MED are

$$\dot{m}_{f,i} = \dot{m}_{fw,fl,i} + \dot{m}_{b,fl,i} \tag{5.33}$$

$$\dot{m}_{b,fl} = \dot{m}_{fw,bo,i} + \dot{m}_{b,i} \tag{5.34}$$

$$\dot{m}_{fw,i} = \dot{m}_{fw,fl,i} + \dot{m}_{fw,bo,i} \tag{5.35}$$

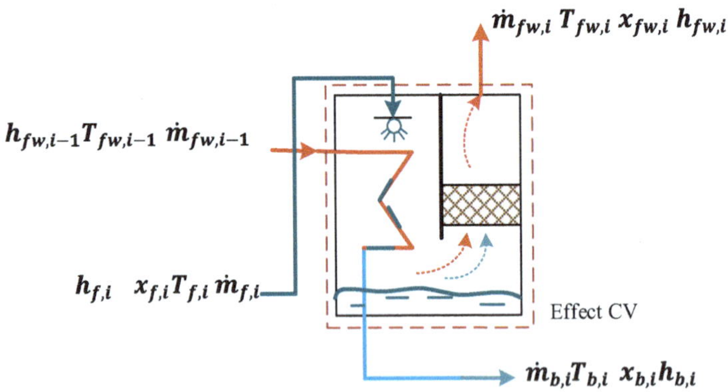

Fig. 5.5 ***Inlet and outlet streams interacting within the nth effect of MED.***

Combining Eqs. (5.33)–(5.35) gives the overall mass rate balance for the ith effect:

$$\dot{m}_{f,i} = \dot{m}_{fw,i} + \dot{m}_{b,i}. \tag{5.36}$$

A mass rate balance for salt, assuming that the vapor is pure water (i.e., $x_{fw_{f\&b}} = 0$), can be written as follows:

$$\dot{m}_{f,i} x_{f,i} = \dot{m}_{b_{f,i}} x_{b_{f,i}}. \tag{5.37}$$

The enthalpy change through the condensation of vapor from the previous effect to vaporize a part of feed at the ith effect to freshwater vapor and brine streams can be determined by the energy rate balance of that effect (Fig. 5.5):

$$\dot{m}_{fw_{i-1}} \Delta h_{fw_{i-1}} = \dot{m}_{fw,i} h_{fw,i} + \dot{m}_{b,i} h_{b,i} - \dot{m}_{f,i} h_{f,i} \tag{5.38}$$

where the value of $\Delta h_{fw_{i-1}}$ is different for the first through nth effects. This is discussed further below.

In the forward MED configuration, shown in Fig. 5.4, there are two major differences between the first effect and other effects. First, the feed seawater enters the first effect at a temperature below the saturation temperature, while in the other effects, the feed enters the effect above the saturation temperature (i.e., in a superheated state). So, flashing does not happen in the first effect ($\dot{m}_{fw_{f,1}} = 0$). Second, in the first effect, the external steam heat source with a specified mass flow rate of $\dot{m}_s$ is used to vaporize a portion of the feed, while in the ith effect ($i \neq 1$) the generated vapor from the previous effect is used to vaporize a portion of the feed. So, for the first effect, instead of $\dot{m}_{fw,i-1} \Delta h_{fw_{i-1}}$, the term $\dot{m}_s h_{fg,s}$ should be used in Eq. (5.38).

In addition, in the second through nth effects, the generated freshwater is condensed at the saturation temperature in the following preheater and effect. But the boiling point of feed seawater exceeds that of pure water, due to the dissolved salt, which elevates the boiling point of seawater (i.e., creates a BPE). So, the vapor generated in the effect is superheated (to a temperature of $T_{v,fw}$) by a BPE value greater than the pure vapor saturation temperature ($T_{v,fw,sat}$). Thus, the freshwater condensation temperature is

$$T_{v,fw,sat} = T_{v,fw} - BPE \tag{5.39}$$

The freshwater vapor generated from the previous effect first enters the preheater in a superheated state, and then enters the current effect in a superheated, saturated vapor, or two-phase saturated mixture state. Therefore, the specific enthalpy change during the complete condensation process of the vapor in the ith effect can be written in terms of the specific enthalpy of the vapor at the entrance of the condensing tube ($h_{fw_{i-1}}$), as

$$\Delta h_{fw_{i-1}} = h_{fw_{i-1}} - h_{sat,fw_{i-1}} \tag{5.40}$$

In the second to nth effects, another energy rate balance is applied to account for the flash process within the effect

$$\dot{m}_{f,i} h_{f,i} = \dot{m}_{fw_{f,i}} h_{fw_{f,i}} + \dot{m}_{b_{f,i}} h_{b_{f,i}} \tag{5.41}$$

To determine the heat transfer area within the effect to satisfy the complete condensation of vapor while evaporating a part of feed stream (i.e., with phase change on both sides of tube), Newton's law of cooling is considered for determining the heat transfer within the MED effect

$$\dot{m}_{fw,i-1} \Delta h_{fw_{i-1}} = A_i U_i \left(T_{v,fw,sat,i-1} - T_{v,fw,i} \right) \tag{5.42}$$

where the specific enthalpy change during phase change ($\Delta h_{fw_{i-1}}$) occurs at the saturation temperature of the previous effect ($T_{v,fw,sat,\ i-1}$). The overall heat transfer coefficient in the ith effect (U_i (W/m^2.K)) can be calculated through the correlation proposed by (El-Dessouky and Ettouney, 2002)

$$\begin{aligned} U_i &= 1939.1 + 1.40562\left(T_{v,fw,sat,i-1}\right) - 0.0207525\left(T_{v,fw,sat,i-1}\right)^2 \\ &\quad + 0.0023186\left(T_{v,fw,sat,i-1}\right)^3 \end{aligned} \tag{5.43}$$

where $T_{v,fw,sat,i-1}$ is in °C. For an actual desalination plant where a particular geometry may exist, the overall heat transfer coefficient can be evaluated through a heat transfer analysis of that geometry.

The freshwater vapor and brine temperatures are identical to the effect temperature. Physical properties of the other streams are quantified as a function of pressure, temperature, and salinity in Section 5.2. In the thermodynamic property evaluation, seawater is assumed to be an incompressible fluid and its properties are taken to be a function of temperature and salinity. The pure liquid water is modeled with zero salinity. The property package developed by IAPWS 1995 (Wagner and Pruß, 2002) is used by engineering equation solver for vapor properties, whereas JACOBIAN uses IAPWS 1997 (Wagner and Pruß, 2002). Values of properties from these two formulations are almost identical.

Flash chamber/box: The condensed distillate freshwater is transferred into the flash chamber/box where it experiences depressurization and partially turns into the vapor. The distillate from the previous effects $\left(\sum_{i=1}^{k-1} \dot{m}_{fw,i}\right)$ also flashes once it enters the *k*th flash box (as the pressure of *k*th flash box is less than that of the (*k*-1)th effect). The produced vapor in the flash chamber ($\dot{m}_{fw,fl,k}$) is transferred to the feed preheater and combined with the distillate ($\dot{m}_{fw,i-1}$), before passing to the next effect. Thus, according to Fig. 5.6, the mass and energy rate balances, respectively, can be written for the *i*th flash chamber as follows:

$$\sum_{i=1}^{k-1} \dot{m}_{fw,k} + \dot{m}_{fw,k} = \dot{m}_{fw,T,k} + \dot{m}_{fw_{fl},k} \tag{5.44}$$

$$\sum_{i=1}^{k-1} \dot{m}_{fw,i} h_{fw,i} + \dot{m}_{fw,k} h_{fw,k} = \dot{m}_{fw,T,k} h_{fw,T,k} + \dot{m}_{fw_{fl},k} h_{fw_{fl},k} \tag{5.45}$$

Feed preheaters: In the feed preheaters, where the heat of freshwater vapor ($\dot{m}_{fw_{ph},i-1}$ which is the summation of $\dot{m}_{fw,i-1}$ and $\dot{m}_{fw_{fl},i-1}$) is recovered to preheat the feed stream, an energy rate balance for the feed preheater can be expressed as

$$\dot{m}_{fw_{ph},i-1}\left(h_{fw_{ph},i-1,in} - h_{fw_{ph},i-1,out}\right) = \dot{m}_{f,T}\left(h_{f,T,out} - h_{f,T,in}\right). \tag{5.46}$$

Applying the heat transfer correlations using the log-mean-temperature-difference method, we obtain

$$\dot{m}_{fw_{ph},i-1}\left(h_{fw_{ph},i-1,in} - h_{fw_{ph},i-1,out}\right) = A_{ph} U_{ph} \left(\frac{T_{f,T,in} - T_{f,T,out}}{\ln\left(\dfrac{T_{fw,sat,i-1} - T_{f,T,out}}{T_{fw,sat,i-1} - T_{f,T,in}}\right)} \right). \tag{5.47}$$

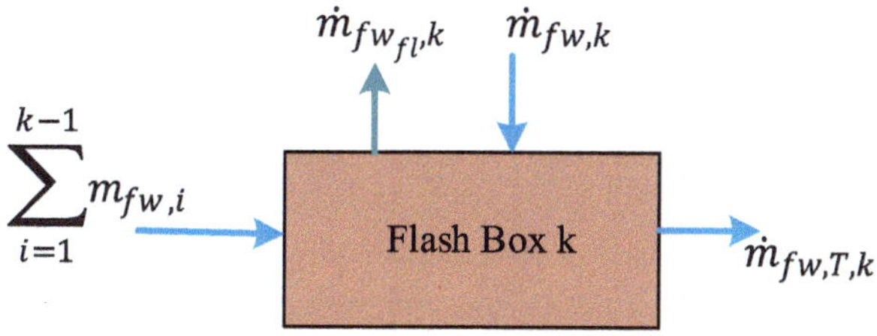

Fig. 5.6 ***Inlet and outlet streams of the kth flash chamber in the MED system.***

The ε − NTU (number of transfer units) method also leads to the same result, as the preheater is the single stream heat exchanger. The overall heat transfer coefficient (U_{ph}) based on El-Dessouky and Ettouney (2002) can be expressed as follows:

$$U_{ph} = 1617.5 + 0.1537\left(T_{fw,sat,i-1}\right) + 0.1825\left(T_{fw,sat,i-1}\right)^2 - 0.00008026\left(T_{fw,sat,i-1}\right)^3 \quad (5.48)$$

where $T_{fw,sat,i-1}$ is the saturation temperature of freshwater vapor in °C.

MED condenser: The vapor distillate from the last effect ($\dot{m}_{fw,n}$) and flash chamber ($\dot{m}_{fw_{fl},n}$) is condensed in the MED condenser (Fig. 5.7). The cooling seawater, with a mass flow rate $\dot{m}_{cw}$, is used in the condensation process for the distillate to adjust the desired feed stream temperature. When the cooling water is exhausted by the condenser, it is returned to the sea. An energy rate balance for the MED condenser can be written as follows:

$$\dot{m}_{fw,C}\Delta h_{fw,C} = \left(\dot{m}_{fw_T} + \dot{m}_{cw}\right)\left(h_{sw,out} - h_{sw,in}\right). \quad (5.49)$$

Applying the heat transfer equations using the log-mean-temperature-difference method yields

$$\left(\dot{m}_{fw_T} + \dot{m}_{cw}\right)\left(h_{sw,out} - h_{sw,in}\right) = A_c U_c \left(\frac{T_{sw,out} - T_{sw,in}}{\ln\left(\dfrac{T_{v,fw} - T_{sw,in}}{T_{v,fw} - T_{sw,out}}\right)} \right) \quad (5.50)$$

where the overall heat transfer coefficient U_c (kW/m^2.K) for the condenser, in terms of $T_{v,fw}$ in °C, can be written as (El-Dessouky and Ettouney, 2002)

$$U_c = 1617.5 + 0.1537\left(T_{v,fw}\right) + 0.1825\left(T_{v,fw}\right)^2 - 0.00008026\left(T_{v,fw}\right)^3. \quad (5.51)$$

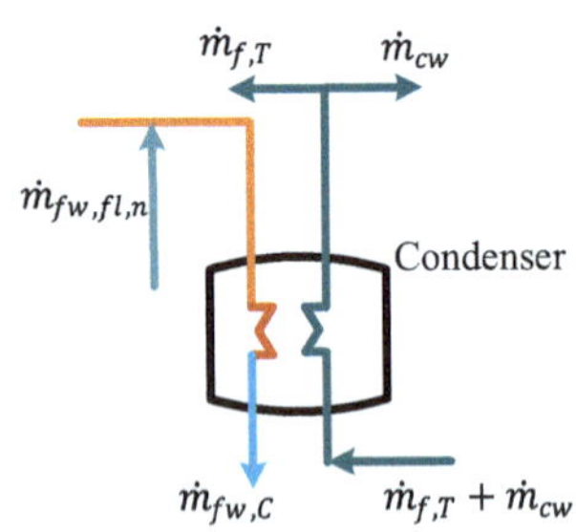

Fig. 5.7 ***Inlet and outlet streams of the MED condenser.***

5.5 MSF desalination model

The MSF desalination technique desalts seawater through flashing a portion of seawater to vapor along a series of stages, each of which is basically countercurrent heat exchangers. In MSF desalination systems, seawater is pumped, heated in a brine heater, and discharged to a flash chamber which is kept below the water saturation pressure. An MSF desalination system is shown in Fig. 5.8. The simplifying assumption of identical saturation temperature difference between the sequential stages of the flashing chamber is considered in the thermodynamic model of the MSF system, so as to avoid an iterative sequential solving process. Similar to MED, the inlet and outlet conditions of each component (feed heater, brine heater, and flash evaporators) are evaluated by applying mass and energy rate balances for each such component simultaneously. The thermodynamic properties of pure water and seawater at varying salinities are formulated and discussed in Section 5.2.

The temperature profile in the MSF desalination system can be expressed in terms of feed seawater temperature (T_f), steam temperature (T_s), top brine temperature (TBT), that is, brine leaving the preheater, and brine temperature leaving the last stage ($T_{b,n}$). A linear temperature profile for the flashing brine and seawater flowing through the condenser is assumed. Then, the temperature drop per stage (ΔT) of an MSF with n stages and the temperature of stage i ($T_{b,i}$), respectively, can be expressed as follows:

$$\Delta T = \left(TBT - T_{b,n}\right) / n \tag{5.52}$$

$$T_{b,i} = TBT - i\Delta T. \tag{5.53}$$

The feed seawater temperature flowing into the condenser tube increases as it receives the heat of the flashed vapor. From an energy rate balance on stage i (Fig. 5.9),

$$\left(\dot{m}_{fw,i}h_{fw,i} - \dot{m}_{fw,i+1}h_{fw,i+1}\right) + \left(\dot{m}_{b,i}h_{b,i} - \dot{m}_{b,i+1}h_{b,i+1}\right) = \dot{m}_f\left(h_{f,i} - h_{f,i+1}\right) \tag{5.54}$$

where subscripts fw, b, and f represent freshwater, brine, and feed. The terms of $\dot{m}$ and h, respectively, denote mass flow rate in kg/s and specific enthalpy in kJ/kg. Substituting the specific enthalpy of each stream in terms of corresponding specific heat at constant pressure and temperature as $h = c_pT$,

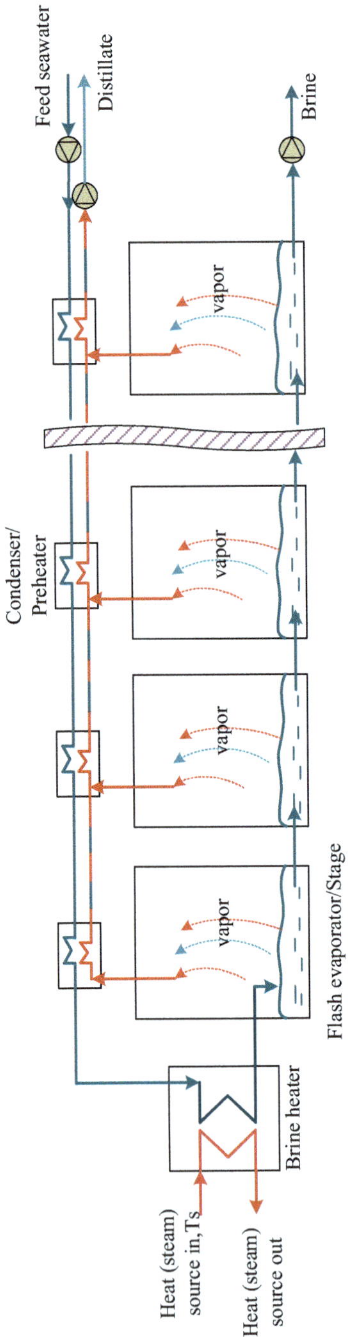

Fig. 5.8 ***Typical multistage flash desalination system.***

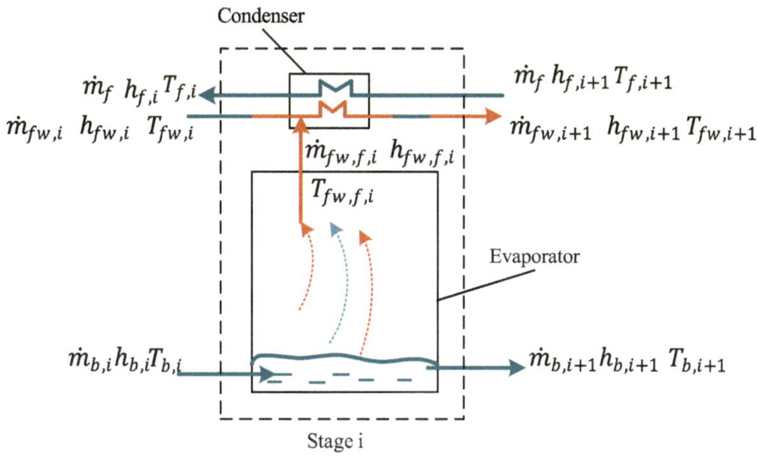

Fig. 5.9 ***Inlet and outlet stream interactions within stage i of an MSF desalination system.***

and assuming negligible temperature difference between the freshwater and brine (i.e., $T_{fw,i} - T_{b,i}$ is small), the energy rate balance (Eq. 5.53) becomes

$$\left(\dot{m}_{b,i} + \dot{m}_{fw,i}\right) c_p T_{b,i} - \left(\dot{m}_{b,i+1} + \dot{m}_{fw,i+1}\right) c_p T_{b,i+1} = \dot{m}_f c_p \left(T_{fi} - T_{f,i+1}\right). \quad (5.55)$$

Recalling $\dot{m}_{f,i} = \dot{m}_{fw,i} + \dot{m}_{b,i}$ implies that the temperature increase in the feed seawater through the condensing tube ($T_{f,i} - T_{f,i+1}$) is equal to the temperature decrease of the brine stream ($T_{b,i} - T_{b,i+1}$). Using Eqs. (5.52) and (5.53), the general expression for the seawater temperature leaving the condenser of stage i becomes

$$T_{f,i} = T_f + \left(n - (i-1)\right) \Delta T. \quad (5.56)$$

Evaporator: Under a steady-state condition, mass, salinity, and energy rate balance equations for the ith flash evaporator can be written as follows:

$$\dot{m}_{b,i} = \dot{m}_{b,i+1} + \dot{m}_{fw,f,i} \quad (5.57)$$

$$\dot{m}_{b,i} x_{b,i} = \dot{m}_{b,i+1} x_{b,i+1} \quad (5.58)$$

$$\dot{m}_{b,i} h_{b,i} = \dot{m}_{b,i+1} h_{b,i+1} + \dot{m}_{fw,f,i} h_{fw,f,i} \quad (5.59)$$

where $\dot{m}_{fw,f,i}$ is the mass flow rate of vapor flashed in the ith stage in kg/s.

Feed heater/condenser: For the *i*th feed heater/condenser, the mass and energy rate balances can be written as

$$\dot{m}_{fw,i} = \dot{m}_{fw,f,i} + \dot{m}_{fw,i+1} \tag{5.60}$$

$$\dot{m}_f \left(h_{f,i} - h_{f,i+1}\right) = \dot{m}_{fw,i} h_{fw,i} + \dot{m}_{fw,f,i} h_{fw,f,i} - \dot{m}_{fw,i+1} h_{fw,i+1}. \tag{5.61}$$

Brine heater: An energy rate balance equation for the brine heater, where feed seawater exchanges heat with the steam heat source, can be written as

$$\dot{m}_s h_{fg,s} = \dot{m}_f \left(h_{f,out} - h_{f,in}\right). \tag{5.62}$$

Equal heat transfer areas for the condensers of the MSF system are considered here, following El- Dessouky and Ettouney (2002). Then, the condenser heat transfer area in the first stage can be obtained with the log-mean-temperature-difference method as

$$\dot{m}_f c_p \left(T_{f,1} - T_{f,2}\right) = A_c U_c (\frac{\left(T_{v1} - T_1\right) - \left(T_{v1} - T_2\right)}{\ln\left(\frac{\left(T_{v1} - T_1\right)}{\left(T_{v1} - T_2\right)}\right)} \tag{5.63}$$

where T_{v1} denotes the condensing vapor temperature in the first stage, which can be expressed as

$$T_{v1} = T_1 - BPE_1. \tag{5.64}$$

The condenser overall heat transfer coefficient U_c (kW/m^2.K) can be expressed based on external source (El-Dessouky and Ettouney, 2002) as

$$U_c = 1.7194 + 3.2063 \times 10^{-3} T_{v1} + 1.5971 \times 10^{-5} T_{v1}^{\ 2} - 1.9918 \times 10^{-7} T_{v1}^{\ 3}. \tag{5.65}$$

Also, the heat transfer area of the brine heater (A_{bh}) can be determined with

$$\dot{m}_f h_{fg,s} = A_{bh} U_{bh} (\frac{\left(T_s - TBT\right) - \left(T_s - t_1\right)}{\ln\left(\frac{\left(T_s - TBT\right)}{\left(T_s - t_1\right)}\right)} \tag{5.66}$$

and U_{bh} can be expressed as (El-Dessouky and Ettouney, 2002):

$$U_{bh} = 1.7194 + 3.2063 \times 10^{-3} T_s + 1.5971 \times 10^{-5} T_s^2 - 1.9918 \times 10^{-7} T_s^3 \quad (5.67)$$

where both T_{v1} and T_s are in °C in Eqs. (5.65) and (5.67).

5.6 DCMD model

DCMD is a thermal desalination technique used for small-scale freshwater production (i.e., on the order of 0.1 m^3/day) (Lee et al., 2011; Martínez-Díez and Florido-Díaz, 2001; Rodríguez-Maroto and Martínez, 2005; Song et al., 2007). The process of a typical DCMD system is shown in Fig. 5.10. With this technology, the feed stream is heated by the heater and then passes across a hydrophobic microporous membrane. Both feed and permeate solutions are in direct contact with the membrane. Thus, the vapor is transferred through the membrane and is directly condensed inside the membrane chamber. The resulting vapor pressure difference is due to the temperature difference between the water streams. This temperature difference can be expressed as a function of the membrane distillation

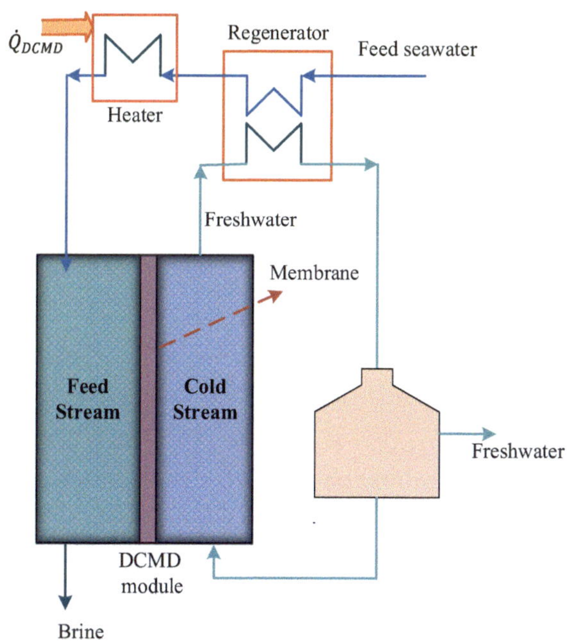

Fig. 5.10 ***Direct contact membrane desalination (DCMD) system.***

coefficient (B) and the vapor pressure difference. The membrane distillation coefficient depends significantly on the membrane structure (e.g., porosity, thickness, and pore size) and represents the diffusion resistance through the pores. It weakly depends on the temperature, pressure, and membrane composition.

The heat and mass transfer mechanism within the DCMD shown in Fig. 5.10 is an important aspect of developing a solid model for the system, and subsequently for evaluation of the system performance. In the DCMD system, there are three main heat transfer regions (Fig. 5.11A).

- Heat transfer at the hot side of the membrane, including the convective heat transfer rate in the feed boundary layer ($\dot{Q}_{f,conv}$) and the heat transfer rate due to mass transfer across the feed thermal boundary ($\dot{Q}_{f,MT}$).
- Heat transfer across the membrane, including the heat transfer rate from vapor movement through the pores of the membrane ($\dot{Q}_{m,MT}$) and conduction across the membrane ($\dot{Q}_{m,cond}$).
- Heat transfer at the cold side of the membrane (permeate side), consisting of convection in the permeate ($\dot{Q}_{p,conv}$) and the heat transfer rate due to the mass transfer along the permeate boundary ($\dot{Q}_{P,MT}$).

The heat flux through the feed side boundary can be written as

$$\dot{q}_f = \dot{Q}_f / A = \dot{q}_{f,conv} + \dot{q}_{f,MT} = h_{htc,f}\left(T_{bf} - T_{mf}\right) + J_w h_{L,f} \quad (5.68)$$

where $h_{htc,f}$ is the heat transfer coefficient of the feed boundary layer, while T_{bf} and T_{mf} denote the temperatures of the bulk feed and the hydrophobic surface of the membrane at the feed side. Also, $h_{L,f}$ is the specific enthalpy of the feed liquid-phase at the average temperature of the feed and the membrane/feed interface $\left(\frac{T_{bf}+T_{mf}}{2}\right)$, and J_w (in kg/(m^2.s)) is the water mass flux through the membrane permeate.

The heat flux through the membrane layer can be written as

$$\dot{q}_m = \dot{Q}_m / A = \dot{q}_{m,cond} + \dot{q}_{m,MT} = h_{htc,m}\left(T_{mf} - T_{mp}\right) + J_w h_v \quad (5.69)$$

where $h_{htc,m}$ is the heat transfer coefficient of the membrane, T_{mp} denotes the temperature at the hydrophobic interface of the membrane at the permeate side, and h_v is the specific enthalpy of produced vapor. The heat flux through the permeate solution thermal boundary is

$$\dot{q}_p = \dot{Q}_p / A = \dot{q}_{p,conv} + \dot{q}_{p,MT} = h_{htc,p}\left(T_{mp} - T_{bp}\right) + J_w h_{L,p} \quad (5.70)$$

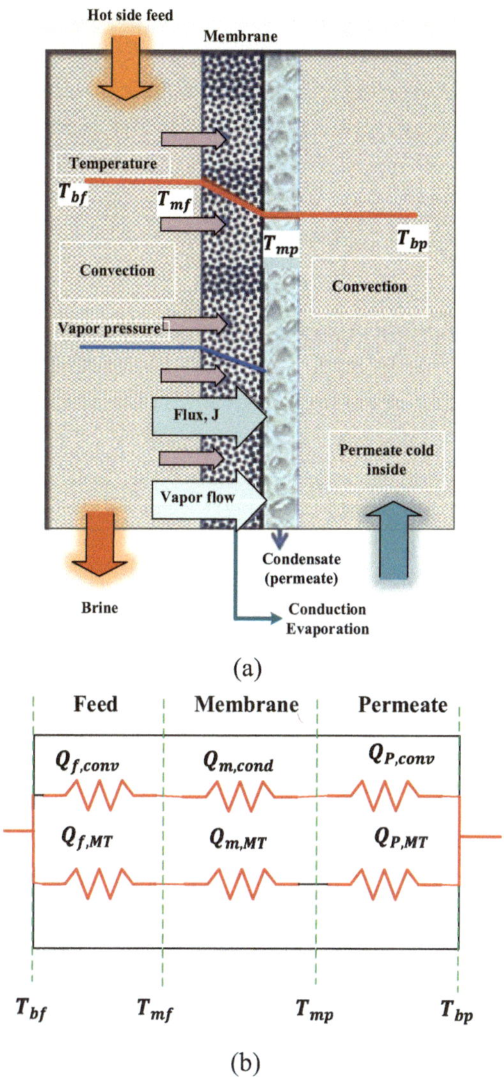

Fig. 5.11 ***(A) Heat and mass transfer mechanism within a typical direct contact membrane desalination (DCMD) module; (B) corresponding heat transfer model.***

where $h_{htc,p}$ is the heat transfer coefficient of the permeate, T_{bp} is the bulk permeate temperature, and $h_{L,p}$ is the specific enthalpy of the permeate solution at the average temperature of the permeate and the membrane/permeate interface $\left(\frac{T_{mp}+T_{bp}}{2}\right)$.

Qtaishat et al. (2008) showed that the effect of mass transfer on the heat transfer in the permeate and feed boundaries ($\dot{Q}_{f,MT}$ and $\dot{Q}_{p,MT}$) is negligible. For steady-state operation of the DCMD system, the heat transfer rates across these three regions are identical, so the overall heat transfer rate through the DCMD can be determined by

$$\dot{Q}_{inside} = \dot{Q}_{DCMD} = UA_m\left(T_{bf} - T_{bp}\right) \tag{5.71}$$

where A_m denotes the membrane filtration area, and T_{bp} and T_{bf} are, respectively, the bulk permeate temperature and bulk feed temperature. The overall heat transfer coefficient for the DCMD system (U) can be obtained using the electrical circuit analogy of heat transfer as follows:

$$U = \left(\frac{1}{\left(\frac{1}{h_f}\right) + \left(\frac{1}{h_m + \frac{J_w h_{fg}}{T_{mf} - T_{mp}}}\right) + \left(\frac{1}{h_p}\right)} \right) \tag{5.72}$$

where h_{fg} is the specific enthalpy of vaporization at the average temperature of the vapor, and the parameters h_m, h_p, and h_f denote the heat transfer coefficients at the hydrophobic membrane, the permeate boundary layer, and the feed boundary layer, respectively. The partial pressure through the membrane and the heat transfer coefficient formulation at different regions can be found in other studies (Qtaishat et al., 2008; Zuo et al., 2011). The heat transfer resistances in the DCMD system, obtained via the electrical analogy of heat transfer, are shown in Fig. 5.11B.

There are several mass transport mechanism models within the membrane, including the viscous model, the Knudsen model, the ordinary-diffusion model as well as a combination of these. The water mass flux across the hydrophobic membrane can be determined by the Antoine equation as follows:

$$J_w = B_m\left(P_{mf} - P_{mp}\right) \tag{5.73}$$

where a linear relationship is commonly observed between the vapor partial pressure difference at the permeate/feed sides and the mass flux across the hydrophobic membrane (Qtaishat et al., 2008). The membrane mass transfer coefficient (B_m) is a strong function of feed temperature and can be evaluated both experimentally and theoretically (Fig. 5.12). The partial pressure difference of vapor between the feed and permeate sides of the

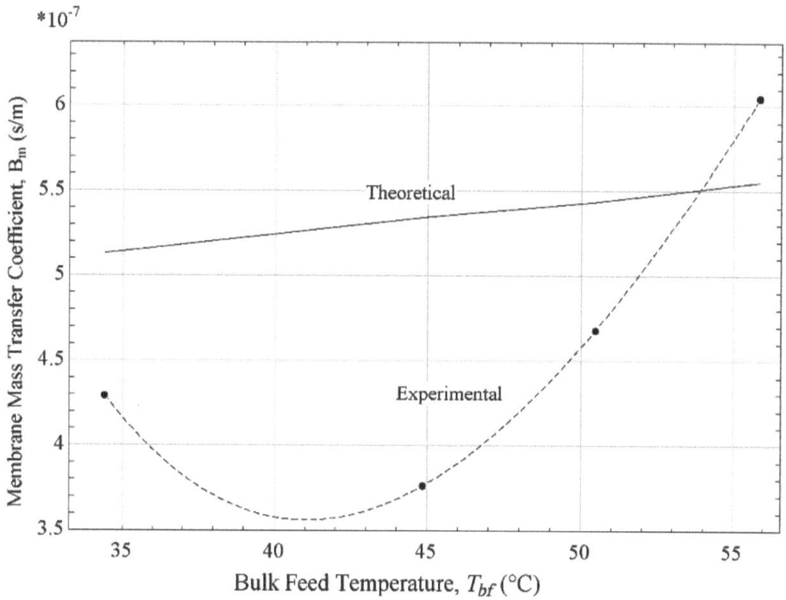

Fig. 5.12 ***Experimental and theoretical variations of the membrane mass transfer coefficient with the bulk feed temperature.*** *From Qtaishat et al. (2008).*

membrane can be expressed with the empirical equation proposed by Qtaishat et al. (2008)

$$P_{mf} - P_{mp} = \exp\left(23.328 - \frac{3841}{T_{mf} - 45}\right) - \exp\left(23.328 - \frac{3841}{T_{mp} - 45}\right). \quad (5.74)$$

Mass transfer through the membrane depends on the mean free path of the transferring species and the pore size. Mass transfer occurs at three primary regions: Knudsen, continuum, and transition. Detailed formulations of these three regions are addressed in other studies (Qtaishat et al., 2008; Zuo et al., 2011).

5.7 MVC model

A typical MVC for desalination is shown in Fig. 5.13. MVC is a widely used seawater desalination process commonly applied for high-salinity feed seawater at various capacities. Research has been carried out to improve the performance of MVC, for example, via enhancing the heat

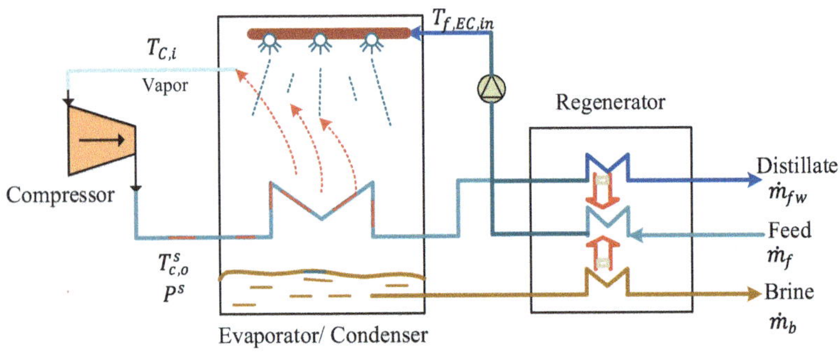

Fig. 5.13 ***Mechanical vapor compression desalination system.***

transfer coefficients for the evaporation and condensation mechanisms within the system. With an increase in the heat transfer coefficient, the driving temperature difference within the evaporation and condensation processes decreases, which results in a lower compression ratio and consequently lower corresponding power required for compression per unit of distillate. The condensation/evaporation heat transfer coefficient can be improved significantly in the high pressure and high temperature MVC system.

In the MVC system in Fig. 5.13, the preheated feed stream is sprayed over the evaporator/condenser (EC) module, and the resulting vapor is compressed to high pressure returning to the EC module. In that module, the high-temperature vapor is condensed, providing heat for the evaporation process. The relatively higher temperatures of brine and freshwater are used to preheat the feed stream in the regenerator (Fig. 5.13).

A model is developed for the MVC desalination system, assuming there is no salt in the product freshwater and a constant heating driving force (equal to the evaporation and condensation temperature difference). Mass and energy rate balances are written as part of the modeling. For the regenerator heat exchanger, with a portion of the input feed seawater receiving heat from the brine stream and the remaining portion receiving heat from the freshwater distillate stream (Fig. 5.14A), an energy rate balance can be written as follows:

$$\dot{m}_f\left(h_{f,EC,in}-h_{f,0}\right)=\dot{m}_b\left(h_b-h_{b,0}\right)+\dot{m}_{fw}\left(h_c-h_{fw,0}\right) \tag{5.75}$$

where subscripts *f*, *b*, and *fw*, respectively, indicate feed, brine, and freshwater. The subscript 0 denotes the reference environment conditions, so h_0 is the

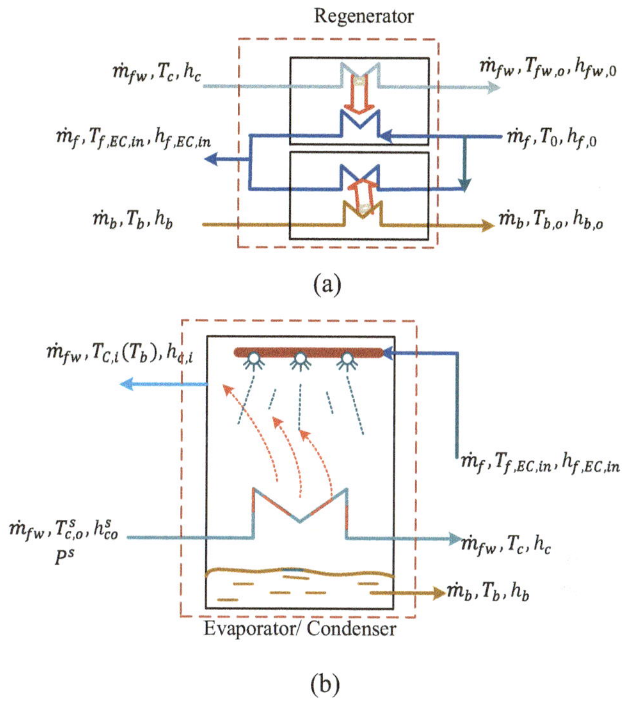

Fig. 5.14 ***Inlet and outlet streams within the (A) regenerator; (B) evaporator/condenser of the MVC desalination system.***

specific enthalpy at the ambient temperature and pressure. The term h_c is the specific enthalpy of freshwater leaving the EC unit.

Similarly, an energy rate balance can be written for the EC module (Fig. 5.14B)

$$\dot{m}_f\left(h_b - h_{f,EC,in}\right) + \dot{m}_{fw}\left(\Delta h_{boiling}\right) = \dot{m}_{fw}\left(\Delta h_{cond}\right) + \dot{m}_{fw}\left(h^s_{c,o} - h_c\right) \tag{5.76}$$

where $\Delta h_{boiling}$ and Δh_{cond}, respectively, are the latent heats of vaporization and condensation for the vapor. The term $h^s_{c,o}$ is the specific enthalpy at the saturation temperature of the distillate at the compressor outlet $T^s_{c,o}$. Note that the specific enthalpy of each stream can be written in terms of its corresponding specific heat at constant temperature and temperature as $h = c_pT$.

The heat transfer area (A_e) required for the evaporator of the MVC system is dependent on the thermal load of the evaporator (i.e., the latent heat of the produced vapor at T_b and the sensible heat to bring feed temperature from $T_{f,EC,in}$ to T_b), the driving force for heat transfer based on the condensation and evaporation temperature difference ($T^s_{c,o}$ - T_b), and the overall heat transfer coefficient (U_e). These terms are related as follows:

$$A_e U_e T^s_{c,o} - T_b) = \dot{m}_{fw} \Delta h_{boiling} + \dot{m}_f c_{pf} \left(T_b - T_{f,EC,in}\right) = \dot{m}_{fw} \left(\Delta h_{cond}\right) + \dot{m}_{fw} c_{p,fw} \left(T^s_{c,o} - T_c\right) \tag{5.77}$$

where the overall heat transfer coefficient U_e(kW/m^2.K) can be expressed as (El-Dessouky and Ettouney, 2002)

$$U_e = 1939.1 + 1.40562\left(T_b\right) - 0.0207525\left(T_b\right)^2 + 0.0023186\left(T_b\right)^3 \tag{5.78}$$

Here, T_b is in °C.

The electrical power required for the MVC system is mostly associated with the compression process rather than pumping. The required electrical power for the compressor per unit of produced freshwater can be expressed in terms of the compression pressure ratio PR (where $PR = P_s$ (compressed vapor pressure at $T^s_{c,o}$) / P_v (vapor pressure at T_b)), the compression efficiency (η_{com}), the specific heat at constant pressure of vapor ($c_{p,v}$), the compressor inlet temperature ($T_{c,i}$), and the ratio of specific heats (k_{SH}), as follows (Thiel et al., 2015):

$$\frac{\dot{W}_c}{\dot{m}_{fw}} = \frac{\text{isentropic work of compression}}{\text{compression efficiency}} = \frac{T_{c,i} c_{p,v} \left(PR^{\frac{k_{SH}-1}{k_{SH}}} - 1\right)}{\eta_{com}}. \tag{5.79}$$

From the Clausius–Clapeyron correlation for an ideal gas, the pressure ratio is a function of the specific enthalpy of vaporization (h_{fg}), the saturation temperatures of pure water at the compressor inlet and outlet ($T^s_{c,i}$ and $T^s_{c,o}$), and the specific gas constant of vapor (R_v = R/M_w). These terms are related as follows:

$$PR = \exp\left[\frac{h_{fg}}{R_v}\left(\frac{1}{T^s_{C,o}} - \frac{1}{T^s_{C,i}}\right)\right]. \tag{5.80}$$

The feed inlet to the EC is at the saturated liquid state ($T_{f,EC,in}$), which exceeds the normal saturation temperature ($T^s_{c,i}$) by the BPE of the feed (BPE_f). So, the vapor state before compression is superheated at the saturation pressure (P^s) that corresponds to the compressor inlet temperature $T^s_{c,i} \left(= T_{f,EC,in} - BPE_f\right)$.

When the feed stream temperature in the EC reaches the brine salinity temperature (i.e., the temperature increases by $BPE_b - BPE_f$), the corresponding enthalpy through the evaporation process can be determined at the average temperature of the feed inlet ($T_{f,EC,in}$) and the brine outlet ($T_{C,i}$) in the EC module, instead of basing it on the bulk enthalpy averaged over the vapor evaporation process. As formulation of an expression for the vapor fraction as a function of temperature requires a heat transfer model and system design characteristics as dimensions, the following approximations are considered for the compressor inlet and outlet temperatures. The first simple average temperature approximation can be written as follows:

$$T_{C,i} = T_{f,EC,in} + 0.5\left(BPE_b - BPE_f\right). \tag{5.81}$$

The compressor outlet pressure needs to be set adequately high so that the temperature of the evaporating stream at its highest salinity (brine) remains above the temperature of the condensing stream by ΔT_t. In addition, the increase in the feed stream temperature at the evaporator for attaining the brine salinity (i.e., $BPE_b - BPE_f$) needs to be considered in the compressor outlet temperature determination. Therefore, the compressor outlet temperature, from which the compressor outlet pressure is derivable, can be written as

$$T_{c,o}^{s} = T_{f,EC,in} + \Delta T_t + \left(BPE_b - BPE_f\right). \tag{5.82}$$

Substituting Eqs. (5.81) and (5.82) into Eq. (5.79) yields the following equation for the electrical power required by the compressor, as a function of terminal temperature difference in the EC, the pressure of the EC, the feed and brine salinities and the compressor efficiency:

$$\frac{\dot{W}_c}{\dot{m}_{fw}} = \left(T_{f,EC,in} + 0.5\left(BPE_b - BPE_f\right)\right)\frac{T_{c,i}c_{p,v}}{\eta_{com}} \times \left\{\exp\left[-\frac{h_{fg}}{c_{p,v}}\left(\frac{1}{T_{f,EC,in} + \Delta T_t + \left(BPE_b - BPE_f\right)} - \frac{1}{T_{f,EC,in} - BPE_f}\right)\right] - 1\right\}. \tag{5.83}$$

Here, for high recovery ratios (i.e., a high value of $\dot{m}_{fw} / \dot{m}_f$), the term $BPE_b - BPE_f$ can potentially be identical to ΔT_t. The rise of pressure ratio due to the salinity change indicates the effect of salinity on the compressor energy consumption. In fact, the higher the salinity, the less is the evaporation pressure and the higher is the discharge pressure to maintain a constant ΔT_t.

For an incomplete water vaporization process in EC, some liquid water content remains within the vapor leaving the EC module and entering the compressor. A higher liquid content leads to an isothermal compression process, which reduces the corresponding required work. The common centrifugal compressor can be damaged when working with liquid water due to the high speed of its blades, while the gerotor positive displacement compressor operates at a relatively lower blade speed and consequently allows for liquid-phase water injection (Lara et al., 2008). In addition, this type of compressor is less expensive, more efficient over a large range of operation, and more easily sized to the intended compression ratio. Therefore, for the case where liquid water exists in the vapor leaving the EC module, the required compression work rate is

$$\dot{W}_{comp} = \frac{(1+x)H_{vapor,out} - \left(H_{vapor,in} + H_{liq,in}\right)}{\eta_{isen,comp}} \tag{5.84}$$

where H is the enthalpy in kJ and x is the fraction of the injection water that evaporates in the compressor. This fraction can be expressed in terms of the corresponding entropy of the interacting streams as

$$x = \frac{S_{vapor,in} - S_{vapor,out}}{S_{vapor,out} - S_{liq,in}}. \tag{5.85}$$

5.8 Reverse osmosis model

In desalination by RO, an osmotic membrane is used through which water passes at relatively higher rate than the dissolved salt. The higher is the applied pressure over the osmotic pressure, the higher is the rate of permeate (low saline water) produced. An RO desalination system is shown in Fig. 5.15. The simple model developed here for the RO system is based on several assumptions: steady-state and isothermal operation, no dependence of salt ions/water permeability coefficient on concentration and temperature, negligible salt flow rate through the membrane, and linear variation of salt concentration in the feed compartment along the membrane area.

The rate of water transport through the semipermeable membrane can be written in terms of the osmotic pressure difference through the

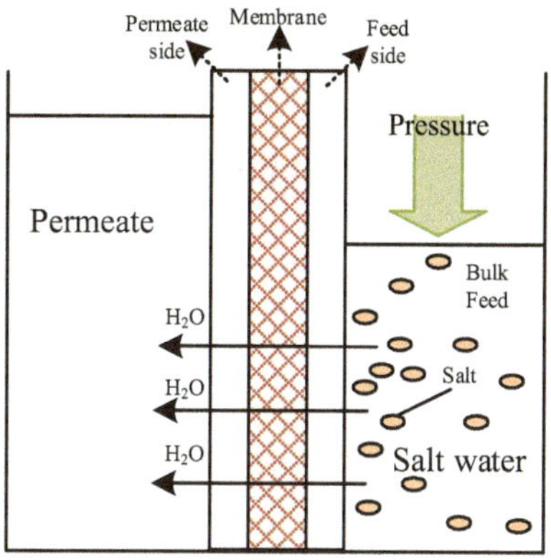

Fig. 5.15 ***Reverse osmosis desalination system.***

membrane ($\Delta\pi$ (kPa)), the hydraulic pressure difference (ΔP (kPa)), and the water permeability coefficient (k_w (m/s.kPa)), which is a property of the membrane characteristics and membrane area (A (m^2)). Thus, the RO permeate flux can be written as

$$J_p = \dot{V}_p / A = k_w \left(\underbrace{\left(\frac{P_f + P_b}{2} - P_p \right)}_{\Delta P} - \underbrace{\left(\frac{\pi_f + \pi_b}{2} - \pi_p \right)}_{\Delta \pi} \right) \tag{5.86}$$

where $\dot{V}_p$ is the volumetric flow rate of permeate (m^3/s); ΔP and $\Delta\pi$ are based on the permeate hydraulic and osmotic pressures (P_p and π_p) and the average hydraulic and osmotic pressures on the feed stream and brine stream ($\overline{P}$ and $\overline{\pi}$). The driving force for pure water transfer from the lower salt concentration (i.e., higher water concentration) to the higher salt concentration (i.e., lower water concentration) is the difference in the water chemical potential. This difference determines the work needed to stop the transport of water. So, the osmotic pressure can be written in terms of the chemical potential of pure water and salt water as

$$\pi = \rho_p \left(\mu_s - \mu_p \right) \tag{5.87}$$

where ρ_p is the density of pure water (kg/m^3), and μ_s and μ_p, respectively, are the chemical potentials of salt water and pure water in kJ/kg.

The salt transfer rate across the membrane is a function of membrane permeability for salt (k_s (m/s)), permeate total dissolved solid concentration

(γ_P (kg/m^3), feed and brine concentrations (γ_f and γ_b), and membrane area (A). So, the salt flux can be expressed as follows:

$$J_s = \dot{m}_{salt} / A = k_s \left(\left(\frac{\dot{m}_f \gamma_f + \dot{m}_b \gamma_b}{\dot{m}_f + \dot{m}_b} \right) - \gamma_P \right) \tag{5.88}$$

where $\dot{m}_{salt}$ is the mass flow rate of salt in kg/s. Note that the water and salt transfer rates through the membrane are not the same. This leads to the salt rejection phenomena. From Eq. (5.86), we see that, with a rise in operating pressure, the water flow increases without changing the salt flow and consequently yields a lower permeate salinity.

Many empirical models have been developed for the mass transport mechanism within the membrane layers of reverse/forward osmosis desalination. These include solution-diffusion, diffusion-viscous flow, finely porous, preferential adsorption capillary flow, and friction models. All of these models are a special case of the statistical-mechanical model proposed by (Mason and Lonsdale, 1990). Such models are nonlinear and require iterative solutions to yield the membrane salt concentration, permeate flow rate, and salt rejection rate. Developing a detailed mass transfer model for RO/forward osmosis (FO) is beyond the scope of this chapter. Instead, the basic mass transfer expressions based on (Mason and Lonsdale, 1990) are applied here to a general reverse osmosis system. Based on this model, the salt concentration in the membrane wall (γ_w), the permeate flux (J_p), and the salt rejection (SR), respectively, can be expressed as follows:

$$\gamma_w = \gamma_{bf} + \left(\gamma_{bf} - \gamma_p\right)\left(e^{\frac{J_w}{k}} - 1\right) \tag{5.89}$$

$$J_p = \left(\Delta P - \sigma \Delta \pi\right)\left(B_1 \gamma_w + B_2\right) \tag{5.90}$$

$$SR = \frac{1}{\left(\frac{E_1}{J_w} + E_2\right)} \tag{5.91}$$

where γ_{bf} denotes the salt concentration in the bulk of the feed compartment and γ_p the salt concentration in the permeate stream. Also, $B_1, B_2, E_1,$ and E_2 are empirical constants, values for which are based on experimental data and can be found elsewhere (Bartman et al., 2009; Roy and Lienhard, 2020; Tow, 2017), ΔP is the pressure difference across the membrane (kPa), $\Delta\pi$ is the osmotic pressure difference across the membrane (kPa), k is the mass transfer coefficient (m/s), and σ is the reflection coefficient.

Concentration polarization refers to the increase in salt concentration (being more than that in the bulk solution) which forms a boundary layer near the membrane surface. Salt polarization adversely affects the performance of reverse osmosis as it increases the osmotic pressure and consequently reduces the net driving pressure difference through the membrane ($\Delta P - \Delta\pi$), increases the salt mass flow across the membrane, and reduces the water flow across the membrane.

The ratio between the concentration of salt at the membrane surface (γ_w) and the bulk surface (γ_{bf}) is a criterion for concentration polarization, and referred to as the concentration polarization factor ($CPF = \gamma_w/\gamma_{bf}$). The rate of ion transport increases with permeate flux, so the membrane surface concentration increases. With increasing feed flow rate, the turbulence close to the membrane surface increases, reducing the thickness of the high-concentration layer. Thus, the CPF can be expressed in terms of permeate flow and feed flow rates as

$$CPF = D.\exp\left(\dot{m}_p / \dot{m}_f\right) \tag{5.92}$$

where D is the proportionality constant, which is dependent on the geometry of system.

5.9 Discussion

The models presented for different desalination techniques help to develop an understanding of desalination technologies and their operation, as well as means for improving them. A proper model accounts for the design and operating parameters when analyzing a desalination system. The thermal model developed for an MED desalination system relies on fewer simplifications and assumptions, and can be solved through a simultaneous equation solver and with no need for developing a solution algorithm. The models developed for the membrane-based desalination systems (i.e., RO and DCMD) are generally straightforward and simple, based on the principles of membrane separation processes. With the presented models, the primary parameters in membrane analysis, which include the required feed pressure, the pressure difference across the membrane and the area of membrane, can be determined so as to yield the desired freshwater production rate.

5.10 Closing remarks

The modeling of various desalination technologies is described in this chapter. Thermodynamic property relations are also developed for seawater and pure water, as the thermodynamic states of streams are necessary in modeling desalination processes. The model for an MED system is simple, flexible to implement, and appropriate for optimization purposes. The modular method used in the MED model, in which each component is modeled individually then pieced together to complete the model, allows for studying different MED configurations like parallel, backward, and forward feed). Several simplifying assumptions are used for the analysis procedure of MSF desalination systems. The mathematical model developed for an MSF desalination system is generally based on mass, salt, and energy rate balances, as well as the heat transfer rate equations used for specifying heat transfer areas. A heat and mass transfer model is presented for the DCMD system by considering three main heat transfer regions, which are the hot side, inside, and cold side of the membrane. In the MVC desalination system, the required electrical work for the compressor is formulated as a function of compression efficiency, compression ratio, and the thermodynamic state of the inlet vapor. The required electrical power for the liquid-vapor mixture of water in an incomplete vaporization of the hot feed stream is also addressed. Finally, for an RO desalination system, a numerical mass transfer model is developed using several simplifying assumptions. In addition, a basic experimental mass transfer model based on the statistical-mechanical method is presented for the RO desalination system. With these models, the required membrane area and feed pressure can be determined for a given freshwater production rate.

Models for desalination systems are used for various purposes, including predictions of performance, design, analysis, and optimization. Many such desalination investigations have been reported, along with elements of the modeling utilized to facilitate the investigations. These include thermodynamic analyses of desalination (Akbari Kordlar et al., 2014; Farsi and Dincer 2019; Orhan et al., 2010; Hamut et al., 2014; Khalid et al., 2016a, 2016b; Hamrang et al., 2020; Chitgar et al., 2019), optimizations (Maleki et al., 2016; Dincer et al., 2017; Wu et al., 2018; Salehi et al., 2018; Peng et al., 2018), and environmental and economic analyses (Keshavarzzadeh et al., 2020; Haghghi et al., 2019).

Nomenclature

A area (m^2)
B_m membrane mass transfer coefficient (s/m)
BPE boiling point elevation (°C or K)
c_p specific heat at constant pressure (J/kg.K)
CPF concentration polarization factor
ΔT_{pp} pinch point temperature difference (°C or K)
g specific Gibbs free energy (kJ/kg)
GOR gained output ratio
h specific enthalpy (J/kg)
h_{gh} latent heat of vaporization/condensation (kJ/kg)
$\dot{H}$ enthalpy rate (kW)
J flux (kg/(m^2.s); m^3/(s.m^2))
k_s membrane permeability for salt (m/s)
k_{SH} ratio of specific heats
k_w water permeability coefficient (m/s.kPa)
M molar mass (kg/mol)
$\dot{m}$ mass flow rate (kg/s)
m_r mass flow rate ratio
n number of effects/stages
P pressure (kPa)
PR compression pressure ratio
$\dot{q}$ heat flux (kW/m^2)
$\dot{Q}$ heat transfer rate (kW)
R Universal gas constant (J/K.mol)
s specific entropy (kJ/kg.K)
S entropy (kJ/K)
SR salt rejection
U overall heat transfer area (W/m^2.K)
v specific volume (m^3/kg)
$\dot{V}$ volumetric flow rate (m^3/s)
w_s salt concentration (kg_s/kg_{sw})

Greek letters

γ concentration (kg/m^3)
η efficiency
μ chemical potential (kJ/kg)
π osmotic pressure (kPa)
ρ density (kg/m^3)
σ reflection coefficient
ε effectiveness
ω humidity ratio

Subscripts

0	reference environment
a	air
B	bottom
b	brine
bh	brine heater
bo	boiling
c	compressor
con	condensation, condenser
cond	conduction
conv	convection
cw	cooling water
da	dry air
DH	dehumidifier
e	evaporator
EC	evaporator/condenser
fl	flash
fw	freshwater
H	humidifier
htc	heat transfer coefficient
m	membrane
ma	moist air
MT	mass transfer
ph	preheater
s	salt, saturation (sat)
T	top
v	vapor
w	water
sw	seawater
wb	wet bulb

Superscript

s	saturation

Acronyms

DCMD	direct contact membrane desalination
EC	evaporator/condenser
HDH	humidification–dehumidification
MED	multieffect desalination
MSF	multistage flash
MVC	mechanical vapor compression
RO	reverse osmosis

References

Akbari Kordlar, M., Mahmoudi, S.M.S., Yari, M., Rosen, M.A., 2014. Energy and exergy analyses of a new combined cycle for producing electricity and desalinated water using geothermal energy. Sustainability 6 (4), 1796–1820.

Bartman, A.R., Christofides, P.D., Cohen, Y., 2009. Nonlinear model-based control of an experimental reverse-osmosis water desalination system. Ind. Eng. Chem. Res. 48 (13), 6126–6136.

Chitgar, N., Emadi, M.A., Chitsaz, A., Rosen, M.A., 2019. Investigation of a novel multi-generation system driven by a SOFC for electricity and fresh water production. Energy Convers. Manage. 196, 296–310.

Darwish, M.A., Al-Juwayhel, F., Abdulraheim, H.K., 2006. Multi-effect boiling systems from an energy viewpoint. Desalination 194 (1-3), 22–39.

Dincer, I., Rosen, M.A., Ahmadi, P., 2017. Optimization of Energy Systems. Wiley, London.

El-Dessouky, H.T., Ettouney, H.M., 2002. Fundamentals of Salt Water Desalination. Elsevier, Amsterdam, The Netherlands.

El-Sayed, Y.M., Silver, R.S., 1980. Fundamentals of distillation. In: Laird, A.D., Spiegler, K.S. (Eds.), Principles of Desalination Second ed. Academic Press, Elsevier.

Farsi, A., Dincer, I., 2019. Development and evaluation of an integrated MED/membrane desalination system. Desalination 463, 55–68.

Farsi, A., Mohammadi, S.H., Ameri, M., 2017. Thermo-economic comparison of three configurations of combined supercritical CO_2 refrigeration and multi-effect desalination systems. Appl. Therm. Eng. 112, 855–870.

Haghghi, M.A., Shamsaiee, M., Holagh, S.G., Chitsaz, A., Rosen, M.A., 2019. Thermodynamic, exergoeconomic, and environmental evaluation of a new multi-generation system driven by a molten carbonate fuel cell for production of cooling, heating, electricity, and freshwater. Energy Convers. Manage. 199, 112040.

Hamrang, F., Shokri, A., Mahmoudi, S.M.S., Ehghaghi, B., Rosen, M.A., 2020. Performance analysis of a new electricity and freshwater production system based on an integrated gasification combined cycle and multi-Effect desalination. Sustainability 12 (19), 7996.

Hamut, H.S., Dincer, I., Rosen, M.A., 2014. Assessment of desalination technologies integrated with renewable energy sources in Turkey. In: Dincer, I., Midilli, A., Kucuk, H. (Eds.), Progress in Exergy, Energy, and the Environment. Springer, New York, pp. 279–292.

Keshavarzzadeh, A.H., Ahmadi, P., Rosen, M.A., 2020. Technoeconomic and environmental optimization of a solar tower integrated energy system for freshwater production. J. Clean. Prod. 270, 121760.

Khalid, F., Dincer, I., Rosen, M.A., 2016a. Analysis and assessment of a gas turbine-modular helium reactor for nuclear desalination. J. Nucl. Eng. Radiat. Sci. 2 (3), 031014-031014-6.

Khalid, F., Dincer, I., Rosen, M.A., 2016b. Comparative assessment of CANDU 6 and sodium-cooled fast reactors for nuclear desalination. Desalination 379, 182–192.

Lara, J.R., Noyes, G., Holtzapple, M.T., 2008. An investigation of high operating temperatures in mechanical vapor-compression desalination. Desalination 227 (1-3), 217–232.

Lee, H., He, F., Song, L., Gilron, J., Sirkar, K.K., 2011. Desalination with a cascade of cross-flow hollow fiber membrane distillation devices integrated with a heat exchanger. AIChE J. 57 (7), 1780–1795.

Maleki, A., Gholipour Khajeh, M., Rosen, M.A., 2016. Weather forecasting for optimization of a hybrid solar-wind-powered reverse osmosis water desalination system using a novel optimizer approach. Energy 114, 1120–1134.

Martinez-Diez, L., Florido-Diaz, F.J., 2001. Desalination of brines by membrane distillation. Desalination 137 (1-3), 267–273.

Mason, E.A., Lonsdale, H.K., 1990. Statistical-mechanical theory of membrane transport. J. Membr. Sci. 51, 1–81.

Millero, F.J., Feistel, R., Wright, D.G., McDougall, T.J., 2008. The composition of standard seawater and the definition of the Reference-Composition Salinity Scale. Deep Sea Res. Part I 55 (1), 50–72.
Mistry, K.H., 2010. Second law analysis and optimization of humidification-dehumidification desalination cycles. (Doctoral diss.). Massachusetts Institute of Technology, Cambridge, US.
Moran, M.J., 1989. Availability Analysis: A Guide to Efficient Energy Use. ASME press, New York.
Orhan, M.F., Dincer, I., Naterer, G.F., Rosen, M.A., 2010. Coupling of copper-chloride hybrid thermochemical water splitting cycle with a desalination plant for hydrogen production from nuclear energy. Int. J. Hydrogen Energy 35, 1560–1574.
Peng, W., Maleki, A., Rosen, M.A., Azarikhah, P., 2018. Optimization of a hybrid system solar-wind-based water desalination by reverse osmosis: comparison of approaches. Desalination 442, 16–31.
Qtaishat, M., Matsuura, T., Kruczek, B., Khayet, M., 2008. Heat and mass transfer analysis in direct contact membrane distillation. Desalination 219 (1-3), 272–292.
Rodríguez-Maroto, J.M., Martinez, L., 2005. Bulk and measured temperatures in direct contact membrane distillation. J. Membr. Sci. 250 (1-2), 141–149.
Roy, Y., 2020. On the presence of solute-solvent transport coupling in reverse osmosis. J. Membr. Sci. 611, 118272.
Salehi, S., Mahmoudi, S.M.S., Yari, M., Rosen, M.A., 2018. Multi-objective optimization of two double-flash geothermal power plants integrated with absorption heat transformation and water desalination. J. Cleaner Prod. 195, 796–809.
Sharqawy, M.H., Lienhard, J.H., Zubair, S.M., 2010. Thermophysical properties of seawater: a review of existing correlations and data. Desalin. Water Treat. 16 (1-3), 354–380.
Song, L., Li, B., Sirkar, K.K., Gilron, J.L., 2007. Direct contact membrane distillation-based desalination: novel membranes, devices, larger-scale studies, and a model. Ind. Eng. Chem. Res. 46 (8), 2307–2323.
Thiel, G.P., Tow, E.W., Banchik, L.D., Chung, H.W., 2015. Energy consumption in desalinating produced water from shale oil and gas extraction. Desalination 366, 94–112.
Tow, E.W., 2017. Organic fouling of desalination membranes. (Doctoral diss.). Massachusetts Institute of Technology, Cambridge, US.
Wagner, W., Pruß, A., 2002. The IAPWS formulation 1995 for the thermodynamic properties of ordinary water substance for general and scientific use. J. Phys. Chem. Ref. Data 31 (2), 387–535.
Wu, B., Maleki, A., Pourfayaz, F., Rosen, M.A., 2018. Optimal design of stand-alone reverse osmosis desalination driven by a photovoltaic and diesel generator hybrid system. Sol. Energy 163, 91–103.
Zuo, G., Wang, R., Field, R., Fane, A.G., 2011. Energy efficiency evaluation and economic analyses of direct contact membrane distillation system using Aspen Plus. Desalination 283, 237–244.

CHAPTER SIX

Application of energy and exergy methods for assessing seawater desalination systems

6.1 Introduction

Generally, seawater desalination technologies are driven by electricity, heat, fuel, or some combination thereof. Assessing the energy requirements of desalination systems becomes more complex when they are combined with other energy systems such as power plants and cogeneration systems. To characterize the performance of desalination systems and to compare the energy costs of diverse system configurations, it is necessary to understand the fundamental thermodynamic limits for the energy requirements of these systems.

Ideal desalination is represented by reversible separation. In such a process, all inlet and outlet streams are in thermal and physical equilibrium with the environment and the dissolved compounds are completely or partially removed from a solution by applying work and heat. The term "ideal desalination" is used here as the heat and work required are at their least for reversible separation. The least work rate ($\dot{W}_l$) and the least heat rate ($\dot{Q}_l$) are defined for making the desalination systems comparable, somewhat in the same way as the Carnot efficiency is the ideal criterion against which all power plants are compared. Thus, the least work and least heat can be defined as the least work and heat needed for the reversible separation of a stream into its components.

The least work and least heat of desalting parameters are useful for a system driven by a single energy source. However, many widely used desalination units, such as multistage flash (MSF) and multieffect desalination (MED), require both mechanical and thermal energy inputs, while some other separation process are driven by chemical energy. So, the least energy benchmark is more useful for considering simultaneously mechanical, thermal, and chemical energy sources.

In this chapter, exergy analysis, described at length in chapter 4, is revisited for a general open control volume (CV). Then, the first and second

Sustainable Energy Technologies for Seawater Desalination
DOI: https://doi.org/10.1016/B978-0-323-99872-7.00001-2

laws of thermodynamics are applied to a generalized desalination system to establish a benchmark for the least energy (work, heat, and fuel) required for desalination. An example is provided to illustrate how the thermodynamic laws are applied for a desalination process. Finally, it is shown how, by introducing one or more additional streams to the feed stream, the reversible work of desalting can be reduced.

It is noted that in recent years many assessments of desalination processes with exergy and exergy-based methods have been reported. These include thermodynamic assessments of desalination (Akbari Kordlar et al., 2014; Farsi and Dincer, 2019; Farsi et al., 2017; Orhan et al., 2010; Hamut et al., 2014; Khalid et al., 2016a, 2016b; Hamrang et al., 2020; Chitgar et al., 2019). Studies have also presented exergy-based economic assessments (Keshavarzzadeh et al., 2020; Haghghi et al., 2019) and exergy-based environmental assessments (Keshavarzzadeh et al., 2020; Haghghi et al., 2019). Finally, optimization procedures utilizing exergy factors have also been reported (Maleki et al., 2016; Dincer et al., 2017; Wu et al., 2018; Salehi et al., 2018; Peng et al., 2018).

6.2 Reversible and irreversible desalting processes

The general exergy relations for a system in a CV considering all possible interactions are used to derive the general exergy relation for a desalination system. Fig. 6.1 shows a CV of an arbitrary system which has q inlet streams and r outlet streams, each different in terms of pressure, temperature, and composition. The environmental temperature, pressure and chemical potential, respectively, are denoted T_0, P_0, and μ_0. The system is able to exchange mass, heat, and work with the environment through the system boundary. The system can also be in thermal contact with the heat reservoirs and can receive or deliver work. For simplicity, potential and kinetic energy terms are taken to be negligible, which is valid in many processes. Energy and entropy rate balances for an open system experiencing heat, work, and mass interactions through a number of inlets and outlets and via multiple heat reservoirs (Fig. 6.1) can be written as follows (Bejan, 2016; Dincer and Rosen, 2021):

$$\frac{dU}{dt} = \sum_{l=0}^{p} \dot{Q}_l - \dot{W} + \sum_{j=1}^{q} \dot{N}_j \bar{h}_j - \sum_{k=1}^{r} \dot{N}_k \bar{h}_k + \sum_{i=1}^{n} \dot{N}_{0,i} \bar{h}_{0,i} - P_0 \frac{dV}{dt} \quad (6.1)$$

$$\frac{dS}{dt} = \sum_{l=0}^{p} \frac{\dot{Q}_l}{T_l} + 0 + \sum_{j=1}^{q} \dot{N}_j \bar{s}_j - \sum_{k=1}^{r} \dot{N}_k \bar{s}_k + \sum_{i=1}^{n} \dot{N}_{0,i} \bar{s}_{0,i} + \dot{S}_{gen} \quad (6.2)$$

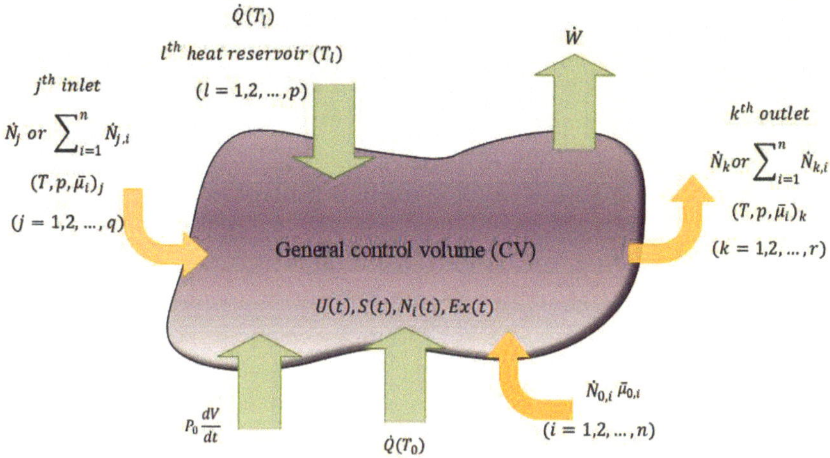

Fig. 6.1 ***An open system in a control volume.*** Possible mass/mole and energy interactions with the environment across the system boundary (i.e., $\dot{N}_{0,i}$, $\dot{Q}_0$ and $P_0 dV/dt$).

where subscripts j and k denote, respectively, the input across port j and exiting across port k; U is the internal energy; $\dot{N}$ is the mole flow rate; $\bar{h}$ is the molal enthalpy; $\dot{Q}_l$ is the heat transfer rate across region l on the control surface (for exchanging heat with the environment $l = 0$ and the heat transfer rate becomes $\dot{Q}_0$); $\dot{W}$ is the work transfer rate out of the CV; $P_0(dV/dt)$ is the work exchange with the environment; $\bar{s}$ is the molal entropy; and $\dot{S}_{gen}$ is the entropy generation rate in the CV.

A word on sign convention is called for here. In the heat exchange process between the CV and reservoir or the environment, heat gain by the CV is defined as positive and heat loss as negative. For work exchanges, the work produced by the CV is defined as positive and work input to or done on the CV as negative. In general, the subscript 0 refers to the environmental condition. In Eq. (6.2) the value of zero is included for the entropy rate associated with the work rate, as no entropy is associated with work interactions. The entropy rate associated with the heat transfer process with a reservoir at temperature (T_l) is $\dot{Q}_l/T_l$.

The general exergy rate balance for the CV of any open system can be obtained through the elimination of the heat transfer with the environment ($\dot{Q}_0$) by combining the first and second laws of thermodynamics (i.e., Eqs. 6.1 and 6.2), along with other manipulations, as follows:

$$\frac{dEx}{dt} = \sum_{l=1}^{p} \dot{Q}_l \left(1 - \frac{T_0}{T_l}\right) - \dot{W} + \sum_{j=1}^{q} \dot{N}_j \overline{ex}_j - \sum_{k=1}^{r} \dot{N}_k \overline{ex}_k - T_0 \dot{S}_{gen} \qquad (6.3)$$

where Ex denotes exergy (kJ) and $\overline{ex}$ molal exergy (kJ/mol). In Eq. (6.3), the first term on the right-hand side represents the equivalent thermal exergy rate associated with the heat transfer process. The second term on the right-hand side is the exergy rate of work rate; these quantities are identical. The third and fourth terms on the right-hand side indicate the inlet and outlet streams for the CV of the open system. The last term on the right-hand represents the entropy generation rate accounting for system irreversibilities (i.e., exergy destructions). The parameter $\overline{ex}$ is the total molal exergy of a material flow. The total exergy of a general material flow is the summation of its physical (or thermomechanical) exergy and chemical exergy as follows:

$$\overline{ex} = \overline{ex}_{ph} + \overline{ex}_{chem} \tag{6.4}$$

The molal physical flow exergy ($\overline{ex}_{ph}$) indicates the maximum useful work that can be transferred to an external user as the flow reaches thermal and mechanical equilibrium with the environment (i.e., the restricted dead state). The molal physical exergy of a material flow can be written as:

$$\overline{ex}_{ph} = \overline{h} - \overline{h}^{\star} - T_0\left(\overline{s} - \overline{s}^{\star}\right) \tag{6.5}$$

where properties with a superscript $\star$ are quantified at the environment temperature (T_0) and pressure (P_0), but at the same concentration or composition as the initial state. This is known as the restricted dead state, which corresponds to thermal and mechanical equilibrium with the environment. Also, the molal enthalpy ($\overline{h}$) and molal entropy ($\overline{s}$) of the overall flowing mixture with i constituents can be written as the summation of the corresponding properties for each constituent as

$$\overline{h} = \sum_{i=1}^{n} \overline{h}_i x_i \tag{6.6}$$

$$\overline{s} = \sum_{i=1}^{n} \overline{s}_i x_i \tag{6.7}$$

Here, x_i denotes the mole fraction of the ith constituent, expressible as

$$x_i = \frac{N_i}{N} \tag{6.8}$$

If the flow mixture with n constituents, this equation can rearranged to show that

$$N = \sum_{i=1}^{n} N_i \tag{6.9}$$

Similarly, the molal enthalpy and molal entropy at the restricted dead state ($\overline{h}^\star$ and $\overline{s}^\star$) can be written as

$$\overline{h}^\star = \sum_{i=1}^{n} \overline{h}_i^\star x_i \tag{6.10}$$

$$\overline{s}^\star = \sum_{i=1}^{n} \overline{s}_i^\star x_i \tag{6.11}$$

The molal chemical flow exergy ($\overline{ex}_{chem}$) is the maximum work obtainable during the transition of a flow from the restricted dead state to the total (or global) dead state. This state is achieved when the temperature, pressure, and constituent concentrations take on the values of these parameters for the reference environment. In fact, at the total dead state condition, the concentration of each constituent in the system takes on its concentration in the environment at T_0 and P_0. Thus, the molal chemical exergy of a mixture can be expressed as

$$\overline{ex}_{chem} = \sum_{i=1}^{n} (\overline{\mu}_i^\star - \overline{\mu}_{0,i}) x_i \tag{6.12}$$

where $\overline{\mu}_i^\star$ is the molal chemical potential of constituent i at the temperature and pressure of the environment, and $\overline{\mu}_{0,i}$ is the molal chemical potential of constituent i at the temperature, pressure, and constituent concentrations of the environment. In fact, when combining the first and second laws of thermodynamics, as expressed in Eqs. (6.1) and (6.2), the term $\overline{\mu}_{0,i}$ is written instead of $\overline{h}_0 - T_0 \overline{s}_0$. Note that for a pure substance such as pure water, the chemical exergy term is zero, while for a multiconstituent substance, such as exhaust gases or seawater, the chemical exergy is not zero and must be accounted for in the analysis.

The general exergy rate balance (Eq. 6.3) can be applied to the general desalination system (Fig. 6.2) operating under steady-state conditions. As shown in Fig. 6.2, in the distant CV the temperature and pressure of streams reach the environment temperature and pressure (restricted dead state), while in the close CV the system streams are not necessarily at the restricted dead state. A separation system can be driven by thermal energy (through heat transfer), and/or mechanical energy (through work), and/or chemical energy (through fuel). The direct material streams are any streams directly involved in the system. For a desalination system, for instance, the direct material streams include input seawater (feed), low saline water (freshwater product), and concentrated brine (waste). Therefore, an exergy

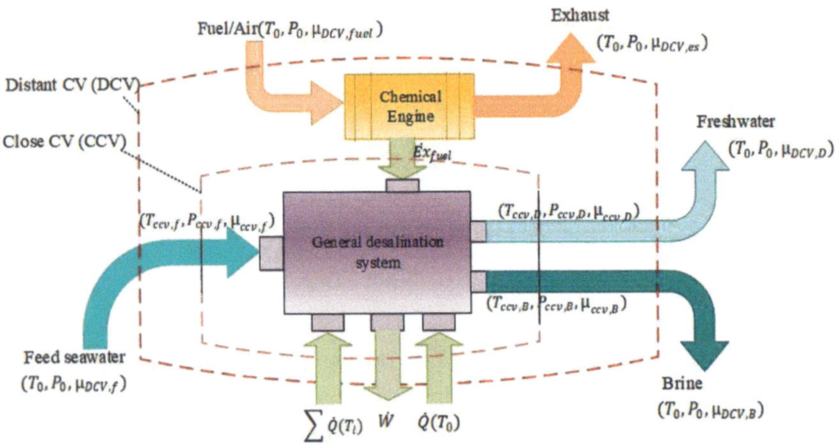

Fig. 6.2 ***A general desalination system, with close and distant control volumes, showing streams.*** For the distant control volume, the streams are in thermal and mechanical equilibrium with the environment (i.e., at the restricted dead state), while for the close control volume, the system streams are not necessarily at the restricted dead state.

rate balance for the distant CV for a general desalination system shown in Fig. 6.2 can be written as

$$0=\sum \dot{Q}_l\left(1-\frac{T_0}{T_l}\right)-\dot{W}+\sum_{fuel,in}\dot{N}_{in}\overline{ex}_{in}-\sum_{exhuast,out}\dot{N}_{out}\overline{ex}_{out}$$
$$+\sum_{dms,in}\dot{N}_{in}\overline{ex}_{in}-\sum_{dms,out}\dot{N}_{out}\overline{ex}_{out}-T_0\dot{S}_{gen} \tag{6.13}$$

where subscript *dms* denotes the direct material stream including the freshwater, brine, and feed streams. The above equation represents the exergy rate balance expression for a general desalination process in which the input and output temperature and pressure of direct material streams are identical to the environmental temperature and pressure. Note that, as all input and output streams are assumed to enter and exit the CV at the environment temperature, a heat transfer rate ($\dot{Q}_0$) must occur with the environment to satisfy the first and second laws of thermodynamics.

As the purpose of a seawater purification system is to produce freshwater (not freshwater at a high temperature), the exergy contents of outlet streams are mainly due to the thermal disequilibrium with the environment. In fact, the exergy contents associated with the outlet streams at the close CV shown in Fig. 6.2 may not be at the environment temperature (T_0), which is regarded as thermal disequilibrium that is not harnessed. Here, entropy is generated as those streams reach thermal equilibrium with the environment. Similarly, the pressure disequilibrium leads to an additional entropy generation. The concentration difference between the system streams indicates the chemical disequilibrium that can be harnessed for producing work. However, as the aim of seawater desalination is to separate impurities from water, the outlet streams are not brought into chemical equilibrium with the environment. As noted before the least energy including least work, least heat, and least fuel are associated with the reversible desalination process in which the entropy generation rate is zero and all streams are at the restricted dead state. In the following, the least energy required for a reversible desalting process is described for three distinct cases in which only one form of the input energy (work, heat, or chemical energy (fuel)) is used for a desalination system.

6.3 Least electrical, thermal, and chemical energy required in desalting process

To compare different desalination systems in terms of energy input required, the least energy is defined as the ideal criteria, somewhat in the same way that the Carnot cycle is used as an ideal heat engine for comparing thermal power cycles. The least energy, which can be the least work, the least heat, or the least chemical energy (fuel), can be defined for a reversible separation process in which all input and output material streams are at the environmental temperature and pressure. The material streams include all direct material streams (feed, freshwater and brine), and input fuel and exhaust streams. This means that all heat and work exchanges with the environment are embodied in the least energy.

For example, consider a thermal desalination using a heat rate $\dot{Q}_D$ to produce a given mass flow rate of freshwater. It is assumed that the intake feed seawater is at the dead state condition. The brine and freshwater output streams with relatively higher temperatures are brought to the environmental temperature. This means that the desalination system experiences a heat transfer rate to environment ($\dot{Q}_0$) through which the brine and freshwater temperature reach the environment temperature. So, the least heat rate required for this desalination system would be $\dot{Q}_{least} = \dot{Q}_D + \dot{Q}_0$. Note that the sign for each heat transfer rate must be considered when summing the heat transfer rates; heat input to the system is taken to be positive and heat output from the system is taken to be negative.

The general expression for the least energy rate of a desalting process can be derived from the energy and entropy rate balances. Consider the general desalination system shown in Fig. 6.2. If the heat transfer rate with the environment and all other reservoirs are embodied in $\dot{Q}$, and also $\dot{W}$ is a general representation for the work exchange which embodies the work exchange with the environment, the energy and entropy rate balances can be written as

$$0 = \dot{Q} - \dot{W} + \sum_{in} \dot{N}_{in} \bar{h}_{in} - \sum_{out} \dot{N}_{out} \bar{h}_{out} \tag{6.14}$$

$$0 = \frac{\dot{Q}}{T} + \sum_{in} \dot{N}_{in} \bar{s}_{in} - \sum_{out} \dot{N}_{out} \bar{s}_{out} + \dot{S}_{gen} \tag{6.15}$$

From Eq. (6.2), $\dot{Q}$ can be expressed as

$$\dot{Q} = T\left(\sum_{out} \dot{N}_{out} \bar{s}_{out} - \sum_{in} \dot{N}_{in} \bar{s}_{in} - \dot{S}_{gen} \right) \tag{6.16}$$

Substituting $\dot{Q}$ from Eq. (6.16) into Eq. (6.14) yields

$$\dot{W} = \left(\sum_{out} \dot{N}_{out} \bar{h}_{out} - \sum_{in} \dot{N}_{in} \bar{h}_{in} \right) - T\left(\sum_{out} \dot{N}_{out} \bar{s}_{out} - \sum_{in} \dot{N}_{in} \bar{s}_{in} \right) + T\dot{S}_{gen} \tag{6.17}$$

As noted before, the least energy rate is defined for the reversible desalting process, so the entropy generation rate term would be zero in Eq. (6.17). So, the least energy rate of separation can be expressed as

$$\dot{W}_{rev} = \dot{E}_{least} = \left(\sum_{out} \dot{N}_{out} \bar{h}_{out} - \sum_{in} \dot{N}_{in} \bar{h}_{in} \right) - T\left(\sum_{out} \dot{N}_{out} \bar{s}_{out} - \sum_{in} \dot{N}_{in} \bar{s}_{in} \right) \tag{6.18}$$

From the above equation (Eq. 6.18) and definition of specific Gibbs free energy as $g = h - Ts$, the least energy rate required for desalination can be expressed as changes in the Gibbs free energy rate of direct material streams. In other words, as all inlet and outlet material streams (including direct material streams, fuel, exhausts, etc.) are at the environmental temperature and pressure, the least energy rate that can be transformed reversibly is equal to the change in the Gibbs free energy rates associated with the direct material streams. So, the right-hand side of Eq. (6.18) can be written in terms of Gibbs free energy changes for the desalination process as

$$\dot{W}_{rev} = \dot{E}_{least} = \sum_{out} \dot{N}_{out} \bar{g}_{out} - \sum_{in} \dot{N}_{in} \bar{g}_{in} \tag{6.19}$$

Chapter 5 (Section 5.2) provides property functions based on the seawater model provided by the International Association for the Properties of Water and Steam (IAPWS) (Wagner and Pruß, 2002). In this model, the specific Gibbs free energy, specific chemical potential, specific entropy, and specific enthalpy of seawater and pure water are presented as a function of salinity and temperature. So, the specific Gibbs free energy of each stream can be evaluated as a function of salinity and temperature. The relations are available for salinities between 0 g/kg and 120 g/kg and temperatures between 0°C and 120°C.

Eq. (6.19) indicates the general least energy rate required for a desalination system, which depends on the changes in the Gibbs free energy of the streams interacting in the system. In the following discussion, the least energy rate expressions are presented for various desalting processes driven by work, heat, or fuel.

For desalination techniques that are driven by work only, such as reverse osmosis (RO), electrodialysis, and mechanical vapor compression (MVC) technologies, Eq. (6.19) can be applied to yield the least work rate required for separation of feed (*f*) into brine (*B*) and distillate (*D*) streams. That is

$$\dot{W}_{D,least} = \dot{W}_{rev} = \dot{N}_D \bar{g}_D + \dot{N}_B \bar{g}_B - \dot{N}_f \bar{g}_f \tag{6.20}$$

Some desalination technologies, such as MSF, MED, humidification–dehumidification (HDH), and membrane desalination (MD), employ only thermal energy to produce freshwater (neglecting the required work of pumps). For these thermal desalination systems, the net heat rate input ($\dot{Q}$) to the desalination system includes the heat exchanges with one or more heat sources at T_H and the environment at T_0 (i.e., $\dot{Q} = \dot{Q}_D + \dot{Q}_0$). Applying energy and entropy rate balances, respectively, for this thermal desalination system yields:

$$\dot{Q}_D + \dot{Q}_0 + \dot{N}_f \bar{h}_f - \dot{N}_B \bar{h}_B - \dot{N}_D \bar{h}_D = 0 \tag{6.21}$$

$$\frac{\dot{Q}_D}{T_H} + \frac{\dot{Q}_0}{T_0} + \dot{N}_f \bar{s}_f - \dot{N}_B \bar{s}_B - \dot{N}_D \bar{s}_D + \dot{S}_{gen} = 0 \tag{6.22}$$

Combining Eqs. (6.21) and (6.22) by multiplying Eq. (6.22) by T_0 and then subtracting it from Eq. (6.21) yields

$$\left(1 - \tfrac{T_0}{T_H}\right)\dot{Q}_D = \left(\dot{N}_B \bar{h}_B + \dot{N}_D \bar{h}_D - \dot{N}_f \bar{h}_f\right) - T_0\left(\dot{N}_B \bar{s}_B + \dot{N}_D \bar{s}_D - \dot{N}_f \bar{s}_f\right) - T_0\,\dot{S}_{gen}$$

So, the least thermal energy rate required for a reversible separation process of feed into brine and distillate can be written as

$$\dot{Q}_{D,least} = \frac{\dot{N}_{fw}\overline{g}_{fw} + \dot{N}_{b}\overline{g}_{b} - \dot{N}_{f}\overline{g}_{f}}{\left(1 - \dfrac{T_0}{T_H}\right)} = \frac{\dot{W}_{rev}}{\left(1 - \dfrac{T_0}{T_H}\right)} \tag{6.23}$$

Several desalination technologies employ chemical energy in the form of a fuel as the sole energy input. The fuel is introduced to the desalination system at the restricted dead state condition, so the only energy input to the system that drives the desalination process is the chemical energy or exergy of the fuel. In fact, the chemical energy or exergy content of a fuel is released through the combustion process in the presence of air/oxygen. So, the least chemical energy rate required for desalination is

$$\dot{W}_{rev} = \dot{E}x_{fuel}^{chem} = \left(\sum_{fuel,in} \dot{N}_{in}\overline{g}_{in} - \sum_{\substack{exhaust\\ streams,out}} \dot{N}_{out}\overline{g}_{out} \right) \tag{6.24}$$
$$= \left(\dot{N}_{D}\overline{g}_{D} + \dot{N}_{B}\overline{g}_{B} - \dot{N}_{f}\overline{g}_{f}\right)$$

However, some fuels such as compressed natural gas (CNG) and liquefied natural gas (LNG) are not inherently at the restricted dead state. For evaluating the least chemical energy of a desalination process driven only by a fuel that is not at the restricted dead state, the mechanical and thermal exergy must also be considered in addition to the chemical exergy.

An illustrative example is presented in the following section to provide a better understanding of the application of the exergy method to desalination systems driven by various types of energy such as electricity, heat, and fuel.

6.4 Example of thermodynamic assessment of an MVC–Rankine system

Consider a MVC desalination system driven by electrical energy supplied by a Rankine thermal power generation cycle. Fig. 6.3 illustrates the combined MVC–Rankine system and its distant CVs (dashed lines). The distant CVs are considered to be sufficiently far from the system so that all of the inlet and outlet streams are at the environmental temperature (T_0) and pressure (P_0), but at different chemical compositions. The heat transfer rate into the system occurs at the environmental temperature (i.e., $\dot{Q}_0$) and a portion of produced work rate from the Rankine cycle is used for the

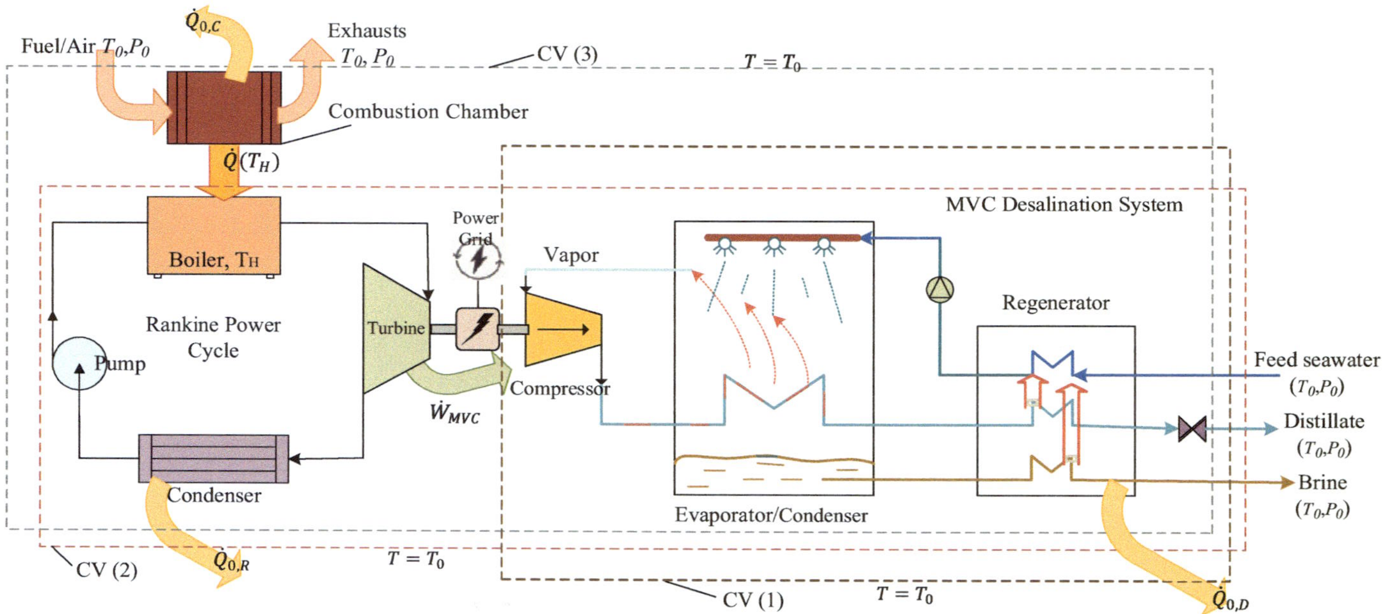

Fig. 6.3 ***MVC desalination and Rankine cycle combined system considered in the example.*** Three control volumes (CV (1), CV (2) and CV (3)) are included, located sufficiently far away from the physical system so that all inlet and outlet streams are at the environmental pressure and temperature.

desalting process in the MVC system (i.e., $\dot{W}_{MVC}$). The feed stream enters the system and is separated into the low saline product/freshwater and brine/concentrated streams using $\dot{W}_{MVC}$. Consider the following steady-state operating conditions:

1. Determine the work and reversible (least) work rate required for the compressor of the MVC desalination system.
2. Determine the heat rate and least heat rate required for the Rankine cycle to generate the corresponding work rate and least work rate for the MVC desalination system. Consider the CV to be drawn for the overall MVC–Rankine combined system in solving the problem.
3. If a combustion chamber is utilized to provide thermal energy for the Rankine power generation cycle, determine the total chemical energy (fuel) and least amount of fuel required for the desalting process.

Solution: In parts (1), (2) and (3), distinct CVs are defined as CV (1), CV (2) and CV (3). The main aim of defining distinct CVs is to apply mass, energy, and entropy rate balances for each of the CVs and then to specify the energy inputs for each CV. For example, for CV (1), the work rate input can be determined, while for CV (2) the heat rate input and for CV(3) the fuel rate input can be determined.

1. To determine the least work rate required for the MVC system, mass, salt mass, energy, and entropy rate balances, respectively, are written for the CV (1) of the system in Fig. 6.3:

$$\dot{m}_f = \dot{m}_D + \dot{m}_B \tag{6.25}$$

$$\dot{m}_f w_f = \dot{m}_D w_D + \dot{m}_B w_B \tag{6.26}$$

$$\dot{W}_{MVC} + \dot{Q}_{0,D} + \dot{m}_f h_f = \dot{m}_D h_D + \dot{m}_B h_B \tag{6.27}$$

$$\frac{\dot{Q}_{0,D}}{T_0} + \dot{m}_f s_f = \dot{m}_D s_D + \dot{m}_B s_B - \dot{S}_{gen} \tag{6.28}$$

Here, $\dot{m}$, x, h, s denote mass flow rate, salt concentration, specific enthalpy, and specific entropy. These terms are applied to the feed (f), distillate (D), and brine (B) streams. The terms $\dot{Q}_{0,D}$ and $\dot{S}_{gen}$ denote, respectively, the heat exchange rate of the desalination system with the environment and the entropy generation rate within the CV (1). The energy and entropy rate balances (Eqs. 6.27 and 6.28) are combined by multiplying Eq. (6.28) by T_0 and then subtracting it from Eq. (6.27). Then the resulting equation can be written in terms of the specific Gibbs free energy ($g = h - Ts$) of the direct material streams (i.e., feed, brine and freshwater) enter and exit the CV (1) at environmental temperature and pressure. So, the work rate required for the MVC desalination system can be written as

$$\dot{W}_{MVC} = \dot{m}_D g_D + \dot{m}_B g_B - \dot{m}_f g_f + T_0 \dot{S}_{gen} \tag{6.29}$$

In the case of the reversible process, the entropy generation term is zero. Then the separation work rate $\dot{W}_{MVC}$ becomes the reversible work rate of separation

$$\dot{W}_{MVC}^{rev} = \dot{m}_D g_D + \dot{m}_B g_B - \dot{m}_f g_f \tag{6.30}$$

As a representative calculation, the generation of 14.5 kg/s salt-free freshwater ($w_D = 0$) from 32.2 kg/s feed seawater input with a salinity of 35 g/kg can be considered. Then, the least work rate required from the Rankine cycle can be found in the following manner.

From Eq. (6.25), we can write

$$32.2\ \text{kg/s} = 14.5\ \text{kg/s} + \dot{m}_B \tag{6.31}$$

Solving leads to $\dot{m}_B = 17.7$ kg/s.

Similarly, with Eq. (6.26), we can write

$$32.2 \times 35 = 17.7 \times w_B \tag{6.32}$$

which yields $w_B = 63.7$ kg/s.

The specific Gibbs free energy of feed, freshwater, and brine streams can be determined either from the seawater property package in Engineering Equation Solver (EES) software or property functions presented in Chapter 5 (Section 5.2). Assuming the environment pressure and temperature are 25°C and 101.13 kPa, the specific Gibbs free energy of the feed, distillate, and brine streams, respectively are:

$$g_f(25°\text{C}, 101.13\ \text{kPa}, 35\ \text{g/kg}) = -4.43\ \text{kJ/kg} \tag{6.33}$$

$$g_D(25°\text{C}, 101.13\ \text{kPa}, 0\ \text{g/kg}) = -4.56\ \text{kJ/kg} \tag{6.34}$$

$$g_B(25°\text{C}, 101.13\ \text{kPa}, 63.67\ \text{g/kg}) = -1.45\ \text{kJ/kg} \tag{6.35}$$

Then, the least work rate produced by the Rankine cycle and used in the MVC system can be found using Eq. (6.30)

$$\dot{W}_{MVC}^{rev} = \dot{W}_{least,MVC} = 14.5 \times (-4.56) + 17.7 \times (-1.45) - 32.2 \times (-4.43) = 50.95\ \text{kW} \tag{6.36}$$

2. To determine the thermal energy rate required for the Rankine cycle to produce the electrical power required by the MVC system (see CV (2) in Fig. 6.3), the energy rate balance for the Rankine cycle can be written as

$$\dot{Q}_{Rankine} - \dot{W}_{Rankine} - \dot{Q}_{0,R} = 0 \tag{6.37}$$

If the Rankine cycle operates between a high-temperature heat source at temperature T_H and the environmental temperature T_0, the entropy rate balance for this cycle can be expressed as

$$\frac{\dot{Q}_{Rankine}}{T_H} - \frac{\dot{Q}_{0,R}}{T_0} - \dot{S}_{gen} = 0 \tag{6.38}$$

Combining Eqs. (6.37) and (6.38) by multiplying Eq. (6.38) by T_0 and then subtracting it from Eq. (6.37) yields

$$\dot{Q}_{Rankine} = \left(1 - \frac{T_0}{T_H}\right)^{-1} \left(\dot{W}_{Rankine} + T_0\dot{S}_{gen}\right) \tag{6.39}$$

where the work produced by the Rankine cycle can be written as $\dot{W}_{Rankine} = \dot{W}^{rev}_{Rankine} + T_0\dot{S}_{gen}$. The least heat input rate to the Rankine cycle is determined through the thermodynamic limit of reversible operation. So, the least thermal energy rate input to the Rankine cycle can be written as

$$\dot{Q}_{least,Rankine} = \left(1 - \frac{T_0}{T_H}\right)^{-1} \dot{W}^{rev}_{Rankine} \tag{6.40}$$

where $\dot{W}^{rev}_{Rankine}$ expression is presented in Eq. (6.30).

For the representative numerical calculation in section (1), the least work rate required from the Rankine cycle and used in MVC system is determined as 50.95 kW to produce 14.5 kg/s freshwater from 32.2 kg/s feed. Then, the least heat rate input (at T_H=130°C) to the Rankine power cycle to generate 50.95 kW electricity would be

$$\dot{Q}_{least,Rankine} = \left(1 - \frac{298}{403}\right)^{-1} \times 50.95 = 195.96 \text{ kW} \tag{6.41}$$

More generally, it is pointed out that the conditions specified in part (2) can be applied for any desalination technique which directly uses thermal energy to separate impurities from water (e.g., MED, MSF, HDH, and MD). In other words, to formulate the heat required for a thermal desalination system, it can be assumed that instead of separation work, heat is used to drive the separation system. From this perspective, if a reversible heat engine is employed to provide the separation work rate from a high-temperature reservoir, the heat and least heat rates required for separation can be obtained with Eqs. (6.37)–(6.40). So, from solution procedures discussed in parts (1) and (2), the least work and least heat required are formulated for a work/heat-driven desalination system. From the desalination least work and least heat rate formulations, it can be seen that entropy generation plays a key role in evaluating the performance of a desalination system. This idea is expanded upon in Chapter 7 (Section 7.6).

The variation of the required least work rate for a general desalination system with the salinity of the feed and product streams and the recovery ratio ($r = \dot{m}_D / \dot{m}_f$) is shown in Fig. 6.4 at a constant feed temperature of 25°C. Eq. (6.30) can be also written based on the normalized least work rate as a function of recovery ratio as $\dot{W}_{least} / \dot{m}_D = g_D + \left(\left(1/r\right) - 1\right) g_B - \left(1/r\right) g_f$. It is seen that the least work rate required for desalination depends greatly on the feed salinity and recovery ratio and not as much on the salinity of the freshwater product. It is also seen that the least work rate approaches a minimum value as the recovery ratio approaches zero. In fact, when the recovery ratio is very close to zero, the energy change is associated with the distillate stream only. For a nonzero value of recovery ratio, work is required to provide the chemical exergy difference between the inlet and outlet streams. For an infinitesimal amount of product (i.e., recovery ratio approaches zero), the least work rate of desalting becomes a minimum. That is, $\dot{W}_{least}^{min} = \lim_{r \to 0} \dot{W}_{least}$.

Besides the main factors affecting the least heat required for desalination (salinity of streams, recovery ratio, and environmental temperature), the reservoir temperature (T_H) also affects the least heat requirement. Based on the Carnot heat engine efficiency definition, the higher is the temperature of the heat source, the lower is the least heat required for separation. This effect is also observed in Fig. 6.4, where a heat-driven desalination system is in thermal contact with thermal reservoirs at temperatures of 50°C and 100°C. A reservoir with the sun's temperature (5527°C) is also considered to be in thermal contact with the thermal desalination to establish the upper bound of heat exchange. The Carnot efficiency approaches unity as the reservoir temperature approaches the sun's temperature (upper bound). In fact, at this condition, the least heat rate and least work rate of separation are almost identical.

3. To specify the amount of fuel and least amount of fuel required to produce the required thermal energy in the boiler of the Rankine cycle, which supplies electrical power to the MVC system, the maximum

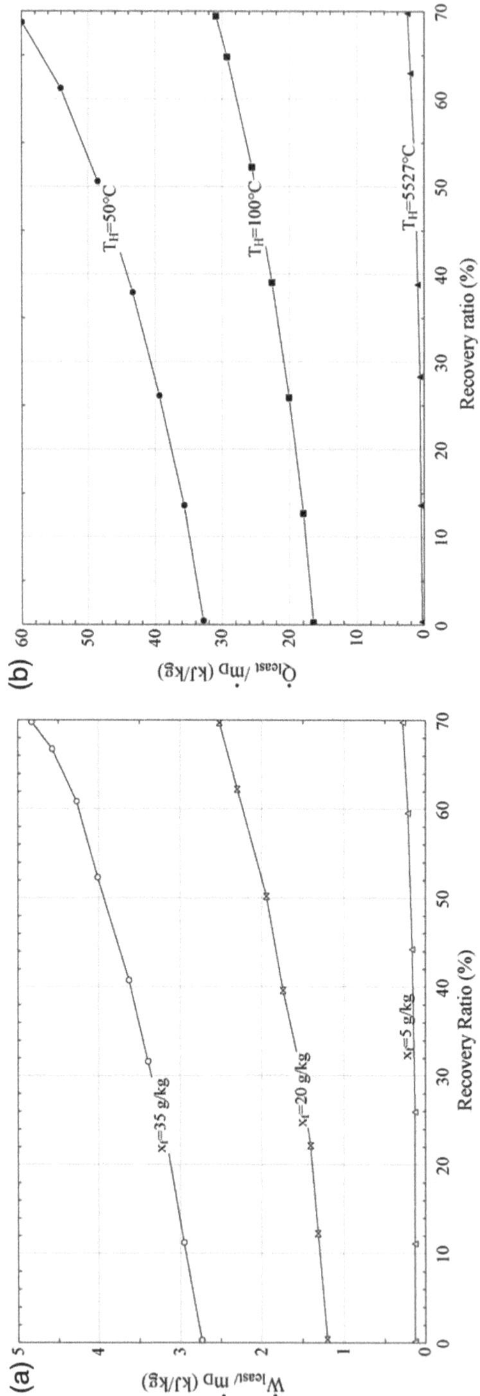

Fig. 6.4 (a): Least work rate required for desalination as a function of recovery ratio and feed salinity (x_f) at a feed temperature of 25 °C. (b): Least heat rate required for desalination as a function of recovery ratio and reservoir temperature (T_H) at a feed temperature of 25 °C.

energy that can be extracted from a fuel needs to be determined. CV (3) in Fig. 6.3 representing the CV includes the combustion chamber, the Rankine cycle, and the MVC desalination system.

As stated before, to calculate the least chemical energy, all material streams, including brine, distillate, feed, fuel, and exhaust, should be at the temperature and pressure of environment. In the example, the fuel combustion process modeling follows the treatments of Bejan (2016) and Moran (1989), in which all reactants (except fuel itself) and products are considered to be at the environmental pressure and temperature and to have the chemical potentials that exist in the environment (i.e., the materials are at the overall dead state).

The fuel is a single component hydrocarbon stream that enters the combustion chamber at steady flow at the restricted dead state. Other species such as O_2, CO_2, and H_2O also enter or leave the combustion chamber at the overall dead state. The following chemical equation expresses the complete combustion of one mole of pure hydrocarbon with oxygen in the absence of nitrogen as

$$C_zH_y + \left(z + \frac{y}{4}\right)O_2 \rightarrow zCO_2 + \frac{y}{2}H_2O_{vapor} \tag{6.42}$$

Using Eq. (6.24), the least chemical energy rate required for the overall desalination-power cycle combined system (i.e., the system in CV (3)) can be written as

$$\begin{aligned} \dot{W}_{rev} &= \dot{Ex}_{fuel}^{chem} = \left(\dot{N}_{C_zH_y}\overline{g}_{C_zH_y} + \dot{N}_{O_2}\overline{g}_{O_2}\right) - \left(\dot{N}_{CO_2}\overline{g}_{CO_2} + \dot{N}_{H_2O_{vapor}}\overline{g}_{H_2O_{vapor}}\right) \\ &= \left(\dot{N}_D\overline{g}_D + \dot{N}_B\overline{g}_B - \dot{N}_f\overline{g}_f\right) \end{aligned} \tag{6.43}$$

Here $\dot{Ex}_{fuel}^{chem}$ denotes the chemical exergy of fuel C_zH_y which can be written in terms of the mole flow rate of fuel and the molal chemical exergy of fuel as

$$\dot{Ex}_{fuel}^{chem} = \dot{N}_{C_zH_y}\overline{ex}_{C_zH_y}^{chem} \tag{6.44}$$

As stated earlier, the fuel enters the combustion chamber at the environmental temperature and pressure, so its molal Gibbs free energy is identical to the molal chemical potential, that is, $\overline{g}_{C_zH_y}(T_0, P_0) = \overline{\mu}_{C_zH_y}$. For the other components, the molal Gibbs free energy of each species can be determined as follows:

$$\begin{aligned} &\overline{g}_i(T_0, P_0, x_i) = \overline{h}_i(T_0, P_0, x_i) - T_0\,\overline{s}_i(T_0, P_0, x_i); \\ &i = O_2, CO_2, \text{ and } H_2O_{vapor} \end{aligned} \tag{6.45}$$

If the constituents O_2, CO_2, and H_2O are considered ideal gases, the specific molar enthalpy $\bar{h}$ is function only of temperature (i.e., $\bar{h}_i(T_0)$) and the specific molar entropy is

$$\bar{s}_i(T_0, P_0, x_i) = \bar{s}_i(T_0, P_0) - \bar{R}\ln x_i \tag{6.46}$$

Then, Eq. (6.45) can be written as follows:

$$\begin{aligned}\bar{g}_i(T_0, P_0, x_i) &= \bar{h}_i(T_0) - T_0\,\bar{s}_i(T_0, P_0) + \bar{R}T_0\ln x_i \\ &= \bar{g}_i(T_0, P_0) + \bar{R}T_0\ln x_i\end{aligned} \tag{6.47}$$

where x_i is the mole fraction of constituent i. The term x_{O_2} is unity, as oxygen is the only oxidant. Substituting Eq. (6.47) into Eq. (6.43) yields the molal exergy content of fuel, which is equivalent to the reversible work per mole of fuel, as follows:

$$\dot{W}_{rev} / \dot{N}_{C_zH_y} = \overline{ex}^{chem}_{C_zH_y} = -\overline{\Delta g}_r(T_0, P_0) + \bar{R}T_0\ln\frac{x_{O_2}^{z+\frac{y}{4}}}{x_{CO_2}^{z}\, x_{H_2O_{vapor}}^{\frac{y}{2}}} \tag{6.48}$$

where $\overline{\Delta g}_r$ is the molar Gibbs free energy of reaction for the fuel, which can be written as

$$\begin{aligned}\overline{\Delta g} = &\left(z\bar{g}_{CO_2}(T_0, P_0) + \frac{y}{2}\bar{g}_{H_2O}(T_0, P_0)\right) \\ &-\left(\bar{g}_{C_zH_y}(T_0, P_0) + \left(z + \frac{y}{4}\right)\bar{g}_{O_z}(T_0, P_0)\right)\end{aligned} \tag{6.49}$$

Using Eqs. (6.43) and (6.44) and $\dot{m} = \dot{N}M$, the least fuel mass flow rate required to drive the Rankine cycle that provides electrical power to the MVC system can be written as follows:

$$\dot{m}_{fuel,least} = \frac{\dot{m}_D g_D + \dot{m}_B g_B - \dot{m}_f g_f}{\overline{ex}_{fuel}} M_{fuel} \tag{6.50}$$

where M denotes molecular weight (kg/mol). The molal chemical exergy of fuel $\overline{ex}_{fuel}$ when C_zH_y is used as fuel can be determined from Eq. (6.50). The total mass flow rate of fuel required for the nonreversible operation of the water-power combined system can then be written as

$$\dot{m}_{fuel} = \frac{\dot{m}_D g_D + \dot{m}_B g_B - \dot{m}_f g_f - T_0\dot{S}_{gen}}{\overline{ex}_{fuel}} M_{fuel} \tag{6.51}$$

Table 6.1 lists the molal-based and mass-based chemical exergy of several fuels. Also, Table 6.2 compares systems using the selected fuels in

Table 6.1 Molal-based and mass-based chemical exergy values of several fuels.

Fuel	M_{fuel} (kg/mol)	$\overline{ex}_{fuel}$ (kJ/mol)[a]	ex_{fuel} (MJ/kg)
Hydrogen, H_2 (g)	2.016×10^{-3}	235.2	116.674
Methane, CH_4 (g)	1.604×10^{-2}	830.2	51.750
Octane, C_3H_{18} (l)	1.142×10^{-1}	5408.7	47.350
Carbon, C (s)	1.120×10^{-2}	410.5	34.178

[a]Data from Bejan (2016).

Table 6.2 Least flow of selected fuels per kilogram of produced freshwater for several values of recovery ratio (r)[a].

Fuel	$\dot{m}_{fuel,least} / \dot{m}_D$ (mg/kg)		
	$r^b = 0$	$r = 35$	$r = 70$
Hydrogen, H_2 (g)	20	30	40
Methane, CH_4 (g)	55	67	75
Octane, C_3H_{18} (l)	60	75	100
Carbon, C (s)	78	102	145

[a]Data extracted from Mistry and Lienhard (2013) and calculated based on reference temperature of 25° C, reference pressure of 1 atm and feed salinity of 35 g/kg.
[b]Recovery ratio is defined as $r = \dot{m}_D / \dot{m}_f$.

terms of the least amount of fuel needed for a desalting process, for several values of recovery ratio. It is observed that the higher the energy density of the fuel, the lower is the quantity of fuel required for the separation process.

Additional notes: It is noted that here we consider the reversible work of the reaction in which all products and reactant are a single component at T_0 and P_0. So, the term $\overline{g}_i(T_0, P_0)$ for each component is identical to the corresponding molal chemical potential at the environment temperature and pressure (i.e., $\overline{g}_i(T_0, P_0) = \overline{\mu}_i(T_0, P_0)$) (Bejan, 2016). It is also observed that the chemical exergy of fuel and Gibbs free energy of reaction are comparable with the lower heating value of fuel. Moran (1989) and Kotas (1986) present expressions for the exergy rates of the fuel, air, and combustion gases. The exergy rate for the fuel at the restricted dead state which is only associated with the chemical exergy of the fuel (i.e., its physical exergy is zero) can be written as

$$\dot{E}x_{fuel} = \dot{E}x_{fuel}^{chem} = \dot{m}_{fuel}.ex_{fuel}^{chem} \tag{6.52}$$

$$ex_{fuel}^{chem} = \gamma.LHV \tag{6.53}$$

where *LHV* is the lower heating value of fuel in kJ/kg, and γ is the chemical exergy/energy ratio, which can be expressed in terms of the ratio of the mass fraction of hydrogen (a) to the mass fraction of carbon (b)

$$\gamma = 1.0437 + 0.1882\left(\frac{a}{b}\right) \tag{6.54}$$

For the air inputs to the combustion chamber, the exergy rate for each component can be expressed as

$$\begin{gathered}\dot{E}x_{air} = \dot{E}x_{air}^{ph} = \sum_j \dot{m}_j\left(h_j - h_0\right) - T_0\left(s_j - s_0\right);\\ j = \text{air species}\left(N_2, O_2, CO_2, H_2O\right)\end{gathered} \tag{6.55}$$

Note that if the air enters the combustion chamber at the restricted dead state, the physical exergy is zero, and if it enters at the total dead state, both physical and chemical exergies are zero.

For the combustion gas produced in the combustion chamber, the exergy rate for each component of the flue gas can be written as

$$\dot{E}x_{flue\ gas} = \dot{E}x_{flue\ gas}^{ph} + \dot{E}x_{flue\ gas}^{chem} \tag{6.56}$$

$$\begin{gathered}\dot{E}x_{flue\ gas}^{ph} = \sum_k \dot{m}_k\left(h_k - h_0\right) - T_\infty\left(s_k - s_0\right);\\ k = \text{flue gas constituent}\left(\mathrm{N_2, O_2, CO_2, H_2O, SO_2}\right)\end{gathered} \tag{6.57}$$

$$\dot{E}x_{flue\ gas}^{chem} = \sum_k \dot{m}_k\left(\frac{\overline{ex}_k^{chem}}{M_k}\right) \tag{6.58}$$

where

$$\overline{ex}^{chem} = \sum x_k.\overline{ex}_k^{\circ chem}\bar{R}T_0 x_k \ln\left(x_k\right) \tag{6.59}$$

Here, $\overline{ex}^{\circ,chem}$ is the standard molal chemical exergy of constituent k in kJ/mol. Note that all gases in air and flue gas are considered to be ideal gases.

From the previous representative numerical calculations for least work rate (50.95 kW) and least heat rate (195.96 kW) produced by Rankine cycle and used in the MVC system, the least methane mass flow rate in the combustion chamber to yield these least work and heat rates would be

$$\begin{aligned}ex_{CH_4}^{chem} &= \gamma.LHV = 1.0437 + 0.1882\left(\frac{0.25}{0.75}\right) \times 50{,}000\ \mathrm{kJ/kg_{CH_4}}\\ &= 55321.6\ \mathrm{kJ/kg_{CH_4}}\end{aligned} \tag{6.60}$$

$$\dot{m}_{fuel,\,CH_4} = \frac{\dot{W}_{least,MVC}}{ex^{chem}_{CH_4}} = \frac{50.95\text{ kW}}{55321.6\text{kJ / kg}_{CH_4}} = 9.21\times 10^{-4}\text{kg}_{CH_4}/\text{s} = 0.921\text{ g}_{CH_4}/\text{s} \quad (6.61)$$

In addition to using combustion to exploit the chemical energy of a fuel, the chemical energy of a solution also can be utilized in theory. For this, the chemical energy of a solution can be harvested while bringing it into the chemical equilibrium with the environment. For instance, the chemical energy of a given solution due to it being in chemical disequilibrium with the environment can potentially be exploited if the reactions are exothermic. This energy can be used to drive a desalting process. However using the energy of a solvent is not a practical means of supplying energy for desalination, as large amounts of salt (or other chemicals) and water are required, which usually renders the process not cost-effective (Mistry and Lienhard, 2013). For example, 7 kg of Ag_2SO_4 and 15 kg of $CaSO_4$ are needed to produce 1 kg of freshwater from exploiting the chemical energy of Ag_2SO_4 and $CaSO_4$ solutions (Mistry and Lienhard, 2013). This indicates that dissolution reactions are not practical for exploitation for large-scale desalination.

6.5 Performance improvement of desalination system using assisted stream

To reduce the amount of energy required for desalination and consequently to improve the overall separation performance, an assisted stream (additional stream) can be used. Fig. 6.5 shows a general work-driven desalination system that reversibly separates the feed stream into low-saline product and brine streams, and also employs a reversible energy recovery device (ERD) to recover work. The work can be recovered through taking in an assist inflow (e.g., wastewater after secondary treatment) and rejecting diluted brine. An amount of water is transferred from the assist stream inflow to the brine stream at the reversible energy recovery device (ERD), also known as the salinity gradient engine. To have feasible work recovery, the salinity of the assist stream at the inlet is less than that of the brine stream. Ideally, the assist stream output from the ERD contains all the salt of its inlet (assist inlet), indicating ideal salt rejection by the desalting process.

The combination of a single-pass RO system and a pressure exchanger with a forward osmosis (FO) and a turbine is an example of a work recovery process through an assist stream and ERD (Figs. 6.5 and 6.6). The feed saline water is split into two streams. One of these is transferred to the high pressure pump and is then pressurized corresponding to the required pressure

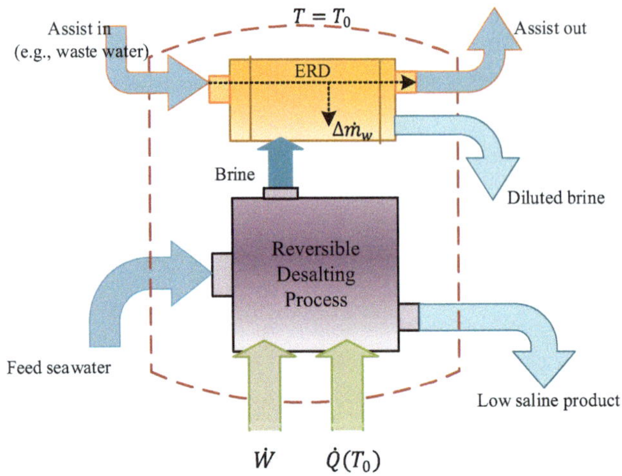

Fig. 6.5 ***Reversible work-consuming desalination system assisted by a reversible energy recovery device (ERD).***

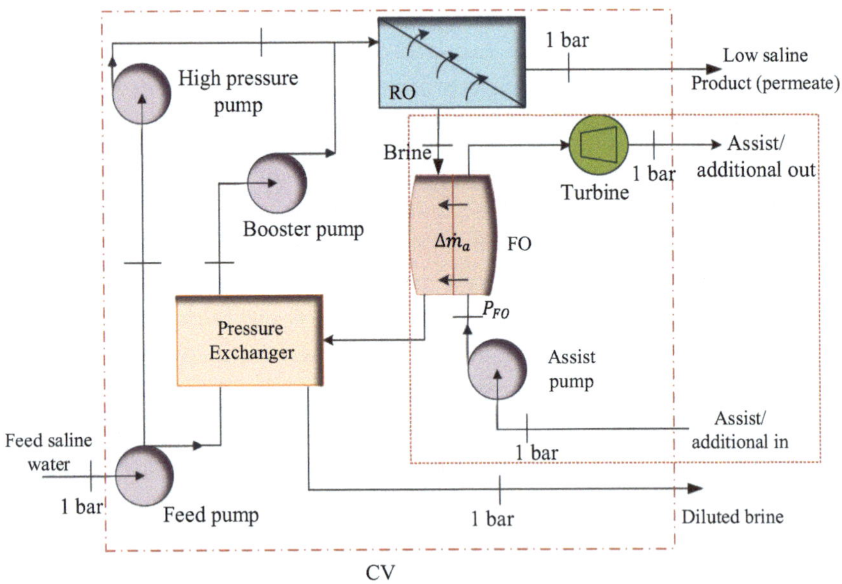

Fig. 6.6 ***Reverse osmosis unit assisted by a pressure exchanger combined with a forward osmosis unit and a turbine for energy recovery.***

for the RO module. The other part of feed enters the pressure exchanger where it is pressurized by the booster pump to an intermediate pressure. The outlet stream from the booster pump is then mixed with the outlet stream from the high pressure pump before entering the RO module. The distillate stream is produced at the low pressure side of the RO module, which is usually at atmospheric pressure. The remaining brine stream exits the high pressure side of the RO unit. The brine stream leaving the RO module is delivered to the FO device, which is treated as an ERD, where water from the additional stream (with a different salt concentration) is passed through the FO membrane to the brine stream. The additional water transferred to the brine results in a higher volumetric flow rate within the pressure exchanger, which reduces the electrical power required for pressurizing of feed stream at the high-pressure pump before it enters the RO system. The ERD of pressure exchanger transfers the pressure of high-pressure concentrate stream to the low-pressure feed seawater. In some cases of using the ERD, the chemical energy difference between the feed and assist streams might exceed the required energy for the desalting process, resulting in work production by the system. Note that there are other kinds of salinity gradient modules that do aim to generate electricity in the system, such as reverse electrodialysis and pressure retarded osmosis (Post et al., 2007).

We now apply mass rate balances to the solutions and salts in the reversible ERD and reversible desalting systems in Fig. 6.6. The following streams interact with the CV: feed (f), assist inflow and outflow (a_{in} and a_{out}), diluted brine (DB), and distillate product (D). So, the mass rate balances for the solutions and salts can be written as

$$\dot{m}_f = \dot{m}_D + \dot{m}_B \tag{6.62}$$

$$\dot{m}_f w_f = \dot{m}_D w_D + \dot{m}_B w_B \tag{6.63}$$

$$\dot{m}_{a,in} + \dot{m}_B = \dot{m}_{a,out} \tag{6.64}$$

$$\dot{m}_{a,in} w_{a,in} = \dot{m}_B w_B + \dot{m}_{a,out} w_{a,out} \tag{6.65}$$

where $\dot{m}$ denotes mass flow rate (kg/s) and w salinity (kg_s/kg_{sw}).

The following assumptions are invoked when applying the first and second laws of thermodynamics to the system in Fig. 6.6:

- The system is open and operates under steady-state conditions.
- The gravitational potential and kinetic energy values for all streams are negligible.
- The system is in thermal contact with the environment, and heat transfer to the system occurs at the environment temperature.

With these assumptions, energy and entropy rate balances can be written, respectively, as follows:

$$\dot{W} + \dot{Q}(T_0) = \sum_{out} \dot{m}h - \sum_{in} \dot{m}h \tag{6.66}$$

$$0 + \frac{\dot{Q}(T_0)}{T_0} + \dot{S}_{gen} = \sum_{out} \dot{m}s - \sum_{in} \dot{m}s \tag{6.67}$$

where $\dot{W}$ and $\dot{Q}$, respectively, are work rate and heat transfer rate in kW; h is the specific enthalpy in kJ/kg; s is the specific entropy in kJ/(kg.K); T_0 is the environment temperature in K; and $\dot{S}_{gen}$ is the entropy generation rate in kW/K. Multiplying Eq. (6.67) by T_0 then subtracting from Eq. (6.66) gives

$$\dot{W} = \left(\sum_{out} \dot{m}h - \sum_{in} \dot{m}h \right) - T_0 \left(\sum_{out} \dot{m}s - \sum_{in} \dot{m}s \right) + T_0 \dot{S}_{gen} \tag{6.68}$$

For a reversible process ($\dot{S}_{gen} = 0$) and assuming all inflow and outflow streams are at the environmental temperature, the first and second terms in the right-hand side of Eq. (6.68) can be written as a Gibbs free energy balance as follows:

$$\dot{W}_{least} = \dot{W}^{rev} = \left(\sum_{out} \dot{m}g - \sum_{in} \dot{m}g \right) \tag{6.69}$$

where $\dot{W}^{rev}$ is the reversible work rate in kW, $\dot{E}x_{least}$ is the least exergy rate in kW, and g is the specific Gibbs free energy in kJ/kg.

In addition to the recovery ratio ($r = \dot{m}_D / \dot{m}_f$), the value for which is selected to achieve good performance and to avoid scaling or precipitation in the desalination system, another nondimensional parameter is introduced: the permeation ratio pr. This ratio is defined for the ERD as the portion of permeated brine from RO in the ERD system. That is,

$$pr = \frac{\dot{m}_{a,in} - \dot{m}_{a,out}}{\dot{m}_{a,in}} \tag{6.70}$$

where $\dot{m}_{a,in}$ and $\dot{m}_{a,out}$, respectively, are the assist inflow and outflow mass flow rates in kg/s. The value of this parameter is greater than or equal to zero and it is limited to no more than unity to avoid salt precipitation in the FO unit.

The mass flow rate ratio r_{ERD} also is defined as the ratio of the mass flow rate of brine from the RO module entering the ERD to the brine rejected from the ERD. That is,

$$r_{ERD} = \frac{\dot{m}_{a,in}}{\dot{m}_B} \tag{6.71}$$

With the definitions in Eqs. (6.69) and (6.70), Eq. (6.69) can be rewritten as a function of the recovery ratio, the ratio of mass flow rates of assist and feed, as follows:

$$\frac{\dot{W}_{least}}{\dot{m}_D} = r_{EDR}\left(\frac{1}{r}-1\right)\left[(1-pr)\left(g_{a,out}-g_{DB}\right)+\left(g_{DB}-g_{a,in}\right)\right] + \left(\frac{1}{r}-1\right)g_{DB}+g_D-\frac{1}{r}g_f \tag{6.72}$$

We now define the recovered least work in the ERD to assess the performance of this module. The recovered work in the ERD per a kilogram of produced distilled water can be expressed as

$$\frac{\dot{W}_{rec,least}}{\dot{m}_D} = \frac{\dot{W}_{least\ with\ ERD}}{\dot{m}_D} - \frac{\dot{W}_{least\ without\ ERD}}{\dot{m}_D} \tag{6.73}$$

where $\dot{W}_{rec,least}$ is the recovered least work rate in kW. Note that a value of $\dot{W}_{rec,least}$ greater than zero means that the ERD is advantageous for work recovery.

The formulations described above can be applied to a desalination system driven by heat or work or chemical energy assisted by an additional stream. In fact, Eq. (6.72) provides the general least energy required for a desalination system combined with an ERD and an additional stream.

The variation of reversible recovered work per kilogram of produced distilled water with permeation ratio is shown in Fig. 6.7. It is seen that, for lower assist stream inlet salinities, the permeation ratio approaches unity, which corresponds to an assist stream outlet salinity of 120 g/kg (the maximum salinity). From Fig. 6.7 it is observed that the recovered least work increases as the assist stream inlet salinity decreases and mass flow rate ratio (r_{EDR}) increases. As the permeation ratio approaches zero, the recovered least work becomes approximately zero, indicating that the ERD utilization is not advantageous. Also, for assist salinities higher than the feed salinity, a corresponding recovery ratio exists at which the desalinating least work is same as the least amount of desalination work without an ERD module. This suggests that the use of the ERD device for recovery ratios close to this point is not significantly practical in terms of saving energy. The optimum permeation value involves a compromise between diluting the brine stream exiting the RO system at the expense of rejecting the assist stream with the higher concentration.

In summary, it is shown in this section how introducing an assist stream can improve the performance of a separation process. It is also demonstrated

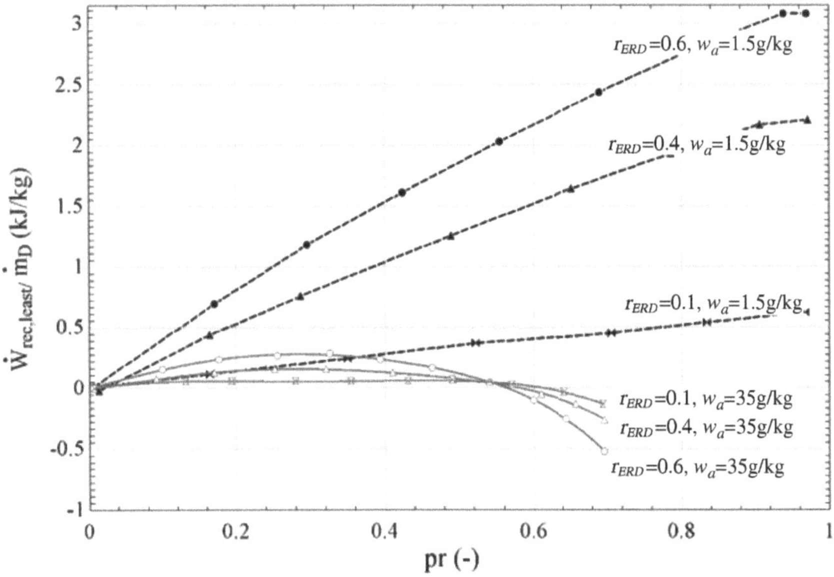

Fig. 6.7 ***Variation with permeation ratio pr of reversible recovered work required in a desalting process assisted by a reversible ERD for several values of salinity of assist stream w_a and mass flow rate ratio r_{ERD}.*** The recovery ratio, feed temperature, and feed salinity are 0.4, 25° C and 35 g/kg, respectively.

that, under specific conditions, the least required separation work can be a negative value in a separation system assisted by an additional stream. In the case of a negative exergy value for the least required separation work, work is produced by the system. Note that the production of work is not feasible for a system that is not supported by an assisted device/stream. To determine the amount of work recovered through an irreversible separation assisted with an ERD, a comprehensive thermodynamic analysis of the irreversible desalination process is necessary. For this, the pump work, the turbine work associated with expansion, the pressure of the feed stream exiting the pressure exchanger, and the mass flow rate through each section of the RO/FO unit need to be examined to determine the work recovered in an irreversible process.

6.6 Closing remarks

In this chapter, energy and exergy analyses are described for a general desalination system. Then, the first and second laws of thermodynamics are applied to a general desalination system to provide a benchmark for the least energy (heat, work, and chemical energy (fuel)) needed for a desalting process. Introducing least work, heat, and fuel assists in comparing desalination techniques. The least energy is defined as the energy required for a reversible desalting process of a stream into its constituents. It is shown that, for conditions in which all direct material streams (commonly feed, brine and freshwater) are at the environmental temperature and pressure, the least exergy of a desalting process is the change in the Gibbs free energy of all direct material streams. The input exergies can be heat, work, fuel, or combination of them. It is also shown that the least exergy of a desalting process does not depend on the specific desalination process used. Instead, it depends on the recovery ratio, the feed and freshwater salinities, and the environmental pressure and temperature. In the case of using an assisted stream, the least exergy of desalting also depends on the assisted salinity. It is demonstrated that the least work is equivalent to the least exergy of the desalting process and the least heat of a desalting process greatly depends on the temperature of the heat source. The least amount of fuel required for desalination depends inversely on the chemical exergy of the fuel. Finally, it is shown that using an additional (assisted) stream can improve the desalination system performance through the recovered least work from the ERD. The combination of single-pass RO with a FO-based ERD is an example of such process.

Nomenclature

Symbols

a	mass fraction of hydrogen
b	mass fraction of carbon
$\overline{ex}$	molal exergy (kJ/mol)
Ex	exergy (kJ)
$\dot{Ex}$	exergy rate (kW)
$\overline{g}$	molal Gibbs free energy (kJ/mol)
G	Gibbs free energy (kJ)
$\overline{h}$	molal enthalpy (kJ/mol)
LHV	lower heating value (kJ/kg)
M	molecular weight (g/mol)
$\dot{N}$	mole flow rate (mol/s)
P	pressure (kPa)
pr	permeation ratio
$\dot{Q}$	heat rate (kW)
r	recovery ratio
$\overline{R}$	universal gas constant (J/mol.K)
$\overline{s}$	molal entropy (kJ/mol. K)

$\dot{S}_{gen}$ entropy generation rate (kW/K)
T temperature (°C, K)
T_0 environmental temperature (°C, K)
U internal energy (kJ)
V volume (m^3)
w mass fraction (kg_s/kg_{sw})
$\dot{W}$ work rate (kW)
x mole fraction

Greek letters

$\bar{\mu}$ molal chemical potential (kJ/mol)
γ chemical exergy/energy ratio

Subscripts

B brine
D distillate
ERD energy recovery device
$chem$ chemical
f feed
ph physical
dms direct material stream
rev reversible
s salt, saturation
sw seawater
w water

Superscripts

$chem$ chemical
ph physical
rev reversible

Acronyms

CNG Compressed natural gas
CV Control volume
ERD Energy recovery device
FO Forward osmosis
HDH Humidification–dehumidification
LNG Liquefied natural gas
MD Membrane desalination
MED Multieffect desalination
MSF Multistage flash
MVC Mechanical vapor compression
RO Reverse osmosis

References

Akbari Kordlar, M., Mahmoudi, S.M.S., Yari, M., Rosen, M.A., 2014. Energy and exergy analyses of a new combined cycle for producing electricity and desalinated water using geothermal energy. Sustainability 6 (4), 1796–1820.

Bejan, A., 2016. Advanced Engineering Thermodynamics, fourth ed. John Wiley & Sons, NJ.

Chitgar, N., Emadi, M.A., Chitsaz, A., Rosen, M.A., 2019. Investigation of a novel multigeneration system driven by a SOFC for electricity and fresh water production. Energy Convers. Manage. 196, 296–310.

Dincer, I., Rosen, M.A, 2021. Exergy: Energy, Environment and Sustainable Development, Third ed. Elsevier, Oxford, UK.

Dincer, I., Rosen, M.A., Ahmadi, P., 2017. Optimization of Energy Systems. Wiley, London.

Farsi, A., Dincer, I., 2019. Development and evaluation of an integrated MED/membrane desalination system. Desalination 463, 55–68.

Farsi, A., Mohammadi, S.H., Ameri, M., 2017. Thermo-economic comparison of three configurations of combined supercritical CO_2 refrigeration and multi-effect desalination systems. Appl. Therm. Eng. 112, 855–870.

Haghghi, M.A., Shamsaiee, M., Holagh, S.G., Chitsaz, A., Rosen, M.A., 2019. Thermodynamic, exergoeconomic, and environmental evaluation of a new multi-generation system driven by a molten carbonate fuel cell for production of cooling, heating, electricity, and freshwater. Energy Convers. Manage. 199, 112040.

Hamrang, F., Shokri, A., Mahmoudi, S.M.S., Ehghaghi, B., Rosen, M.A., 2020. Performance analysis of a new electricity and freshwater production system based on an integrated gasification combined cycle and multi-Effect desalination. Sustainability 12 (19), 7996.

Hamut, H.S., Dincer, I., Rosen, M.A., 2014. Assessment of desalination technologies integrated with renewable energy sources in Turkey. In: Dincer, I., Midilli, A., Kucuk, H. (Eds.), Progress in Exergy, Energy, and the Environment. Springer, New York, pp. 279–292.

Keshavarzzadeh, A.H., Ahmadi, P., Rosen, M.A., 2020. Technoeconomic and environmental optimization of a solar tower integrated energy system for freshwater production. J. Clean. Prod. 270, 121760.

Khalid, F., Dincer, I., Rosen, M.A., 2016a. Analysis and assessment of a gas turbine-modular helium reactor for nuclear desalination. J. Nucl. Eng. Radiat. Sci. 2 (3), 031014.

Khalid, F., Dincer, I., Rosen, M.A., 2016b. Comparative assessment of CANDU 6 and sodium-cooled fast reactors for nuclear desalination. Desalination 379, 182–192.

Kotas, T.J., 1986. Exergy Method of Thermal and Chemical Plant Analysis. Elsevier, London, UK.

Maleki, A., Gholipour Khajeh, M., Rosen, M.A., 2016. Weather forecasting for optimization of a hybrid solar-wind-powered reverse osmosis water desalination system using a novel optimizer approach. Energy 114, 1120–1134.

Mistry, K.H., Lienhard, J.H., 2013. Generalized least energy of separation for desalination and other chemical separation processes. Entropy 15 (6), 2046–2080.

Moran, M., 1989. Availability Analysis: A Guide to Efficient Energy Use. ASME Press, New York.

Orhan, M.F., Dincer, I., Naterer, G.F., Rosen, M.A., 2010. Coupling of copper-chloride hybrid thermochemical water splitting cycle with a desalination plant for hydrogen production from nuclear energy. Int. J. Hydrogen Energy 35, 1560–1574.

Peng, W., Maleki, A., Rosen, M.A., Azarikhah, P., 2018. Optimization of a hybrid system solar-wind-based water desalination by reverse osmosis: comparison of approaches. Desalination 442, 16–31.

Post, J.W., Veerman, J., Hamelers, H.V., Euverink, G.J., Metz, S.J., Nymeijer, K., Buisman, C.J., 2007. Salinity-gradient power: evaluation of pressure-retarded osmosis and reverse electrodialysis. J. Membr. Sci. 288 (1-2), 218–230.

Salehi, S., Mahmoudi, S.M.S., Yari, M., Rosen, M.A., 2018. Multi-objective optimization of two double-flash geothermal power plants integrated with absorption heat transformation and water desalination. J. Clean. Prod. 195, 796–809.

Wagner, W., Pruß, A., 2002. The IAPWS formulation 1995 for the thermodynamic properties of ordinary water substance for general and scientific use. J. Phys. Chem. Ref. Data 31 (2), 387–535.

Wu, B., Maleki, A., Pourfayaz, F., Rosen, M.A., 2018. Optimal design of stand-alone reverse osmosis desalination driven by a photovoltaic and diesel generator hybrid system. Sol. Energy 163, 91–103.

CHAPTER SEVEN

Second law analysis of desalination systems

7.1 Introduction

Second law or exergy analysis is a powerful thermodynamic assessment tool. Applying exergy methods to desalination systems is useful, as there are notable differences between the theoretical and practical minimum energy required for a desalting process due to the irreversibilities in real systems. The use of exergy methods in desalination systems helps to identify the locations and components in a process with high entropy generation or exergy destruction. Reducing the sources of irreversibilities within a desalination system helps to improve it. For cases in which the energy costs associated with the irreversibilities dominate the total cost for freshwater production, thermodynamic improvement of desalination system is highly advantageous.

To identify the sources of irreversible losses for desalination systems, exergy analysis is performed for the main desalination techniques discussed earlier in this book, including multistage flash (MSF) desalination, multieffect desalination (MED), mechanical vapor compression (MVC) desalination, direct contact membrane desalination (DCMD), humidification–dehumidification (HDH) desalination, and reverse osmosis (RO).

This chapter presents a consistent definition of exergy efficiency for a general desalination system to provide a benchmark for comparing the thermodynamic performance of desalination systems. The approach is analogous to how the Carnot heat engine and its efficiency is used for comparing thermal power plants. A consistent definition of exergy efficiency is presented that is applicable to desalination systems driven by any combination of thermal, mechanical, and chemical energies. Additionally, the exergy analysis of a desalination system operating as a part of a cogeneration plant is presented and the heat recovery potential for various desalination technologies is discussed. Finally, the entropy generation associated with the thermodynamic processes within desalination systems is formulated based on exergy methods to assess the irreversibilities associated with a desalting

Sustainable Energy Technologies for Seawater Desalination
DOI: https://doi.org/10.1016/B978-0-323-99872-7.00003-6

process. These methodologies are applied to commonly used desalination technologies to evaluate and compare their thermodynamic efficiencies.

7.2 Thermodynamic performance parameters

7.2.1 Energy efficiency definition of desalination systems

Energy-based performance parameters are commonly used to assess desalination systems. The commonly used evaluating parameters are gained output ratio (GOR), performance ratio (PR), and specific electricity consumption (SEC).

The dimensionless parameter GOR is applicable for thermally driven desalination technologies. It represents the ratio of the enthalpy of vaporization (or latent heat of vaporization) of pure water at the inlet water temperature to the thermal energy input to the desalination system. GOR basically indicates how much thermal energy is consumed for distillate (or freshwater) production. For a flow process, GOR can be expressed as

$$GOR = \frac{\dot{m}_D h_{fg}}{\dot{Q}_D} \tag{7.1}$$

where $\dot{m}_D$ is the mass flow rate of produced distillate (kg/s), h_{fg} is the latent heat of vaporization (kJ/kg), and $\dot{Q}_D$ is the input heat rate to the desalination system (kW). The higher the GOR of a thermally driven desalination system, the better is its efficiency. The value of GOR typically varies between 0.3 and 8.1 depending on the system design and operating conditions. Using heat recovery devices, effective insulation, improved module design, optimized permeate flux, and more recovery stages can improve GOR.

In MED and MSF desalination systems, if steam is used as the heat source (i.e., $\dot{Q}_D = \dot{m}_{steam} . \Delta h$) and if the steam condenses at the same temperature as the first effect/stage of the MED/MSF desalination systems (i.e., $\Delta h = h_{fg}$), the enthalpy of vaporization in the numerator and denominator of Eq. (7.1) cancels. The remaining terms are of the ratio of the mass flow rate of produced freshwater to the mass flow rate of heating steam, which is known as PR. This parameter is used for thermal desalination systems such as MED and MSF which are driven by the heat released from steam as it condenses. So, for a flow process, the performance parameter PR can be written as

$$PR = \frac{\dot{m}_D}{\dot{m}_{steam}} \tag{7.2}$$

For a work-driven desalination process, which consumes electrical energy, the performance parameter SEC is useful. For a flow process, SEC is defined as the ratio of the work rate required for desalination to the mass flow rate of produced freshwater

$$SEC = \frac{\dot{W}_D}{\dot{m}_D} \tag{7.3}$$

Note that, to compare values of SEC with those of GOR or PR, a proper thermal-to-electrical conversion factor must be applied as the performance assessment parameters are otherwise inconsistent.

7.2.2 Exergy efficiency definition of a desalination system

In contrast to the energy efficiency that measures the ratio of useful energy to the input energy, the exergy efficiency reveals the extent to which a process is reversible. An exergy efficiency of unity indicates a reversible process, while the energy efficiency is less than 1 for real processes. In general, the exergy (or second law) efficiency can be defined as the exergy rate of useful process outputs $\left(\dot{E}x_{out,useful}\right)$ to the exergy rate of the process inputs $\left(\dot{E}x_{in}\right)$

$$\Psi = \frac{\dot{E}x_{out,useful}}{\dot{E}x_{in}} = 1 - \frac{\dot{E}x_{Des} + \dot{E}x_{loss}}{\dot{E}x_{in}} \tag{7.4}$$

Note that the exergy rate of useful process outputs is equal to the input exergy rate $\left(\dot{E}x_{in}\right)$ less the exergy destruction rate $\left(\dot{E}x_{Des}\right)$ and the exergy loss rate $\left(\dot{E}x_{loss}\right)$

$$\dot{E}x_{out,useful} = \dot{E}x_{in} - \left(\dot{E}x_{Des} + \dot{E}x_{loss}\right) \tag{7.5}$$

The destroyed exergy represents the amount of available work that is lost due to system irreversibilities, while the exergy loss represents the amount of available work that is lost with outgoing waste streams. This latter term is often equivalent to the exergy loss due to bringing the direct material streams to the environmental state. The direct material streams refer to all streams directly involved in the system. For a desalination system, for instance, the direct material streams include feed, freshwater, and brine streams. Therefore, the exergy efficiency relation (Eq. 7.4) can be also written as $\Psi = 1 - \left(\dot{E}x_{Des} + \dot{E}x_{loss}\right) / \dot{E}x_{in}$.

In a desalination process, the exergy rate of useful process output is the minimum work rate required to separate a feed seawater stream into freshwater and brine streams. The minimum work required for a desalination system can be defined when all input and output streams, except freshwater, are at the total dead state (i.e., environmental temperature, pressure, and constituent concentrations), while the freshwater stream is at the restricted dead state (ambient temperature and pressure). Then, the minimum work rate of separation ($\dot{W}_{min}$), which can be referred to as the reversible work rate, can be determined using the exergy rate of the inlet and outlet direct material streams for the system. For example, consider a general reversible desalination system that separates a feed (f) stream into the brine (B) and distillate (D) streams. The work rate required for this reversible separation process, that is, the minimum work rate, would be

$$\dot{W}_{min} = \dot{m}_D ex_D + \dot{m}_B ex_B - \dot{m}_f ex_f \tag{7.6}$$

Note that the minimum work rate required for a desalting process is used for determination of the exergy efficiency rather than the least work rate. The least work rate required for the desalting process is introduced as an ideal criterion through which desalination systems can be compared. It can be defined as the required work rate for a reversible desalting process in which all streams including feed, distillate, and brine are at the restricted dead state. In other words, the least work rate can be expressed in terms of the changes in the Gibbs free energy rates of the streams. The concept of least energy of desalting is described in detail in Chapter 6. It is pointed out that, in the wastewater treatment processes where both distillate and brine streams are treated as useful products, the chemical composition of the brine is not brought to chemical equilibrium with the environment. So, the least required work rate should be used in the exergy efficiency expression for wastewater treatment processes.

One type of exergy efficiency of a desalination system can be defined as the ratio of the minimum work rate required for desalination (i.e., the exergy rate of freshwater, $\dot{W}_{min}$) to the actual exergy input rate $\left(\dot{E}x_{input}\right)$, as follows:

$$\Psi = \frac{\dot{W}_{min}}{\dot{E}x_{input}} \tag{7.7}$$

The expression for exergy rate input (denominator in Eq. 7.7) for a standalone desalination system operated by a combination of work rate ($\dot{W}_D$), heat rate ($\dot{Q}_D$), and fuel flow rate ($\dot{m}_{fuel}$) is the summation of the exergy rate inputs

$$\dot{E}x_{input} = \dot{W}_D + \left(1 - \frac{T_0}{T_H}\right)\dot{Q}_D + ex_{fuel}\dot{m}_{fuel} \tag{7.8}$$

It can be seen from Eq. (7.8) that $\dot{E}x_D$ is a strong function of reservoir temperature (T_H). Furthermore, the minimum work of separation is related to the exergy flow of streams, which is determined by the salinities. So, the exergy efficiency (Ψ) is a strong function of T_H and the feed salinity x_f.

There are several methods for calculating thermodynamic properties of substances involved in desalination systems, including ideal mixtures, sodium chloride aqueous solution, and seawater model for aqueous sodium chloride solution. It is found that the ideal mixture model gives values for exergy flow and consequently minimum work of separation with more than 50% error from the actual values (Cerci et al., 2003). The other two models properly predict the minimum separation work (Bromley et al., 1974; Sharqawy et al., 2010; Spiegler and El-Sayed, 2001). So, the ideal mixture model of sodium chloride solution should not be used in the exergy analysis of seawater desalination systems. In Chapter 5 of this book, thermodynamic properties of seawater and pure water are presented based on the up-to-date seawater model. In the exergy analysis of desalination systems, the intake seawater conditions are usually taken as the environmental dead state. The intake seawater conditions including temperature, atmospheric pressure, and salinity differ in general with geographic region. Fig. 7.1 shows the specific flow exergy of

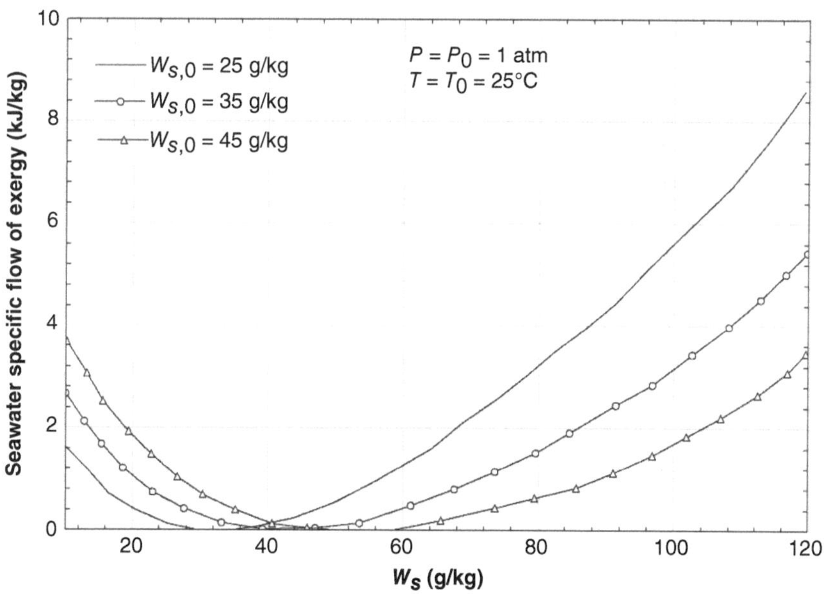

Fig. 7.1 ***Variation in specific exergy of seawater with salt concentration at the dead state temperature and pressure.***

seawater as a function of salt concentration of intake seawater (initial state) when the pressure and temperature are equal to the dead state values (i.e., P_0, T_0). It is seen that the specific exergy of seawater is always positive and it is zero only at the dead state salt concentration. This is true for any salt concentration value taken as the dead state condition. Similarly, for the temperatures and pressures other than dead state values when the salt concentration is equal to dead state, the specific exergy of seawater is always positive and it is zero at the corresponding temperature/pressure dead state value.

In the following, an illustrative example is provided to clarify how to practically employ the minimum exergy and input exergy concepts for evaluating the exergy efficiency of a desalination system.

7.3 Example of forward osmosis desalination with a thermal regeneration system

Consider a typical forward osmosis (FO) brine concentration system shown in Fig. 7.2. The system includes an FO unit and a thermal regeneration system. By applying thermal draw regeneration, the FO system can desalinate a high-salinity input to produce freshwater more efficiently than a RO system, as FO avoids fouling conditions during operation more reliably than RO. In the FO unit, the driving force for separating dissolved solutes from water is the osmotic pressure gradient. The draw solution, which is a high concentration solution with high osmotic pressure, triggers

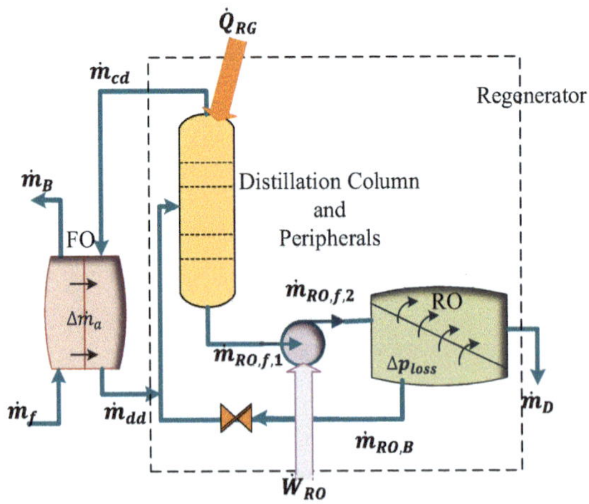

Fig. 7.2 *Forward osmosis (FO) system for desalination with thermal draw regeneration including a distillation column and a reverse osmosis (RO) unit.*

water transfer across the membrane, consequently separating water from its solutes. Through this process, the feed stream becomes more concentrated and the draw concentrated stream becomes diluted. Thermal regeneration plays an important role in the case of high-salinity wastewater (or produced water from industry). Determine the exergy efficiency expression for the FO system with thermal draw regeneration.

Solution:

System description: In the FO module, water is drawn from the feed stream through a saline high-osmotic pressure draw solution. The feed flow becomes more concentrated and consequently the draw solution becomes diluted. The reconcentration of the draw solution occurs in the distillation column (e.g., a stripping column, which is a type of distillation column). The high temperature adversely effects gas solubility allowing for regeneration through a simple distillation column. In other words, the draw solution is regenerated through the distillation columns. The dilute bottom product of the distillation column is delivered to the RO unit to eliminate remaining impurities. The concentrates stream from RO, which is relatively low in salt concentration compared to the draw streams, is returned to the diluted draw solution instead of the concentrate draw solution to reduce the irreversibility. Thermophysical properties of the draw solution can be found using the UNIQUAC model (Darde et al., 2010; Wagman et al., 1982). Note that this model is available in numerous software packages, including the Aspen Plus process simulator (Aspen Plus, 2013).

Methodology: The FO brine concentration exergy efficiency can be written in terms of the exergy efficiency of each system component, that is, the FO exchanger and the draw regenerator (Fig. 7.2). The regenerator (or reconcentrator) is responsible for separation of the dilute draw (*dd*) solution into the concentrated draw (*cd*) and distillate (*D*) streams. The minimum work required for this module is identical to the mixing work associated with combining the concentrated draw and dilute streams at the restricted dead state (i.e., environmental temperature and pressure). So, the minimum work rate required for the draw regenerator can be written as

$$\dot{W}_{min,RG} = \dot{m}_D ex_D + \dot{m}_{cd} ex_{cd} - \dot{m}_{dd} ex_{dd} \tag{7.9}$$

The actual exergy input rate to the regenerator in the denominator of the exergy efficiency definition can be expressed as

$$\dot{E}x_{act,RG} = \dot{Q}_{RG}\left(1 - \frac{T_0}{T_s}\right) + \dot{W}_{RO} \tag{7.10}$$

So, the exergy efficiency of the regenerator is

$$\Psi_{RG} = \frac{\dot{W}_{min,RG}}{\dot{E}x_{act,RG}} = \frac{\dot{m}_D ex_D + \dot{m}_{cd} ex_{cd} - \dot{m}_{dd} ex_{dd}}{\dot{Q}_{RG}\left(1 - \frac{T_0}{T_s}\right) + \dot{W}_{RO}} \tag{7.11}$$

In the FO unit, the water passage across the membrane due to the chemical potential difference is a spontaneous irreversible process. The exergy efficiency of the FO exchanger can be expressed based on the minimum work rate required for removing water from the feed to the minimum work rate required for the same water removal process from the draw, plus the parasitic power consumption by the FO exchanger for overcoming hydraulic losses. That is,

$$\Psi_{FO} = \frac{\dot{W}_{min,\,FO}}{\dot{W}_{min,\,RG} + \dot{W}_{FO}} = \frac{\dot{m}_D ex_D + \dot{m}_B ex_B - \dot{m}_f ex_f}{\dot{m}_D ex_D + \dot{m}_{cd} ex_{cd} - \dot{m}_{dd} ex_{dd} + \dot{W}_{FO}} \tag{7.12}$$

A comparison of Eqs. (7.11) and (7.12) when neglecting the parasitic exchanger power consumption ($\dot{W}_{FO}$) indicates that the denominator of Eq. (7.12) is same as the numerator of Eq. (7.11). The FO exchanger consumes a small amount of electrical power to drive the flow through the channels. So, according to the exergy efficiency expressions for the regenerator and the FO exchanger modules and considering that the hydraulic pressure drop across the FO channels is negligible compared to the osmotic pressure of the draw, the overall exergy efficiency of the FO brine concentration system can be written as follows:

$$\Psi_{FO,RG} = \Psi_{FO} \Psi_{RG} \tag{7.13}$$

$$\Psi_{FO,RG} = \frac{\dot{m}_D ex_D + \dot{m}_B ex_B - \dot{m}_f ex_f}{\dot{Q}_{RG}\left(1 - \frac{T_0}{T_s}\right) + \dot{W}_{RO}} \tag{7.14}$$

It is clear, therefore, that both the FO exchanger and the regenerator influence the efficiency of the FO brine concentration system. The draw streams falls inside the control volume surrounding all system modules, which explains why they do not appear in Eq. (7.14).

Supplementary note: The exergy efficiency of the RO pump within the regenerator can be defined as the ratio of minimum work of desalting to the actual desalting work (i.e., $\dot{W}_{min,RO} / \dot{W}_{act,RO}$). The minimum work rate for RO desalination system can be defined as the required work rate for reversible desalting (i.e., a process with zero exergy destruction), which is equivalent to the exergy rate changes in the inlet and outlet streams interacting with the RO module. That is,

$$\dot{W}_{min,RO} = \dot{m}_{RO,f} ex_{RO,f,2} - \dot{m}_D ex_D - \dot{m}_{RO,D} ex_{RO,D} \tag{7.15}$$

The actual work rate required for the RO module is the exergy rate input to the pump divided by the pump efficiency (η_{pump}):

$$\dot{W}_{act,RO} = \frac{\dot{E}x_{input,pump}}{\eta_{pump}} = \frac{\dot{E}x_{RO,f,2} - \dot{E}x_{RO,f,1}}{\eta_{pump}} \tag{7.16}$$

The electrical power utilized by the RO module also can be estimated through the operating pressure of the RO-pump (P_{RO}), mass flow rate of feed ($\dot{m}_f$), the density of the feed (ρ_f), and the isentropic efficiency of the pump (η_{pump}), as follows:

$$\dot{W}_{act,RO} = \frac{\dot{m}_f P_{RO}}{\eta_{pump} \rho_f} \tag{7.17}$$

7.4 Exergy efficiency evaluation of a desalination system operating as a part of a cogeneration plant

The low-pressure steam requirement in MSF and multieffect evaporators has been investigated for many years (Farsi and Dincer, 2019; Rahimi et al., 2017; Moghimi et al., 2018; Moradi et al., 2019). As a result, a dual-purpose combined system with an electrical generation plant and low-pressure steam desalination is often preferred to stand-alone production. The decision of whether to have a single purpose (stand-alone) or dual purpose (combined) system depends on the economic feasibility of the complete system and other factors. Many industrial large-scale desalination systems are combined with other energy systems, usually electrical generation plants. Determination of the additional energy in the forms of heat and work can be advantageous for performance assessments of such combined systems.

A temperature-entropy (T-s) property diagram of a dual-purpose electricity-water system is shown in Fig. 7.3. The steam at the turbine outlet is considered to be dry saturated. In other words, that steam has a quality of unity (i.e., $x = 1$). The single-purpose power plant cycle is specified by "1-2-3-4-1," while line "5-6" shows the single-purpose desalination

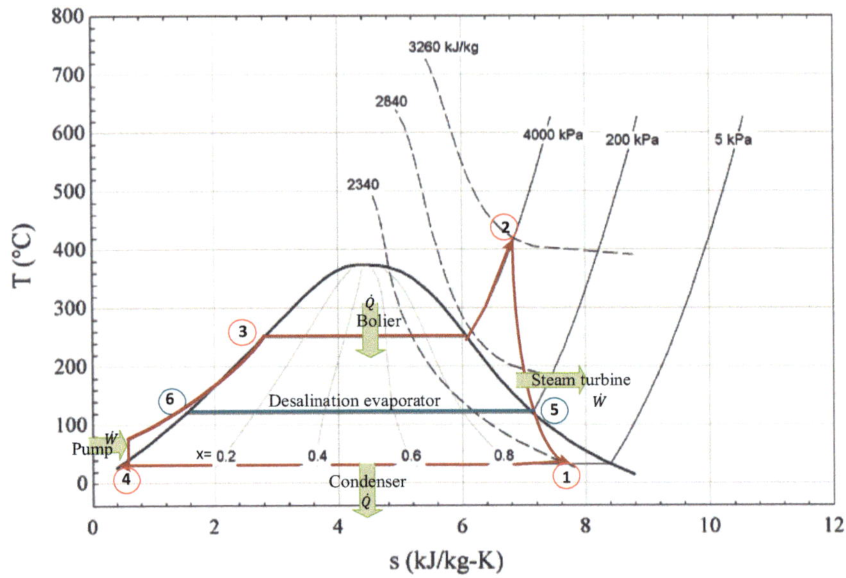

Fig. 7.3 ***Temperature–entropy diagram for the stand-alone and combined electricity-water cogeneration systems operating at the specified design conditions.***

process. In the dual-purpose combined system, the steam leaves the turbine at a relatively higher pressure and temperature (state 5) compared to the stand-alone power plant, for which steam exits at state 1. Then the turbine outlet steam (state 5) is delivered into the evaporator of the desalination system to supply heat for evaporation in the desalination module. Finally, the partially condensed vapor is fully condensed in the condenser. In the combined electricity-water system ("1-2-3-6-5-4-1"), however, the latent heat of the steam is released at the desalination evaporator, which provides the thermal energy required for the desalination process.

Some desalination processes are operated in cogeneration or multigeneration schemes in which a primary energy source (commonly a fuel) is used to produce freshwater, electricity, heat, etc. The exergy efficiency for such cases—a desalination system integrated with other energy systems—should be determined in terms of the primary energy inputs to the overall system rather than the only the desalination unit. In fact, by selecting a proper control volume for the overall cogeneration/multigeneration system, all of the irreversibilities and thermodynamic losses are accounted for, specifically those in the associated energy conversion process (e.g., electricity generation from thermal energy).

A combined power plant cycle and desalination system is shown in Fig. 7.4. It is seen that the heat reservoir with temperature T_H supplies

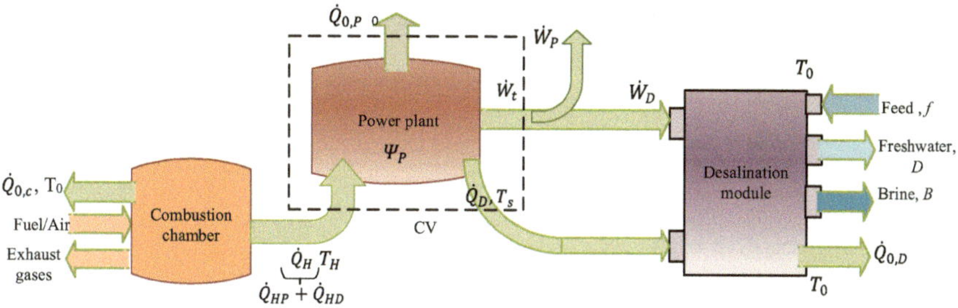

Fig. 7.4 ***A general desalination system combined with a power plant.*** **The desalination module is powered by work rate ($\dot{W}_D$) and heat rate ($\dot{Q}_D (\uparrow \dot{Q}_{HD})$) produced by the power plant, which itself is driven by the heat rate of $\dot{Q}_H$ at T_H.**

the heat rate for the combined system. If the total heat input rate to the combined system is $\dot{Q}_H$, a part of this total heat rate (i.e., $\dot{Q}_{HP}$) is used for electricity production in the power plant cycle and the remaining part (i.e., $\dot{Q}_{HD}$) is used for freshwater production in the desalination system. The energy and entropy rate balances can be written for the power cycle control volume shown in Fig. 7.4 as follows:

$$\dot{Q}_H - \dot{Q}_D - \dot{Q}_{0,P} - \dot{W}_P - \dot{W}_D = 0 \tag{7.18}$$

$$\frac{\dot{Q}_H}{T_H} - \frac{\dot{Q}_D}{T_s} - \frac{\dot{Q}_{0,P}}{T_0} + \dot{S}_{gen} = 0 \tag{7.19}$$

To eliminate the term representing the heat transfer rate with the environment ($\dot{Q}_{0,P}$), Eq. (7.19) is multiplied by T_0 and then subtracted from Eq. (7.18). This yields

$$\dot{Q}_H\left(1 - \frac{T_0}{T_H}\right) - \dot{Q}_D\left(1 - \frac{T_0}{T_s}\right) - T_0\dot{S}_{gen} = \dot{W}_P + \dot{W}_D \tag{7.20}$$

The net heat input rate to the power plant can be defined as the total heat input rate from the heat source ($\dot{Q}_H$) minus the heat input rate to the desalination system ($\dot{Q}_D$). As $\dot{Q}_D$ is recovered in the desalination system, it should not be treated as a heat loss. Thus, the exergy efficiency of the power plant cycle can be written as

$$\Psi_p = \frac{\dot{W}_P + \dot{W}_D}{\left(1 - \frac{T_0}{T_H}\right)\dot{Q}_H - \dot{Q}_D\left(1 - \frac{T_0}{T_s}\right)} \tag{7.21}$$

From the above equation, the total power produced by the power plant can be expressed as

$$\dot{W}_P + \dot{W}_D = \Psi_p\left(\left(1 - \frac{T_0}{T_H}\right)\dot{Q}_H - \dot{Q}_D\left(1 - \frac{T_0}{T_s}\right)\right) \tag{7.22}$$

Substituting the term $(\dot{W}_P + \dot{W}_D)$ in Eq. (7.22) into Eq. (7.20) yields

$$T_0\dot{S}_{gen} = \left(1 - \Psi_p\right)\left(\dot{Q}_H\left(1 - \frac{T_0}{T_H}\right) - \dot{Q}_D\left(1 - \frac{T_0}{T_s}\right)\right) \tag{7.23}$$

Recall that the total heat input rate to the combined system from the heat source includes the heat rate required for the power plant and the desalination system, as follows:

$$\dot{Q}_H = \dot{Q}_{HP} + \dot{Q}_{HD} \tag{7.24}$$

To determine the part of total heat input rate for the desalination system (i.e., $\dot{Q}_{HD}$), it can be assumed that the power plant is not combined with the desalination system. The power plant cycle receives $\dot{Q}_{HD}$ from the heat source and produces an electrical energy rate $\dot{W}_P$ (i.e., $\dot{Q}_{HD}$, $\dot{Q}_D$ and $\dot{W}_D$ are zero). Then, the exergy efficiency of this power cycle can be expressed as

$$\Psi_p = \frac{\dot{W}_P}{\left(1 - \frac{T_0}{T_H}\right)\dot{Q}_{HP}} \tag{7.25}$$

Finding $\dot{Q}_{HP}$ from Eq. (7.25) and substituting it into Eq. (7.24) and then substituting the obtained relation for $\dot{Q}_H$ into Eq. (7.22), yields the part of total heat input rate to the combined system that is used for the desalination system. That is,

$$\dot{Q}_{HD} = \frac{\dot{W}_D}{\Psi_p\left(1 - \frac{T_0}{T_H}\right)} + \dot{Q}_D\frac{\left(1 - \frac{T_0}{T_s}\right)}{\left(1 - \frac{T_0}{T_H}\right)} \tag{7.26}$$

For typical values of temperatures T_H, T_s, and T_0 of 1000 K, 380 K, and 300 K, respectively, and neglecting the work consumed by the desalination pumps (i.e., $\dot{W}_D = \dot{W}_{pump} = 0$), it can be shown that

$$\dot{Q}_{HD} = 0.3\dot{Q}_D \tag{7.27}$$

Eq. (7.27) indicates that the portion of the heat rate used for the distillation unit of the combined electricity-water plant can be only about one-third of that normally used by a stand-alone distillation system with the same design operating conditions.

To evaluate the exergy efficiency of the desalination system in Fig. 7.4, the desalination work rate $\dot{W}_D$ and heat rate $\dot{Q}_D$ are not properly representative of the energy inputs to the desalination process, as these two quantities are applicable for a stand-alone desalination system, not one linked to a cogeneration system. To properly determine the energy rates (heat and work rates) required for the desalination system combined with a power plant, the extra thermal energy supplied by the power plant (i.e., $\dot{Q}_{HD}$ from Eq. 7.26) is included in the exergy efficiency, as follows:

$$\Psi_D = \frac{\dot{W}_{min}}{\dot{Q}_{HD}\left(1-\frac{T_0}{T_H}\right)} = \frac{\dot{W}_{min}}{\left(\frac{\dot{W}_D}{\Psi_p}\right)+\dot{Q}_D\left(1-\frac{T_0}{T_s}\right)} \tag{7.28}$$

In the thermal desalination plants where $\dot{W}_D = 0$, the exergy efficiency can be expressed as

$$\Psi_D = \frac{\dot{W}_{min}}{\dot{Q}_D\left(1-\frac{T_0}{T_s}\right)} \tag{7.29}$$

In a similar way, for work-driven desalination systems, Eq. (7.28) reduces to

$$\Psi_D = \Psi_P\frac{\dot{W}_{min}}{\dot{W}_D} \tag{7.30}$$

The exergy efficiency of the power plant is seen to be one of the contributors in determining the exergy efficiency of the desalination plant, as $\dot{W}_D$ is produced from the power plant through an irreversible process. If the power plant operates reversibly, the exergy efficiency of power plant is obviously unity ($\Psi_P = 1$). If the combustion chamber had been also included in this analysis, both $\dot{W}_D$ and $\dot{Q}_D$ would be divided by the exergy efficiency of the combustion chamber, which would reduce the exergy efficiency of the desalination system. Note that, as the combustion chamber provides the primary energy for the power plant by producing heat from a fuel, both $\dot{W}_D$ and $\dot{Q}_D$ are affected similarly by the losses associated with the electrical power generation process.

Generally, in the case where multiple energy sources are input to a desalination system, the source of each form of primary energy input needs to be recognized, much like $\dot{W}_D$ and $\dot{Q}_D$ that stem from $\dot{Q}_{HD}$. Then the exergy of each of the energy source inputs, after tracing them back to their primary source, should be accounted for in the exergy rate balance.

For the power-freshwater cogeneration system in which the desalination plant is driven by work only, the exergy efficiency of the desalination module in Eq. (7.30) cannot be 100% in the case of irreversible operation of the power plant, even if the desalination system operates reversibly. This is because the cogeneration system pays an energy penalty for energy conversion of heat to power through an irreversible process. If the combustor is located within the control volume of the cogeneration system, in addition to Ψ_P, the exergy efficiency of the combustion process (Ψ_C) also contributes to the calculation of the desalination exergy efficiency.

7.5 Thermodynamic performance comparison of desalination systems

Theoretically, a reversible thermal desalination plant can achieve an exergy efficiency of 100% because, unlike work-driven desalination, the energy penalty associated with the heat-to-electricity conversion in the power plant cycle is not attributed to these systems. This is verified in Fig. 7.5, where an exergy efficiency of 100% is attained when $\dot{Q}_D$ approaches $\dot{Q}_{least}$, regardless of the steam temperature T_s. To achieve such an exergy efficiency, an infinite heat exchanger area, and an infinite number of stages are required, which is impossible. At a fixed required heat rate for desalination process, the exergy efficiency decreases with increasing steam temperature. In fact, for a constant desalting required heat rate, the exergy of the steam is not utilized efficiently as the temperature of steam increases. From Fig. 7.5, the exergy efficiency of a thermal desalination plant is seen to increase sharply as the heat rate required for the desalination process approaches its lowest value.

MSF systems typically operate at relatively higher temperatures (around 100 °C) than MED systems (around 80 °C). MSF systems generally have GOR values between 6 and 9 and an exergy efficiency of about 5%, while MED systems generally have GOR values between 10 and 13 and an exergy efficiency between 7% and 9% (Miller, 2003). The exergy efficiency and the operating and capital costs associated with both MSF and MED systems always require a compromise. As the steam temperature increases,

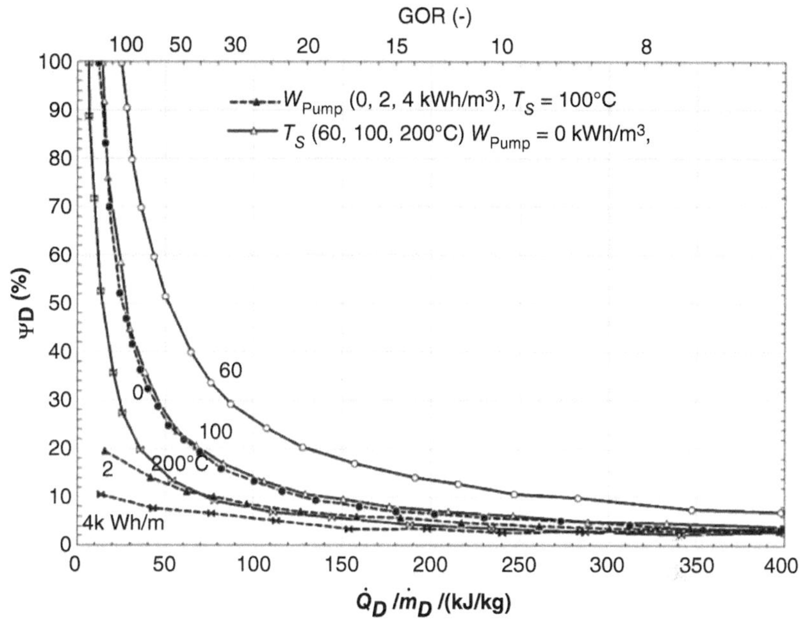

Fig. 7.5 ***Variation in exergy efficiency and GOR of a thermal desalination system driven by heat supplied from a power plant with the heat required for the desalination per kilogram of distillate for several values of steam temperature and required pump work.*** The power plant uses thermal energy as the primary energy source. $T_0 = 25$ °C and $w_f = 35$ g/kg. *Data extracted from Mistry and Lienhard (2013).*

the exergy efficiency rises, but more stages are required (i.e., the heat transfer area increases) and consequently the capital costs increases. But as steam temperature decreases, the temperature pinches of each stage decrease leading to an additional heat transfer area and higher associated capital costs. As a result, the lower the steam temperature, the lower are the number of stages and the overall exergy efficiency for MSF and MED systems.

The exergy efficiency of a typical RO system varies between 10% and 20%. The feed stream properties and the characteristics of the system design are the main factors affecting the exergy efficiency of an RO system (Fritzmann et al., 2007; Stover, 2007). When comparing heat-driven and work-driven desalination systems from the thermodynamic perspective, it can be found that the work-driven desalination systems are more efficient than those driven by heat. But it is noted that the considerable amount of work for pumping a large amount of seawater into the thermal desalination system is not considered in this comparison. If this were to be done, the advantages of the work-driven systems over the heat-driven systems would be greater. Fig. 7.5 shows the exergy efficiency of a steam-driven desalination plant accounting for pump work ranging from 0 to 4 kWh. When the required pump work is eliminated (i.e., $W_{pump} = 0$), an exergy efficiency of 100% is obtained. But for a practical system in which a pump is used to drive seawater into the system, the exergy efficiency is considerably lower. For instance, at a steam temperature of 100 °C, the exergy efficiency of the desalination system drops about 80% when counting the electrical pump work of 2 kWh per m^3 of produced freshwater.

Work-driven desalination systems exhibit higher exergy efficiencies than thermal desalination systems. Recent advances in the membrane desalination (MD) modules have made them competitive with work-driven systems in terms of exergy efficiency. Furthermore, important factors such as quality of feed seawater and produced freshwater, capital and operating costs, expertise and infrastructure can make thermal desalination plants more beneficial for a given application and/or location. Specifically, in the case of heat recovery from power plants, a thermal desalination system can be integrated properly with a system for power/freshwater cogeneration. The selection of thermal desalination technology for integrating with other energy systems requires considerations of the unused temperature reduction of waste heat. For example, when a high-temperature energy source (over 120 °C) is available from a power plant, selecting MED to be combined with the power plant could increase the unused temperature as it is reduced to meet the MED operating temperature (around 70 °C). Instead, regardless of cost and exergy efficiency, an MSF system with a relatively higher operating temperature (around 110 °C) can be a better choice for integrating with the power plant.

7.6 Irreversibilities in seawater desalination technologies

Exergy destructions and losses in desalination systems are mostly associated with several typical thermodynamic processes such as pressure exchange, heat transfer, and bringing concentration and temperature into the equilibrium environmental conditions. Determining the exergy destruction in each desalination component/process provides useful information for identifying the sources of inefficiencies and irreversibilities and how to improve a desalting process. In an assessment of a desalination system when employing ideal gas and incompressible fluid models, a reasonable selection of a reference state is necessary for the determination of physical properties of seawater and pure water. Selection of a reference state is discussed further in the following section. The entropy generation of the aforementioned thermodynamic processes can be determined using the control volume containing the entire physical process, then applying the second law of thermodynamics.

7.6.1 Entropy generation expressions for common thermodynamic processes in desalination systems

Entropy generation rate expressions for common thermodynamic processes in diverse desalination methods are presented now.

Heat exchange: In the operation of an actual heat exchanger, viscous forces contribute to pressure drops. For a heat exchanger in which a single-flow is exchanging, a thermodynamically optimum geometry can be achieved through a compromise between the entropy generation due to pressure drop in a finite flow volume and heat transfer across a finite temperature difference (Bejan, 1978). However, the pressure drop is not significant in desalination systems. Hence, the generated entropy within the heat exchangers can be determined based on the final temperatures only. Then, the general entropy generation rate through a heat exchanger can be expressed as

$$\dot{S}_{gen}^{Heat\ exchanger} = \left[\dot{m}\left(s_{out} - s_{in}\right)\right]_{stream1} + \left[\dot{m}\left(s_{out} - s_{in}\right)\right]_{stream2} \quad (7.31)$$

where $\dot{m}$ denotes mass flow rate (kg/s); s is specific entropy (kJ/(kg.K)) and $\dot{S}_{gen}^{Heat\ exchanger}$ the entropy generation rate in the heat exchanger (kW/K). The specific entropy change for a phase change process at a saturation condition (i.e., evaporation and condensation) at constant pressure can be written as a function of quality x of the exit stream as follows:

$$s_{out} - s_{in} = \begin{cases} xs_{fg} = x\left(s^{gas} - s^{fluid}\right) & \text{for evaporation} \\ (x-1)s_{fg} = (x-1)\left(s^{gas} - s^{fluid}\right) & \text{for condensation} \end{cases} \quad (7.32)$$

Also, the change in specific entropy of an incompressible flow and ideal gas at an approximately constant pressure can be written, respectively, as

$$s_{out} - s_{in} = c \ln \frac{T_{out}}{T_{in}} \quad (7.33)$$

$$s_{out} - s_{in} = c_p \ln \frac{T_{out}}{T_{in}} \quad (7.34)$$

where T_{in} and T_{out}, respectively, are the inlet and outlet temperatures in K, and c and c_p, respectively, are the specific heat of an incompressible fluid and specific heat at constant pressure (kJ/(kg.K)). Note that the specific entropy change expression in Eq. (7.33), includes a single type of specific heat (i.e., c) as the specific heat at constant pressure and volume are identical for an incompressible fluid (i.e., $c_p = c_v$).

Bringing products temperature to thermal equilibrium: Consider exit streams (e.g., freshwater and brine) from a control volume that is in mechanical equilibrium but not thermal equilibrium with the environment (Fig. 7.6). The entropy generation rate associated with the thermal disequilibrium when the exit stream i at a temperature $T_{i,out1}$ is brought to the environmental temperature (so that $T_{i,out2} = T_0$) can be written as

$$\dot{S}_{gen}^{T\ disequilibrium} = \dot{m}_i \left\{ \left(s_{i,out2} - s_{i,out1} \right) \right\} - \frac{\dot{Q}}{T_0}$$
$$= \dot{m}_i \left\{ \left(s_{i,out2} - s_{i,out1} \right) - \frac{h_{i,out2} - h_{i,out1}}{T_0} \right\} \quad (7.35)$$

where $\dot{Q}$ is the heat transfer rate with the environment and h_i is the specific enthalpy of stream i. To bring an incompressible fluid to thermal equilibrium, the terms of specific entropy change (i.e., $(s_{i,out2} - s_{i,out1})$) and specific enthalpy change (i.e., $(h_{i,out2} - h_{i,out1})$) in Eq. (7.35) can be written as $c_i \ln T_0/T_i$ and $c_i(T_0 - T_i)$, respectively. Then, the entropy generation rate associated with bringing an incompressible flow to the environmental temperature can be expressed as

$$\dot{S}_{gen}^{T\ disequilibrium} = \dot{m}c \left[\frac{T_i}{T_0} - 1 + \ln\left(\frac{T_0}{T_i} \right) \right]. \quad (7.36)$$

Bringing brine concentration to chemical equilibrium: The brine stream of a desalination process is typically treated as a waste output that is discharged

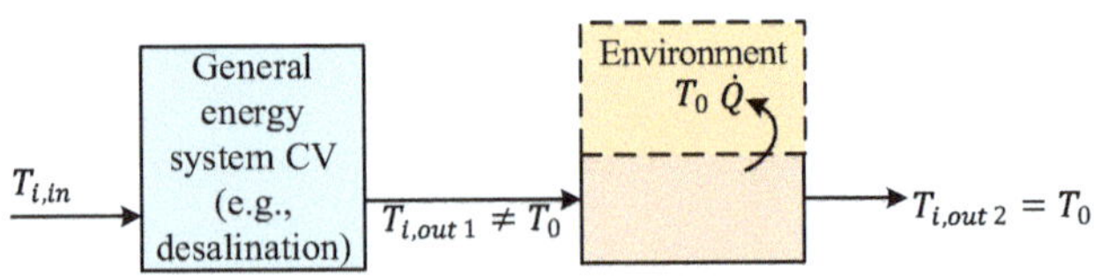

Fig. 7.6 ***Process for bringing outlet steam from the system control volume temperature ($T_{i,out1}$) to the environment temperature (T_0).***

back to the sea. Entropy is generated through restoring the brine stream with a relatively higher salinity to the seawater salinity. The entropy generated through bringing brine to chemical equilibrium with the environment is determined either through the exergy destruction method due to irreversible mixing or the least work method for desalting, both of which were proposed by Bejan (2016). The entropy rate generated through mixing the brine stream (B) at the restricted dead state (i.e., ambient temperature and pressure) with a sufficiently large seawater reservoir so that is in chemical equilibrium with the seawater (i.e., $x_{B,out} = x_{sw}$) and reaches the total dead state (i.e., $g_{out} = g_{sw}$) can be expressed as

$$\dot{S}_{gen}^{chem\ disequilibrium} = \frac{\dot{E}x_D^{mixing}}{T_0} = \frac{\dot{E}x_{out} - \dot{E}x_{in}}{T_0} \tag{7.37}$$

where the change in the input and output exergy rates can be expressed in terms of the Gibbs free energy change of the inlet and outlet streams, as

$$\dot{E}x_{out} - \dot{E}x_{in} = \left(\dot{m}_{sw,reservoir} + \dot{m}_B\right) g_{out} - \left(\dot{m}_B g_{out} + \dot{m}_{sw,reservoir} g_{sw}\right) \tag{7.38}$$

Here, $\dot{E}x$ is the exergy rate (kW), g is the specific Gibbs free energy (kJ/kg), and x is the molal concentration.

Irreversible expansion: In expansion devices without phase change a pressure drop occurs across flow constrictions such as pipes, throttles, membranes, expanders, etc. The entropy generation rate due to irreversible expansion of an *incompressible fluid* can be expressed as a function of the isentropic efficiency of expansion devices ($\eta_{isen,\ exp}$), as follows:

$$\dot{S}_{gen}^{expansion\ fluid} = \dot{m}c\ ln\left[\frac{v}{cT_{in}}\left(P_{in} - P_{out}\right)\left(1 - \eta_{isen}\right) + 1\right] \approx \dot{m}\left(\frac{v}{T_{in}}\left(P_{in} - P_{out}\right)\right) \tag{7.39}$$

where v denotes specific volume (m^3/kg) and P pressure (kPa). The isentropic efficiency of expansion relates the work produced per unit mass through the actual (w) and isentropic (w_s) processes. That is,

$$\eta_{isen,\ exp} = \frac{w}{w_s} = \frac{h_{in} - h_{out}}{h_{in} - h_{out}^{isen}} \tag{7.40}$$

where h_{out}^{isen} is the specific enthalpy of the outlet stream for the isentropic expansion process. Note that if a complete irreversible pressure drop occurs (e.g., through a throttle process) where no work is produced, the isentropic efficiency becomes zero.

The generated entropy rate for the reversible expansion of an *ideal gas* can be expressed as

$$\dot{S}_{gen}^{expansion\ gas} = \dot{m}\left(c_p \ln\left[1+\eta_{isen}\left\{\left(\frac{P_{out}}{P_{in}}\right)^{\frac{R}{c_p}}-1\right\}\right]-R\ln\left(\frac{P_{out}}{P_{in}}\right)\right) \tag{7.41}$$

From Eqs. (7.39) and (7.41), it can be seen that the entropy generation rate for an incompressible fluid is a function of the pressure difference, while for an ideal gas it is a function of the pressure ratio.

Flashing: In the flashing process in desalination, saturated liquid water (or water near the saturation state) undergoes a rapid pressure drop in a device such as a throttling valve. As a result, a portion of the liquid is converted to vapor. The specific entropy generated in the mixture of liquid/gas ($s_{gen}^{flashing}$) can be modeled based on an incompressible fluid for an ideal gas in terms of quality (x), specific heat at constant pressure (c_p), specific heat of an incompressible fluid (c), specific volume of liquid (ν), ideal gas constant for vapor (R), and the specific enthalpy and entropy of the vapor (h_v°, s_v°) and the fluid (h_{lw}°, s_{lw}°) at the reference state. Then,

$$\begin{aligned} s_{gen}^{flashing} &= x\left\{\left(c_p - c\right)\ln T_{out} - R\ln P_{out}\right\} \\ &\quad + x\left\{\left[s_v^\circ - s_{lw}^\circ - \left(c_p - c\right)\ln T^\circ + R\ln P^\circ\right]\right\} + c\ln\frac{T_{out}}{T_{in}} \end{aligned} \tag{7.42}$$

or, rearranged for the quality as

$$x = \frac{c\left(T_{in} - T_{out}\right) + \nu\left(P_{in} - P_{out}\right)}{\left(c_p - c\right)T_{out} - \nu P_{out} + \left[h_v^\circ - h_{lw}^\circ - \left(c_p - c\right)T_{ref} + \nu P^\circ\right]}. \tag{7.43}$$

Note that the standard specific enthalpy, specific entropy, specific heat, and specific volume should be determined at the reference state. The saturation state at the average temperature of the inlet and outlet streams can be a proper reference state for pure water. For seawater, however, in addition to the average temperature, the average salinity should also be considered. But, for the processes in which the compositions significantly change, the average salinity is not applicable for specifying heat capacity. Bringing the brine concentration to the equilibrium concentration is an example of composition change. In this case, the specific heat significantly decreases as salinity increases. Then, instead of average salinity, the Gibbs free energy should be used. This concept is used for bringing the brine concentration to the environmental constituent concentration, as is done in Eq. (7.38).

Compression and pumping: The entropy generation rate associated with compression and pumping processes can be expressed in terms of the isentropic efficiency of compression and pumping as (Bejan, 2016)

$$\dot{S}_{gen}^{compressing} = \dot{m}\left(c_p \ln\left(1 - \frac{1}{\eta_{isen,c}}\left[1 - \left(\frac{P_{out}}{P_{in}}\right)^{R/c_p}\right]\right) - R\ln\left(\frac{P_{out}}{P_{in}}\right)\right) \tag{7.44}$$

$$\dot{S}_{gen}^{pumping} = \dot{m}\left(c\ln\left[1 + \frac{v}{cT_{in}}\left(P_{out} - P_{in}\right)\left(\frac{1}{\eta_{isen,p}} - 1\right)\right]\right)$$

$$\approx \dot{m}\left(\frac{v}{T_{in}}\left(P_{out} - P_{in}\right)\left(\frac{1}{\eta_{isen.p}} - 1\right)\right) \tag{7.45}$$

where $\eta_{isen,c}$ and $\eta_{isen,p}$ are the isentropic efficiencies of the compressor and the pump, respectively, which can be expressed as

$$\eta_{isen,p\&c} = \frac{w_s}{w} = \frac{h_{out}^{isen} - h_{in}}{h_{out} - h_{in}}. \tag{7.46}$$

With the above formulas for the entropy generation rates of the various processes that commonly occur in desalination systems, the extent of irreversibilities in a system can be determined. As noted before, the exergy efficiency can be defined as the usfuel exergy rate divided by the exergy rate input to the system. As the useful exergy rate is equal to the exergy input rates minus the exergy destruction and exergy loss rates, the exergy efficiency can be expressed as a function of the ratio of the summation of the exergy destruction and exergy loss rates to the total input exergy rate, as follows:

$$\Psi_D = \frac{\dot{E}x_{useful}}{\dot{E}x_{in}} = 1 - \frac{\sum \dot{E}x_{Des,i} + \sum \dot{E}x_{loss,i}}{\dot{E}x_{in}} \tag{7.47}$$

where $\dot{E}x_D$ and $\dot{E}x_{loss}$, respectively, are exergy destruction and exergy loss rates. The exergy destruction rate indicates the lost available work rate due to the irreversibilities (i.e., $\sum \dot{E}x_{Des} = \sum T_0\dot{S}_{gen}$) and the exergy loss rate indicates the lost available work rate due to releasing waste streams to the environment.

7.6.2 Exergy destruction in desalination technologies

In this section, exergy destruction values based on past studies are provided for the thermodynamic processes commonly occuring in several types of desalination systems. Then, the identified sources of irreversibilities are compared and discussed.

HDH desalination: In the HDH desalination system, air as a carrier gas is used to evaporate water from the feed seawater and generate distillate through subsequent condensation. The HDH desalination system usually operates at atmospheric pressure and relatively low temperature (i.e., 50–65 °C). Each HDH system has at least three components: heater, humidifier, and dehumidifier. Depending on the arrangements of these three components various configurations can be formed. One of the primary configurations is the closed air open water (CAOW) HDH class. Fig. 7.7 shows a typical CAOW water-heated HDH desalination system.

In a CAOW configuration of a HDH system, air always circulates between the dehumidifier and humidifier. From Fig. 7.7, the feed seawater input to the dehumidifier at a temperature of 30 °C and a salinity of 35 g/kg condenses the moist hot air stream. Then the preheated feed stream is further heated to the top brine temperature of 70 °C in the water heater and then sprayed into the humidifier. In the humidifier, a portion of the sprayed hot feed stream is vaporized and absorbed by the dry air to form the hot and moist air. The remaining concentrated seawater exits the system as the brine stream.

The exergy destruction rates for various processes of a typical CAOW HDH desalination system, operating at specific design conditions, are listed

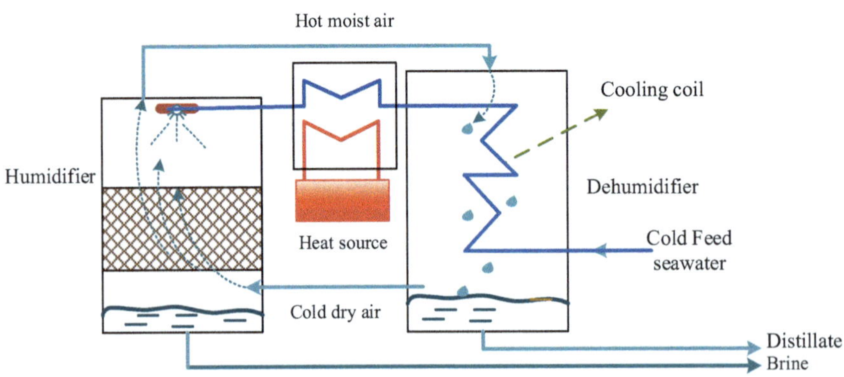

Fig. 7.7 ***Closed air open water (CAOW) water-heated humidified-dehumidified desalination system.***

Table 7.1 Typical operating design characteristics, exergy destruction rates, and performance parameters of the CAOW water-heated HDH system in Fig. 7.7.

Parameter	Value
Design parameters	
Seawater salinity	35,000 ppm
Seawater temperature	30 °C
Top brine temperature	70 °C
Mass flow rate ratio for seawater to dry air	3
$\dot{Q}_{Heater}/\dot{m}_D$	962 $kJ/kg_{Distillate}$
Normalized exergy destruction rates	
Chemical disequilibrium of brine	0.11 $kJ/kg_{Distillate}$
Temperature disequilibrium of freshwater	3.3 $kJ/kg_{Distillate}$
Temperature disequilibrium of brine	14.11 $kJ/kg_{Distillate}$
Heater	19.07 $kJ/kg_{Distillate}$
Humidifier	14.3 $kJ/kg_{Distillate}$
Dehumidifier	59.1 $kJ/kg_{Distillate}$
Performance parameters	
GOR	2.5
Ψ_{HDH}	2.4%

Data modified from Mistry et al. (2011), Mistry and Mitsos (2011), Mistry and Zubair (2010).

in Table 7.1. The overall system operates at atmospheric pressure so the pumping work is negligible. The entropy generation rate expressions for the heat exchanger (Eqs. 7.31–7.34) are used for the humidifier, dehumidifier, and heater.

In Mistry et al. (2011), the normalized entropy generation rates associated with components of the CAOW HDH desalination system are presented. It can be observed from Table 7.1 that the greatest exergy destruction occurs in the dehumidifier module. It is seen that the exergy destruction rate associated with the brine salinity disequilibrium with the environment is relatively small. This means that the distillate production from the input seawater is minor, which results in a small increase in the brine salinity. So, the relatively low irreversibility associated with the salinity disequilibrium of reject brine indicates that the CAOW HDH desalination system has a small recovery ratio ($\dot{m}_D / \dot{m}_f$).

It is indicated in Table 7.1 that the GOR of the CAOW water-heated HDH desalination system is 2.5. Generally, the GOR of an HDH desalination system depends on various factors, such as humidity ratio of air at the exit of the humidifier, top temperature at the humidifier, humidifier/dehumidifier effectiveness, and mass flow rate ratio. The mass flow rate ratio can be defined as the ratio of the mass flow rate of water to the humidifier to

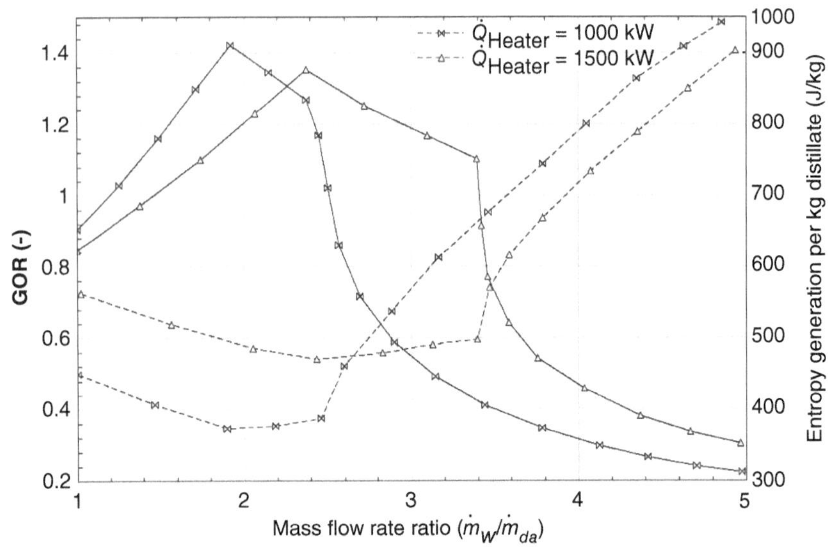

Fig. 7.8 ***Variation of GOR (solid lines) and entropy generation per kg of distillate (dashed lines) with mass flow rate ratio, for two heat input rates to the CAOW water-heated HDH desalination system.*** *Data from Mistry and Zubair (2010).*

the mass flow rate of air through that humidifier. Fig. 7.8 shows the variation of GOR and entropy generation per kilogram of distillate produced ($s_{gen} = \dot{S}_{gen,total} / \dot{m}_D$) with the mass flow rate ratio at several heat input rates for the desalting process ($\dot{Q}_{Heater}$), for a CAOW water-heated HDH desalination system. It can be seen that the maximum GOR corresponds to the minimum entropy generation per kilogram of distillate produced. This means that there is an optimum operating condition for which GOR is a maximum and entropy generation per kilogram of distillate is a minimum. Note that the trend in Fig. 7.8 does not appear when plotting total entropy generation rate or entropy generation per kilogram of distillate against GOR. If GOR is plotted based on the variations of entropy generation per kilogram of distillate (s_{gen}) or total entropy generation rate ($\dot{S}_{gen,total}$) in the system, the GOR is not a single-valued function of both s_{gen} and $\dot{S}_{gen}$. Instead, GOR is a strong function of mass flow rate ratio, as seen in Fig. 7.8.

Multieffect desalination (MED): MED desalination systems generally consist of multiple effects in which seawater is heated by steam tubes. An external steam heat source is used in the first effect and it vaporizes a portion of sprayed seawater. The generated vapor is transferred from the first effect to the second effect to vaporize a portion of the sprayed seawater at the second effect. This process is successively repeated in the following effects with lower temperature and pressure after each effect. In essence, each effect recovers the energy from the previous effect.

There are several MED configurations depending on the feed flow pattern, such as backward-feed, forward-feed, and parallel-feed. Fig. 7.9 shows a typical forward-feed MED desalination system with six effects driven by a steam heat source at a temperature of 70 °C. Design parameters for a typical MED system are presented in Table 7.2. The results of an MED thermodynamic analysis are also provided in Table 7.2. They are obtained utilizing the following simplifying assumptions:

- The produced distillate in all effects is almost the same. This demonstrates the impact of raising latent heat as the temperature of effect decreases.
- A portion of the distillate produced from the flashing process in each effect (excluding the first effect) is determined by the mass flow rate of brine from the previous effect, as $\dot{m}_{b,i-1}c_{p,i}\Delta T / h_{fg,i}$. The remaining part of the distillate produced in each effect is obtained from partial evaporation of the feed in that effect.
- Flashing does not occur in the first effect.
- The temperature difference between sequential effects is equal, at $\Delta T_i = \Delta T_{i+1} = \Delta T$, where $\Delta T = (T_{steam} - T_n)/n$ and n denotes number of effects.
- The temperature difference between the evaporating brine and the condensing vapor (ΔT) is the same as the temperature rise through the feed heaters.

The highest exergy destruction rates for the MED forward-feed desalination system are seen in Table 7.2 to be associated with the heat exchanger modules, specifically the effects units (responsible for 56.5% of the total exergy destruction rate), the condenser (21.8%), and the feed heater (12.3%). This is mainly due to the fact that heat transfer processes are dominant in the MED system. The destructions of exergy in the flash boxes are relatively small (0.6%). The exergy destructions attributed to the disequilibrium discharge of the brine and distillate account for about 8.7% of total exergy destruction rate. Note that the exergy destruction rates attributable to the pumps are not included as they make up less than 1% of total of exergy destruction rate. The exergy destruction rates in each of the effects of MED are almost the same. So, on an individual basis, the condenser has a higher exergy destruction rate than any of the individual MED effects. The large exergy destruction rate of the condenser is due to the condensation of vapor using a large flow rate of water, even though the component involves

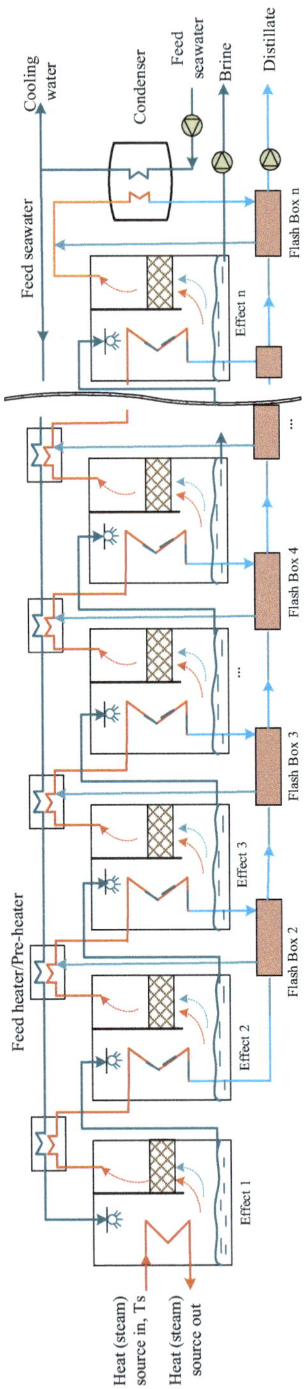

Fig. 7.9 ***Forward-feed MED desalination system.***

Table 7.2 Typical design parameters, exergy destruction rates, and performance parameters for the MED forward feed desalination system in Fig. 7.9.

Parameter	Value
Design parameters[a]	
Number of effects, n	6
Steam temperature, T_S	70 °C
Last effect temperature, T_6	40 °C
Seawater temperature (environment temperature), T_0	25 °C
Mass flow rate of produced water, $\dot{m}_D$	25 °C
Feed seawater salinity	42,000 ppm
Maximum salinity	70,000 ppm
Distillate salinity	0.0 ppm
Normalized exergy destruction rates	
Effects	25.8068 $kJ/kg_{Distillate}$
Feed heaters	8.8804 $kJ/kg_{Distillate}$
Condenser	12.665 $kJ/kg_{Distillate}$
Flash boxes	0.350 $kJ/kg_{Distillate}$
Temperature disequilibrium of brine	1.788 $kJ/kg_{Distillate}$
Temperature disequilibrium of freshwater	1.639 $kJ/kg_{Distillate}$
Chemical disequilibrium	0.894 $kJ/kg_{Distillate}$
Performance parameters	
GOR	5.4
Ψ_{MED}	5.9%

[a]Data modified from Darwish et al. (2006) and El-Dessouky and Ettouney (2002).

a relatively low-temperature heat transfer process. Using thermal vapor compressor (TVC) reduces the generated entropy within the condenser as a portion of vapor is entrained from the last effect and then reinjected into the first effect. This serves to improve the performance of the MED system significantly but, as substantial exergy is destroyed by the TVC module, the total exergy destruction may not be reduced considerably.

Multistage flash (MSF) desalination: In the MSF desalination system shown in Fig. 7.10, the intake feed seawater at 25 °C is pumped to the brine heater where it exchanges heat with high-temperature steam at a temperature of 116 °C. In the brine heater, the feed seawater is heated and then transferred to the first stage that is maintained at a pressure below the saturation temperature of water. After the flashing process, the generated vapor is condensed at the condenser and preheats the input feed seawater. The remaining seawater exits the first stage as brine and flashes in the second stage, which is maintained at a pressure lower than that of first stage. This process repeats in the successive stages of the MSF desalination system. The brine from the last effect exits the stage at a temperature of 40 °C.

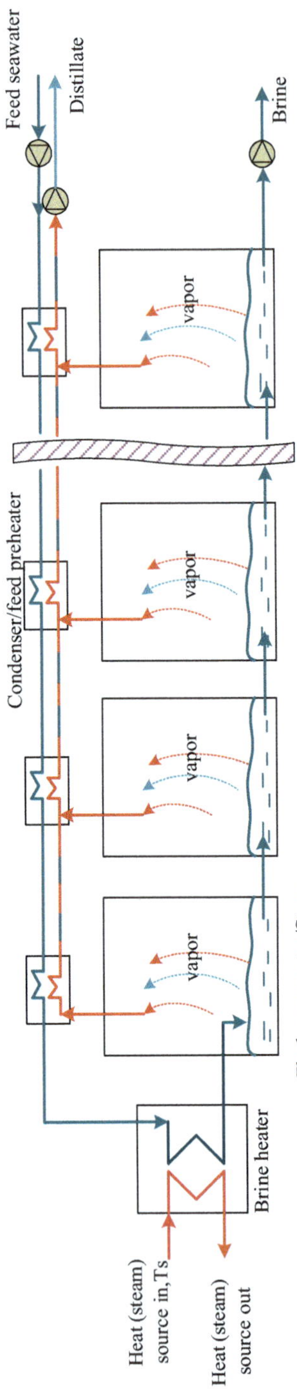

Fig. 7.10 ***Typical multistage flash desalination system.***

In the MSF model, the difference in saturation temperatures of sequential stages can be assumed constant (El-Dessouky and Ettouney, 2002). The energy rate balance for the brine heater using steam as a heat source can be expressed as

$$\dot{m}_s h_{fg} = \dot{m}_f \left(h_{f,TBT} - h_{f,ph} \right) \tag{7.48}$$

where $\dot{m}_s$ and $\dot{m}_f$, respectively, are the mass flow rates of the steam and feed streams, while $h_{f,TBT}$ and $h_{f,ph}$, respectively, are the specific enthalpies of the feed at the top brine temperature and at the last preheater temperature. An energy rate balance for stage i can be written as

$$\dot{m}_{b,i-1} h_{b,i-1} = \dot{m}_{b,i} h_{b,i} + \dot{m}_{D,i} h_{D,i} \tag{7.49}$$

where $\dot{m}_{b,i}$ and $\dot{m}_{D,i}$, respectively, are the brine and distillate mass flow rates in effect i.

Table 7.3 provides the MSF desalination design and performance parameters and exergy destruction rates for the system components. The MSF desalination system in Fig. 7.10 consists of 24 stages, 24 flash chambers, 24 feed heaters, and a brine heater. The exergy efficiency of MSF

Table 7.3 Typical design parameters, exergy destruction rates, and performance parameters for the MSF desalination system in Fig. 7.10.

Parameter	Value
Design parameters[a]	
Steam temperature	116 °C
Feed temperature	25 °C
Brine reject temperature from last effect	40 °C
Feed mass flow rate	3.384×10^3 kg/s
Feed seawater salinity	42 g/kg
Distillate mass flow rate	378.8 kg/s
Normalized exergy destruction rates	
Flash chambers	1.430 $kJ/kg_{Distillate}$
Feed heaters	93.035 $kJ/kg_{Distillate}$
Brine heater	16.092 $kJ/kg_{Distillate}$
Temperature disequilibrium of brine	11.427 $kJ/kg_{Distillate}$
Temperature disequilibrium of freshwater	1.192 $kJ/kg_{Distillate}$
Chemical disequilibrium	0.298 $kJ/kg_{Distillate}$
Performance parameters	
GOR	4.6
Ψ_{MSF}	2.9%

[a]Data modified from Dessouky and Ettouney (2002).

desalination is 2.9%. This value is of the same order as the exergy efficiency for the MED system discussed previously. The greatest exergy destructions occur in the feed heaters due to the large mass of thermal brine circulating at a low recovery ratio for the heat transfer process. The exergy destruction distribution between the feed heaters is almost the same for both processes. On an individual basis, the brine heater has the greatest exergy destruction rate. In this heat exchanger, a large exergy destruction rate is incurred through the condensation of high-quality steam. The use of low-quality and low-temperature steam can reduce the exergy destruction rate. The specific exergy destruction is relatively small in each of the flash evaporators (0.06 $kJ/kg_{Distillate}$) due to the isothermal evaporation of a small quantity of vapor, which is a consequence of the small recovery ratio. According to Table 7.3, the specific exergy destruction associated with the temperature disequilibrium of the brine and distillate streams, respectively, are 11.427 $kJ/kg_{Distillate}$ and 1.192 $kJ/kg_{Distillate}$. The total contribution of temperature disequilibrium to the total exergy destruction rate is 10%. The contribution of temperature disequilibrium of the brine stream is significant here because of the low recovery in the MSF system and the large rejection of brine at a relatively high temperature compared to the environmental temperature.

Direct contact membrane desalination (DCMD): DCMD is the simplest configuration of MD system and widely used. Fig. 7.11 shows a DCMD system. The feed seawater is preheated in the regenerator and then heated further in the heater module. The hot feed stream enters the hot side of membrane where it partially evaporates at the membrane interface. The pressure difference across the membrane and permeate side moves the generated vapor to the membrane for condensation and consequently distillate production. The hydrophobic feature of the membrane does not allow the feed stream to pass across the membrane. Instead only the gas phase (vapor) exists in the pores of membrane.

Table 7.4 presents input design parameters for a typical DCMD system. The outlet temperatures and permeate flux in the DCMD presented here are based on the model developed by Bui et al. (2010) and Lee et al. (2011). The resulting recovery ratio for a flat sheet membrane configuration was determined about 4.4%. Also, it was found that the pressure drop across the thin channel within the membrane is the main source of entropy generation and irreversibility. In Mistry et al. (2011), the normalized entropy generation rates associated with components of the DCMD system are presented. The exergy destruction of processes per kilogram of the produced

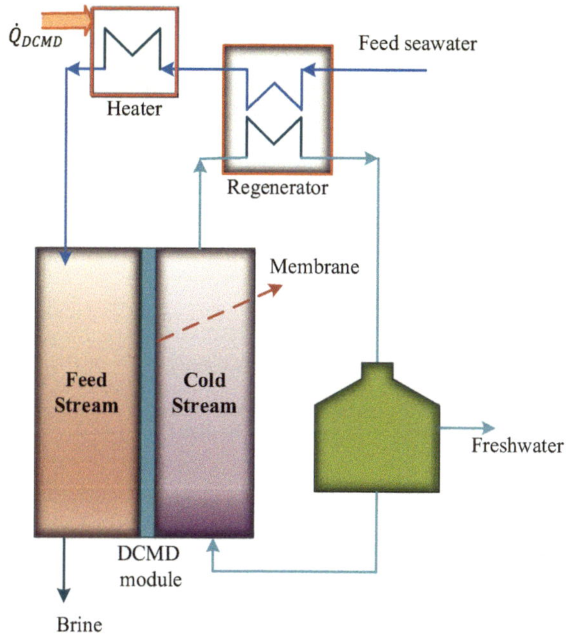

Fig. 7.11 ***Direct contact membrane desalination system.***

freshwater and the GOR and exergy efficiency of DCMD system are also presented in Table 7.4.

It can be observed in Table 7.4 that the DCMD module is the greatest contributor to the total exergy destruction rate of the DCMD system. This is mainly due to the diffusion across the pores and losses associated with the heat conduction in the thin membrane separator between the cold and hot streams. The substantial diffusion resistance is attributable to the very small pore size, which is 1000 times smaller than the thickness of the membrane. The exergy destruction in the heater occurs mainly because of the large temperature difference between the steam temperature, treated as the energy source, and the heater inlet stream. The smaller temperature difference and consequently the lower heat exchange rate in the regenerator results in a relatively lower exergy destruction rate for this component. Note that the specific exergy destruction increases with evaporation temperature. The exergy destruction rate of the pump is negligible, as for the other desalting methods discussed here. The lower temperature of the brine stream leaving the DCMD module compared to other thermal desalination methods results in a correspondingly lower entropy generation due to temperature disequilibrium. Also, the minor recovery process, which yields

Table 7.4 Typical design parameters, exergy destruction rates, and exergy efficiency for the DCMD system in Fig. 7.11.

Parameter	Value
DCMD module details[a]	
Length	10 m
Width	0.7 m
Flow channel depth	4 mm
Membrane coefficient	16×10^{-7} s/m
Feed stream details[b]	
Initial feed seawater temperature	27 °C
Feed mass flow rate	1 kg/s
Brine recovery mass flow rate	1 kg/s
Feed salinity	35 g/kg
Terminal temperature difference of regenerator heat exchanger	3 °C
Feed inlet temperature	85 °C
Heat source temperature	90 °C
Normalized exergy destruction rates[c]	
DCMD module	95.1 $kJ/kg_{Distillate}$
Heater	72.5 $kJ/kg_{Distillate}$
Temperature disequilibrium	63.4 $kJ/kg_{Distillate}$
Regenerator	44.9 $kJ/kg_{Distillate}$
Performance parameters	
Ψ_{DCMD}	1.0%

[a]Data from Bui et al. (2010), Fath et al. (2008), and Lee et al. (2011).
[b]Data from Bui et al. (2010) and Lee et al. (2011).
[c]Data modified from Mistry et al. (2011).

little distillate production from the feed input, leads to a negligible chemical disequilibrium of the brine. Accounting for all sources of irreversibilities, the relatively low exergy efficiency of the DCMD system (about 1.0%) is observed to be mainly attributable to the low recovery ratio and high discharge temperature of the DCMD system.

MVC desalination: A single-effect MVC desalination system is shown in Fig. 7.12 and mainly consists of a mechanical vapor compressor, a regenerator, and an evaporator/condenser heat exchanger. The generated vapor in the evaporator/condenser unit is compressed in the compressor, causing the saturation temperature of the vapor to increase to the superheated state. The hot compressed vapor enters the evaporator/condenser unit, which is a falling film shell and tube heat exchanger. In this unit, the sprayed feed stream is partially vaporized by condensing the hot compressed vapor inside the tubes, which yields the distillate water. The produced distillate

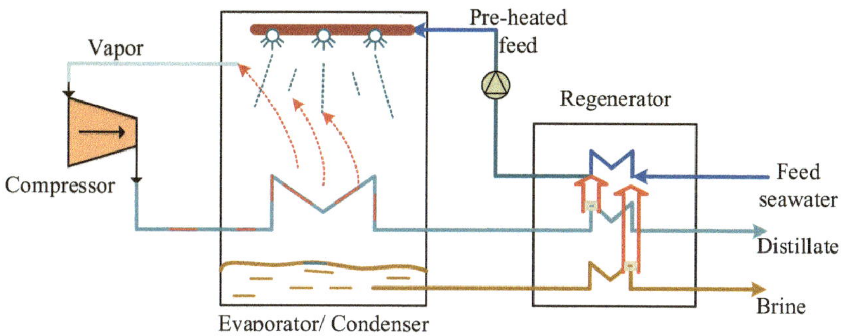

Fig. 7.12 ***Single effect mechanical vapor compression desalination system.***

and brine streams exit the evaporator/condenser unit and enter the regenerator to preheat the feed seawater input. The feed stream, which enters at a temperature of 25 °C and salinity of 35 g/kg, is split into two parts to receive heat from both the distillate and the brine, which are both at a temperature of 60 °C. The preheated feed streams are mixed and then enter the evaporator/condenser unit.

Typical input design characteristics for the MVC desalination system and the exergy destruction rate for its thermodynamic processes are tabulated in Table 7.5. The normalized entropy generation rates associated with components of the MVC desalination system are listed in Mistry et al. (2011). Several simplifying assumptions are considered in Table 7.5, for example, the average saturation pressure at the feed and brine salinities for the compressor inlet pressure, and a constant temperature difference between the brine and feed and between the freshwater and feed.

The compressor requires 8840Wh of electrical energy to produce 1000 L of freshwater. From Table 7.5, the temperature difference between the condensation and the evaporation processes is the main contributor to the exergy destruction. Sensible heat exchange within the regenerator is the main cause of exergy destruction in this component. The relatively smaller exergy destruction in the regenerator compared to the evaporator/condenser module is expected as entropy generation associated with the phase change evaporation/condensation process is much more than that associated with sensible heat transfer in the regenerator. The relatively high recovery ratio leads to notable irreversibilities associated with returning the concentrated brine to the seawater salinity. Also, as the regenerator performs effectively in bringing the outlet streams to the environmental temperature, the entropy generation attributable to the temperature disequilibrium is

Table 7.5 Typical design parameters, exergy destruction rates, and exergy efficiency of the MVC desalination system in Fig. 7.12.

Parameter	Value
Design parameters[a]	
Top brine temperature	60 °C
Seawater inlet temperature	25 °C
Brine salinity	53,330 ppm
Seawater inlet salinity	35,000 ppm
Product freshwater salinity	0.0 ppm
Compressor inlet pressure	19,400 Pa
Compression ratio	1.15
Recovery ratio	40%
Pinch point temperature difference in evaporator and condenser	3 °C
$\eta_{isen,comp}$	70%
Specific work required for compressor	8.84 kWh/m^3
Normalized exergy destruction rates[b]	
Evaporator/condenser	16.7 $kJ/kg_{Distillate}$
Compressor	8.176 $kJ/kg_{Distillate}$
Regenerator	3.212 $kJ/kg_{Distillate}$
Chemical disequilibrium	0.905 $kJ/kg_{Distillate}$
Temperature disequilibrium	0.2 $kJ/kg_{Distillate}$
Performance parameters	
Ψ_{MVC}	8.5%

[a]Data from Aly (1999) and Veza (1995).
[b]Data modified from Mistry et al. (2011).

relatively small. Note that the exergy destruction of the compressor strongly depends on its isentropic efficiency and compression ratio.

RO desalination: Fig. 7.13 shows a single-pass RO unit with a pressure exchanger. The feed seawater input at 25 °C and a mass flow rate of 1 kg/s is raised by feed pump to a pressure of 2 atm. Then the feed stream is split into two streams. One of these streams (40% of the feed) is transferred to the high-pressure pump, where it is pressurized further to 69 atm, which corresponds to the required pressure for the RO module. The other part of feed (60%) passes to the pressure exchanger and is then pressurized by the booster pump to an intermediate pressure of 65.4 atm. The outlet stream from the booster pump is then mixed with the outlet stream from the high-pressure pump before entering the RO module. Through the RO process, the distillate stream is produced at the low-pressure side of the RO module, which is at atmospheric pressure. The remaining brine stream exits the high-pressure side of the RO unit at a pressure of 67 atm rather than

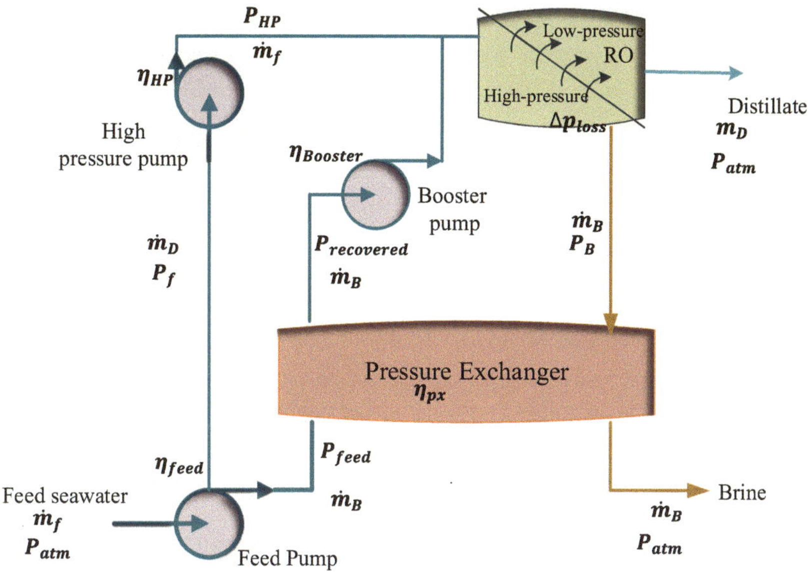

Fig. 7.13 ***RO desalination system assisted by a pressure exchanger module.*** *From Mistry et al. (2011).*

69 atm. This pressure difference accommodates the pressure drop due to viscous losses in the RO unit. Then the brine stream is transferred to the pressure exchanger, which is designed to permit two streams with the same volume flow rate to enter and exit.

The pressure exchanger device pressurizes a portion of feed utilizing the work rate produced from the expansion of the brine stream. In the pressure exchanger, the energy of the brine stream is transferred hydraulically to the feed stream, with an efficiency of 96%. The work rate produced through expansion and the work rate required for compression can be related by incorporating the isentropic efficiencies of expansion and pumping (Eqs. 7.39 and 7.41) for incompressible flows through the pressure exchanger. Then, the recovered pressure within the pressure exchanger can be determined as

$$p_{recovered} = P_f - \eta_{expansion}\eta_{pump}\left(\frac{\rho_f}{\rho_B}\right)(P_B - P_{atm}) \tag{7.50}$$

where ρ_f and ρ_B are the feed and brine densities, respectively; P_f, P_B, and P_{atm}, respectively, are the feed, brine, and atmospheric pressures; and $\eta_{expansion}$ and η_{pump}, respectively, are the isentropic efficiencies of the expansion and pumping processes.

For an RO desalination system operating under the conditions listed in Table 7.6, considering the pressurization work required by the pumps leads to accurate results. In fact, the temperature rise and the pressure loss corresponding to pump inefficiencies are very small. The analysis in this section draws extensively from the work of Mistry et al. (2011). The input design characteristics and assumptions are provided in Table 7.6. Seawater and pure water properties from section 5.2 are utilized.

For determination of exergy destruction rates via entropy generation rates in the RO module, both chemical and mechanical states of seawater need to be considered. This is because both are changing during the process. To account for those states, the process can be divided into two

Table 7.6 Design parameters, assumptions, exergy destructions, and exergy efficiency of the RO system in Fig. 7.13.

Parameter	Value
Design parameters	
Inlet feed seawater temperature, T_f	25 °C
Inlet feed seawater mass flow rate, $\dot{m}_f$	1 kg/s
Inlet feed seawater pressure, P_f	1 atm
Inlet feed seawater salinity, w_f	35,000 ppm
Distillate salinity, w_D	0.0 ppm
Recovery ratio	0.4
High pressure pump pressure, P_{HP}	69 atm
Brine pressure after RO unit, P_B	67 atm
Distillate pressure, P_D	1 atm
Pressure drop of from hydraulic friction, ΔP_{loss}	2 atm
Normalized exergy destruction rates	
RO module	3.17 $kJ/kg_{Distillate}$
Chemical disequilibrium	0.92 $kJ/kg_{Distillate}$
High pressure pump	1.156 $kJ/kg_{Distillate}$
Feed through pressure exchanger	0.185 $kJ/kg_{Distillate}$
Brine through pressure exchanger	0.19 $kJ/kg_{Distillate}$
Booster pump	0.12 $kJ/kg_{Distillate}$
Feed Pump	0.046 $kJ/kg_{Distillate}$
Assumptions	
Pump efficiency (for high pressure, boost and feed pumps)	85%
Isentropic efficiency of compression and expansion processes	98%
Performance parameters	
Ψ_{RO}	32%

Data modified from Banchik and Lienhard (2012) and Mistry et al. (2011).

subprocesses involving isobaric and isothermal separation into two streams with different compositions. Then, those two streams are expanded at constant salinity to find the membrane diffusion pressure drop (68 atm) and brine hydraulic pressure drop (2 atm). Consequently, the entropy generation rate within the RO module can be expressed as

$$\Delta \dot{S}_{gen}^{RO} = \Delta \dot{S}_{chem\ comp} + \Delta \dot{S}_{expansion} \tag{7.51}$$

where $\Delta \dot{S}_{chem\ comp}$ and $\Delta \dot{S}_{expansion}$, respectively, are the entropy rate changes associated with the chemical composition change and expansion at constant salinity.

The entropy rate change associated with the chemical composition change represents the least work rate required for creating this composition change. So, the entropy rate change for the chemical composition change can be written as follows:

$$T_0 \Delta \dot{S}_{chem\ comp} = \dot{W}_{least} = \Delta \dot{G}_{chem\ comp} \tag{7.52}$$

where the change in the Gibbs free energy rate can be expressed as

$$\Delta \dot{G}_{chem\ comp} = \Delta \dot{H}_{chem\ comp} - T_0 \Delta \dot{S}_{chem\ comp} \tag{7.53}$$

When applying Eq. (7.53) to seawater, the dependency on the pressure is eliminated as the seawater is an incompressible liquid. The entropy generation rate associated with the separation can be assessed by formulating thermophysical properties of seawater as a function of pressure, temperature, and salinity. For simplicity, it is assumed that the change in chemical concentration (x) occurs at the environmental temperature (T_0) and a constant high pressure (P_{HP}). This assumption yields the following equation for the entropy rate change associated with the chemical composition change:

$$\begin{aligned} \Delta \dot{S}_{chem\ comp} &= \dot{m}_D s_D \left(T_0, P_{HP}, x_D\right) + \dot{m}_B s_D \left(T_0, P_{HP}, x_D\right) \\ &\quad - \dot{m}_f s_f \left(T_0, P_{HP}, x_f\right). \end{aligned} \tag{7.54}$$

Then, using the entropy generation rate expression (Eq. 7.40) for the expansion of brine and distillate streams, and considering an isentropic efficiency of zero for a completely irreversible pressure drop, we can write the entropy rate change associated with expansion as follows:

$$\Delta \dot{S}_{i,expansion} \approx \dot{m}_i \left(\frac{v_i}{T_{i,in}} \left(P_{i,in} - P_{i,out}\right) \right) \tag{7.55}$$

where ν_i is the specific volume of stream i (brine and distillate), $T_{i,in}$ is the inlet temperature of brine/distillate in the expansion process for stream i, and $P_{i,in}$ and $P_{i,in}$, respectively, are the inlet and outlet pressures of the brine/distillate in the expansion process for stream i. The summation of the entropy rate change through the chemical composition change (Eq. 7.54) and the entropy generation rate associated with the depressurization of the brine and product streams (Eq. 7.55) gives the total entropy generation rate of the RO module and consequently the exergy destruction rate in this module (i.e., $\dot{E}x_{Des} = T_0\dot{S}_{gen}$).

Normalized entropy generation rates associated with components of the RO desalination system in Fig. 7.13 are given elsewhere (Mistry et al., 2011). Also, the specific exergy destructions for the RO desalination system are given in Table 7.6. It is seen that the RO module has the highest exergy destruction rate. The entropy generation rate is significantly greater for the depressurization of the freshwater stream rather than the brine stream. This indicates that the greatest source of exergy destruction is the diffusion of water across the membrane, which is mainly due to the large pressure drop of 68 atm. Although the same flow rate of water passes through the same pressure difference in the high-pressure pump, its relatively high isentropic efficiency of 85% leads to lower irreversibilities than those for the flow passage across the membrane.

The actual and minimum required work input, respectively, for the RO system are 8.5 kJ and 2.71 kJ per kilogram of distillate (or freshwater) produced. So, the exergy efficiency of the RO desalination system (i.e., the ratio of the minimum work to the actual work on a percentage basis) is 32%, which is higher than the other thermal desalination technologies considered in this section. Furthermore, reducing the recovery ratio or employing an osmotic power recovery device on the brine stream exiting the system can potentially reduce the irreversibilities associated with brine chemical disequilibrium.

Employing two or more stages in the RO desalination system significantly improves the efficiency of the system. A two-pass RO module in which each pass consists of two stages is shown in Fig. 7.14. In the first pass, brine from the first stage enters the second stage, where its concentration is reduced to yield potable water. The permeate from the first pass enters the first stage of the second pass to further remove salt, and then the brine stream enters the second stage of second pass to yield the final permeate and brine streams. The first stage operates at a relatively smaller recovery ratio, resulting in a lower concentration of the brine stream. Furthermore,

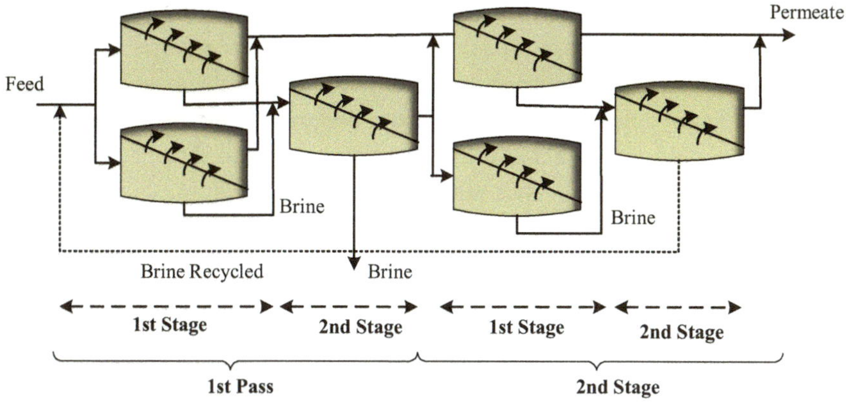

Fig. 7.14 ***Two-pass reverse osmosis process, where each pass contains two stages.***

the lower recovery ratio leads to a lower operating pressure at the first stage, as the osmotic pressure and feed salinity specify the pressure required for this stage. Then, in the second stage, additional water is extracted through the further pressurization of the brine stream from the first stage. The lower seawater flow rates through the high-pressure pump in the two-stage RO system compared to the flow rate in the single-stage RO system result in a lower entropy generation rate for the two-stage system (Elimelech and Phillip, 2011). In addition, the volume of water required for the pressurization process can be reduced through the batch processing of seawater. This approach is developed elsewhere (Desalitech Ltd, 2016).

7.6.3 Discussion

The exergy destruction rates and the exergy efficiencies of various desalination techniques are determined and discussed. The relatively low exergy efficiency of desalination systems, especially thermal desalination systems, indicates the major sources of lost work. Comparing the exergy efficiencies of desalination technologies demonstrates which system is closer to reversible operation. More importantly, perhaps, comparing the exergy efficiencies of a single desalination system at several operating conditions provides useful information for engineers to determine the irreversibilities and potentials for improvements.

For thermal desalination systems, heat transfer across the temperature difference in heat exchangers (such as evaporators, condensers, and heaters) is the main reason for the destruction of exergy. In these systems, for better recovery and utilization of exergy, heat recovery from brine discharges can be employed, and in some instances this can be coupled with heat pumps (or other energy systems). In all desalination technologies, after exergy destruction due to the temperature difference across the heat exchanger, the other main contributors to the exergy destructions from highest to lowest are: bringing brine and distillate temperatures to thermal equilibrium, compression, flashing, RO, throttling/expansion, pumping, and bringing brine concentration to chemical equilibrium. In the MSF and MED systems, increasing the number of stages/effects could increase the exergy efficiency but at a higher capital cost that may not be viable. The possible use of low-temperature and low-quality steam can reduce the exergy losses attributable to the current use of high-quality steam in the MSF and MED systems. For membrane-based desalination systems (such as RO and MD), the membrane is the main contributor to the exergy destruction. Increasing the permeability of the membrane can improve the exergy efficiency of these systems.

7.7 Closing remarks

The application of second law analysis to desalination systems, mainly through the use of exergy analysis, is described in this chapter. The exergy efficiency of a desalting process is often defined as the ratio of useful work output to the total work input to the control volume. It is shown that the useful work output from a desalination system is determined when all input and output streams, except freshwater, are at the total dead state (i.e., environmental temperature, pressure, and constituent concentration), while the freshwater product is in the restricted dead state (ambient temperature and pressure). The exergy input to a desalting process can be in the form of work, heat, fuel, or combination of them. In the combined operation of a desalination system with a thermal power plant or any other energy system, the exergy input to the desalination system is considered to be the exergy of each form of energy input after tracing it back to the primary source. However, work-driven desalination technologies have higher exergy efficiencies than thermally driven desalination systems, some factors such as existing infrastructure, feed quality, cost, and robustness, to varying conditions, make thermal desalination more favorable. Examining the entropy generation of a desalting process provides useful information for determining the main sources of lost work. To provide an accurate entropy-generation determination for a desalting process, it is necessary to take into account the entropy generation due to the temperature and chemical disequilibrium of product streams (i.e., brine and freshwater) with the environment when considering the irreversibilities in the desalting process. It is shown that the entropy generation associated with chemical disequilibrium of product streams is notable in desalination techniques with high recovery ratios, whereas the entropy generation associated with temperature disequilibrium of the product streams is significant in thermal desalination systems. Recovering the potential energy difference between the product streams and the environment can potentially reduce the energy input to the desalination system.

The relatively low exergy efficiencies of desalination technologies implies that there are significant sources of exergy destructions and losses. When understood, these can help in developing potential improvements in the thermodynamic operation of these systems. Adapting desalination technologies for low temperature heat sources, identifying heat recovery potentials (e.g., heat recovery from brine before rejecting it to the environment), recirculation of brine, and integration of desalination with other energy systems can potentially improve the exergy efficiencies of desalination systems.

The material presented in this chapter complements the numerous studies of desalination processes based on the second law analysis, particularly using exergy methods. These include assessments of desalination based on the second law analysis (Akbari Kordlar et al., 2014; Farsi and Dincer 2019; Orhan et al., 2010; Hamut et al., 2014; Khalid et al., 2016a, 2016b; Hamrang et al., 2020; Chitgar et al., 2019), in addition to investigations using economic and environmental assessments based on exergy (Keshavarzzadeh et al., 2020; Haghghi et al., 2019), as well as exergy-based optimization applications (Maleki et al., 2016; Dincer et al., 2017; Wu et al., 2018; Salehi et al., 2018; Peng et al., 2018).

Nomenclature

c	specific heat of an incompressible fluid (kJ/(kg.K))
c_p	specific heat at constant pressure (kJ/(kg.K))
ex	specific exergy (kJ/kg)
Ex	exergy rate
g	specific Gibbs free energy (kJ/kg)
GOR	gained output ratio
$\bar{h}$	molal enthalpy (kJ/mol)
h_{fg}	specific latent heat of vaporization (kJ/kg)
$\dot{H}$	Enthalpy rate (kW)
M	molecular weight (g/mol)
$\dot{m}$	mass flow rate (kg/s)
$\dot{N}$	mole flow rate (mol/s)
P	pressure (kPa)

PR	performance ratio
$\dot{Q}$	heat transfer rate (kW)
R	gas constant (kJ/mol. K)
s	specific entropy (kJ/kg. K)
s_{fg}	specific entropy change of vaporization (kJ/kg. K)
$\dot{S}$	entropy rate (kW)
$\dot{S}_{gen}$	entropy generation rate (kW/K)
SEC	specific electricity consumption (kW/m^3)
T	temperature (°C, K)
T_0	reference environment temperature (°C, K)
v	specific volume (m^3/kg)
W	work energy (kJ)
$\dot{W}$	work rate (kW)
w_s	mass fraction of salt (g/kg, ppm)
x	mole fraction

Greek letters

η_{isen}	isentropic efficiency
Ψ	exergy efficiency
ρ	density (kg/m^3)

Subscripts

act	actual
B	brine
D	distillate, desalination
dd	dilute draw solution
Des	destruction
cd	concentrated draw
$comp$	compressor
$chem$	chemical
f	feed
gen	generator, generation
lw	liquid water
p	permeate
RG	regenerator
sw	seawater
TDS	total dead state
v	vapor
w	water

Superscripts

T	temperature
$chem$	chemical

Acronyms

CAOW	Closed air open water
DCMD	Direct contact membrane desalination
FO	Forward osmosis
HDH	Humidification–dehumidification
MD	Membrane desalination
MED	Multi-effect desalination
MSF	Multistage flash
MVC	Mechanical vapor compression
RO	Reverse osmosis
SEC	Specific electricity consumption
TVC	Thermal vapor compressor

References

Akbari Kordlar, M., Mahmoudi, S.M.S., Yari, M., Rosen, M.A., 2014. Energy and exergy analyses of a new combined cycle for producing electricity and desalinated water using geothermal energy. Sustainability 6 (4), 1796–1820.

Aly, S.E., 1999. Gas turbine total energy vapour compression desalination system. Energy Convers. Manage. 40 (7), 729–741.

Banchik, L.D., Lienhard, J.H., 2012. Thermodynamic analysis of a reverse osmosis desalination system using forward osmosis for energy recovery. In: ASME International Mechanical Engineering Congress and Exposition 45226. American Society of Mechanical Engineers, pp. 1017–1029.

Bejan, A., 1978. General criterion for rating heat-exchanger performance. Int. J. Heat Mass Transfer 21 (5), 655–658.

Bejan, A., 2016. Advanced Engineering Thermodynamics, Fourth ed. John Wiley & Sons, NJ.

Bromley, L.A., Singh, D., Ray, P., Sridhar, S., Read, S.M., 1974. Thermodynamic properties of sea salt solutions. AIChE J. 20 (2), 326–335.

Bui, V.A., Vu, L.T., Nguyen, M.H., 2010. Modelling the simultaneous heat and mass transfer of direct contact membrane distillation in hollow fibre modules. J. Membr. Sci. 353 (1-2), 85–93.

Cerci, Y., Cengel, Y., Wood, B., Kahraman, N., Karakas, E.S., 2003. Improving the thermodynamics and economic of desalination plants: minimum work required for desalination and case studies of four working plants. National Technical Information Service (NTIS), U.S. Department of the Interior. Report No. 78.

Chitgar, N., Emadi, M.A., Chitsaz, A., Rosen, M.A., 2019. Investigation of a novel multigeneration system driven by a SOFC for electricity and fresh water production. Energy Convers. Manage. 196, 296–310.

Darde, V., van Well, W.J., Stenby, E.H., Thomsen, K., 2010. Modeling of carbon dioxide absorption by aqueous ammonia solutions using the extended UNIQUAC model. Ind. Eng. Chem. Res. 49 (24), 12663–12674.

Darwish, M.A., Al-Juwayhel, F., Abdulraheim, H.K., 2006. Multi-effect boiling systems from an energy viewpoint. Desalination 194 (1-3), 22–39.

Desalitech Ltd, 2016. Doing away with RO energy-recovery devices, water desalination reuse. https://www.desalination.biz/desalination/doing-away-with-ro-energy-recovery-devices/ (Accessed 16 June 2021).

Dincer, I., Rosen, M.A., Ahmadi, P., 2017. Optimization of Energy Systems. Wiley, London.

El-Dessouky, H.T., Ettouney, H.M., 2002. Fundamentals of Salt Water Desalination. Elsevier, Amsterdam, The Netherlands.

Elimelech, M., Phillip, W.A., 2011. The future of seawater desalination: energy, technology, and the environment. Science 333 (6043), 712–717.
Farsi, A., Dincer, I., 2019. Development and evaluation of an integrated MED/membrane desalination system. Desalination 463, 55–68.
Fath, H.E., Elsherbiny, S.M., Hassan, A.A., Rommel, M., Wieghaus, M., Koschikowski, J., Vatansever, M., 2008. PV and thermally driven small-scale, stand-alone solar desalination systems with very low maintenance needs. Desalination 225 (1-3), 58–69.
Fritzmann, C., Löwenberg, J., Wintgens, T., Melin, T., 2007. State-of-the-art of reverse osmosis desalination. Desalination 216 (1-3), 1–76.
Haghghi, M.A., Shamsaiee, M., Holagh, S.G., Chitsaz, A., Rosen, M.A., 2019. Thermodynamic, exergoeconomic, and environmental evaluation of a new multi-generation system driven by a molten carbonate fuel cell for production of cooling, heating, electricity, and freshwater. Energy Convers. Manage. 199, 112040.
Hamrang, F., Shokri, A., Mahmoudi, S.M.S., Ehghaghi, B., Rosen, M.A., 2020. Performance analysis of a new electricity and freshwater production system based on an integrated gasification combined cycle and multi-Effect desalination. Sustainability 12 (19), 7996.
Hamut, H.S., Dincer, I., Rosen, M.A., 2014. Assessment of desalination technologies integrated with renewable energy sources in Turkey. In: Dincer, I., Midilli, A., Kucuk, H. (Eds.), Progress in Exergy, Energy, and the Environment. Springer, New York, pp. 279–292.
Keshavarzzadeh, A.H., Ahmadi, P., Rosen, M.A., 2020. Technoeconomic and environmental optimization of a solar tower integrated energy system for freshwater production. J. Clean. Prod. 270, 121760.
Khalid, F., Dincer, I., Rosen, M.A., 2016a. Analysis and assessment of a gas turbine-modular helium reactor for nuclear desalination. J. Nucl. Eng. Radiat. Sci. 2 (3), 031014.
Khalid, F., Dincer, I., Rosen, M.A., 2016b. Comparative assessment of CANDU 6 and sodium-cooled fast reactors for nuclear desalination. Desalination 379, 182–192.
Lee, H., He, F., Song, L., Gilron, J., Sirkar, K.K., 2011. Desalination with a cascade of cross-flow hollow fiber membrane distillation devices integrated with a heat exchanger. AIChE J. 57 (7), 1780–1795.
Maleki, A., Gholipour Khajeh, M., Rosen, M.A., 2016. Weather forecasting for optimization of a hybrid solar-wind-powered reverse osmosis water desalination system using a novel optimizer approach. Energy 114, 1120–1134.
Miller, J.E., 2003. Review of Water Resources and Desalination Technologies. Sandia National Laboratories, Albuquerque, 2003-0800.
Mistry, K.H., Lienhard, J.H., 2013. Generalized least energy of separation for desalination and other chemical separation processes. Entropy 15 (6), 2046–2080.
Mistry, K.H., McGovern, R.K., Thiel, G.P., Summers, E.K., Zubair, S.M., Lienhard, J.H., 2011. Entropy generation analysis of desalination technologies. Entropy 13 (10), 1829–1864.
Mistry, K.H., Mitsos, A., 2011. Optimal operating conditions and configurations for humidification–dehumidification desalination cycles. Int. J. Therm. Sci. 50 (5), 779–789.
Mistry, K.H., Zubair, S.M., 2010. Effect of entropy generation on the performance of humidification-dehumidification desalination cycles. Int. J. Therm. Sci. 49 (9), 1837–1847.
Moghimi, M., Emadi, M., Akbarpoor, A.M., Mollaei, M., 2018. Energy and exergy investigation of a combined cooling, heating, power generation, and seawater desalination system. Appl. Therm. Eng. 140, 814–827.
Moradi, M., Ghorbani, B., Shirmohammadi, R., Mehrpooya, M., Hamedi, M.H., 2019. Developing of an integrated hybrid power generation system combined with a multi-effect desalination unit. Sustain. Energy Technol. Assess. 32, 71–82.
Orhan, M.F., Dincer, I., Naterer, G.F., Rosen, M.A., 2010. Coupling of copper-chloride hybrid thermochemical water splitting cycle with a desalination plant for hydrogen production from nuclear energy. Int. J. Hydrogen Energy 35, 1560–1574.

Peng, W., Maleki, A., Rosen, M.A., Azarikhah, P., 2018. Optimization of a hybrid system solar-wind-based water desalination by reverse osmosis: comparison of approaches. Desalination 442, 16–31.
Plus, A., 2013. Aspen Physical Property System: Physical Property Methods, Aspen Technology Inc., http://profsite.um.ac.ir/~fanaei/_private/Property%20Methods%20 8_4.pdf (Accessed 16 June 2021).
Rahimi, M., Hamedi, M.H., Amidpour, M., 2017. Thermodynamic simulation and economic modeling and optimization of a multi generation system partially fed with synthetic gas from gasification plant. Gas Process. J. 5 (2), 49–68.
Salehi, S., Mahmoudi, S.M.S., Yari, M., Rosen, M.A., 2018. Multi-objective optimization of two double-flash geothermal power plants integrated with absorption heat transformation and water desalination. J. Clean. Prod. 195, 796–809.
Sharqawy, M.H., Lienhard, J.H., Zubair, S.M., 2010. Formulation of seawater flow exergy using accurate thermodynamic data, Proc. ASME International Mechanical Engineering Congress and Exposition, 44298. American Society of Mechanical Engineering, pp. 675–682.
Spiegler, K.S., El-Sayed, Y.M., 2001. The energetics of desalination processes. Desalination 134 (1-3), 109–128.
Stover, R.L., 2007. Seawater reverse osmosis with isobaric energy recovery devices. Desalination 203 (1-3), 168–175.
Veza, J., 1995. Mechanical vapour compression desalination plants: a case study. Desalination 101 (1), 1–10.
Wagman, D.D., Evans, W.H., Parker, V.B., Schumm, R.H., Halow, I., 1982. The NBS tables of chemical thermodynamic properties. Selected values for inorganic and C1 and C2 organic substances in SI units. National Standard Reference Data System, J. Phys. Chem. Ref. Data 11 (2), 2–391.
Wu, B., Maleki, A., Pourfayaz, F., Rosen, M.A., 2018. Optimal design of stand-alone reverse osmosis desalination driven by a photovoltaic and diesel generator hybrid system. Sol. Energy 163, 91–103.

CHAPTER EIGHT

Seawater desalination systems using sustainable energy technologies

8.1 Introduction

To address energy and water issues as well as concerns about environmental impacts, especially climate change, practices for employing sustainable energy technologies, particularly those based on renewable energy, are growing in importance. The utilization of sustainable energy technologies with water desalination systems is of interest to many. The combination of sustainable energy sources with technologies for reuse, recovery, and recycle of water and energy can provide an environmentally beneficial and sustainable approach to meeting energy and water demands.

Many areas related to sustainable energy have been the subject of investigations and advances in recent years. For instance, Kilkis et al. (2020) and Rezaie and Rosen (2020) have investigated the importance of water systems and their integration with energy and environment systems. Renewable energy options and their applications have been reviewed (Koohi-Fayegh and Rosen, 2020a; El Haj Assad and Rosen, 2021). Specific studies have been undertaken of solar thermal energy (Taslimi et al., 2021 expected) and photovoltaics (PVs) (Yu et al., 2021), geothermal energy based on hot sources (Jalili et al., 2021) and ambient ground sources (Rosen and Koohi-Fayegh, 2017; Alavy et al., 2021), biomass energy (Cuomo et al., 2018; Habibollahzade et al., 2021), and nuclear energy (Rosen, 2021). Technologies that facilitate the use of sustainable energy sources have also received much attention, such as energy storage (Koohi-Fayegh and Rosen, 2020b; Dincer and Rosen, 2021) and cogeneration and district energy (Rosen and Koohi-Fayegh, 2016).

In this chapter, industrial desalination technologies, including membrane-based and phase-change processes, are integrated with one or more sustainable energy sources. Sources such as solar, wind, geothermal, and nuclear energies are considered. Various configurations of sustainable desalination processes are presented, compared, and discussed in terms of

Sustainable Energy Technologies for Seawater Desalination
DOI: https://doi.org/10.1016/B978-0-323-99872-7.00010-3

operational conditions as well as performance and efficiency. A number of pilot renewable energy–driven desalination systems from various parts of the world are examined. Furthermore, an illustrative example is provided in which a renewable energy–driven desalination system is analyzed from thermodynamic and environmental viewpoints. The example demonstrates how thermodynamic principles (i.e., energy and exergy concepts) permit the behavior to be determined for a renewable energy–driven desalination system and how the behavior relates to environmental impact assessments.

Many studies integrating desalination with sustainable energy technologies have been reported in recent years, reinforcing the importance of this approach. These include investigations of desalination based on solar energy (Wu et al., 2018; Keshavarzzadeh et al., 2020), biomass (Hamrang et al., 2020), geothermal energy (Akbari Kordlar et al., 2014; Salehi et al., 2018), and nuclear energy (Orhan et al., 2010; Hamut et al., 2014; Khalid et al., 2016a, 2016b; Rosen, 2021). In addition, studies have been undertaken of hybrid desalination systems that use more than one energy source, such systems based on solar and wind (Maleki et al., 2016; Peng et al., 2018), of integrated systems using fuel cells ((Haghghi et al., 2019; Chitgar et al., 2019), and of multigeneration systems based on biomass using multieffect desalination (MED) (section 10.5 of Dincer et al. (2017)) and reverse osmosis (RO) (section 11.3.1 of Dincer et al. (2017)).

8.2 Solar energy

Today, there are numerous industrial-scale solar plants with capacities on the order of megawatts for heat and electricity production purposes. The first commercial, large-scale solar plant was established in New Mexico, United States in 1979; it included more than 220 heliostats and had a thermal energy production capacity of 5 MW (Kalogirou, 2005). The capacity became larger, at 35 MW of thermal output, for a second large-scale solar plant in California, United States (Kalogirou, 2005, 2004). The high-temperature superheated steam (at 400°C) produced from such large-scale solar plants can be utilized in heat- or work-driven desalination systems to produce freshwater in conjunction with electricity and heat, simultaneously.

A global map of daily and annual direct normal irradiance (DNI) values, based on long-term averaged data, is shown in Fig. 8.1. It can be seen that several regions, including the Middle East, the southwest United States, North Africa, Australia, India, Spain, Chile, and southern China, experience

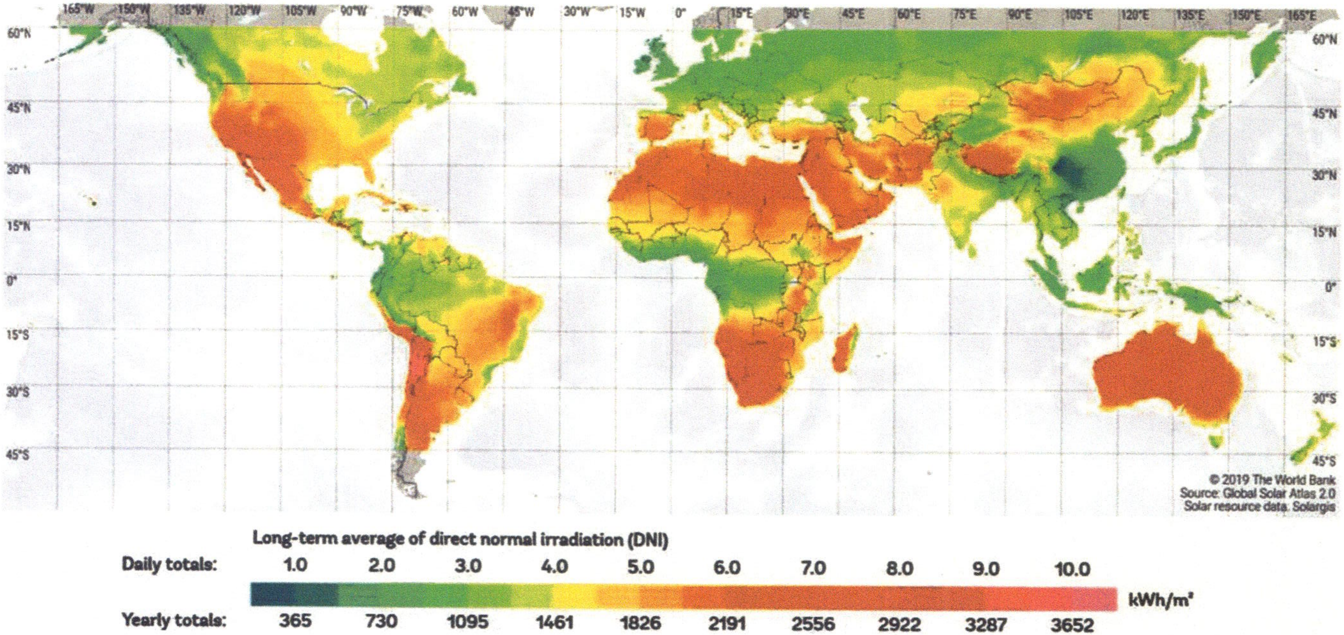

Fig. 8.1 ***Global solar atlas for daily and annual direct normal irradiance (kWh/m^2) based on long-term averaged data.*** Obtained from Global Solar Atlas 2.0, a free, web-based application developed and operated by the company Solargis s.r.o. on behalf of the World Bank Group, utilizing Solargis data, with funding provided by the Energy Sector Management Assistance Program (ESMAP, 2019). For additional information: https://globalsolaratlas.info.

relatively high annual values of DNI, ranging from 2000 kWh/m^2 to over 3000 kWh/m^2. Most of these regions are desert or semidesert areas, often having water scarcities. So, the installation of solar desalination facilities in these regions can be advantageous for meeting freshwater demands.

Solar energy can be utilized in seawater desalination systems either by generating electrical energy (mainly through solar thermal power plants and/or PV cells) to operate a work-driven distillation process (such as RO, mechanical vapor compression (MVC), and electrodialysis (ED)), or by producing heat (through solar collectors and/or solar ponds) to drive phase-change desalting processes (such as MED and multistage flash (MSF) desalination, humidification–dehumidification (HDH) desalination, and membrane desalination (MD)). Solar desalination systems can be categorized as direct and indirect collection systems. In direct collection systems, freshwater is produced directly in the solar collector, while in indirect collection systems two separate subsystems are used, one for collecting solar energy (e.g., PV, solar collector) and the other for delivering the collected energy to the desalination system.

In addition, direct collection systems use solar energy when it is available, whereas indirect collection systems collect solar energy in the form of steam, hot water, or another heated substance, which is treated as an indirect medium. Energy storages can be employed to store energy for use during periods of low insolation, but they sometimes can be expensive. Electricity from the grid or backup boilers are alternatives to energy storage. In a solar desalination system, storage devices, solar energy collectors and desalination subsystems are the main contributors to the cost, and strongly affect the optimum performance condition for the integrated system.

The use of multigeneration for on-site production has been extensively investigated. This can be observed in the Middle East and North Africa (MENA) regions, for instance, where there are extensive demands for cooling and water and 50% of the electrical load is used for air-conditioning due to the hot and arid weather conditions. For example, consider the Aqaba Solar Water Project (Trieb and Muller-Steinhagen, 2009). This integrated system includes a power plant, compression and absorption chillers, and a thermal desalination system, and is intended for on-site production. The absorption chiller provides cooling for hot seasons and the compression chiller provides cooling for peak periods and intermittent demands. The power plant and desalination systems provide, respectively, electricity and freshwater. It is shown for this system that the application of solar energy reduces the fuel input to the overall system by up to 35%. Fig. 8.2

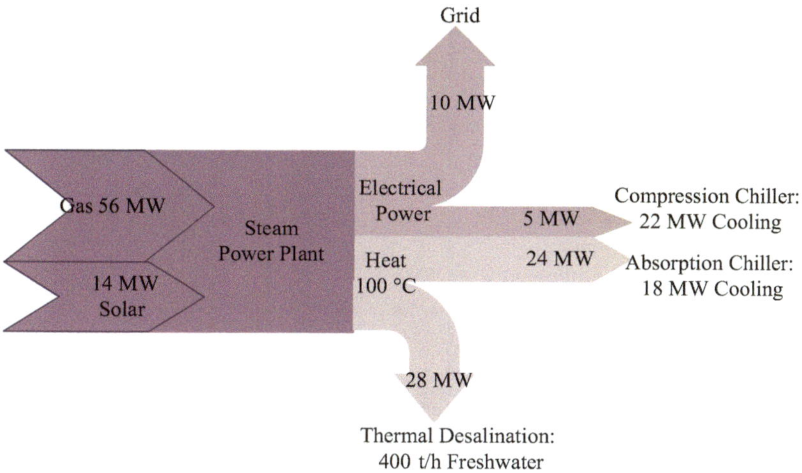

Fig. 8.2 ***Process flow diagram for the multigeneration of electricity, cooling and freshwater through an integrated system comprised a steam power plant, absorption and compression chillers, and thermal desalination systems.*** *Modified from (Trieb and Muller-Steinhagen, 2009).*

shows an energy flow diagram of this multigeneration system, highlighting the electrical, cooling, and freshwater products, for an on-site application.

8.2.1 Solar thermal collectors

Solar thermal collectors receive solar radiation, convert it to the thermal energy, and transform it to the internal energy of a transport medium (commonly water, air, or oil) circulating through the collectors. The collected thermal energy is used directly for a determined application, commonly space heating or hot water, or is stored in thermal energy storage (TES), typically for recovery during periods of low solar radiation such as cloudy days and nights.

Generally, solar collectors can be classified into concentrating and nonconcentrating (often stationary) collectors. Table 8.1 lists various commercial solar thermal collectors. For nonconcentrating collectors, the intercepting and absorbing areas for solar radiation are identical. Concentrating solar collectors normally provide higher temperatures than flat collectors. For concentrating collectors, the energy delivery temperature is usually enhanced by reducing the heat loss area. The receiver absorbs the radiation and converts it to another form of energy (usually thermal energy). It consists of an absorber, insulation and cover materials. The optical system

Table 8.1 Commercial solar thermal collectors.

Collector category	Collector type	Absorber type/ configuration	Concentration ratio	Provided temperature range
Nonconcentrating collector				
	Evacuated tube collector	Flat	1	50–200 °C
	Flat plate collector	Flat	1	30–80 °C
Concentrating collector				
	Linear Fresnel reflector	Tubular	10–40	60–250 °C
	Compound parabolic collector	Tubular	1–15	60–300 °C
	Cylindrical through collector	Tubular	10–50	60–300 °C
	Parabolic through collector	Tubular	15–45	60–300 °C
	Heliostat field collector	Point	100–1500	150–2000 °C
	Parabolic dish reflector	Point	100–1000	100–500 °C

Data from Kalogirou (2013).

or concentrator conveys radiation to the receiver. Concentrators can be refractor or reflector types, surfaces of revolution or cylindrical, and segmented or continuous. Receivers can be covered or uncovered, as well as flat, convex or concave. Concentrators yield higher temperatures and can operate in conjunction with storages. Concentrators typically require solar trackers to ensure radiation is properly directed onto the absorber, and special maintenance to maintain the optical system quality and to avoid oxidization as well as the buildup of dirt and corrosive atmospheric components. The relatively higher operating and maintenance costs associated with solar concentrators often limit their utility compared to flat-plate collectors (Duffie et al., 2013). Fig. 8.3 shows the main concentrating solar collectors typically used in electricity production and desalination plants.

8.2.2 Solar collector thermal performance

For solar thermal collectors, the conversion of incoming solar energy to useful energy can be assessed through an energy rate balance and used for

Fig. 8.3 ***Main types of concentrating solar collectors for high-temperature production electricity generation and desalination.*** parabolic trough (upper left) (DOE, 2016a), solar tower (upper right) (DOE, 2016b), linear Fresnel (bottom left) (DOE, 2021), and dish Stirling (bottom right) (DOE, 2021).

determining the solar collector efficiency and performance. The incident radiation less optical losses is referred to the absorbed solar radiation. These terms are often provided per unit area of absorber surface, for example, the absorbed solar radiation rate per unit area is $S(\mathrm{W/m^2})$. A portion of solar thermal energy is lost from the collector to the environment through conduction, convection, and infrared radiation (Fig. 8.4). This energy loss rate can be assessed using an overall heat transfer coefficient $U_L(\mathrm{W/m^2K})$. So, under steady-state conditions, the useful output thermal energy rate from a collector with the surface area A_c, expressed as the difference between the solar absorbed radiation rate and the thermal loss rate, can be written as (Duffie et al., 2013)

$$\dot{Q}_{useful} = A_c\left(S - U_L\left(\overline{T_p} - T_0\right)\right) \tag{8.1}$$

where T_0 and $\overline{T_p}$, respectively, are the environmental temperature and mean absorber plate temperature in °C or K. The mean absorber plate temperature ($\overline{T_p}$) is a function of the incident solar radiation, the collector design, and the inlet working fluid condition. As evaluation of the absorber temperature is not simple, Eq. (8.1) can be written in terms of fluid inlet

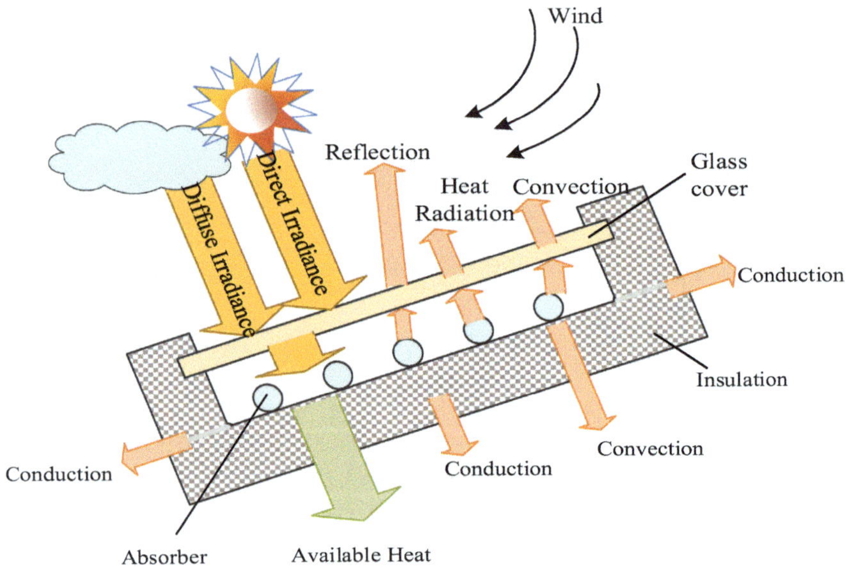

Fig. 8.4 ***Typical flat plate solar collector and thermal interactions.***

temperature and collector heat removal factor. The collector heat removal factor is used to allow for substitution of inlet working fluid temperature for absorber plate temperature. It accounts for the flow rate of fluid, inherent properties of the transfer fluid, and the collector-to-fluid interface. The collector removal factor is determined through basic principles and experimental measurements, as addressed by Duffie et al. (2013).

A proper thermal loss model for concentrating collectors varies for geometry of each receiver. As an example of calculating the loss coefficient and thermal losses, consider a receiver with an uncovered cylindrical absorbing tube. Assuming no temperature difference around the tube of the receiver, the thermal loss rate per unit area of receiver can be expressed in terms of conduction through the support structure, and convection and radiation from the surface. That is,

$$\frac{\dot{Q}_{loss}}{A_r} = U_L\left(T_r - T_0\right) = h_w\left(T_r - T_0\right) + \varepsilon\sigma\left(T_r^4 - T_{sky}^4\right) + U_{cond}\left(T_r - T_0\right) \quad (8.2)$$

where T_r is the receiver temperature, ε is the collector emissivity, and σ is the Stefan–Boltzmann constant. The parameter h_w is the wind convection coefficient that can be determined through fluid mechanic principles. For a flow of air across a single tube exposed to the environment, the Nusselt number ($Nu = h_w L/k$) can be expressed by the modified McAdams equation as (McAdams, 1942)

$$Nu = \begin{cases} 0.4 + 0.54Re^{0.52} & \text{for } 0.1 < Re < 1000 \\ 0.3Re^{0.6} & \text{for } 1000 < Re < 50{,}000 \end{cases} \tag{8.3}$$

where L is the length of tube in m, k is the thermal conductivity of the fluid in W/(m.K), and Re is the Reynolds number.

The thermal efficiency of the solar collector can be defined as the ratio of the useful gained thermal energy rate over a specific period to the incident solar energy rate (G_T (W/m^2)) for that period, as follows:

$$\eta = \frac{\int \dot{Q}_{useful}}{A_c \int G_T dt} \tag{8.4}$$

8.2.3 Concentrating solar power plants for desalination

A concentrating solar power (CSP) plant can provide high-temperature thermal energy for various uses, including electricity generation via an electrical generation plant. Often CSP systems use mirrors (or heliostats) that track the sun's position to provide solar concentration. In CSP technologies, the receiver is usually designed in a way that avoids excessive thermal losses when the sunlight is concentrated on the receiver area. The collected thermal energy is transferred to a steam power plant, when that is its destination, via a heat transfer fluid (e.g., oil, water, molten salt, air). That energy is used to raise high temperature and pressure steam for steam turbine.

CSP facilities using solar tower, parabolic trough, and linear Fresnel collectors can potentially produce 5 MW to over 200 MW of electrical power with a thermal energy efficiency between 30% and 40% (Trieb, 2007). The solar to electricity conversion efficiency in these collectors is normally between 10% and 20% (IEA, 2010). The solar to electrical power efficiency embodies both the solar to heat conversion efficiency and heat to electricity conversion efficiency.

In the daytime, the thermal energy generated by solar collectors often exceeds the heat required for the steam power plant. Then, the additional collected heat can be stored in a TES, which can contain such storage materials as molten salt, concrete, and ceramics. During periods of low solar radiation (e.g., at night and on cloudy days), the stored energy can be recovered to operate the power cycle. However employing TESs in the power plants increases the investment cost, even though they enhance the capacity of electricity production to meet the electricity demand. Often, using TES with a power plant system to adapt to the electricity demand is more economic than increasing the turbine capacity. Moreover, TES permits continuous electrical power production, thereby avoiding technical fluctuations in power plant operation.

Alternatively, fossil fuels and renewable fuels such as biomass can be used as a backup energy supply for the power cycle when needed. The continuous operation of a solar power plant becomes important when combining it with a desalination system, because desalination system cannot easily operate with intermittent energy inputs. Hybrid fossil fuel– and solar energy–driven power cycles are undergoing research to raise the portion of solar thermal energy input to the system and to reduce the capital and maintenance costs of the solar facilities.

Table 8.2 presents total freshwater demand, CSP demand, CSP potential, and CSP-desalination combined system demand in the MENA region. The CSP electricity production potential is calculated based on the annual

Table 8.2 Total freshwater, CSP electrical, and combined CSP-desalination demands in arid countries.

Country	Total freshwater demand 2050 ($\times 10^3$ millions m^3/y)	CSP demand 2050 (TWh/y)	Total CSP potential (TWh/y)	CSP-desalination demand 2050 ($\times 10^3$ millions m^3/y)
Algeria	13.6	120	135,771	2.1
Bahrain	0.4	4	16	0.3
Egypt	92.6	395	57,140	6.7
Iran	114.9	290	32,134	10.6
Iraq	74.2	180	24,657	13.3
Israel	6.4	29	151	4.4
Jordan	2.5	38	5884	1.5
Kuwait	1.3	13	1372	0.7
Lebanon	1.9	5	5	0.3
Libya	6.2	25	82,714	2.8
Morocco	26.1	110	8428	8.5
Oman	3.1	14	14,174	2.0
Qatar	0.6	8	555	0.4
Saudi Arabia	37.2	135	75,832	23.7
Syria	25.3	135	8449	0.9
Tunisia	6.5	34	5673	2.6
UAE	3.3	40	447	2.4
Yemen	24.2	65	8486	13.1

Modified from Moser et al. (2013).

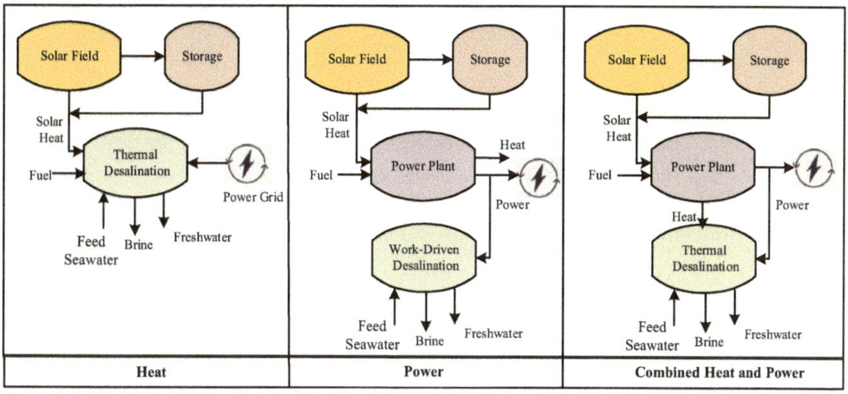

Fig. 8.5 ***Three configurations for thermal/work driven desalination systems powered by a concentrating solar power (CSP) plant.*** Left: the produced heat from solar field is directly used in a thermal desalination plant. Center: the produced power from CSP is used for a work-driven desalination plant. Right: the waste heat from CSP is recovered and used in a thermal desalination plant.

DNI and the average annual electricity-generation efficiency of 15%. The RO, MED, or both desalination techniques can be selected to combine with the CSP plant. It can be seen that the potential of concentrating power technology in the most of the listed countries is sufficiently enough for the future demand of freshwater. Therefore, in most parts of the MENA region, there is sufficient CSP potential to cover the growing demand for electricity and additional energy demand for freshwater production in desalination plants.

Fig. 8.5 shows several configurations for combining a desalination plant with a solar field. On the left side of the figure, the thermal desalination plant directly uses thermal energy produced and stored in the solar field. In the middle, the RO or other work-driven desalination system consumes a part of electricity produced from the solar power plant (commonly a CSP plant for cogeneration of electrical energy and water. On the right side of the figure, the waste heat from the solar power plant is recovered and used in the thermal desalination plant. Note that, in some cases, the desalination system should be large enough to provide sufficient water for both freshwater demand and cooling of the power plant.

Table 8.3 lists selected solar-assisted MED and MSF desalination systems operating in different regions and data for the systems. In the Platform a Solar de Almeria (SPA) site of Spain, a forward-feed MED system with 14 effects and a freshwater production capacity of 3 m^3/h is used (Gregorzewski et al., 1991). In the first phase of this project, the reliability

Table 8.3 Data for a selection of installed solar-assisted MED and MSF desalination plants.

Plant type	Location	Annual average global radiation (kWh/m^2/d)	DNI average radiation (kWh/m^2/d)	Freshwater production capacity (m^3/d)	Solar collector type	Operation temperature of solar collector (°C)	Solar collector area (m^2)
MED	Abu Dhabi, UAE	5.61	6.41	100	ETC[a]-PV	60–80	2500–12,500[b]
MED	Sydney, Australia	4.98	5.93	100	Flat	NA	NA
MED	PSA[c], Spain	4.65	5.6	72	PTC[d]	<70	500[b]
MSF	Tamil Nadu, India	NA	4.97	0.0085	Flat	NA	2
MSF	Suez, Egypt	NA	7.03	0.0025	Flat	67	2.39
MSF	Tianjin, China	NA	5.58	0.3	Flat	78	NA

[a]Evacuated tube collector.
[b]Collector total absorber area.
[c]Platform a Solar de Almeria (SPA).
[d]Parabolic trough collector.
Data extracted from Li et al. (2013).

and technical details of parabolic trough collectors coupled with MED was investigated. In the second phase, a double absorption heat pump was used to enhance the system performance.

In a system in Abu Dhabi, evacuated tube collectors (ETCs) were used to drive an MED desalination system. This plant operated from 1984 to 2002 (El-Nashar, 1993). The ETCs comprise a number of parallel cylindrical transparent glass tubes connected to a header. In each tube, a copper fin is attached to a metal heat pipe. The heat transfer fluid circulates and absorbs heat from the tubes. The solar heat is transferred by the heat pipes to the seawater in the MED desalination system. The deposition of dust could reduce the freshwater production by up to 40% compared to a clean collector. Thus, plant maintenance is required for such systems.

Extraction steam at a temperature of 90 °C to 120 °C from the steam power plant (e.g., Rankine cycle) can operate an MSF desalination system (Fig. 8.6). For an MSF-steam power plant combined system, when

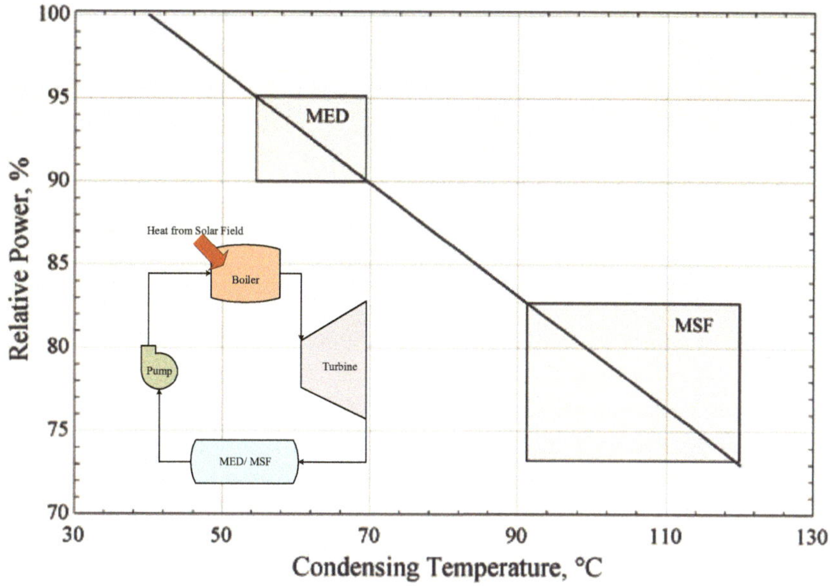

Fig. 8.6 ***Reduction in electrical power capacity of a Rankine cycle with increasing condensation temperature when replacing the condenser of the Rankine cycle by an MED or MSF desalination system.*** The gray squares represent the operating temperatures of MED and MSF desalination systems.

the steam temperature is above 100 °C at atmospheric pressure (i.e., the steam temperature is above the condensation temperature), a back-pressure turbine is required. The back-pressure turbine generates electricity with a relatively small decrease in steam pressure. In the combined scheme, however, the condenser of the steam power cycle is replaced by the desalination system, and the extraction of relatively high-temperature steam from the back-pressure turbine reduces the electrical power production compared to the conventional condensing steam turbine operating at a temperature of 35 °C–40 °C. Therefore, the application of an MED desalination system with a CSP plant offers a better combined operation and efficiency than an MSF desalination plant.

In a low-enthalpy MED system, on the other hand, the operating top brine temperature (i.e., the brine temperature in the first effect of MED) is normally between 70 °C and 55 °C. The relatively lower temperature allows for the use of low-grade waste heat, which reduces corrosion and scale formation. Therefore, if an MED system is combined with a steam power cycle, the reduction in electricity production is much less than for those steam plants combined with MSF systems (Fig. 8.6). In some combined steam power plant–MED applications, conventional condensing turbines can be employed rather than back-pressure turbines.

Extensive research has shown that RO and MED systems, respectively, are at present the most efficient work and heat-driven desalination technologies. So, the best options for combining a solar power plant with a desalination process are RO and MED systems.

In an RO–CSP combined system, the electricity required for the RO process is supplied from the solar power cycle. Fig. 8.7 shows an RO desalination system integrated with a parabolic trough-CSP plant. One of the advantages of RO–CSP over MED–CSP is that the two subsystems (i.e., RO and the solar power plant) are not necessarily in the same location. This is important as coastal areas receive less DNI than areas far from the sea. So, by installing an RO system close to the sea and a solar power plant in a relatively higher solar radiation area, the pumping power required for transferring seawater to the RO module can be kept low. In addition, in the case in which there is a sufficient cooling water source (below 30 °C) for the condenser of the power plant, integration of the RO technology with the power cycle has advantages over other desalination methods (even MED). This is because, in the RO-power cycle, the reduction in electricity production for incomplete expansion in the steam turbine that takes place in the MED/MSF-power system is eliminated. However, discharging a large amount of cooling water to the sea can lead to the adverse impacts on the environment, specifically marine species. In addition, in hot weather conditions, the cooling process of the power plant cannot be accomplished properly due to the relatively high temperature of seawater used for cooling. By coupling an MED system with a power plant instead, the relatively high temperature of the feed seawater in hot weather conditions is able to cool the high-temperature exhaust steam from the turbine, rather than cooling the relatively low-temperature steam at the turbine outlet in the RO-power cycle configuration.

In MED–CSP systems, the exhaust steam from the CSP plant can be either used directly as the thermal energy source for a low-temperature MED system or delivered first to a thermal vapor compressor (TVC) before entering the MED plant (Fig. 8.8A). A TVC is an efficient device that can be used in any thermally driven desalination system (MED, MSF, etc.). Depending on the pressure and quality of the exhaust steam, various configurations of TVC can be considered. For example, as shown in Fig. 8.8B, a portion of the exhaust steam can be transferred to a TVC and the remainder condensed in a wet or dry cooling system. In hot and dry regions, dry cooling systems are more common than wet cooling systems

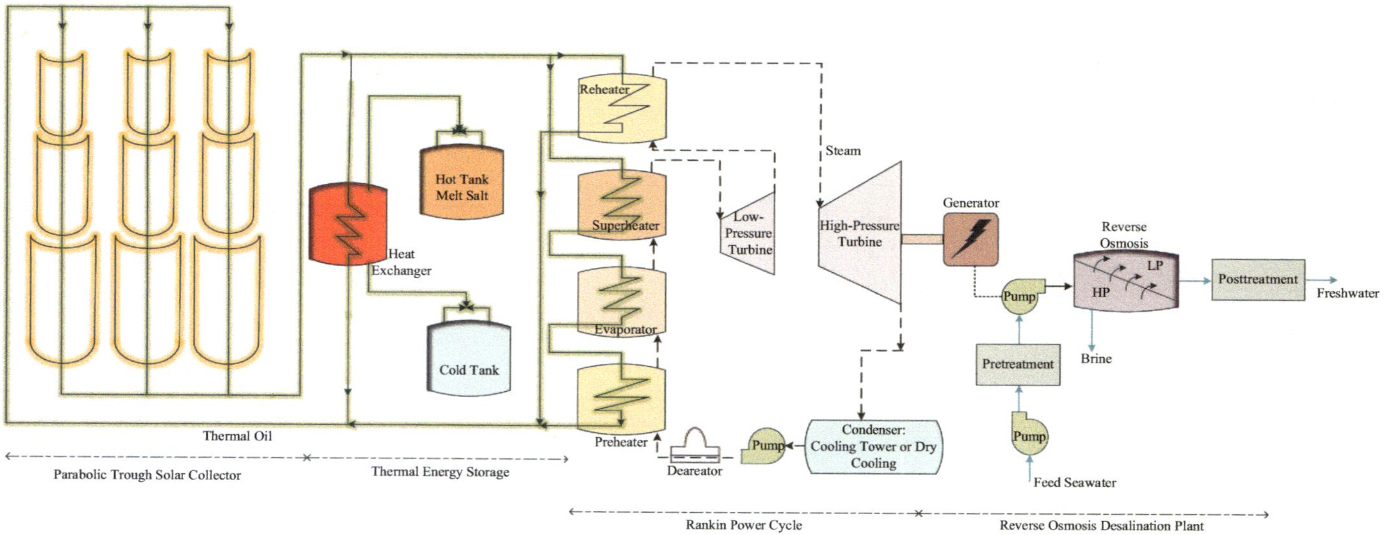

Fig. 8.7 ***Parabolic trough solar concentrator with a thermal energy storage used to drive a Rankine power cycle-RO desalination combined system.***

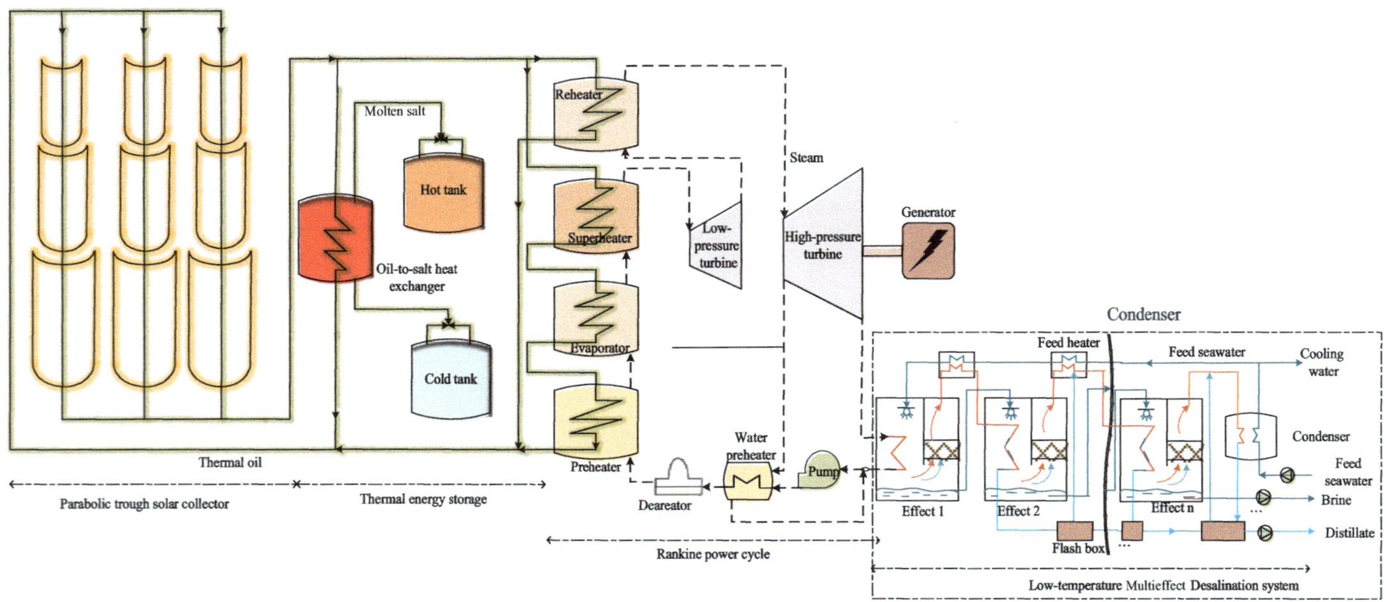

Fig. 8.8 ***Parabolic trough solar concentrator with a thermal energy storage used to drive.*** (a) a combined system including a Rankine power cycle and a low-temperature MED unit; (b) a combined system including a Rankine power cycle and a low-temperature MED-TVC unit.

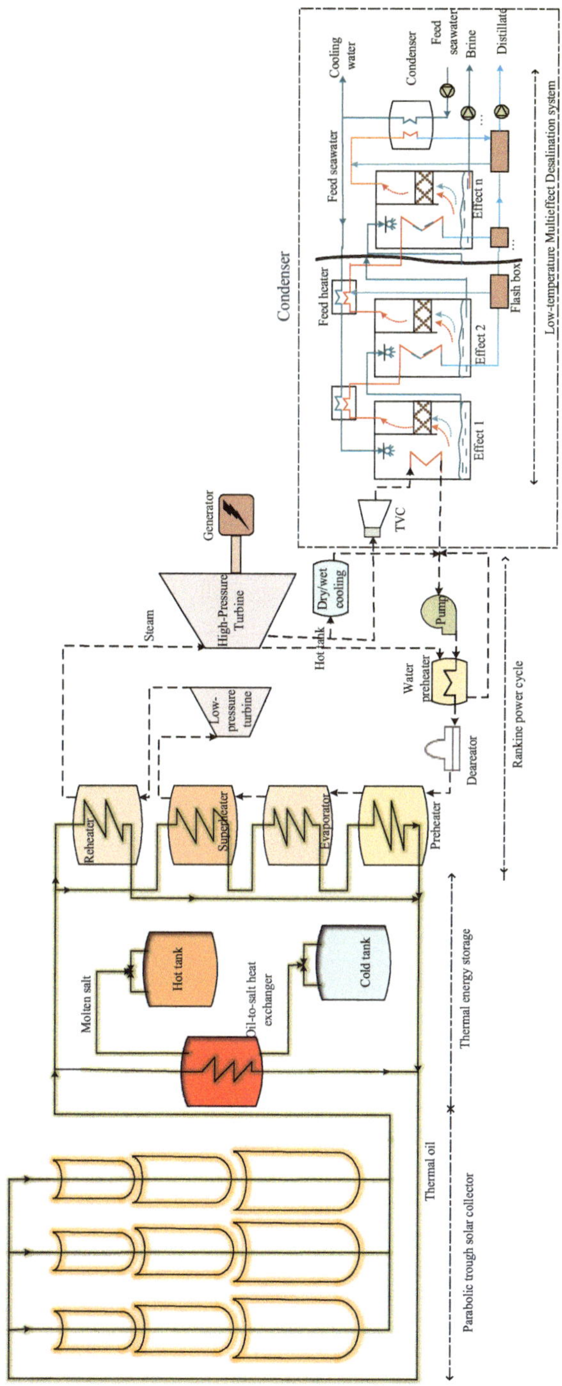

(b)

Fig. 8.8 *(Cont'd)*

(e.g., cooling towers), largely because the latter consume large amounts of water for cooling for the power cycle. For example, 2–3 metric tons of water is typically required per 1 MWh of electrical energy production. The outlet steam from the MED plant is pumped to the water preheater where a portion of the high-pressure turbine steam outlet preheats the incoming water. A deaerator is used to remove dissolved gases from the water. Then, the water working fluid in the Rankine cycle is converted to superheat steam by the preheater, evaporator, and superheater. In the TES subsystem, hot (300–400 °C) and cold (200–300 °C) tanks containing molten salt are used for storing the solar energy collected from the parabolic trough concentrators. Note that a thermal oil (e.g., Monsanto Therminol VP-1) is used as the heat transfer fluid in the parabolic trough solar field. The Monsanto Therminol VP-1 heat transfer fluid is an aromatic hydrocarbon, suitable for temperatures up to 400 °C (Eastman, 2020).

8.2.4 Solar ponds

In a solar pond, a large amount of water essentially is heated via solar energy and stored in the saltwater in the pond. Specifically, a portion of incoming solar beam is absorbed and converted to thermal energy and then stored as dense heated water in the lower part of pond. Lower density heated water rises to the surface of pond and transfers heat to the environment through convection and radiation. Solar ponds can be considered as both solar collectors and TESs. In salt gradient lakes, temperature and salinity naturally increase in the downward direction leading to the formation of three zones: an upper convecting zone, a nonconvecting zone in the middle, and a lower convecting zone.

Small-scale solar ponds can be used for water and space heating. Large solar ponds can be utilized in industrial electrical power generation and desalination systems.

Many studies have been reported for electrical power production from solar ponds (Alcaraz et al., 2018; Kasaeian et al., 2018; Mittal et al., 2021; Mosaffa and Farshi, 2021). The only practical operational plants are installed in Israel (Tabor, 1981). In this plant a 1500 m^2 pond with an operating temperature of 90 °C is used to supply energy for a 6 kW Rankine cycle, and a 7000 m^2 pond supplies 150 kW of peak power. Low-temperature desalination modules can be driven by the salt-gradient solar pond using high-temperature brine as the heat source to vaporize seawater in a multieffect boiling evaporator. This solar pond coupled desalination system is specifically applicable in desert areas near the ocean.

There are numerous solar desalination plants in which solar collectors are integrated with industrial desalination systems. The Margarita de Savoya solar desalination plant in Italy with a freshwater production capacity of 50–60 m^3 per day is an example. In this plant, MSF desalination is coupled with a solar pond system (Delyannis, 1987). The combined system is considered cost-effective and competitive with fossil-fueled desalination systems (Hoffman, 1992). In another example, a solar pond–powered MED plant with a freshwater production capacity of 3000 m^3 per day has been installed at the Dead Sea (CRES, 1998).

Furthermore, extensive studies have been carried out on solar pond applications for desalting. Among those studies, the work conducted by the University of Texas (Lu et al., 2002, 2001) on the solar pond desalination system is noteworthy. This work aimed for zero-discharge desalination, attained by integrating a salinity-gradient solar pond (SGSP) (with 3000 m^2 surface area and a depth of 3.25 m) and multieffect/multistage (MEMS) desalination (with a 63–80 °C operation temperature) in conjunction with other desalination plants (Fig. 8.9). The freshwater production rate varied between 2.3 and 7.2 m^3/day. The brine from the thermal or work-driven desalination system supplies makeup water to the SGSP. In the SGSP the produced brine feeds the MEMS. The high-concentration brine from MEMS desalination is provided as feed to a brine concentrator and recovery system (BCRS). The BCRS uses thermal energy from the SGSP to form a semislurry salt discharge. A portion of the salt is transferred back to the SGSP and the remainder can be processed as chemicals for sale or use. The zero-discharge approach in the proposed system removes the need for

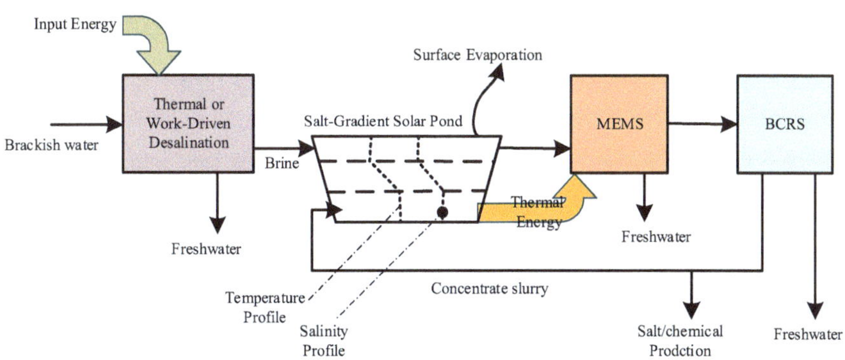

Fig. 8.9 ***Zero-discharge desalination process, integrating a salinity-gradient solar pond (SGSP), a multieffect/multistage (MEMS) desalination system and a brine concentrator and recovery system (BCRS).***

waste disposal, as the brine is reused, and can thereby significantly address environmental concerns associated with inland desalination facilities.

8.2.5 Solar PV systems

PV modules directly convert solar energy to electricity via conducting the excited electrons exposed to sunlight in silicon or other semiconductor materials. Individual PV panels can be connected in series or parallel to create a PV array to produce design current and voltage. The power outputs from a typical PV module, with a surface area of around 15–25 m^2, is 50–80 W. Most PV modules produce direct current (DC) electricity at 12 volts, while most industrial and household appliances operate with alternating current (AC) at 240 V or 120 V or 415 V. An inverter that converts DC voltage to AC is therefore necessary in a PV-electric setup.

The maximum attainable efficiency of PV systems depends on the system characteristics and capacity and the distance to the electricity grid. Typical efficiencies of solar panels are reported to be between 15%–22%, and the best observed efficiency is about 23% (Energy stage, 2021). These electricity-generation systems have several advantages over the conventional fossil fuel–based electricity production systems, such as no direct environmental emissions or impacts, no noise, small degradation in power-generation efficiency throughout the lifetime (less than 1% per year), and low maintenance costs (Al Suleimani and Nair, 2000; NREL, 2018). However, the capital cost of PV cells is relatively high in many jurisdictions, even though their average cost has decreased markedly over past two decades, in some cases by six times. PV modules use has increased greatly in recent years in many countries, particularly where fossil fuels are expensive.

8.2.5.1 Solar PV model

The current generated by a PV module depends on incident solar radiation, operating voltage, and solar cell temperature. These are the main factors in the classic one-diode model of a PV unit. With such a model, the output current from a solar PV unit can be written as (Wenham et al., 2011)

$$I = I_{ph} - I_0\left(\exp\left(\frac{IR_s + V}{\left(\frac{nk_BT_c}{q}\right)}\right) - 1\right) - \frac{IR_s + V}{R_{sh}} \tag{8.5}$$

where I_{ph} is the light-generated current and I_0 is the reverse saturation current, which is affected by temperature. Also, V and I, respectively, denote operating voltage and current; n is the diode ideality factor; q is the charge of an electron; k_B is the Boltzmann constant; T_c is the cell temperature, and R_s and R_{sh}, respectively, are the panel series resistance and shunt resistance. The parameters I_0, R_s, and R_{sh} are dependent to the panel. The light-generated current (I_{ph}) can be expressed in terms of the solar panel area (A_{sp}), the incoming solar radiation normal to the cell (G_{Rad}), and the specific parameters of the panel (C_1 and C_2) as follows:

$$I_{ph} = A_{sp} G_{Rad} \left(C_1 + C_2 T_c \right) \tag{8.6}$$

Solar radiation and cell temperature have a notable effect on the optimal operating condition, that is, the condition at which the electrical power output is a maximum ($P_{max} = I_{max} V_{max}$). For example, a 1 °C increase in cell temperature for a silicon panel reduces the power output by 0.4% (Wenham et al., 2011). Note that the number of PV modules is calculated by dividing the peak power required by the peak power produced from each PV module.

Generally, for renewable energy–based electrical generation systems (such as solar PV panels and wind turbines) applied to a specific load like that of an RO desalination unit, a battery is used to store electrical energy during high power production periods. Such periods usually occur when solar radiation or wind speeds are high. Subsequently, the stored energy is recovered to provide electricity for the load. The storage addresses the intermittent nature of many renewable energy sources and permits continuous energy supply. The operational current used for charging the battery at any time can be written as the difference between the current produced by the solar PV panels (or wind turbines or other technologies) and the load current, as follows:

$$I = \sum_{\substack{Generated\ by \\ PV/wind}} I_i - \sum_{Load} I_o \tag{8.7}$$

The operational current is positive (i.e., $I > 0$) during the charging period and negative (i.e., $I < 0$) during the discharging period. If the charging period is divided into time steps, the charge transfer over a time step can be expressed as

$$Q = \int_{t_i}^{t_{i+1}} |I| \, dt \tag{8.8}$$

The battery state of charge (SOC) during charging and discharging can be expressed as (Sinha and Chandel, 2015)

$$SOC_{ch}(t+1) = SOC(t).\left[\left(1-\sigma(t)\right)\right] + \left[I_B(t).\Delta t.\frac{\eta_{ch}(t)}{C_B}\right] \tag{8.9}$$

$$SOC_{disch}(t+1) = SOC(t).\left[\left(1-\sigma(t)\right)\right] - \left[\frac{I_B(t).\Delta t}{\eta_{disch}(t)C_B}\right]; \tag{8.10}$$

$$\text{with } \left(i - \text{Depth of Discharge (DOD)}\right) \leq SOC(t) \leq 1$$

where σ is the hourly self-discharge rate, which depends on the battery state and usually is constant at around 0.02%; C_B and I_B, respectively, are the nominal battery capacity and operational battery current; η_{ch} is the battery charge efficiency, which depends on charging current and normally varies between 65% and 85%, and η_{disch} is the battery discharge efficiency (usually 100%).

8.2.5.2 PV–RO systems

A PV electricity generation system can be directly connected to an RO desalination system. Both PV and RO are easily scalable and modular. The compatibility of an RO system with PV technology has improved with the advent of reliable and efficient energy recovery devices (ERDs) such as turbines or pressure exchangers. In fact, the energy of the brine produced at the high pressure (HP) side of an RO unit can be harnessed by the ERD to assist in pumping feed water. Fig. 8.10 shows a PV–RO system using

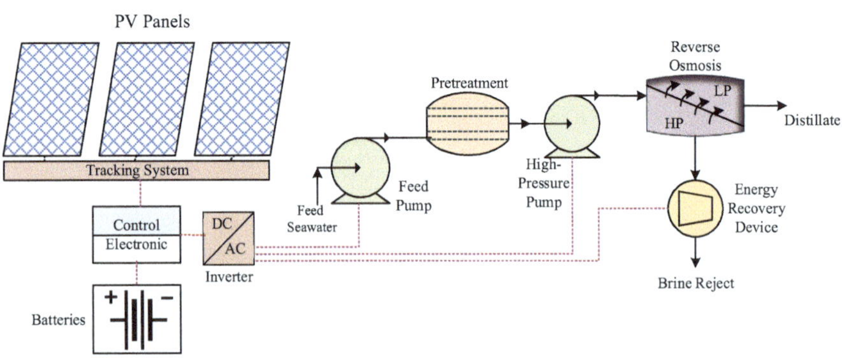

Fig. 8.10 ***Typical PV-RO system using a battery to store electrical energy and an energy recovering device (EDR) to harness energy of the high-pressure brine stream leaving the RO module.***

a battery and an ERD. ERD modules can substantially reduce the power required in the RO process; however, application of such devices for small-scale operations is often not economic.

The membrane-based RO desalination system desalts water using external electrical power to overcome the osmotic pressure and membrane resistance. According to Fig. 8.10, the RO desalination subsystem includes a unit for pretreatment of feed seawater and an RO membrane module. The spiral-wound membranes are one of the common constructions in PV–RO systems. In the RO module, seawater is partially desalted by passing through the membrane. The high-concentration brine stream exits from the high-pressure side of the RO module. The distillate flux across the membrane is related to the pressure difference minus osmotic pressure difference, as follows:

$$J_D = \dot{V}_D / A = k_w \left(\Delta P - \Delta \pi \right) \tag{8.11}$$

where $\Delta\pi$ and ΔP, respectively, are the osmotic pressure difference through the membrane and the hydraulic pressure difference. Also, k_w is the water permeability coefficient that is dependent on the membrane characteristics, and A is the membrane area. It can be shown with Eq. (8.11) that the distillate flow rate increases with an increase in the permeability and area of the membrane and the driving pressure applied across the membrane.

During actual operation of the membrane in an RO module, a quantity of salt is passed through the pores of the membrane. The salt concentration in the produced freshwater (distillate) can then be written as follows:

$$C_D = CPF.k_s.A.TCF.C_{fc} / \dot{V}_D \tag{8.12}$$

Here, CPF denotes concentration polarization factor, k_s the membrane permeability for salt, TCF the water permeability temperature correction factor, and C_{fc} the average concentration of water in the feed channel of the membrane. In design of an RO module, the permeability of salt and the area of the membrane need to be considered as they affect the permeate concentration. Note that a complete model for RO desalination system is presented in Chapter 5 (Section 5.8).

The battery bank provides a constant energy flow during fluctuations in solar electricity production (such as on cloudy days and at nighttime). Fig. 8.11 shows the performance of an RO desalination system driven by solar energy (or wind or hybrid solar–wind energy), with and without a battery bank storage and with a backup diesel generator. In the battery configuration (left side of Fig. 8.11), during the daytime (i.e., from

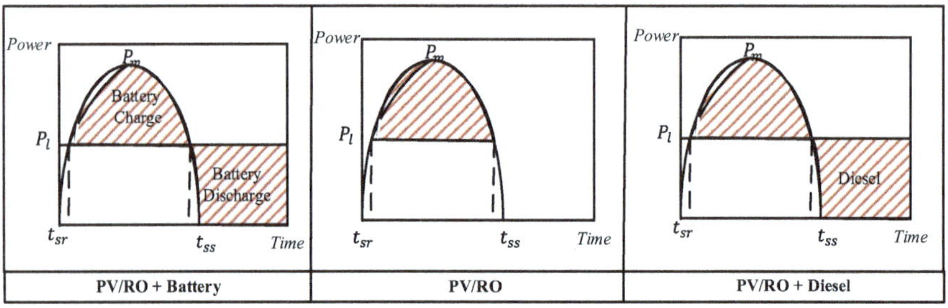

Fig. 8.11 Performance of a PV–RO system using (left) battery electrical energy storage only; (center) no battery and no diesel backup; and (right) diesel generator only.

sunrise (t_{sr}) to sunset (t_{ss})), incoming solar energy is converted to electrical energy by PV modules to supply electrical energy for both the RO unit and the battery bank. When there is no solar radiation, the stored energy in the battery is recovered to permit continuous operation of the RO system by meeting the power requirement throughout the entire day (hatched area). In the nonbattery configuration (center of Fig. 8.11), the RO operation is halted during the night. In the last configuration, where a diesel generator is used as a backup energy supply (right side of Fig. 8.11), the RO operation is continuous throughout the day as the nighttime power requirement for the RO system is supplied by the backup diesel generator.

Rechargeable batteries are expensive and require proper maintenance to avoid a high degradation rate (reduction in the amount of energy a battery can store, or the amount of power it can deliver). Some problems are associated with battery operation in the PV systems and should be considered when the system is set up. For example, leakage in lead-acid batteries reduces their efficiency, while average battery lifetime declines in hot and dry conditions. In addition, some energy is lost from batteries when the chemical-to-electrical energy conversion occurs to supply electricity for pumping. Electronic controls are included in PV–RO systems to determine the start-up and shutdown times and to permit the RO to work at a fixed condition or point. The panel set-point is controlled so as to provide the optimum electrical power to the batteries. Several techniques exist for determining battery start-up and shutdown times (Herold and Neskakis, 2001; Mohamed and Papadakis, 2004).

Electrodialysis (ED) desalination also can be coupled with PV. In the ED system, the ions in an electric field migrate toward the positive electrode. However, combined PV–ED systems are less investigated than

PV–RO systems, ED systems typically are much more cost-effective than RO systems for lower salt concentrations in seawater.

Numerous PV–RO installations exist, especially in rural areas for small desalination capacities, ranging from 1 m^3/day to 5 m^3/day (AlMadani, 2003; Gocht et al., 1998; Herold and Neskakis, 2001; Joyce et al., 2001; Richards and Schäfer, 2003). Table 8.4 presents selected PV–RO systems installed between 1985 and 2016 at various locations.

The electrical conversion efficiency of PV modules declines significantly as temperature rises on the surface of PV panels. Introducing a cooling fluid circulating through the panels can improve the efficiency of PV panels and also yield a high-temperature fluid usable for such applications as domestic hot water production and space heating. When the heat is recovered in this manner, the PV system is referred to as a PV-thermal system. Such produce electrical and thermal energy simultaneously.

8.3 Wind energy

Wind energy is produced from the uneven distribution of solar radiation on the earth's surface, which generates atmospheric pressure differences and results in air movement. In fact, wind energy can be considered indirect solar energy. Of the total solar power reaching the earth (17.3×10^{10} MW), only 12×10^{8} TW (i.e., 0.7%) is harnessed to drive energy systems (Sorensen, 2017).

The theoretical wind-to-mechanical conversion efficiency of wind turbines is 59% (Burton et al., 2011) while, in practice, the most efficient airfoil technologies in wind turbine facilities have the maximum achievable efficiencies of 48% (Burton et al., 2011). In addition, the need for shutting down wind turbines during the presence of very high or low wind speeds and operation at a nonoptimum conditions further decrease wind turbine efficiencies (Beurskens and Jensen, 2013). There are sometimes problems associated with wind power installations, such as significant land requirements, telecommunication interference, audible noise, and adverse impacts on the wildlife and natural habitats. To avoid some of these problems, wind plants can be placed off-shore, although doing so can introduce other challenges. The costs of wind turbine systems can be reduced by decreasing the capital cost (e.g., constructing slender blades), utilizing the best wind sites and employing reliable design methods. The wind turbine industry is in some ways mature, as electrical power generation from wind power is commercially available for a large range of nominal power outputs.

Table 8.4 Selected solar PV-RO plants installed in various geographical areas.

System and location	Freshwater production capacity (m^3/d)	PV electrical generation capacity (kW)	Battery	Energy recovery	Backup power supply
HEI[a], Vancouver, Canada (Keeper et al., 1985)	0.5–1	0.48	Yes	Yes	No
CRESTA[b], Curtin University of Technology, Australia (Al-Alawi et al., 2007)	1	1.2	Yes	No	Diesel
CIEA-ITC[c], Canary Island, Spain (Herold and Neskakis, 2001)	1.24	4.8	Yes	Yes	No
CRES[d], Laviro, Greece (Tzen et al., 2004)	3.12	4	Yes	No	Wind
CREST[e], Loughborough University, UK (Thomson and Infield, 2003; Thomson et al., 2003)	3.9	2.4	No	Yes	No
GECOL[f], Ras Ejder, Libya (Kershman et al., 2005)	300	50	No	Yes	Wind and grid
MIT[g], Cambridge, USA (Bilton et al., 2011)	0.3	0.23	No	Yes	Small batteries
NTUA[h], Athens, Greece (Ntavou et al., 2016)	Three identical units, each of capacity 12.0–16.8 m^3/d	10–20	No	Yes	Axial piston motor

[a]Highquest Engineering Incorporated (HEI).
[b]Center for Renewable Energy Sustainable Technology Australia (CRESTA).
[c]Energy and Water Research Centre of the Canary Islands Technological Institute (CIEA-ITC).
[d]Centre for Renewable Energy Sources (CRES).
[e]Centre for Renewable Energy Systems Technology (CREST).
[f]General Electricity Company of Libya (GECOL).
[g]Massachusetts Institute of Technology (MIT).
[h]National Technical University of Athens (NTUA).

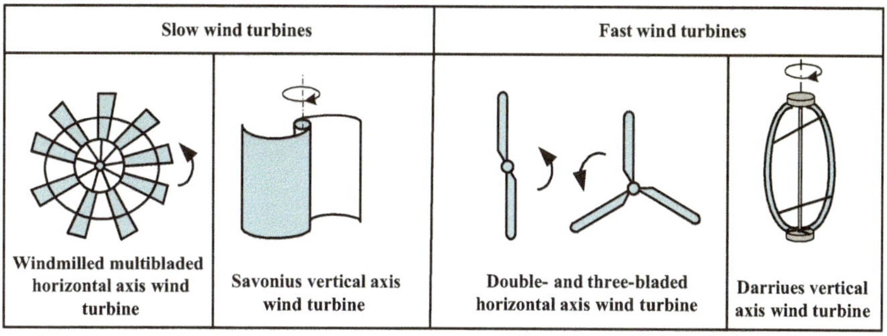

Fig. 8.12 ***Common types of commercial wind turbines.***

Nonetheless much work is ongoing to improve wind turbines, including efforts to improve energy storage and develop new control strategies that can potentially enhance the efficiencies of such systems.

Wind turbines can have blades or other means for harnessing the wind. They can rotate on a vertical axis or horizontal axis. Fig. 8.12 shows several horizontal axis wind turbines and vertical axis wind turbines. All wind turbines convert the kinetic energy of wind to mechanical energy in the form of a rotating rotor.

8.3.1 Wind power modeling

Based on the design, wind turbines can be categorized in three primary types: variable speed wind turbines, fixed speed wind turbines, and slip ring induction motors. The wind turbines discussed here convert the kinetic energy of wind to electrical energy. A general expression for the average mechanical power produced by a wind turbine P_{WT} follows (Burton et al., 2011):

$$P_{WT} = \frac{CP.\rho_{air}.A.\bar{u}^3}{2} \tag{8.13}$$

where ρ_{air} denotes air density, A the swept area of the turbine blades, and $\bar{u}$ the average hourly wind speed. The swept area of blade (A) can be expressed as a function of rotor diameter (D_{wt}), as follows:

$$A = \neq\left(D_{wt}^{\ 2}/4\right) \tag{8.14}$$

The term CP in Eq. (8.13) is the coefficient of performance (also called the maximum power coefficient), which expresses the ratio of the actual wind turbine power output (provided by turbine manufacturers) to the available wind power. The coefficient CP is related to the

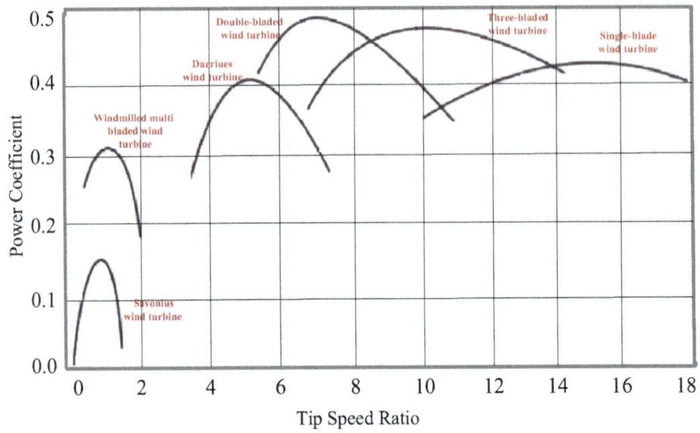

Fig. 8.13 ***Power coefficient (CP) variation with tip speed ratio for several wind turbines.*** *Data from Carta, (2019) and Hau (2013).*

electro-mechanical efficiency of the wind turbine, which can be achieved at the optimal value of tip speed ratio. The tip speed ratio represents the speed ratio of turbine to wind as follows:

$$\lambda = \frac{\omega R}{u} \tag{8.15}$$

where R is the radius of the rotor, ω the rotor angular velocity, and u the incoming wind speed. Fig. 8.13 shows the power coefficient variation with tip speed ratio for various wind turbines. It is seen that there is a nonlinear relationship between CP and tip speed ratio. In addition, slow wind turbines have a relatively smaller CP value than fast wind turbines. However, the starting torques of fast wind turbines are lower than those of slow wind turbines.

From the Betz theory (Burton et al., 2011), the maximum power that can be extracted from wind energy via a wind turbine is 16/27 (=0.593) of the kinetic energy of the wind at the turbine. Thus, the maximum value of CP is 16/27. CP values for advanced manufactured wind turbines are about 70%–80% of the Betz value (Burton et al., 2011). Details of wind generator dynamics are addressed elsewhere (Soetedjo et al., 2011).

8.3.2 Wind-RO system

Wind energy is the most well-established renewable energy source after solar energy for small-scale desalination plants, especially RO systems. Table 8.5 lists several existing RO plants powered by wind turbine plants,

Table 8.5 Selected RO-wind turbine installations for various geographical areas.

System and location	Freshwater production capacity of RO (m^3/d)	Wind turbine nominal power production (kW)	Specific energy consumption (kWh/m^3)
CERST, UK (Miranda and Infield, 2003)	0.5	2.5	NA
AEROGEDESA, Spain (Lai et al., 2016; Loutatidou et al., 2017; Ma and Lu, 2011)	18	15	8.4
ENERCON, Mediterranean (Lai et al., 2016; Loutatidou et al., 2017; Ma and Lu, 2011)	180–360	2.5	2–2.8 (SW[a])/ 0.8–1.3 (BW[b])
Canary Island, Spain (Subiela et al., 2009)	200	42	5
ENERCON, Germany (Paulsen and Hensel, 2007)	175–1400 (SW[a])/ 350–2500 (BW[b])	200	2–2.5

[a]Seawater.
[b]Brackish water.

installed in a variety of regions between 2002 and 2014. A wind-RO desalination plant on Milos Island, Greece is shown in Fig. 8.14. The electrical power produced from the wind turbines (850 kW) supplies the electrical power required for producing 3300 m^3/day freshwater with an RO unit. It is estimated that 1800 MWh of electrical energy per year can be generated from those wind turbines.

Although an MVC desalination plant consumes more energy than an RO system, the former is better able to tolerate intermittent operation. Compared to RO, an MVC system is more compatible for remote areas as

Fig. 8.14 ***Photograph of wind-powered reverse osmosis (RO) plant installed at Milos Island, Greece.*** *Source: ITA (2018).*

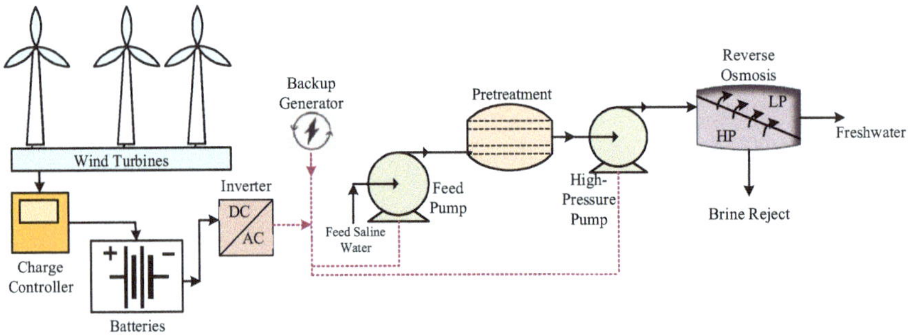

Fig. 8.15 ***Wind-RO desalination system using a battery bank as energy storage.***

it does not require membrane replacement, is more robust, requires fewer skilled worker, and uses less chemicals.

Due to the fluctuations of wind power output, intermediate energy storage such as battery is necessary to provide the continuous and constant electrical power required for an RO plant. Typically, a wind-powered RO setup comprised an RO plant, wind generators, a battery bank, a charge controller, and an inverter (Fig. 8.15). The electrical power generated by the wind turbines drives the RO system and charges the battery bank. The battery bank is discharged when the wind speed is inadequate for generating electricity with the wind turbines needed by the high-pressure pump of the RO system. Batteries are suitable for energy storage for small-scale RO-wind plants, with production capacities of up to 5 m^3/d of freshwater. The charge controller is used to prevent the batteries inside the bank from overcharging. It monitors the charging and discharging levels of batteries so as to drive the pumps of the RO without fluctuations. A diesel generator or grid-connection can be used as a backup energy supply for the RO unit. The backup supply is needed for high-load RO operation (more than 5 m^3/d freshwater production), low battery charge, and insufficient wind speed. Diesel generators are typically operated at 80%–100% of their power rating.

The number of wind turbines required in a wind-RO system can be determined iteratively. First, the required number of wind turbines are calculated as the ratio of the annual energy requirement of the wind-RO system to the annual energy output of a unit wind turbine module (in kWh) as follows:

$$N_{wt} = \frac{E_{wind-RO}}{E_{wt}} \tag{8.16}$$

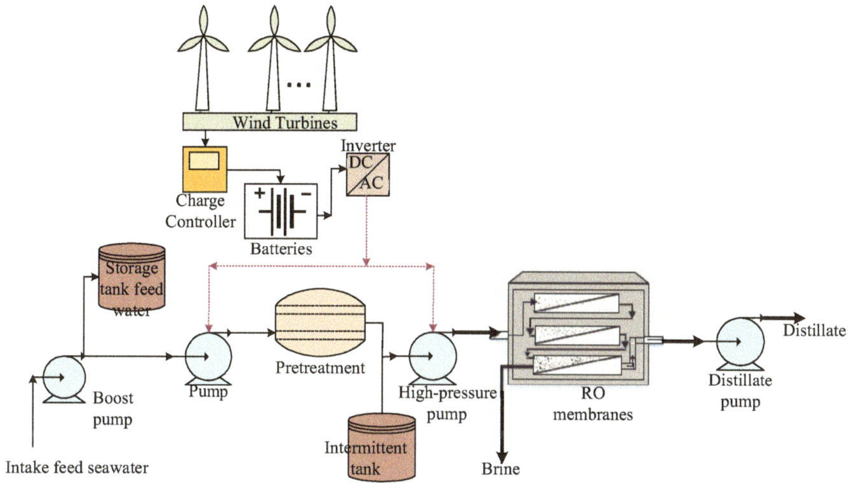

Fig. 8.16 ***Wind-RO system prototype installed in Suderoog Island, Germany.***

Then, the annual freshwater production can be determined. If the calculated value is far from the target annual freshwater production, then a wind turbine is added or subtracted and the freshwater production calculation is repeated until the freshwater production value is sufficiently close to the target freshwater production.

A 6 kW wind turbine prototype was installed in Germany to drive three RO modules with a maximum freshwater capacity of 0.375 m^3/h (Petersen et al., 1979). Fig. 8.16 shows a simplified schematic of the prototype. The original wind turbine was replaced by an Aeroman wind turbine model with a capacity of 11 kW. The control system used in the prototype regulates automatically the speed of wind turbine. This unit allows for direct coupling of the wind turbine with the RO system without a converter and battery. For an indirect connection of the wind turbine and RO modules, the electrical energy stored in a battery is recovered to drive the three pumps used in the system. The authors addressed the need for continuous operation of system with no reduction in distillate production rate despite the intermittent electricity generation from the wind turbines. This indicates the importance of battery storage in wind facilities.

8.3.3 Hybrid wind-solar PV power

Wind energy has some advantages over solar energy, but it also has disadvantages. For instance, fluctuations in wind speed often exceed those in solar radiation. Hence, the use of wind energy often entails greater use of batteries for energy storage or of backup energy supplies. Also, the transmission of electrical power produced by wind turbines from windy regions to users is often more difficult than the transmission of electrical power produced from solar energy, because windy regions are commonly located far from urban locations, making energy delivery more complicated.

By incorporating solar energy with wind energy, the drawback of intermittent power production wind power plants can be somewhat offset. The use of hybrid wind-solar energy systems, such as wind generator-PV systems, provides a potentially beneficial alternative for regions that experience high solar radiation intensities when wind speeds are inadequate, and vice versa. Fig. 8.17 illustrates this concept. For large-scale RO plants (e.g., 300 m^3 per day freshwater production); however, utilizing a diesel engine backup in conjunction with a hybrid wind-solar energy can ensure provision of the required electrical power at peak load conditions.

There are few practical solar-wind hybrid powered RO desalination facilities. This kind of system is still undergoing research and development. Most studies focus on technological developments and trends and the needs for implementing such systems in various regions (Kershman et al., 2005; Mohamed and Papadakis, 2004; Voivontas et al., 2001; Weiner et al., 2001). A hybrid solar-wind powered RO desalination pilot is installed in Cadarache Centre in France (Maurel, 1991). The system has a freshwater production capacity of 0.1 m^3 per day from a compact solar distiller and 6 m^3 per day from RO and ED units. The PV field supplies 4 kW peak electrical power and two wind turbines are used as an assisting energy source (Maurel, 1991).

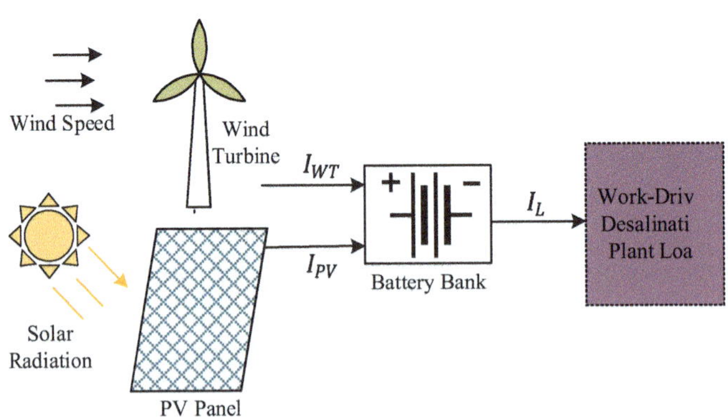

Fig. 8.17 ***Utilization of a hybrid solar/wind energy system for driving a desalination system.***

8.4 Geothermal energy

There are two main categories of geothermal energy: ground-source and high-temperature.

Ground-source geothermal energy exploits the ground's high thermal inertia, which causes the underground temperature to remain almost constant over the year. In the ground surface zone, at depths less than about 1 m, the ground temperature varies seasonally and with weather conditions. In the shallow zone, which contains light dry soil at depths of 1–8 m and moist heavy soil at a depth of 20 m, the temperature of ground is almost constant at the average annual temperature of air. The seasonal variation of ground temperature is based on the cycle climate variations. In the deep zone, below the shallow zone, the temperature of the ground is practically constant (Popiel et al., 2001).

High-temperature geothermal energy sources can be categorized in terms of temperature, as follows: low temperature (below 100 °C), medium temperature (100–150 °C), and high temperature (more than 150 °C). The thermal gradient varies in the range of 15 °C and 75 °C per kilometer. However, the heat flux is dependent on the geographical region. Table 8.6 shows typical values of temperature, enthalpy, and pressure for several geothermal reservoirs.

Ground heat exchangers are commonly used for delivering underground energy to thermal loads. Ground heat exchangers typically are manufactured from polyethylene material, which is durable, and are warranted for as much as 50 years. The tough plastic material equipped with heat fuse joints permits heat to pass. The working fluid can be either water or an environmentally benign antifreeze. Hot water extraction from underground aquifers is quite mature. Another type of underground heat exchanger utilizes copper pipes. The length of the pipes mainly depends on the loop configuration, ground conditions, the thermal load, and the local climate.

Table 8.6 Typical geothermal reservoir characteristics.

Reservoir type	Temperature (°C)	Specific enthalpy (kJ/kg)	Pressure (bar)
Compressed liquid	240	1038	100
Saturated water-steam mixture	235	1351	31
Compressed liquid	210	900	35
Saturated water-steam mixture	205	1146	17

Low-temperature geothermal water sources beneath the ground (i.e., within about 100 m depth from surface) are particularly suitable energy sources for low-temperature desalination processes such as MED. Nonetheless, freshwater production using geothermal energy is not widespread commercially. The cost of freshwater produced from this technology has been demonstrated to be competitive that for freshwater produced from large-scale multieffect cogeneration plants (Ophir, 1982). One important advantage of geothermal energy is that it provides a reliable, noninterruptible and stable heat supply, unlike solar or wind energy. The freshwater production cost from a geothermal source is lower than that from solar/wind sources (Mahmoudi et al., 2017).

Fig. 8.18 shows the global distribution of heat flow rates per unit surface area. It is seen that high heat flow rates are mostly located at many inland and coastal regions such as Japan, Peru, New Zealand, and the western part of the United States (Davies, 2013). Areas located along the Red Sea, southern Europe (specifically Greek islands and Italy) and North Africa also have a high heat flow rates. For instance, underground water temperatures in southern Europe vary between 80 °C and 110 °C in the shallow zone and rise to as high as 325 °C at depths of 70–1400 m. For areas with relatively low water flow rates, higher underground hydraulic conductivities

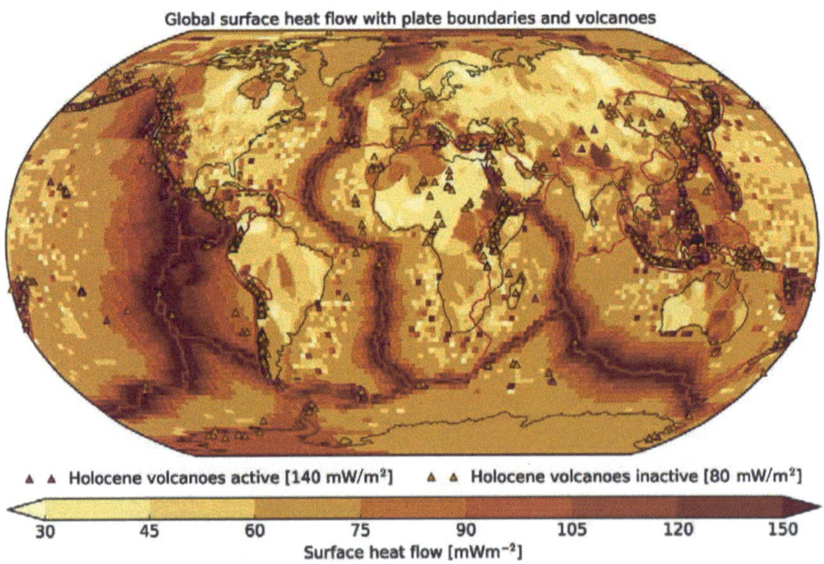

Fig. 8.18 ***Global map of surface heat flow rate per unit surface area from geothermal energy.*** *Source: (Limberger, 2018).*

are dominant due to the hot-rock temperatures. Most of these areas are dry regions that experience freshwater scarcities. So, installation of desalination plants in such regions may help satisfy the demand for freshwater (Ghaffour, 2009).

8.4.1 Geothermal desalination systems

Geothermal desalination systems are typically divided into low-enthalpy (low-temperature) and high-enthalpy (high-temperature) systems. Some geothermal desalination systems utilize naturally hot water or dry steam directly, while others initially generate electrical power that is then used to drive desalination units. Another geothermal desalination system configuration involves coupling with other renewable energy sources to produce freshwater in dual or multigeneration schemes.

Thermal energy from geothermal resources can be extracted via wet-rock geothermal energy production, through which hot water or steam is harnessed. In another approach, called hot dry rock recovery or enhanced geothermal systems, water is injected into wells drilled into the ground, pumped through fractures in the rocks and finally recovered and returned to the ground surface through recovery wells (Axelsson, 1991).

8.4.1.1 Low-enthalpy geothermal-based desalination systems

Hot saline underground or thermal spring discharges can be utilized in low-temperature MD systems (Fig. 8.19). Such processes draw on low-enthalpy geothermal heated water sources. The driving force in the MD system is the partial vapor pressure difference between the hot and cold sides of the membrane. Having hot saline water at the hot side and a cooler seawater at the cold side leads to the diffusion of water vapor through the membrane. The vapor is condensed at the permeate (coolant) side. The temperature difference between the hot and cold sides is typically only between 5 °C and 10 °C. This method is not applicable for other membrane-based desalination processes such as reverse/forward osmosis or ED systems as they require pressure differences or electrical potentials to produce freshwater. Instead, an MD unit employs a microporous and hydrophobic membrane to generate freshwater by a liquid-vapor interface at the pores. Corrosion is not problematic in MD process as no metallic materials are employed. In addition, MD unit is not highly affected by the feed salinity.

Low-enthalpy geothermal heat sources can potentially drive thermally driven HDH desalination systems. In such systems, air is heated in a heat exchanger by a geothermal energy source at a temperature of

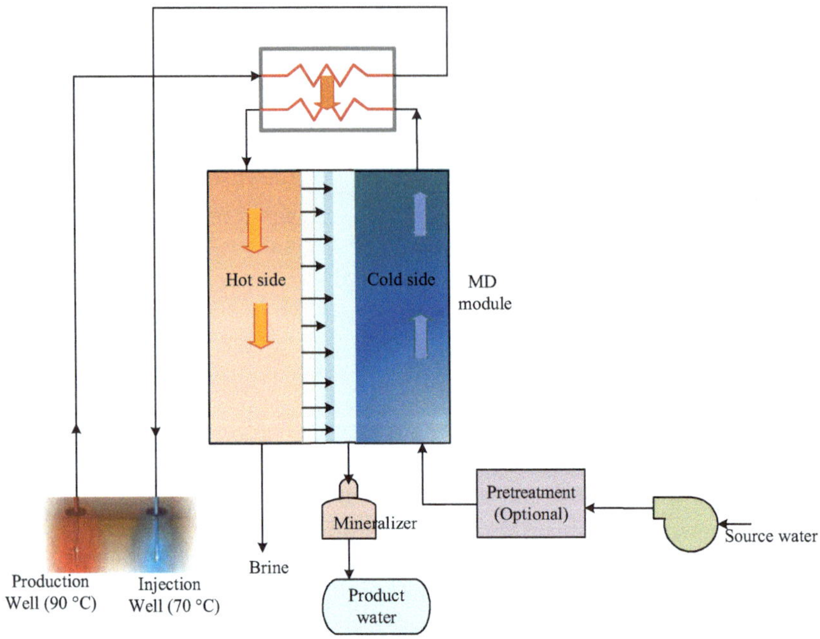

Fig. 8.19 ***Membrane desalination system driven by low-enthalpy geothermal energy.***

80–100 °C (Mohamed and El Minshawy, 2009), (Fig. 8.20). Mohamed and El-Minshawy (2009) showed that the HDH desalination efficiency using low-temperature geothermal energy increases as the geothermal inlet temperature rises, at the optimum water-to-air ratio (1.5–2.5). The HDH system can also operate efficiently with underground heat source temperatures above 100 °C. The coastal areas of California are among the regions where such systems can be implemented efficiently for freshwater production.

8.4.2 Hybrid low-enthalpy geothermal and solar/wind systems using underground TES

As mentioned in preceding sections, solar and wind energy cannot easily operate continuously without thermal or electrical energy storage due to the variability and intermittency of solar radiation and wind speeds. Geothermal energy, however, can continuously operate independent of the weather conditions. So, low-enthalpy geothermal energy facilities can be coupled with solar/wind plants to allow continuous operation, without the need for energy storage. Combined solar-geothermal-MD is an example of such a hybrid system.

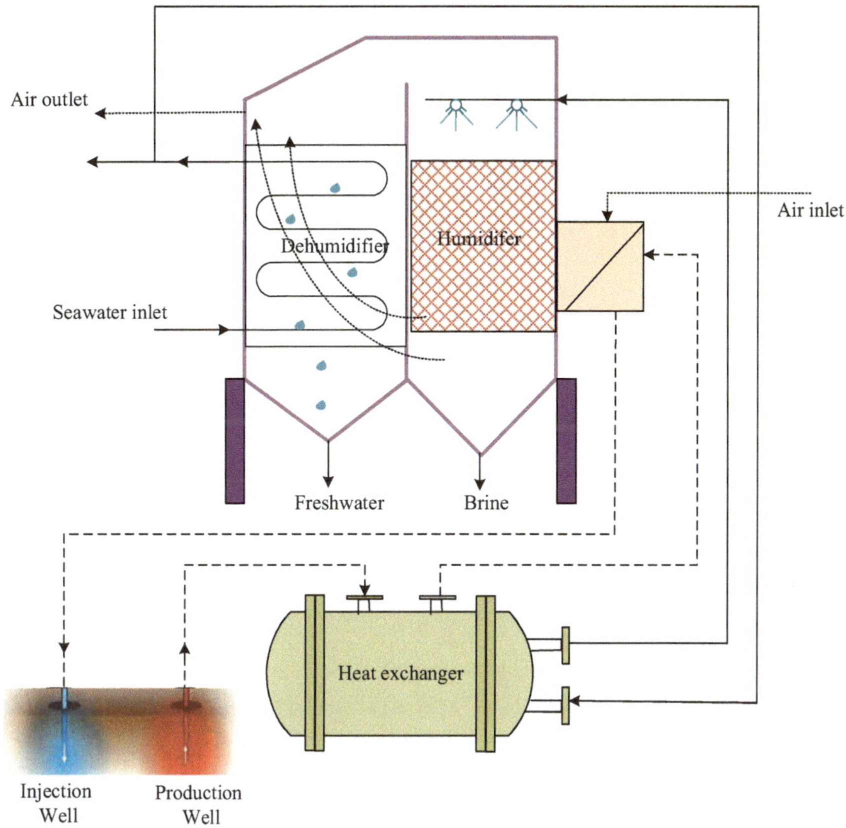

Fig. 8.20 ***Low-enthalpy geothermal unit used for an open-air open-water humidification–dehumidification desalination system.***

Some in-ground TESs can be used in conjunction with solar MD systems. One such storage that can provide short- or long-term energy storage is aquifer TES. In aquifer TES, thermal energy is stored by heating the groundwater in aquifers using groundwater wells. The stored thermal energy can be recovered subsequently when needed. The rock material surrounded the aquifer or borehole/tunnel usually has a low thermal conductivity, allowing it to serve as an insulator with a small amount of occupied space. In the aquifer, water is located in impermeable clay or rock. Aquifer TES uses the water in aquifers as a storage medium; heat is transferred into and out of storage by transmitting water to and from water-bearing rock. The aquifer's volume commonly exceeds millions of cubic meters containing 25% water, often making them a suitable choice for high-capacity systems. The water in an aquifer TES is extracted for a specific application then reinjected at another point in the aquifer for storage and then recovery. Thus, one well for cold storage and one well for hot storage is drilled.

In the charging mode, heated water at temperature T_c and mass flow rate $\dot{m}_{ch}$ is injected into the aquifer over a time period t_{ch} (charging period). In the discharging mode, water is recovered over a period t_{disch} from the aquifer at a mass flow rate of $\dot{m}_{disch}$. The discharge temperature of the water is a function of time (i.e., $T_d = T_d(t)$) with a limiting environmental temperature (i.e., $T_d(\infty) = T_0$). If a linear profile for the discharge temperature decrease with time is followed, from the initial temperature value of $T_d(0)$ to the final value of T_0, after a time period of t_f, the fluid temperature reaches a constant temperature T_0 and remains constant for the rest of the time. That is,

$$T_d(t) = \begin{cases} T_d(0) - \left(T_d(0) - T_0 . \dfrac{t}{t_f} \right); & 0 \leq t \leq t_f \\ T_0; \ t_f \leq t \leq \infty \end{cases} \tag{8.17}$$

Also, there are other underground TES types, including soil and earth bed, borehole, rock cavern, tank and solar pond. These are described in detail elsewhere (Rosen and Koohi-Fayegh, 2017).

8.4.3 Cogeneration and desalination using low-enthalpy geothermal energy

Electrical power generation using a geothermal energy source is a low-emission method for electricity production. In geothermal-driven electricity-generation plants, hot water and steam are extracted from drilled wells and transferred directly/indirectly to the turbine of a power plant. The geothermal energy is converted to electricity by the power plant. The residual fluids from electricity production plants have temperatures between 80 °C and 150 °C that can be used in thermal desalination systems (Bundschuh et al., 2015). The temperature at the first or second pass of a turbine in an electrical power production cycle, such as an organic Rankine or Kalina cycle, is sufficiently high that it can be used to drive a low-temperature desalination process such as MD or MED.

A self-sufficient pilot geothermal-based MED system was established at Kimolos Island, Greece (Fytikas et al., 2005; Karytsas et al., 2002). The system has a freshwater production capacity of 75 m^3/h, with a geothermal energy resource temperature of 61 °C. It is shown that utilizing low-enthalpy geothermal energy can save the equivalent of 453,600 kg (about 500 tons) of oil consumption per year (Karytsas et al., 2002).

The feasibility was examined of a geothermal-powered MED plant in Texas (Birney et al., 2019). Texas has geothermal resources at temperatures below 150 °C (i.e., low-enthalpy geothermal energy), at depths of 3000–4000 m. The results of the thermodynamic assessment of a geothermal MED plant showed that the freshwater production from the proposed system increases from 121 m^3/day to 1132 m^3/day as the geothermal fluid flow rate increases from 30 kg/s to 90 kg/s. This range of freshwater production is sufficient to meet the water demand of 230–2130 people. In addition, employing existing wells can significantly reduce the costs of drilling associated with the geothermal system.

8.4.3.1 High-enthalpy geothermal desalination systems

High-enthalpy geothermal system (at temperature more than 150 °C) can use either steam or hot water to drive desalination systems. After elimination of particulates from the dry steam, it can be used directly to heat the feed seawater and then for other thermal loads. In this type of configuration, a secondary heat exchanger is not required, although the final determination of that depends on the dry steam temperature. In essence, the dry steam is used in place of the boiler that provides steam for thermal desalination systems.

The utilization of hot water from high-enthalpy geothermal sources introduces several complexities, especially regarding temperature and water quality. For high-temperature hot water applications with temperatures above 170 °C, the hot water should be kept under pressure to prevent flashing. Another problem is the precipitation of chemical compounds in the piping downstream of the flashpoint. Thus, an intermediate heat exchanger is necessary to transfer thermal energy to the feed seawater. In addition, geochemical techniques can be used to ensure no precipitation forms in the piping and injection well. It is difficult and expensive to remove silicate precipitate (Stober and Bucher, 2013). Near the shoreline of an ocean, geothermal resources mostly contain hot water at a temperature of 170–300 °C, rather than dry steam. In the application of high-enthalpy geothermal resources near shorelines in desalination systems, a heat exchanger is commonly required.

The hot water in a high-enthalpy geothermal energy source can either directly be input to a thermal desalination system or it may enter as steam for conventional MSF and MED plants, which are operated by steam. The steam is generated in a flash chamber. The number of effects/stages in a thermal desalination plant depends on the hot water temperature and the scaling factor at high temperatures. Note that utilizing a flash chamber to produce steam can potentially lead to significant precipitation of chemical compounds; these compounds need to be filtered from the steam before it is input to the thermal desalination system. Alternatively, a hot dry rock system with a higher recovery pressure and temperature can serve as a direct mechanical device for driving RO desalination. Details of this type of facility are addressed elsewhere (Missimer et al., 2016).

8.4.3.2 High-enthalpy geothermal electricity generation

High-enthalpy geothermal electricity generation systems are typically one of three types: flash steam, dry steam, or binary cycles, as shown in Fig. 8.21. The feed groundwater temperature in all these high-temperature facilities should be between at least 175 °C and more than 300 °C; the latter temperature can be harnessed from high-temperature wells. The configuration shown on the left side of Fig. 8.21 uses a separator to produce steam from the underground water extracted for the Rankine power cycle. In the dry steam configuration (center configuration in Fig. 8.21), the steam from the geothermal source is directly used in the Rankine power cycle. In the binary cycle (right configuration in Fig. 8.21), a heat exchanger is used to transfer heat from the high-temperature geothermal fluid to the working fluid in the electrical power generation system.

The pumping energy requirement to draw underground water depends on several factors: depth of the water source (Z_i (m)), water density (ρ (kg/m^3)), volumetric flow rate of water ($\dot{V}$ (m^3/s)), pipe diameter (d(m)), pipe length (l (m)), friction factor (f), efficiency of pump (η_p), and capacity factor of load system (CF) (typically 90%–100%). That is (Birney et al., 2019),

$$E_{p,i} = \frac{\rho g \dot{V}}{1000.\eta_p.CF} \times \left[Z_i + \frac{\left(\frac{4q_i}{\pi d^2} \right)^2}{2g} \times \frac{f}{d} \times (Z_i + l) \right] \tag{8.18}$$

Note that, when a geothermal water resource originates in a geopressured resource, the underground water moves freely to the surface without the need for pumping. The total mass flow rate extracted from

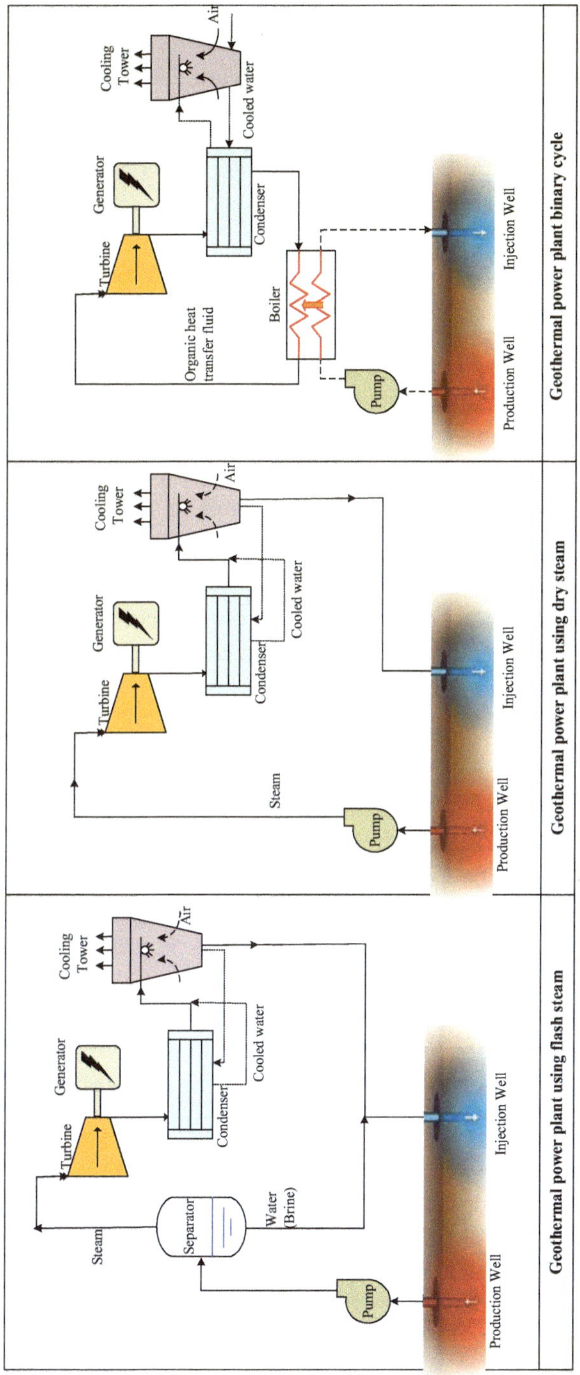

Fig. 8.21 Typical geothermal power plant using (left): flash steam, (middle): dry steam and (right): binary cycle.

a geothermal field is a function of wellhead pressure, that is, the pressure at which the separator operates. The mass production rate of geothermal fluid from a geothermal field can be determined through the following polynomial expression in terms of wellhead pressure (P_w) (DiPippo, 2012):

$$\dot{m}_{geo} = 99.663 - 2.6287P_w + 0.5802P_w^{\ 2} - 0.04212P_w^{\ 3} \qquad (8.19)$$

where $\dot{m}_{geo}$ is in kg/s and P_w in bar.

Coupling a thermal desalination system with any of the three high-enthalpy geothermal electricity-generation schemes in Fig. 8.21 is not complex. In the flash steam scheme, steam is delivered to the turbine and after expansion to the condenser. The hot water (brine) at the separator is also transferred for other thermal applications such as desalination. The dry steam scheme operates in the same manner as a typical electricity generation-thermal desalination (MSF or MED) cogeneration plant, in which the turbine outlet steam is transferred to the first stage/effect of the desalination system. Such configurations are currently operated in the Middle East in oil-fired power plants. The binary cycle geothermal power system offers more flexibility when combined with one or multiple high-temperature desalination systems. The heat exchanger utilized in this scheme allows for distribution of steam to various devices requiring thermal energy such as desalination units.

A geothermal electricity-generation facility can also be coupled with an RO desalination system (Fig. 8.22). Missimer et al. (2014) filed a patent for a system in which a geothermal electricity generator is integrated with aquifer storage and used to drive an RO module. The system also includes thermal MED and MD desalination systems. The RO module uses the electrical power produced by the geothermal recovery process. In fact, when the dry steam recovery facility operates at a steam temperature of 300 °C, the corresponding pressure at the geothermal reservoir is around 6–7 MPa. A portion of the high-temperature and high-pressure steam can be pressurized to generate mechanical work for the RO system. The primary portion of the steam is employed to produce electrical power in an electricity-generation facility. Thus, the geothermal energy source can be applied to both thermal- and work-driven desalination systems. Note that coupling an RO system with such an HP geothermal facility eliminates the need for pumping the intake seawater and for a chemical-dosing system that controls septicity in the wastewater.

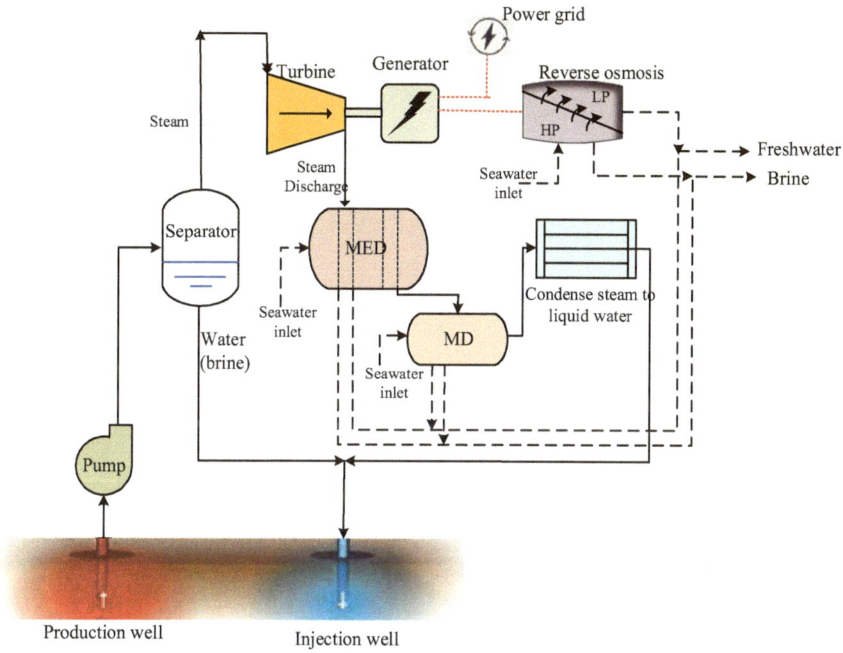

Fig. 8.22 ***Geothermal power plant combined with thermal desalination systems (MED and MD) and a work-driven desalination system (RO).***

8.5 Example: assessment of a geothermal combined system with an organic Rankine cycle and MED desalination

A knowledge of thermodynamics is important for understanding geothermal-based energy systems. Thermodynamic models of geothermal electricity-freshwater production systems can provide practical information to understand the system behavior. Energy and exergy analyses of such systems can predict thermodynamic properties of streams interacting within the system and system behavior. In this section, an illustrative example is presented of a thermodynamic analysis of a geothermal-driven organic Rankine cycle (ORC) and MED desalination system.

Consider the combined electricity-desalination system shown in Fig. 8.23. It consists of an ORC using ammonia as a working fluid, a steam generator, an MED system, and a fish farming unit. The primary heat source of the system is hot water from a geothermal resource. The hot

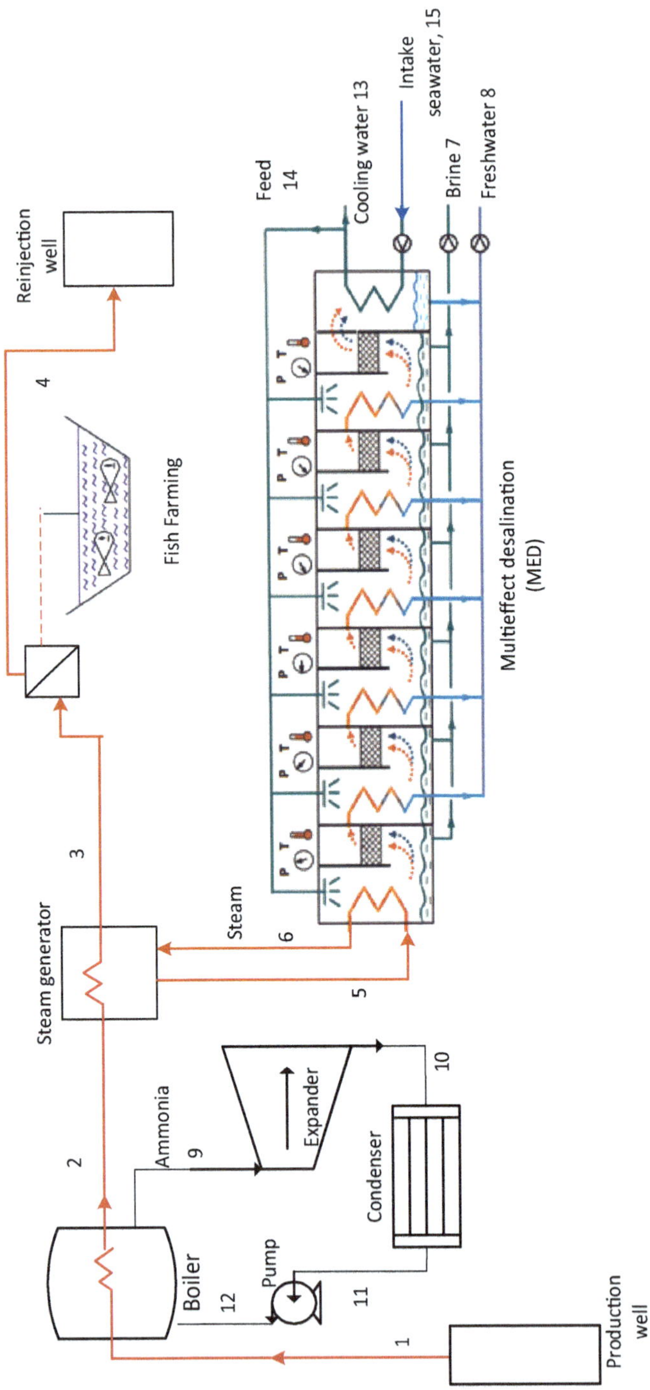

Fig. 8.23 *Geothermal energy source used to drive ORC and MED systems and a fish farm.*

hydrothermal fluid (150 °C) extracted from the production well (stream 1) superheats the ammonia in the boiler (stream 9). The boiler steam outlet is at a sufficiently high temperature (about 85 °C) to drive an MED desalination unit by generating steam in the steam generator and exiting at a temperature of 68 °C (stream 3). The energy in stream 3 can be further used in the fish farming plant, after which it is returned to the ground through the reinjection well (stream 4). Design parameters for the geothermal-based ORC–MED system are given in Table 8.7

Considering the design parameters given in Table 8.7 for this geothermal-based system and assuming no pressure drops and thermal losses in the system components:

- Develop mass, energy, entropy, and exergy rate balances for each component of the geothermal-based system.
- Determine the performance ratio (*PR*) for the MED system, as well as the energy and exergy efficiencies of each system.
- Link the exergy efficiency to environmental impact and sustainability for the combined system.

Solution: Energy and exergy analyses of a generalized system are described in detail in Chapters 4, 6, and 7. General expressions for energy

Table 8.7 Design parameters for the geothermal-based electricity-water production combined system.

Parameter	Value
Geothermal water temperature	150 °C
Environment temperature	25 °C
Environment pressure	101.135 kPa
ORC subsystem	
Expander isentropic efficiency	0.85
Boiler temperature	130 °C
Boiler pressure	3800 kPa
Condenser pressure	1500 kPa
Pump isentropic efficiency	0.85
MED subsystem	
Number of effects	6
Seawater input temperature	20 °C
Top brine temperature (TBT)	60 °C
Input seawater salinity	45 g/kg
Brine salinity	70 g/kg
Distillate salinity	0 g/kg

and exergy analyses are provided here. Neglecting kinetic and potential energy values, general mass, energy, entropy, and exergy rate balances under steady-state conditions for an open system in a control volume can be expressed, respectively, as

$$\sum_{in} \dot{m} - \sum_{out} \dot{m} = 0 \tag{8.20}$$

$$\dot{Q} - \dot{W} + \sum_{in} \dot{m}h - \sum_{out} \dot{m}h = 0 \tag{8.21}$$

$$\frac{\dot{Q}}{T} + \sum_{in} \dot{m}s - \sum_{out} \dot{m}s + \dot{S}_{gen} = 0 \tag{8.22}$$

$$\dot{E}x^{\dot{Q}} - \dot{E}x^{\dot{W}} + \sum_{in} \dot{m}ex - \sum_{out} \dot{m}ex - \dot{E}x_D = 0 \tag{8.23}$$

where $\dot{m}$, h, s and ex, respectively, denote mass flow rate, specific enthalpy, specific entropy, and specific exergy; $\dot{Q}$, $\dot{W}$, $\dot{S}_{gen}$, and $\dot{E}x_D$, respectively, denote heat transfer rate, work rate, entropy generation rate, and exergy destruction rate; and $\dot{E}x^{\dot{Q}}$ and $\dot{E}x^{\dot{W}}$, respectively, denote exergy rates associated with heat transfer and work. The latter two terms can be expressed as follows:

$$\dot{E}x^{\dot{Q}} = \left(1 - \frac{T_0}{T}\right)\dot{Q} \tag{8.24}$$

$$\dot{E}x^{\dot{W}} = \dot{W} \tag{8.25}$$

where T is the temperature (in K) at which the heat transfer process occurs.

General energy and exergy efficiencies for an energy system can be defined as

$$\eta = \frac{\text{Useful product energy rate}}{\text{Input energy rate}} \tag{8.26}$$

$$\Psi = \frac{\text{Useful product exergy rate}}{\text{Input exergy rate}} \tag{8.27}$$

ORC for electrical power generation: From Fig. 8.23, mass, energy, entropy, and exergy rate balances for the ammonia-based ORC components can be expressed, respectively, as follows:

Boiler

$$\dot{m}_{12} = \dot{m}_9 \tag{8.28}$$

$$\dot{Q}_{Boiler} = \dot{m}_{Geo}\left(h_1 - h_2\right) = \dot{m}_9 h_9 - \dot{m}_{12} h_{12} \tag{8.29}$$

$$\dot{m}_{12} s_{12} + \frac{\dot{Q}_{Boiler}}{T_{Boiler}} + \dot{S}_{gen,Boiler} = \dot{m}_9 s_9 \tag{8.30}$$

$$\dot{m}_{12} ex_{12} + \dot{Q}_{Boiler}\left(1 - \frac{T_0}{T_{Boiler}}\right) = \dot{m}_9 ex_9 + \dot{E}x_{D,Boiler} \tag{8.31}$$

Expander (exp)

$$\dot{m}_9 = \dot{m}_{10} \tag{8.32}$$

$$\dot{m}_9 h_9 = \dot{m}_{10} h_{10} + \dot{W}_{exp} \tag{8.33}$$

$$\dot{m}_9 s_9 + \dot{S}_{gen,exp} = \dot{m}_{10} s_{10} \tag{8.34}$$

$$\dot{m}_9 ex_9 = \dot{m}_{10} ex_{10} + \dot{W}_{exp} + \dot{E}x_{D,exp} \tag{8.35}$$

Condenser (cond)

$$\dot{m}_{10} = \dot{m}_{11} \tag{8.36}$$

$$\dot{m}_{10} h_{10} = \dot{m}_{11} h_{11} + \dot{Q}_{cond,ORC} \tag{8.37}$$

$$\dot{m}_{10} s_{10} + \dot{S}_{gen,cond} = \dot{m}_{11} s_{11} + \frac{\dot{Q}_{cond}}{T_{cond}} \tag{8.38}$$

$$\dot{m}_{12} ex_{12} = \dot{m}_{11} ex_{11} + \dot{Q}_{cond}\left(1 - \frac{T_0}{T_{cond}}\right) + \dot{E}x_{D,cond} \tag{8.39}$$

Pump

$$\dot{m}_{11} = \dot{m}_{12} \tag{8.40}$$

$$\dot{m}_{11} h_{11} + \dot{W}_{pump} = \dot{m}_{12} h_{12} \tag{8.41}$$

$$\dot{m}_{11} s_{11} + \dot{S}_{gen,pump} = \dot{m}_{12} s_{12} \tag{8.42}$$

$$\dot{m}_{11} ex_{11} = \dot{m}_{12} ex_{12} + \dot{E}x_{D,pump} \tag{8.43}$$

The energy and exergy efficiencies of the ORC subsystem can be expressed as

$$\eta_{ORC} = \frac{\dot{W}_{net}}{\dot{Q}_{Boiler}} = \frac{\dot{W}_{exp} - \dot{W}_{pump}}{\dot{Q}_{Boiler}} \tag{8.44}$$

$$\Psi_{ORC} = \frac{\dot{W}_{net}}{\dot{Q}_{Boiler}\left(1-\dfrac{T_0}{T_{Boiler}}\right)} = 1 - \frac{\dot{E}x_{D,Boiler} + \dot{E}x_{D,exp} + \dot{E}x_{D,cond} + \dot{E}x_{D,pump}}{\dot{Q}_{Boiler}\left(1-\dfrac{T_0}{T_{Boiler}}\right)} \tag{8.45}$$

MED system: A process flow diagram of a parallel-feed MED is shown in Fig. 8.24. In Chapter 5, a thermodynamic model is developed for an MED system. Seawater and pure water property formulations are also presented in Chapter 5 (Section 5.2). In the example, the work rate required for pumping feed, brine, and distillate streams are not considered. Denoting salt concentration w and using subscripts f, B, and D, respectively, to denote feed, brine, and distillate, mass, energy, entropy, and exergy rate balances for the MED effects, the condenser, and the overall unit can be written, as follows:

First effect

$$w_{sw}\dot{m}_{f1} = w_{B,1}\left(\dot{m}_{f1} - \dot{m}_{D1}\right) \tag{8.46}$$

$$\dot{Q}_{source} = \dot{m}_{f1}c_p\left(T_1 - T_f\right) + \dot{m}_{D1}h_{fg1} \tag{8.47}$$

$$\dot{m}_{14}s_{14} + \frac{\dot{Q}_{source}}{T_S} + \dot{S}_{gen} = \dot{m}_{B1}s_{B1} + \dot{m}_{D1}s_{D1} \tag{8.48}$$

$$\dot{m}_{14}ex_{14} + \dot{Q}_{source}\left(1-\frac{T_0}{T_s}\right) = \dot{m}_{B1}ex_{B1} + \dot{m}_{D1}ex_{D1} + \dot{E}x_{D,effect1} \tag{8.49}$$

Effect i to n

$$w_{sw}\dot{m}_{f1} + w_{B(i-1)}\dot{m}_{Br(i-1)} = w_{B(i)}\dot{m}_{B(i)} \tag{8.50}$$

$$\dot{Q}_{i-1} = \dot{m}_{D,n-1}h_{fgn-1} = \dot{m}_{D,i}h_{fgi} + \dot{m}_{B,i-1}c_{p,w}\left(T_{i-1} - T_i\right) \tag{8.51}$$

$$\dot{m}_i s_i + \frac{\dot{Q}_{i-1}}{T_{i-1}} + \dot{S}_{gen} = \dot{m}_{Bi}s_{B,i} + \dot{m}_{D,i}s_{D,i} \tag{8.52}$$

$$\dot{m}_i ex_i + \dot{Q}_{i-1}\left(1-\frac{T_0}{T_{i-1}}\right) = \dot{m}_{Bi}ex_{Bi} + \dot{m}_{Di}ex_{Di} + \dot{E}x_{D,effecti} \tag{8.53}$$

$$T_i = T_{vs,i} + BPE_i \quad i = 1, 2, \ldots, n \tag{8.54}$$

Here, T_i is the temperature of vapor generated in effect i, which is a function of the pure vapor saturation temperature ($T_{vs,i}$) and the boiling point elevation (BPE).

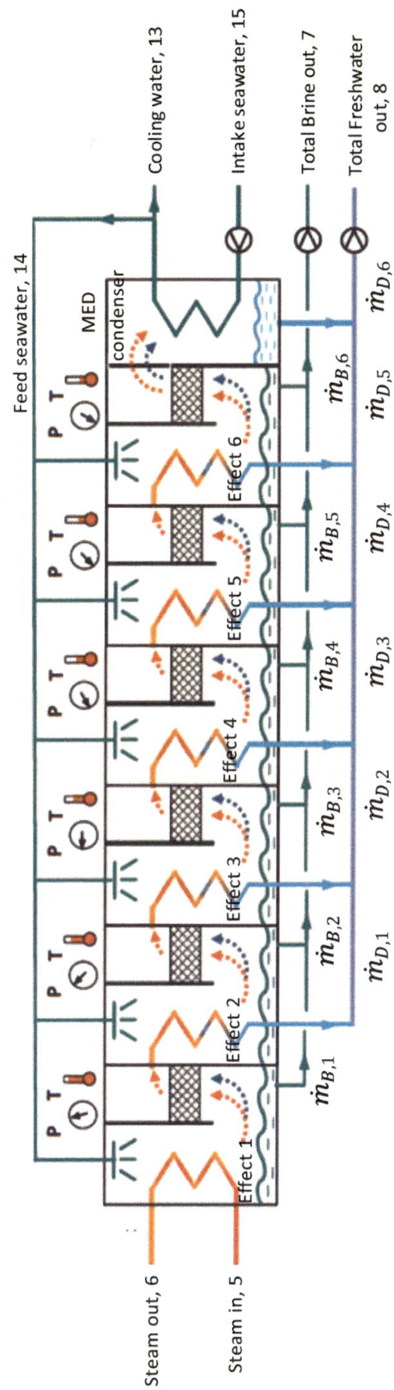

Fig. 8.24 ***Parallel-feed MED desalination unit.***

An energy rate balance for the condenser, in which vapor from the last effect is condensed using cooling water (*cw*), can be written as follows:

$$\dot{Q}_{cond} = \dot{m}_{D,n} h_{fg,n} = \left(\dot{m}_f + \dot{m}_{cw}\right) c_p \left(T_{sea} - T_f\right) \tag{8.55}$$

Steam generator (SG) heat exchanger

$$\dot{m}_{s,in} = \dot{m}_{s,out} \tag{8.56}$$

$$\dot{m}_2 = \dot{m}_3 \tag{8.57}$$

$$\dot{m}_2 h_2 + \dot{m}_{s,out} h_{s,out} = \dot{m}_3 h_3 + \dot{m}_{s,in} h_{s,in} \tag{8.58}$$

$$\dot{m}_2 s_2 + \dot{m}_{s,out} s_{s,out} + \dot{S}_{gen,SG} = \dot{m}_3 s_3 + \dot{m}_{s,in} s_{s,in} \tag{8.59}$$

$$\dot{m}_2 ex_2 + \dot{m}_{s,out} ex_{s,out} + \dot{S}_{gen,SG} = \dot{m}_3 ex_3 + \dot{m}_{s,in} ex_{s,in} + \dot{E}x_{D,SG} \tag{8.60}$$

Effects Block

$$\dot{m}_6 = \dot{m}_7 + \dot{m}_8 + \dot{m}_{13} \tag{8.61}$$

$$\dot{m}_6 h_6 + \dot{Q}_{SG} = \dot{m}_7 h_7 + \dot{m}_8 h_8 + \dot{m}_{13} h_{13} \tag{8.62}$$

$$\dot{m}_6 s_6 + \frac{\dot{Q}_{SG}}{T_S} + \dot{S}_{gen,MED} = \dot{m}_7 s_7 + \dot{m}_8 s_8 + \dot{m}_{13} s_{13} \tag{8.63}$$

$$\dot{m}_6 ex_6 + \dot{Q}_{SG}\left(1 - \frac{T_0}{T_s}\right) = \dot{m}_7 ex_7 + \dot{m}_8 ex_8 + \dot{m}_{13} ex_{13} + \dot{E}x_{D,MED} \tag{8.64}$$

In the analysis of a desalination process, it is necessary to determine the minimum work required for separation of feed into brine and distillate to evaluate the exergy efficiency of the desalting process. The minimum required work for separation can be defined for a reversible desalination operation (i.e., one with an entropy generation rate of zero), in which all input and output process streams (except distillate) are brought to the total dead state (i.e., environmental temperature, pressure and constituent concentrations). The distillate, as the sole useful product, is brought to the restricted dead state (i.e., environmental temperature and pressure). In this example, the process streams interacting with the desalination system are feed, brine, and distillate. So, the minimum work rate required for the desalting process is equal to the sum of the exergy rates corresponding to the process streams, as follows:

$$\dot{W}_{min,MED} = \dot{m}_D ex_D + \dot{m}_B ex_B - \dot{m}_{sw} ex_{sw} \tag{8.65}$$

Note in Eq. (8.65) that the control volume is selected such that all streams (i.e., feed, brine, and distillate) are in thermal and mechanical equilibrium with the environment.

The common energetic *PR* of an MED thermal desalination system can be defined as the ratio of mass flow rate of the distillate product to the mass flow rate of steam as the heating medium in the first effect. So, the *PR* of MED system can be expressed as

$$PR = \frac{\dot{m}_{D,total}}{\dot{m}_s} \tag{8.66}$$

The exergy efficiency of a desalination system is a measure of effectiveness of a desalting process compared to its performance in reversible conditions. Exergy destructions and losses reduce exergy efficiencies. The exergy efficiency of a desalination system is a function of the environment temperature, the feed seawater concentration, the recovery ratio (i.e., $\dot{m}_D / \dot{m}_f$), and the product salinity. One type of exergy efficiency is the ratio of minimum exergy rate required for the desalination process to actual exergy rate input to the system. With this definition, the exergy efficiency of the MED subsystem can be expressed as

$$\Psi_{MED} = \frac{\dot{W}_{min,MED}}{\dot{E}x_{in,MED}} = \frac{\dot{m}_7 ex_7 + \dot{m}_8 ex_8 - \dot{m}_{15} ex_{15}}{\dot{m}_{s,in} ex_{s,in} - \dot{m}_{s,out} ex_{s,out}} \tag{8.67}$$

Also, the exergy efficiency of an MED desalination system can be expressed using Eq. (8.68). As the useful exergy rate is equal to the exergy input rates minus the exergy loss and exergy destruction rates, the exergy efficiency can be expressed based on the ratio of the summation of the exergy loss and exergy destruction rates to the total input exergy rate. That is,

$$\begin{aligned}\Psi_{MED} &= 1 - \frac{\sum \dot{E}x_{Des,i} + \sum \dot{E}x_{loss,i}}{\dot{E}x_{in,MED}} \\ &= 1 - \frac{\sum_{effect\ i=1}^{6} \dot{E}x_{Des,i} + \dot{E}x_{Des,Cond,MED} + \dot{E}x_{Des,T,D} + \dot{E}x_{Des,T\&ch,B} + \dot{E}x_{Des,T\&ch,cw}}{\dot{m}_{s,in} ex_{s,in} - \dot{m}_{s,out} ex_{s,out}}\end{aligned} \tag{8.68}$$

Here, the exergy destruction rate indicates the lost available work rate due to irreversibilities (i.e., $\sum \dot{E}x_{Des} = \sum T_0 \dot{S}_{gen}$) and the exergy loss rate indicates the lost available work rate due to releasing waste streams (i.e., brine and cooling water) to the environment. The term $\dot{E}x_{Des,T,D}$ in Eq. (8.68) is the exergy destruction rate associated with bringing the distillate temperature to the environmental temperature. Then, the exergy destruction due to the temperature disequilibrium for the distillate can be written as

$$\dot{E}x_{Des,T,D} = \dot{m}_8 ex_8 \left(T_8, P_0, w_8\right) - \dot{m}_8 ex_8 \left(T_0, P_0, w_8\right) \tag{8.69}$$

where w_8 is assumed to be zero. Note that, for the brine and cooling water streams, the exergy destruction rates are associated with bringing both temperature and concentration to the environmental temperature and constituent concentration values (i.e., $\dot{E}x_{Des,T\&ch,B}$ and $\dot{E}x_{Des,T\&ch,cw}$). Then, the exergy destruction rates associated with temperature disequilibrium and chemical disequilibrium for the brine and cooling water can be written, respectively, as

$$\dot{E}x_{Des,T\&ch,B} = \dot{m}_7 ex_7\left(T_7, P_0, w_7\right) - \dot{m}_7 ex_7\left(T_0, P_0, w_0\right) \tag{8.70}$$

$$\dot{E}x_{Des,T\&ch,cw} = \dot{m}_{13} ex_{13}\left(T_{13}, P_0, w_{13}\right) - \dot{m}_{13} ex_{13}\left(T_0, P_0, w_0\right) \tag{8.71}$$

It is pointed out that the last terms on the right-hand sides of both Eqs. (8.70) and (8.71) are zero, because they are at the total dead state conditions.

Sustainability analysis: By utilizing sustainable or renewable sources of energy, as well as using efficiently nonrenewable sources such as natural gas, reduced environmental impact and greater sustainability can be attained. The sustainable index (*SI*) is employed here to relate the exergy and environmental impact concepts as follows:

$$SI = \frac{1}{D_p} \tag{8.72}$$

Here, D_P is the depletion number, defined as the ratio of destroyed exergy to exergy input. This equation shows how decreasing the environmental impact of a system can be attained by decreasing the exergy destruction. Therefore, the sustainability index indicates the extent of the contribution of exergy efficiency to sustainable development. The sustainability index can be written as follows:

$$SI = \frac{1}{1-\Psi} \tag{8.73}$$

Simulation results: Mass, energy, entropy, and exergy rate balances are written above for all system components, as are energy and exergy efficiency expressions for each system. To solve this example, the developed thermodynamic model is simulated using the thermodynamic properties of working fluids (i.e., ammonia, seawater, distillate, and brine) at each state point. Engineering equation software (EES) software is utilized to solve all equations simultaneously. Furthermore, input and evaluated values of temperature, pressure, specific enthalpy, specific entropy, and specific exergy for all state points are presented in Table 8.8. In addition, for the MED subsystem, the effect temperature and pressure, specific enthalpy, specific entropy

Table 8.8 Input and evaluated thermodynamic properties at state points in Fig. 8.23.

State no.	Fluid	T°C	P (kPa)	h (kJ/kg)	s (kJ/kg.K)	ex (kJ/kg)	x (kg/kg)
0	Seawater	25	101.135	98.7	0.34	0	0.045
0	Ammonia	25	101.135	1546	6.6	0	-
0	Water	25	101.135	104.3	0.36	0	-
1	Water	150	476.2	632.2	1.84	87.83	-
2	Water	85	461.2	356.3	1.13	22.81	-
3	Water	68	461.2	285	0.93	12.29	-
4	Water	56	461.2	234.8	0.78	6.72	-
5	Steam	65	101.135	2618	7.83	288.8	-
6	Steam	65	101.135	272.1	0.89	10.31	-
7	Brine	52.3	14.46	198.2	0.65	5.07	0.07
8	Distillate	52.1	14.46	217.7	0.73	8.32	0
9	Ammonia	130	3800	1680	5.317	516.5	-
10	Ammonia	60.26	1500	1558	5.38	375.1	-
11	Ammonia	38.7	1500	384.1	1.62	320.8	-
12	Ammonia	39.53	3800	388.8	1.62	324.8	-
13	Seawater	34.44	101.135	136.1	0.46	0.54	0.045
14	Seawater	34.44	101.135	136.1	0.46	0.54	0.045
15	Seawater	25	101.135	98.7	0.34	0	0.045

and specific exergy of brine, and entropy generation and exergy destruction rates for each effect, are provided in Table 8.9.

Discussion: Table 8.10 lists the heat and electrical power loads and the mass flow rate of produced freshwater in each MED effect. Theoretically, the required heat rates for the boiler and steam generator (79,407 kW and 20,523 kW, respectively) are supplied from the geothermal heat source. This indicates that using a renewable geothermal thermal energy resource

Table 8.9 Input and evaluated thermodynamic parameters for the MED effects.

Effect no.	T (°C)	P(kPa)	h_B (kJ/kg)	h_{vapor} (kJ/kg)	s_B (kJ/kg.K)	s_{vapor} (kJ/kg.K)	ex_B (kJ/kg)	$\dot{S}_{gen}$ (kW)	$\dot{E}x_{Des}$ (kW)
1	60	20.49	229.3	2610	0.7464	7.898	8.15	0.1494	44.51
2	56.8	17.65	218.2	2604	0.7153	7.95	6.72	0.6556	195.4
3	53.6	15.16	206.7	2599	0.682	8.003	5.45	0.5698	169.8
4	50.4	12.97	194.9	2593	0.647	8.057	4.32	0.4948	147.5
5	47.2	11.06	182.9	2587	0.6105	8.113	3.32	0.4316	128.6
6	44	9.39	170.7	2582	0.5729	8.17	2.45	0.3811	113.6

Table 8.10 Heat rate, work rate, and total freshwater production rate of the geothermal-based ORC–MED system.

Parameter	Value
η_{ORC} (%)	9.1
Ψ_{ORC} (%)	38.3
PR	4.25
Ψ_{MED} (%)	16
$\Psi_{overall}$ (%)	44.5
$\dot{Q}_{Boiler}$ (kW)	79,407
$\dot{Q}_{Cond,ORC}$ (kW)	72,193
$\dot{Q}_{SG}$ (kW)	20,523
$\dot{Q}_{Cond,MED}$ (kW)	11,907
$\dot{W}_{Expander}$ (kW)	7500
$\dot{W}_{Pump}$ (kW)	285.7
Freshwater production rate in each effect (kg/s)	$\dot{m}_{D1} = 7.82$, $\dot{m}_{D2} = 7.02$, $\dot{m}_{D3} = 6.34$, $\dot{m}_{D4} = 5.77$, $\dot{m}_{D5} = 5.31$, $\dot{m}_{D6} = 4.97$, $\dot{m}_{D,total} = 37.24$
Feed flow rate input to MED (kg/s)	19.56 each effect (equal feed mass flow rate for all effects)
Cooling water mass flow rate in MED (kg/s)	184.3
Steam flow rate input to the MED (kg/s)	8.75
Geothermal underground water mass flow rate (kg/s)	287.9
Ammonia mass flow rate (kg/s)	61.49

can significantly reduce the carbon dioxide emissions that would otherwise occur due to the consumption of conventional fossil fuels. The available mass flow rate of geothermal fluid extracted from the production well and the geothermal source temperature are the main factors that determine the system capacity. In a condition where there is insufficient geothermal energy, a backup renewable or nonrenewable energy source is required. Fig. 8.25 also shows the exergy destruction rate breakdown for all component of the geothermal-based ORC–MED system. As can be seen, the greatest exergy destruction rates are associated with the steam generator followed by the boiler. In general, the irreversibilities that are significant in

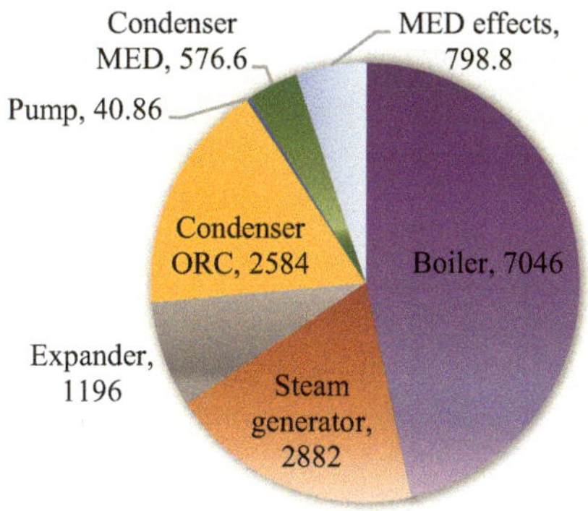

Fig. 8.25 ***Exergy destruction rate (kW) breakdown by component for the geothermal-based ORC-MED system.***

the heat exchangers are primarily due to heat transfer over large temperature differences. The exergy loss rate can be greater in the heat exchangers where phase change occurs or where different capacity rates (i.e., $\dot{m}_f c_p$) exist on either side of heat exchanger, or where a significant change in the specific heat capacity takes place. These conditions result in significant differences between the temperature and enthalpy rates. From a practical viewpoint, an important source of irreversibility in the heat exchanger is attributed to flow friction in the channels of the heat exchangers.

In this example, the high rate of exergy destruction in the steam generator and boiler heat exchangers is mainly due to the high temperature differences between the inlet and outlet streams. For the boiler, the main sources of irreversibilities are the large temperature differences and frictional losses (if pressure loss is counted), whereas for the steam generator the phase-change experienced by the vapor and condensed water, as well as the large temperature difference between the heating mediums, are the main contributors for the exergy destruction rate. An important way to lower the irreversibility in a heat exchanger is to reduce the temperature difference involved.

From the EES simulation, the obtained thermodynamic properties at each state point of the geothermal-based ORC–MED system can be used to calculate the *PR* and energy and exergy efficiencies of the MED and ORC systems. Then, the energy and exergy efficiencies of the ORC system can be calculated using Eqs. (8.44) and (8.45). respectively, as

$$\eta_{ORC} = \frac{\dot{W}_{exp} - \dot{W}_{pump}}{\dot{Q}_{Boiler}} \times 100 = \frac{7500 - 285.7}{79407} \times 100 = 9.1\% \qquad (8.74)$$

$$\Psi_{ORC} = \frac{\dot{W}_{net}}{\dot{Q}_{Boiler}\left(1-\frac{T_0}{T_{Boiler}}\right)}\times 100 = \frac{7500-285.7}{79407\left(1-\frac{298}{390}\right)}\times 100 = 38.31\% \quad (8.75)$$

The *PR* and exergy efficiency of MED system can be obtained using Eqs. (8.66) and (8.68), respectively, as:

$$PR = \frac{\dot{m}_{D,total}}{\dot{m}_s} = \frac{37.24}{8.75} = 4.25 \quad (8.76)$$

$$\Psi_{MED} = 1 - \frac{\left(\sum_{effect\ i=1}^{6} \dot{E}x_{Des,i}\right) + \dot{E}x_{Des,Cond,MED} + \dot{E}x_{Des,T,D} + \dot{E}x_{Des,T\&ch,B} + \dot{E}x_{Des,T\&ch,cw}}{\dot{m}_{s,in}ex_{s,in} - \dot{m}_{s,out}ex_{s,out}}\times 100$$

$$= \frac{(44.51+195.4+169.8+147.5+128.6+113.6)+576.6+179.3+394.7+99.9}{(8.75\times 288.8)-(8.75\times 10.31)}\times 100$$

$$= 15.88\% \quad (8.77)$$

Fig. 8.26 shows the impact of environment temperature on the sustainability index and the exergy destruction rate of the geothermal ORC and MED systems. As the environment temperature increases from 20 °C to 35 °C, the sustainability index for both the ORC and MED systems increases, while the corresponding exergy destruction rates decrease. For instance with an increase in the environment temperature from 20 °C to 35 °C, the sustainable index increases for the ORC and MED systems, respectively, from 1.17 to 1.23 and from 1.57 to 1.75.

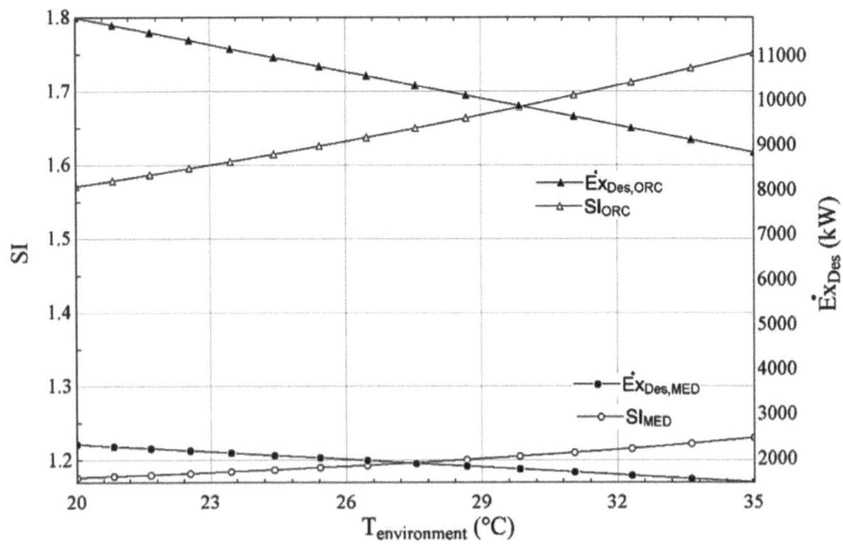

Fig. 8.26 ***Effect of environment temperature on exergy destruction rate and sustainability index for the ORC and MED systems.***

8.6 Nuclear energy

Nuclear energy is considered by many as sustainable energy option, especially for electric power generation. Based on the data reported to the International Atomic Energy Agency (IAEA) for 2019, nuclear power supplies around 10% of the worldwide electricity generation which is around one-third of the overall low-carbon electricity after hydro power (IAEA, 2019). At the end of 2018, 450 nuclear power reactors were being operated around the world, with overall electricity-generation rate of 398.9 GW (IAEA, 2019). Nuclear energy can be used for sustainable freshwater production through coupling nuclear power plants with desalination processes. Nuclear power plants are commonly installed in developed countries where the need for freshwater is not significant, while freshwater needs in many developing countries are vital. Technical knowledge has nonetheless been exchanged between developed and developing countries to allow the latter to benefit from nuclear electrical power and freshwater production.

In terms of the coolant type, nuclear reactors can be divided into two categories: heavy water reactors (HWRs) and light water reactors (LWRs). HWRs include pressurized HWRs (PHWRs), while LWRs include pressurized water reactors (PWRs) and boiling water reactors (BWRs). Other types of nuclear reactors include the high-temperature gas-cooled reactor (HTGR) and the liquid metal fast breeder reactor (LMFBR), which have been operated successfully in several countries. Among the various types of nuclear reactors, LMFBRs and PHWRs are typically more suitable for combining with desalination plants.

Nuclear reactors can also be classified in terms of their power production capacity, as small modular reactors (SMRs) (less than 300 MW net power), medium reactors (between 300 and 700 MW), and large reactors (more than 700 MW). The use of large nuclear reactors with desalination plants needs to consider performance as well as safety and stability factors (Yan et al., 2017).

Nuclear-power generation is the process of harnessing electrical power through the fission-controlled nuclear reactions that are accomplished in three stages. In the first stage, the kinetic energy of the fission fragments produced by splitting the heavier nucleus (U-235) into lighter nuclei yields thermal energy (a heat rate of around 196 MW) and gamma emissions. In the second stage, a working fluid is used in heat exchangers to remove the generated heat from the fuel and to transfer the thermal energy to a gas or steam before it enters a turbine. The working fluids typically can be heavy water (deuterium oxide, D_2O), light water (H_2O), a gas (CO_2 or helium), a liquid metal (sodium or lead/lead-bismuth), and molten salts (fluoroborates, chlorides, fluorides, eutectic compositions, or mixtures). In the last stage, the mechanical energy produced by expansion of the gas/steam fluid in the turbine is converted to electrical energy. The turbine outlet stream is commonly transferred to a condenser. The condensate is then preheated and pumped to the reactor to complete the power cycle.

In the direct nuclear cycle, the working fluid is water that absorbs heat from the fuel and raises steam for the turbine directly. This reactor type is the BWR shown in Fig. 8.27.

In the indirect nuclear cycle, a steam generator is used as a safety barrier heat exchanger to transfer heat from water inside the nuclear reactor to the secondary water-steam for expansion in a turbine. An example of this cycle is the PWR shown in Fig. 8.28. For a PWR, water in a liquid form as the primary coolant in the nuclear reactor flows into thousands of small tubes at a very high temperature and pressure. This high-temperature primary coolant transfers heat to the lower pressure secondary water coolant in the steam generator, a heat exchanger that evaporates the secondary coolant and conveys it to the steam turbine. The PWR is the most common nuclear reactor as of 2019, with an average net electrical power production capacity per reactor unit of 951 MW (Alonso et al., 2020). Table 8.11 provides and compares various characteristic values for BWRs and PWRs.

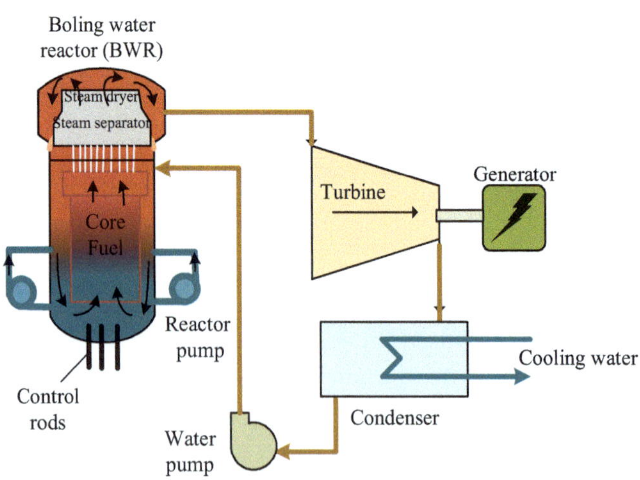

Fig. 8.27 ***Boiling water reactor (BWR) power plant.***

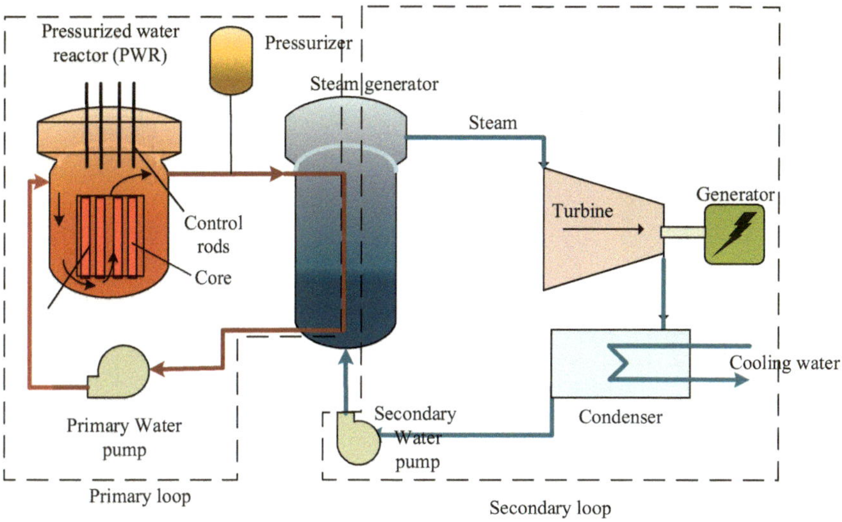

Fig. 8.28 ***Pressurized water reactor (PWR) power plant.***

Table 8.11 BWR and PWR design parameters.

Parameter	BWR	PWR
Power: thermal/electric (MW)	3600/1200	3600/1200
Fuel mass (kg-U)	138,000	96,000
Reactor coolant flow rate (kg/day)	1.89×10^6	2.55×10^6
Reactor coolant pressure (kPa)	7170	15,513.2
Coolant temperature: core inlet/outlet (°C)	276.6/286.1	290.5/326.6
Average heat flux (kW/m^2)	504.6	599.2
Specific power (kW_t/kg-U)	26.1	37.5
Maximum fuel centerline temperature (°C)	1871.1	1926.6

Data adapted from (Murray and Holbert, 2019).

In a PHWR, heavy water is used as the primary coolant and neutron moderator (speed reducer for fast neutrons). As with the PWR, the primary coolant in a PHWR is kept under pressure to avoid boiling. In the steam generator, heat is transferred to the lower pressure, secondary, nonradioactive water medium to generate steam for the turbine. Fig. 8.29 shows a simplified flow diagram of a PHWR power cycle.

In a light-water graphite-moderator reactor shown in Fig. 8.30, light water is used as the primary coolant and carbon graphite as the moderator, enabling nonenriched uranium to be used as nuclear fuel. Pressure tubes are used to keep the light water at high temperature and pressure so as to

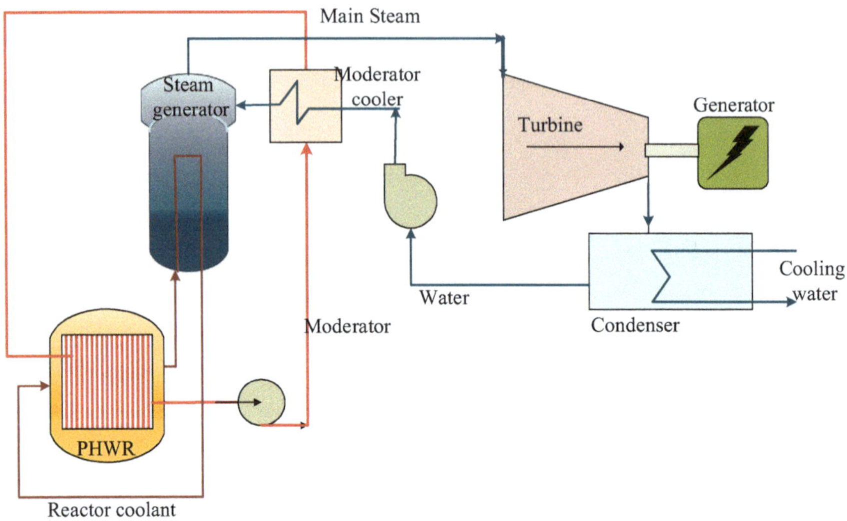

Fig. 8.29 ***Pressurized heavy water reactor (PHWR) power plant.***

prevent steam generation. Again, in the steam generator, heat is transferred to the secondary water coolant to generate steam that is fed to the turbine.

In the HTGR, also known as the very high-temperature reactor, the primary coolant is CO_2 or helium and the moderator is graphite. A very high-temperature coolant (more than 1000 °C) heats the light water in

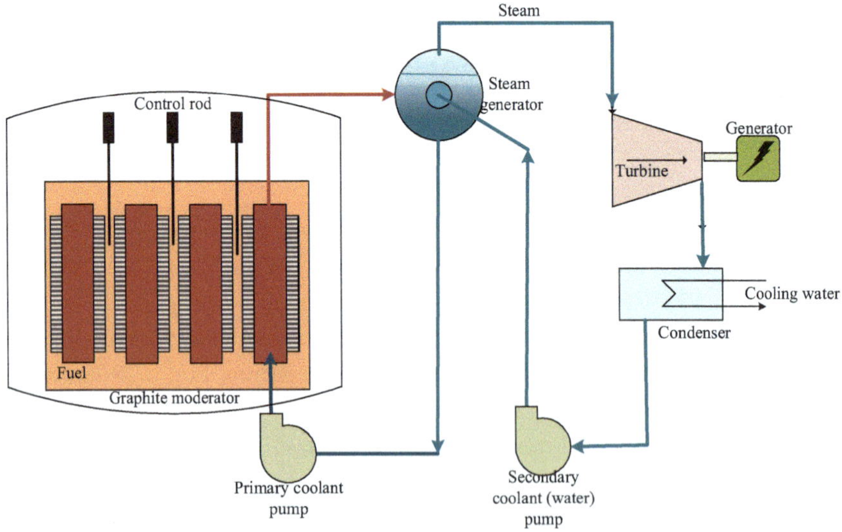

Fig. 8.30 ***Light water graphite-moderator reactor power plant.***

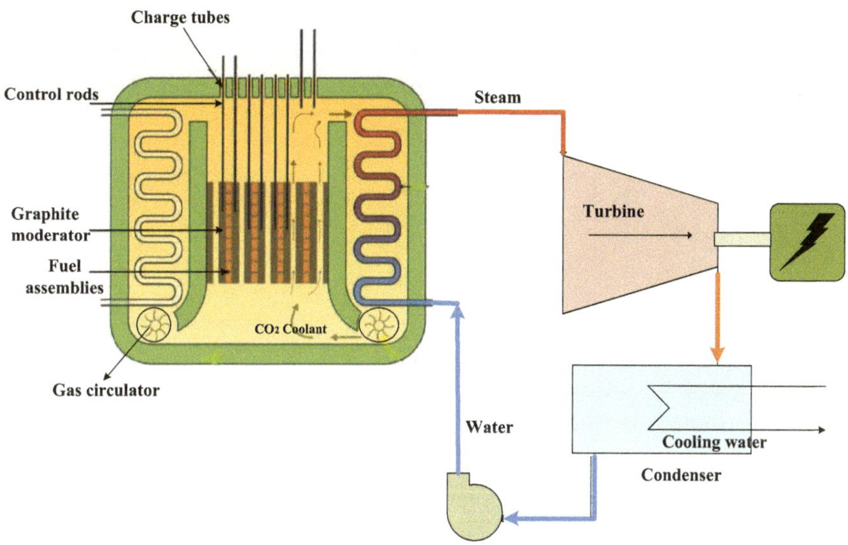

Fig. 8.31 ***High-temperature gas-cooled reactor (HTGR) power plant.***

the steam generator to provide steam for the turbine. Fig. 8.31 shows a HTGR nuclear power reactor, which is considered a Generation IV reactor concept.

Lastly, a liquid LMFBR is shown in Fig. 8.32. In this type of reactor, the primary cooling fluid is high-temperature sodium. The reactor uses fast neutrons instead of slow ones to strike a uranium-238 nucleus and form plutonium-239 (Murray and Holbert, 2014). The high-temperature sodium is passed from the first heat exchanger where the opposing side working fluid is also liquid sodium but with lower pressure. Then in the second heat exchanger, heat is transferred from the secondary liquid sodium to water to generate steam for the turbine. Thus, there are two heat exchangers in the design of this nuclear plant, helping to assure that the radioactive steam from the secondary loop does not enter the primary loop.

8.6.1 Electrical power production in a nuclear power plant

In a typical nuclear reactor, the thermal energy production rate by fission is often considered uniform throughout the fuel rod, as the height H of the rod is long in comparison with its radius R. Thus, the overall heat production rate in a cylindrical fuel rod, which has a cross-sectional area of πR^2, having a constant surface temperature of T_s can be written as (Murray and Holbert, 2014)

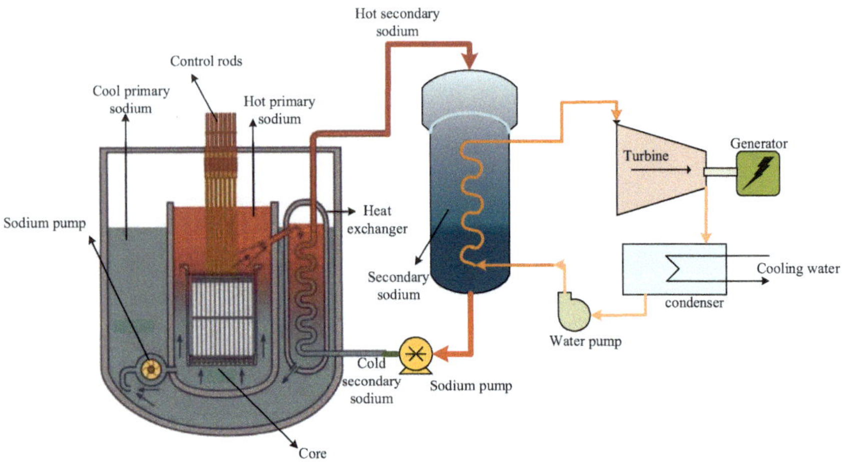

Fig. 8.32 ***Liquid metal fast breeder reactor (LMFBR) power plant.***

$$q = \left(\pi R^2 H\right)\left(q''\right) \tag{8.78}$$

where q'' is the rate of heat generation per unit volume. This term is associated with the nuclear fission process as follows:

$$q'' = w\Sigma_f \Phi \tag{8.79}$$

where $\Sigma_f\Phi$ is the fission reactor rate (related to the neutron flux) and w is the energy released per fission process.

A mass flow rate of coolant, $\dot{m}_c$, with a specific heat c_p and temperature $T_{c,in}$, enters the reactor, flows though the fuel rod channels, and is heated before exiting at temperature $T_{c,out}$. So, the reactor thermal power can be expressed as follows:

$$\dot{Q}_R = \dot{m}_c c_p \left(T_{c,out} - T_{c,in}\right) \tag{8.80}$$

Note that Eq. (8.80) is valid as long as there is no phase-change process, which is normally valid due to pressurization of the coolant. The state of a specific coolant, water, in terms of liquid and vapor phase, is shown in Fig. 8.33. The saturated condition lies on the liquid-vapor saturation curve.

The electrical power generation rate (P_R) in a nuclear power plant can be determined from the thermal efficiency of the power cycle ($\eta_{th,R}$) and thermal energy rate ($\dot{Q}_R$) as follows:

$$P_R = \eta_{th,R}\dot{Q}_R \tag{8.81}$$

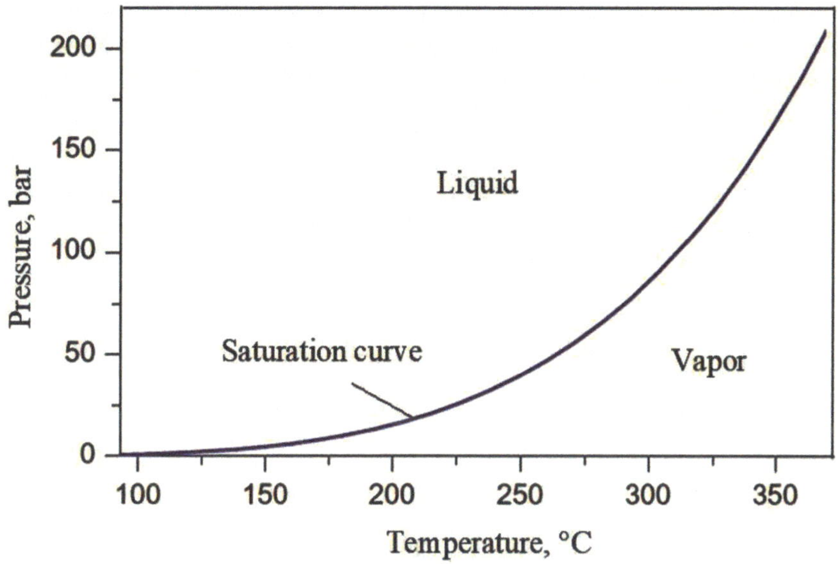

Fig. 8.33 ***Variation in liquid and vapor phases of water with temperature and pressure.***

The waste heat from a nuclear power plant is dissipated through the condenser and the cooling water flowing through it. With an energy rate balance, the rejected heat rate ($\dot{Q}_C$) from the condenser of a nuclear power plant can be written as

$$\dot{Q}_C = \dot{Q}_R - P_R \tag{8.82}$$

As the thermal efficiency of a typical nuclear power plant is about 33%, around 67% of input thermal energy is rejected from the plant via the condenser.

8.6.2 Nuclear desalination plants

Numerous theoretical and practical studies on coupling desalination systems with nuclear power plants for cogeneration of freshwater and electrical power have been reported in many countries, including Japan, Kazakhstan, Pakistan, India, the United States of America, China, Canada, and Russia. Much effort has been expended on eliminating the possibility of radioactive leakage into the product freshwater. Consequently, an intermediate loop consisting of a heat exchanger and a recirculation pump, is often used to connect the generated steam from the nuclear facility to the desalination plant. This loop operates at a higher pressure than the secondary circuit of the nuclear power cycle. The intermediate loop mainly serves as a pressure barrier to prevent radioactive steam leakage into the desalting process in the event of a rupture in the steam generator and heat exchanger tubes.

The intermediate loop is selected based on the type of desalination process. For an MED system, for example, the intermediate loop also includes a flash tank for supplying steam to the first effect of the MED process (Fig. 8.34). For an MSF desalination process, hot water is used in the heat exchanger (brine heater) to heat the feed seawater. Nuclear-based MSF desalination plants can be driven either by heating reactors or electrical generation reactors. Heating reactors normally have lower capital costs. In a combined nuclear power plant-MSF desalination system, the layout is the same as for the fossil fuel–driven power plants; steam from the back-pressure steam turbine is used to heat seawater in the brine heater of the desalting process (Fig. 8.35).

Nuclear desalination systems typically operate at larger plant sizes than renewable energy–based desalination plants (Zarzo and Prats, 2018), largely because nuclear plants have notably higher energy densities. Installing nuclear desalination plants in developing countries, where demands for energy and freshwater are significant and increasing, requires technical, safety, and economic plans to manage with challenges involved in the design and operation of such cogeneration plants.

The IAEA is involved in nuclear desalination projects in many countries (Gowin and Konishi, 1999; IAEA, 2007; IAEA, 2002). In addition, numerous research activities for nuclear desalination have been undertaken by the IAEA. These activities seek to:

- provide practical information on possibilities for harnessing waste heat from a nuclear plant and using it in thermal desalination,
- develop environmentally benign nuclear desalination systems, and
- reduce total facility costs improving plant operating performance.

8.6.3 Established and under-evaluation nuclear desalination plants

Several selected nuclear desalination projects, both established and under evaluation and from various countries, are described in this section.

Japan: Japan is one of the leading countries in nuclear energy, as it supplies 33% of the country's electricity demand. Of the 10 nuclear reactors that were combined with MED/MSF desalination plants, only half are currently working. Several reactors are not operating anymore after the Fukushima accident in 2011. The combined systems are capable of

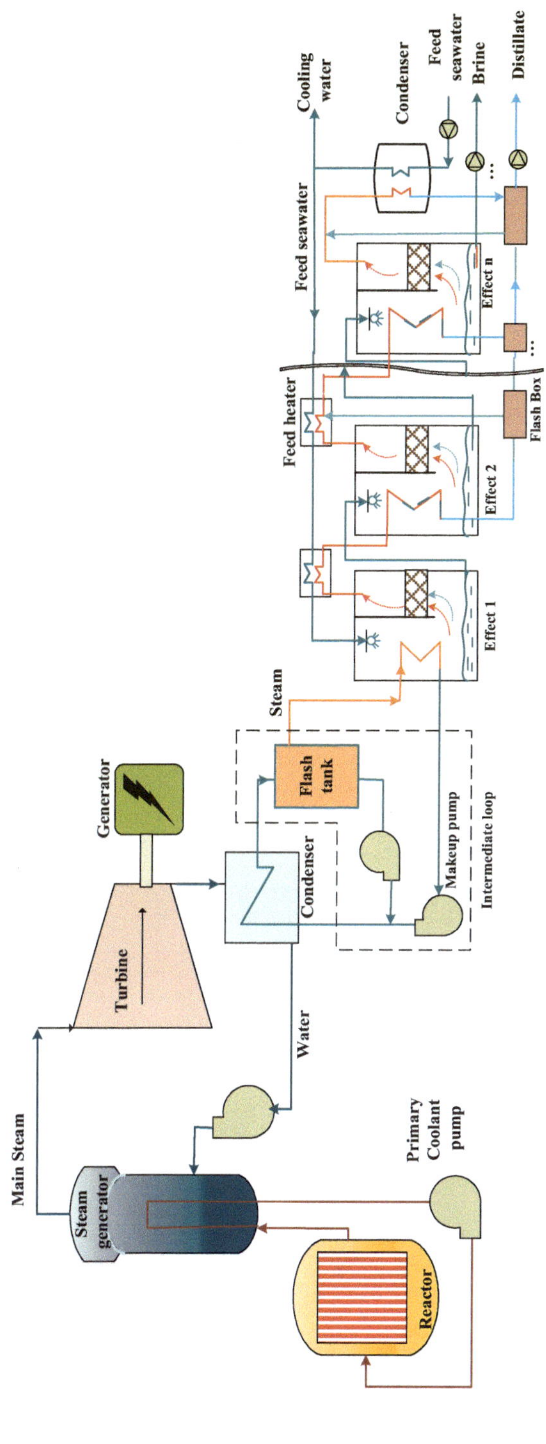

Fig. 8.34 *Nuclear power plant coupled with a multieffect desalination system.*

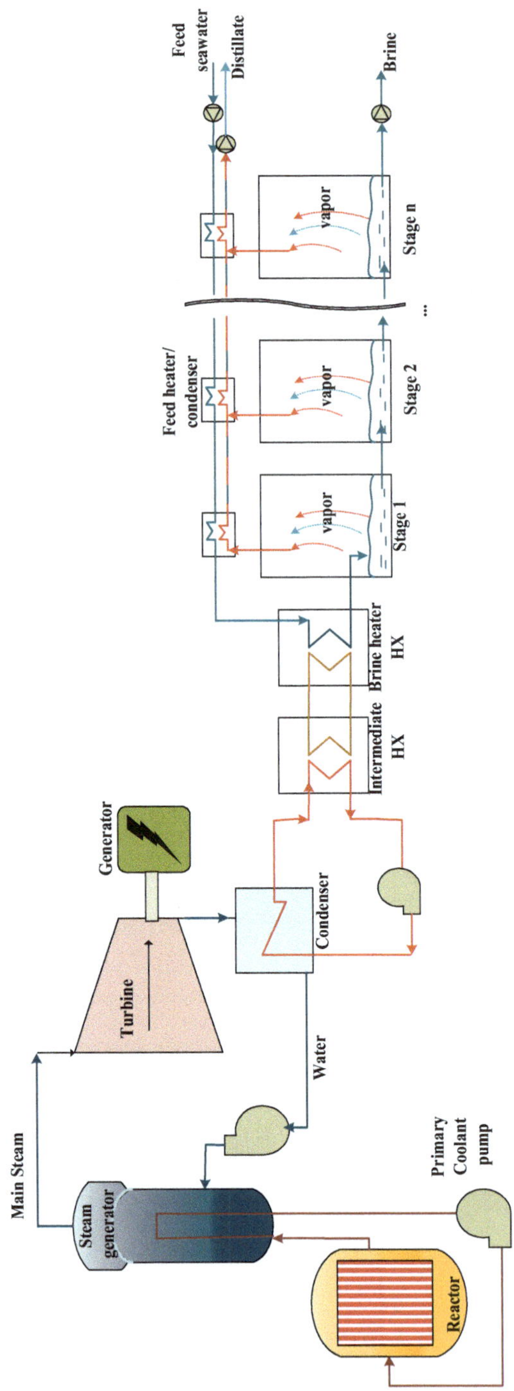

Fig. 8.35 *Nuclear power plant coupled with an MSF desalination system.*

producing electricity and freshwater, simultaneously. The MED/MSF desalination capacities are less for these plants than commercial ones, which typically have capacities around 1000–3000 m^3 per day. Based on operation and maintenance experience, successful operation of combined nuclear-desalination plants without radioactive leakage into the produced freshwater has been reported (IAEA, 2007; IAEA, 2002; Yan et al., 2017).

In 1973, the first PWR was coupled with an MSF desalination facility, using the Ohi-1 and Ohi-2 reactors, which each have a power production capacity of 1175 M. After 3 years of operation, the desalination unit capacity was increased from 1300 m^3/d to 2600 m^3/d. After the Ohi (I and II) experiences, the Ikata-I and Ikata-II nuclear power reactors with electric power capacities of 566 MW each began operation with an MSF desalination facility having a freshwater production capacity of 2000 m^3/d. Finally, in 1985 and 1992, the Kashiwazaki BWR (1100 MW) and the Genkai-IV power reactor (1180 MW) were successfully linked with MSF and MED units, respectively, each with a the freshwater production capacity of 1000 m^3/d.

Furthermore, a membrane-based RO desalination system was combined with the Genkai-III (1180 MW), Ohi-IV (1180 MW), and Ikata-III (890 MW) nuclear reactor plants for producing 1000 m^3, 2600 m^3, and 1000 m^3 of freshwater per day, respectively.

Kazakhstan: In 1973, the first fast breeder nuclear reactor (the BN-350), which has an electric power capacity of 150 MW, was combined with MED facilities for producing 120,000 m^3/d freshwater. A natural gas–based power plant was used as a backup. This cogeneration plant was the main water supplier in the Aktau region, providing 80% of the freshwater demand. By 1990 the plant capacity was upgraded to 10 MED units, with a production capacity for 80,000 m^3/d of distilled water; however, plant operation stopped due to technical and operational problems in both the nuclear reactor and desalination units (Gowin and Konishi, 1999).

India: A pilot plant in Trombay having a combined heavy water CIRUS nuclear reactor and a low-temperature MED facility was investigated in 2004 (Alonso et al., 2020). The MED desalination subsystem is capable of producing 30 m^3/d freshwater. The low-temperature MED system uses nuclear heat at a temperature of 530 °C and produces freshwater for the makeup needs of the reactor. In the CIRUS reactor, light water serves as the cooling fluid, while the moderator is heavy water. A metallic natural uranium is used as fuel. For safety reasons, an extra heat exchanger is employed between the nuclear plant and the desalination facility to prevent radioactive leakage into the produced freshwater.

MED and MSF desalination plants normally require more energy per unit freshwater product than RO systems, and they produce freshwater with a higher quality (i.e., having lower salinity and fewer impurities). Thus, there has been a tendency to couple nuclear reactors with two types of desalination systems in a hybrid manner, commonly involving MED-RO or MSF-RO configurations. The Nuclear Desalination Demonstration Project (NDDP) at Madras Atomic Power Station (MAPS) in India is an example of such a hybrid combined system. In this system, an HWR power plant with a capacity of 170 MW is coupled with a hybrid MSF–RO desalting process producing 6300 m^3/d freshwater. Fig. 8.36 shows a simplified view of this combined facility, including the PHWR power plant with MSF–RO desalination system.

Russia: In Russia, four VVER-1000 PWRs were combined with eight MED modules for a total freshwater production capacity of 9600 m^3/d. Low-grade heat from the power plant turbines is used in the desalination process (Alonso et al., 2020).

Pakistan: A 125 MW nuclear power plant based on a PHWR (KANUPP-1) was linked with an MED unit with a freshwater production capacity of 1600 m^3/d (Alonso et al., 2020). The plant capacity was subsequently upgraded by installing additional MED and RO systems.

United States: The only facility in the United States that produces its own needs for freshwater is the Diablo Canyon nuclear desalination plant in California. This facility is capable of producing 2180 m^3/d of freshwater, which is primarily used for cooling water makeup. RO and ultrafiltration modules are used in this facility. Fig. 8.37 shows the Diablo Canyon nuclear desalination facility, which is situated in a coastal cove.

Canada: In 1993, the CANDESAL enterprise started working actively with Atomic Energy of Canada Limited (AECL) and the IAEA on a potable water program and the employment of the Canada Deuterium Uranium (CANDU) reactor for desalination of seawater (IAEA, 2002). The CANDU nuclear reactor (having unit capacities of 450–668 MW) is a safe and reliable reactor design, making it suitable for integration with desalination. Both MED and RO desalination systems were candidates for coupling with the CANDU reactor. The relatively expensive changes associated with matching the heat required for the MED adversely affected the electrical power generation efficiency. Therefore, the CANDU-RO design with a freshwater production capacity of 12,000–24,000 m^3/d was preferred.

Some of the key features of the CANDU reactor are low-pressure, small diameter horizontal tubes, a low-temperature calandria (tank) containing

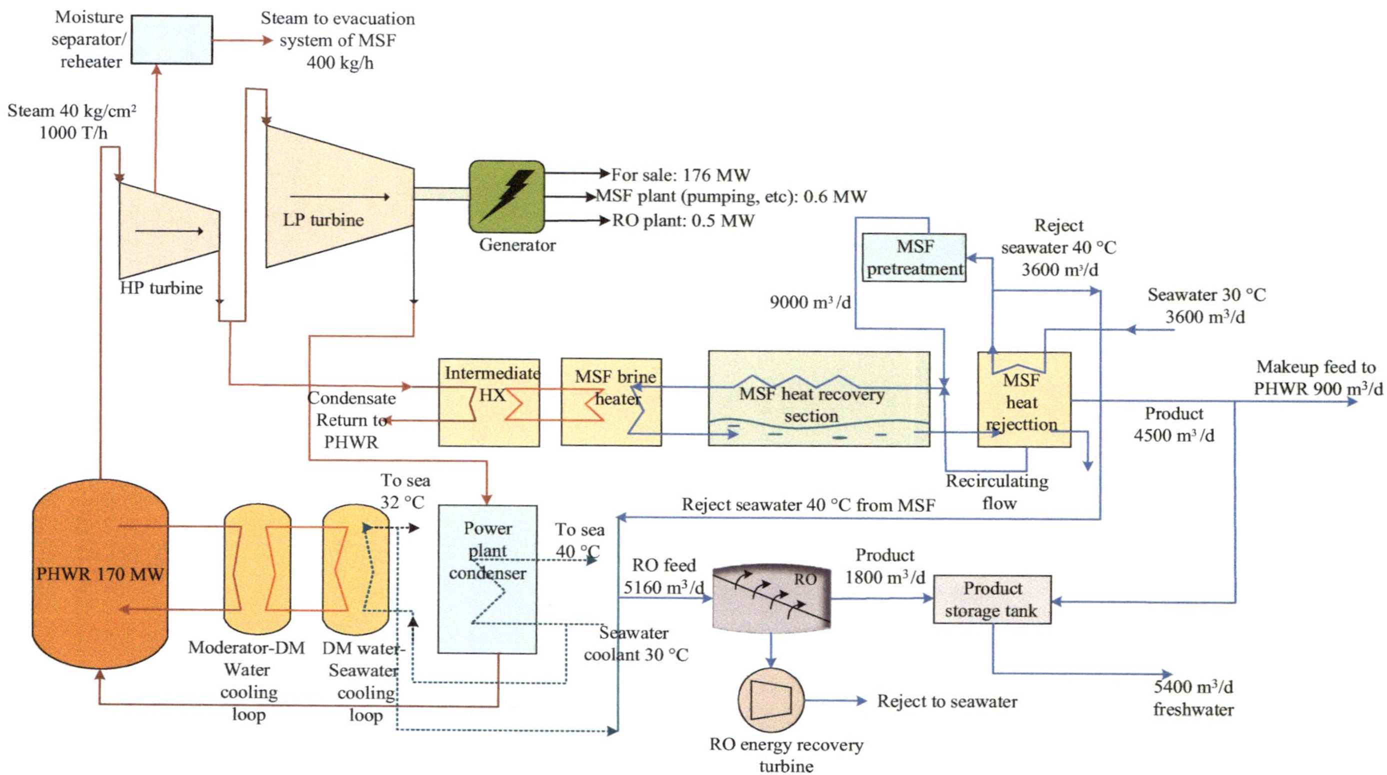

Fig. 8.36 ***PHWR power plant with a capacity of 170 MW combined with an MSF-RO desalination plant with a freshwater production capacity of 6300 m³/d.*** The plant was established by the Bhabha Atomic Research Centre (BARC). *Modified from IAEA, 2002. All rights reserved.*

Fig. 8.37 ***Diablo Canyon nuclear desalination plant with a freshwater production capacity of 2180 m³/d, in California, USA.*** *Source: IAEA, 2010. All rights reserved.*

heavy water as the primary coolant and moderator, operation with low fissile content fuel (e.g., natural uranium), and absence of any chemicals in the reactor coolant for reactivity control.

Fig. 8.38 presents a flow diagram of a PHWR-RO nuclear desalination system based on the CANDU-6 reactor in Canada. The cooling water in the condenser of the plant is used as the preheated feed seawater. This practice significantly improves the desalting and cogeneration efficiency, as it reduces the energy consumption per unit of freshwater produced. In addition, waste heat from the moderator cooling water module is used for additional heating of the feed stream temperature, raising its temperature by up to 9 °C before it enters the RO module. This process is feasible due to the unique design of CANDU reactor, while other reactors without this moderator cooling source of waste heat only have condenser waste heat with which to work. Moreover, ultrafiltration pretreatment is applied for improving the quality of feed stream, as well as for improving the performance of the membranes and extends their lifetimes.

In Table 8.12, selected installed nuclear desalination plants from around the world are summarized.

8.6.4 Environmental impact of nuclear desalination plants

Having the same location for nuclear plants and desalination systems when they are integrated has raised concerns about the impacts on marine, coastal,

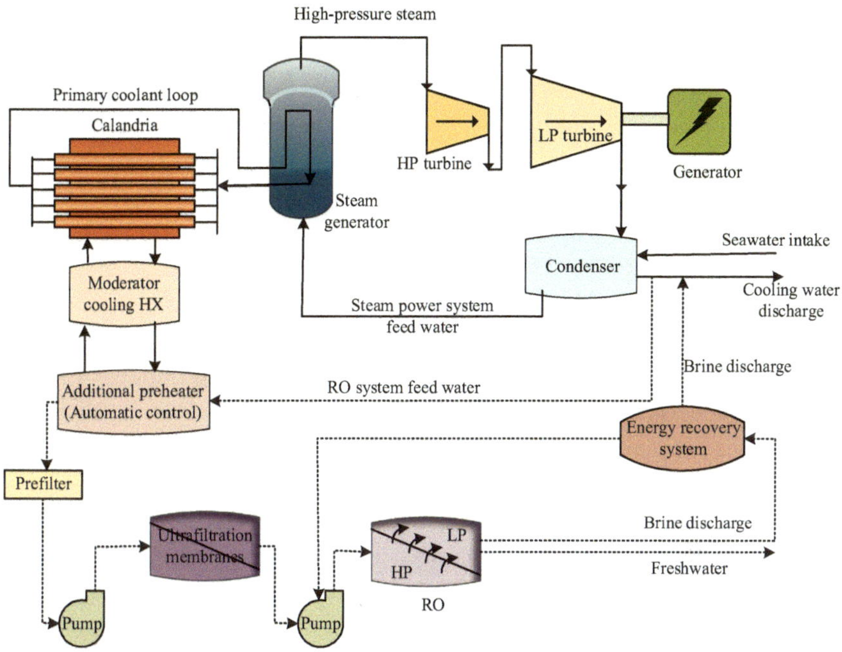

Fig. 8.38 ***CANDU-6 (PWHR)/CANDESAL (RO) nuclear desalination system.*** *Adapted from IAEA, 2002. All rights reserved.*

Table 8.12 Selected nuclear desalination power plants in various geographical regions.

Location	Plant	Type of power/ desalination system	Freshwater production capacity (m^3/d)	Gross power generation (MWe)
Fukui, Japan	Ohi-1,2	PWR/MSF	3900	2 × 1175
Fukui, Japan	Ohi-3,4	PWR/RO	2600	1 × 1180
Ehime, Japan	Ikata-1,2	PWR/MSF	2000	566
Ehime, Japan	Ikata-3	PWR/RO	2000	890
Fukuoka, Japan	Genkai-4	PWR/RO	1000	1180
Fukuoka, Japan	Genkai-3,4	PWR/MED	1000	2 × 1180
Fukui, Japan	Takahama-3,4	PWR/RO	1000	2 × 870
Aktau, Kazakhstan	Shevchenko	LMFR/MED, MSF	80,000–145,000	150
Trombay, India	LTE	PWHR/LTE*	30	40 (MW(t))
Kalpakkam, India	NDDP	PWHR/Hybrid: MSF-RO	6300	170
San Luis Obispo, USA	Diablo Canyon	PWR/RO	2180	2 × 1100

*Low temperature evaporation (LTE) desalination.
Data from IAEA, 2010; IAEA, 2002.

and atmospheric environments resulting from the continuous operation of such cogeneration systems. Indirect intake systems, such as horizontal well and beach well intakes (shown in Fig. 8.39), cannot provide sufficient water for the nuclear plants. It is noted that nuclear plants require more water than fossil fuel plants. Instead, direct intake systems (like open surface intakes) accompanied by pumps and filters are commonly used for nuclear power plants, but these systems result in significant impacts on aquatic life (Peters and Pintó, 2008). The high brine salinity and temperature significantly reduce the dissolved oxygen level, which can adversely affect the submarine species. Also, the chemical compounds present in rejected brine, such as heavy metal, chlorine, coagulants, and corrosion products, can lead to undesirable marine issues. The construction of nuclear desalination facilities requires the use of heavy machinery, which can result in visual and noise disturbances in natural coastal habitats.

Careful considerations of marine impacts of nuclear stations and actions to mitigate those impacts are necessary. Employing indirect intake water, even if it is limited to small plant scales, benefits from porous rocks and sand creates low suction forces and provides barriers for the marine organism. Fig. 8.39 shows two indirect intake designs: horizontal well and beach well. These designs have a specific capacity generally smaller than that for direct intake designs. The beach well consists of perforated covers with many

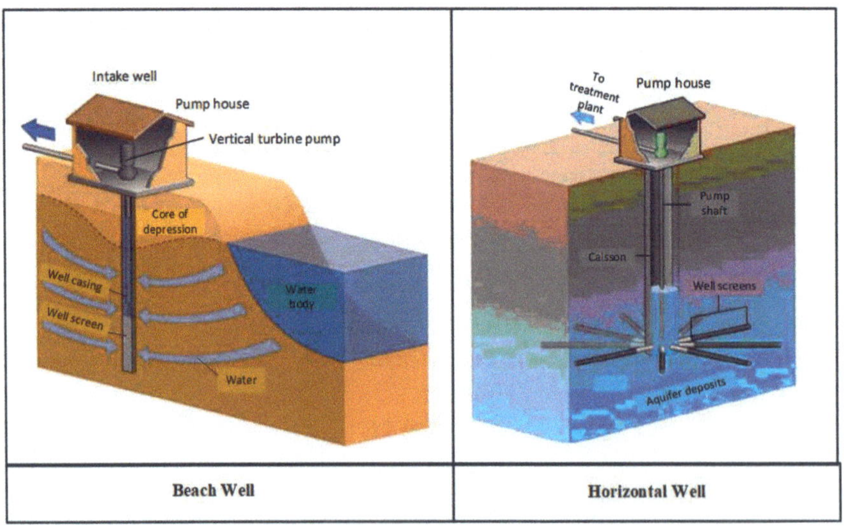

Fig. 8.39 ***Two common indirect intake designs used in nuclear desalination to reduce impingement and entrainment.*** *Modified from IAEA, 2010. All rights reserved.*

slots. These wells are often placed close to the seashore to harvest a high amount of seawater. In horizontal wells, pipes are laid in a radial pattern, which provides high intake seawater capacities. In addition, employing a cooling tower instead of once-through cooling could result in a considerable reduction in the seawater intake quantity and consequently reduce the rate of impingement and entrainment. Impingement and entrainment rates can also be considerably reduced by utilizing a fine mesh screen, mitigation methods, behavior deterrent devices, and Ristroph travelling screens (collection system) technologies. These are discussed in detail elsewhere (IAEA, 2010). Brine disposal can be eliminated through brine recovery, such as salt production processes or diluting of brine by the cooling water cycle prior to discharge.

In spite of the aforementioned potential adverse marine and coastal effects of nuclear desalination plants, these facilities have much lower impacts on the atmosphere compared to equivalent-capacity fossil-fueled desalination plants. In fact, the greenhouse gas emissions of nuclear plants are comparable with those of solar-, wind-, hydro-, and biomass-based desalination facilities (Weisser, 2007). For instance, the carbon dioxide released to the atmosphere from a nuclear-driven RO desalination system is about 10–60 gCO_2-equivalent per cubic meter of freshwater production, while it is 1000–2000 gCO_2-equivalent for the case where natural gas is used to drive the RO desalination plant (IAEA, 2010).

8.6.5 Safety issues in nuclear desalination plants

Although nuclear desalination facilities can be efficient for cogeneration of electricity and freshwater, safety considerations have a substantial influence on the public acceptance of such plants. As a consequence of the damages attributed to the Fukushima and Chernobyl nuclear accidents, nuclear policies are affected and reviewed by many countries to control and elevate safety concerns. Accidents in nuclear reactors have the potential for emissions of radioactive materials from the plant, which can have harmful impacts on humans and the environment. A strong safety culture is necessary in designing, operating, and monitoring nuclear desalination processes, and in particular to prevent accidental transfer of radioactive materials into the freshwater product. The IAEA leads safety activities for nuclear reactors in several countries, for example, Poland, Norway, Turkey, and Jamaica (IAEA, 2001).

A general agreement has been approved for the use of SMRs, due to their significant safety improvements, specifically for the new countries to nuclear energy (Locatelli et al., 2014). SMRs typically have capacities of not more than 300 MW and are considered as Generation III reactor concepts. The cooling process of an SMR after accidents or shutdown (which avoids radioactive material releases) is much easier than for other types of nuclear plants (Budnitz et al., 2018). SMRs offer several advantages, including safer operation, smaller area needs, longer lives, and reduced costs and installation time. These advantages have attracted attention throughout the world. SMRs are well suited for heating applications, especially desalination and district heating. Examples of this approach are the SMART system in Korea, KLT-40 in Russia, CAREM in Argentina, and IRIS in the United States (Khan et al., 2017). In addition to SMRs, off-shore nuclear power plant designs, which offer emergency passive containment cooling, is another approach that is under investigation (Lee et al., 2013).

As mentioned before, employing an intermediate loop between a nuclear plant and an MSF/MED desalination facility can reduce significantly the risk of radioactive material contamination of the freshwater product. In fact, by using an intermediate heat exchanger, if any radioactive leakage occurs, it does not impact feed seawater stream. In addition, for an RO desalination system that consumes electrical power, utilizing a separate steam generator for supplying electrical energy assures that there is no physical contact between the nuclear reactor and the desalination unit.

8.7 Closing remarks

In this chapter, sustainable desalination technologies are described and discussed. Many combinations and configurations exist of renewable energy–based desalination systems. In addition, modeling studies of solar thermal collectors, PV, wind turbines, battery banks, and nuclear reactors are described. Design considerations for hybrid renewable energy technology that drives electricity-water production combined systems are discussed. The efficient and practical selection of a sustainable desalination technology depends on potential and type of accessible sustainable energy source, availability of grid or battery power backup, specific energy consumption of plant, remoteness, technical infrastructure, and feed seawater salinity. In addition to technological and economic feasibility factors for the application of renewable energy in desalination systems, the freshwater demand and the local availability of renewable energy are important factors. Some integrations of renewable energy with desalination are more feasible for large-scale applications (e.g., concentrating solar collector-MED systems and wind-RO systems), while others are better suited to small-scale applications (e.g., PV-RO systems and solar still-MD/HDH systems).

Often abundant solar energy is available in regions with water scarcity challenges. So establishing solar power plants in these regions can be an important driver for the adoption of solar desalination technologies. Recent advances in PV technology are making PV-RO systems commercially viable despite their high initial costs and the intermittent nature of solar energy. Geothermal energy, as a nonintermittent renewable energy resource, can be combined with both high- and low-enthalpy desalination systems, at a variety of freshwater production capacities. The performance of geothermal desalination facilities can be significantly enhanced through innovations and improvements in aquifer thermal storage, harvesting geothermal energy optimally, and designing balanced heat collection to allow for replenishment of heat extracted in a sustainable manner. Relatively low-cost fuel and high-density energy production make nuclear energy a viable option for driving desalination systems. With rigid safety protocols and efforts to reduce marine and coastal effects of the operation of nuclear desalination plants, this technology is comparable to other renewable energies in terms of carbon dioxide emission. Renewable energy–driven desalination systems offer many advantageous over fossil fuel facilities, but extensive research and development is still required. The optimization of renewable energy–based desalination systems is important for the further development of sustainable freshwater production, and this topic is the focus of Chapter 10.

Nomenclature

A	area (m^2)
BPE	boiling point elevation(°C, K)
C_1	PV panel performance constant (V)
C_2	PV panel performance constant (V/K)
c_p	Specific heat at constant pressure (kJ/kg/K)
C_B	battery nominal capacity (Amp. Hour (Ah))
C_D	Salt concentration in product freshwater (mg/L)
CF	capacity factor
CP	power coefficient
C_{fc}	average concentration of water in feed channel of membrane (mg/L)
CPF	concentration polarization factor
D_P	depletion number
D_{wt}	rotor diameter (m)
$\dot{m}$	mass flow rate (kg/s)
E	energy (kJ)

ex	*specific exergy* (kJ/kg)
$\dot{E}x$	exergy rate (kW)
$\dot{E}x_D$	exergy destruction rate (kW)
G	incident solar energy (W/m^2)
GOR	gained output ratio
h	specific enthalpy (kJ/kg); heat transfer coefficient (W/m^2.K)
h_{fg}	latent heat (kJ/kg)
H	height (m)
I	current (amp)
I_{ph}	light-generated current (A)
I_0	reverse situation current (A)
J_D	distillate flux (kg/(m^2.s); m^3/(s.m^2))
k	thermal conductivity (W/m.K)
k_B	Boltzmann constant
k_w	water permeability coefficient (m/s.kPa)
k_s	membrane permeability for salt (m/s)
Nu	Nusselt number
P	pressure (kPa)
PR	Performance ratio
q	charge of electron
q''	rate of heat generation per unit volume (kW/m^3)
$\dot{Q}$	heat transfer rate (kW)
R	electrical resistance (Ω); radius (m)
R_s	panel series resistance (Ω)
R_s	shunt resistance (Ω)
Re	Reynolds number
s	specific entropy (kJ/kg.K)
S	absorbed solar radiation per unit area (W/m^2)
SOC	state of charge (%)
SI	sustainability index
T	temperature (°C, K)
TBT	top brine temperature (°C, K)
TCF	water permeability temperature correction factor
u	speed (m/s)
U	overall heat transfer coefficient (W/m^2.K)
V	potential difference (V)
ν	specific volume (m^3/kg)
$\dot{V}$	volumetric flow rate (m^3/s)
w	energy released per fusion (kJ); salinity (kg/kg)
$\dot{W}$	work rate (kW)
Z_i	depth of water source (m)

Greek letters

η	efficiency
ε	solar collector emissivity
ρ	density (kg/m^3)

σ	Stefan-Boltzmann constant; hourly self-discharge rate
Ψ	exergy efficiency
π	osmotic pressure (kPa)
ω	rotor angular velocity (1/s)

Subscripts

0	reference state
B	brine
ch	chemical
cw	cooling water
cond	condensation, condenser
D	distillate
Des	destruction
exp	expander
f	feed
sep	separator
SG	steam generator
sw	seawater
wt	wind turbine

Acronyms

BCRS	Brine concentrator and recovery system
BWR	Boiling water reactor
CSP	Concentrating solar power
DNI	Direct normal irradiance
ED	Electrodialysis
EES	Engineering equation solver
ERD	Energy recovery device
IAEA	International Atomic Energy Agency
HDH	Humidification–dehumidification
HP	High pressure
HTGR	High-temperature gas-cooled reactor
HWR	Heavy water reactor
LMFBR	Liquid-metal fast breeder reactor
LWR	Light water reactor
MD	Membrane desalination
MED	Multieffect desalination
MEMS	Multieffect multistage
MENA	Middle East and North Africa
MSF	Multistage flash
MVC	Mechanical vapor compression
ORC	Organic Rankine cycle
PHWR	Pressurized heavy water reactor
PV	Photovoltaic
PWR	Pressurized water reactor

RO	Reverse osmosis
SGSP	Salinity gradient solar pond
SMR	Small modular reactor
TES	Thermal energy storage
TVC	Thermal vapor compressor

References

Akbari Kordlar, M., Mahmoudi, S.M.S., Yari, M., Rosen, M.A., 2014. Energy and exergy analyses of a new combined cycle for producing electricity and desalinated water using geothermal energy. Sustainability 6 (4), 1796–1820.

Al-Alawi, A., Al-Alawi, S.M., Islam, S.M., 2007. Predictive control of an integrated PV-diesel water and power supply system using an artificial neural network. Renew. Energy 32 (8), 1426–1439.

Alavy, M., Peiris, M., Wang, J., Rosen, M.A., 2021. Assessment of a novel phase change material-based thermal caisson for geothermal heating and cooling. Energy Convers. Manage. 234, 113928.

Alcaraz, A., Montalà, M., Cortina, J.L., Akbarzadeh, A., Aladjem, C., Farran, A., Valderrama, C., 2018. Design, construction, and operation of the first industrial salinity-gradient solar pond in Europe: an efficiency analysis perspective. Sol. Energy 164, 316–326.

AlMadani, H.M.N., 2003. Water desalination by solar powered electrodialysis process. Renew. Energy 28 (12), 1915–1924.

Alonso, G., Del Valle, E., Ramirez, J.R., 2020. Desalination in Nuclear Power Plants. Elsevier, United Kingdom.

Al-Suleimani, Z., Nair, V.R., 2000. Desalination by solar-powered reverse osmosis in a remote area of the Sultanate of Oman. Appl. Energy 65 (1-4), 367–380.

Axelsson, G., 1991. Reservoir engineering studies of small low-temperature hydrothermal systems in Iceland. Report No. SGP-TR-134-19. National Energy Authority, Grensasvegur, Reykjavik Island.

Beurskens, J., Jensen, P.H., 2013. Wind energyEUREC Agency, The Future for Renewable Energy 2: Prospects and Directions, Second ed. Taylor & Francis Group, Routledge. London and New York.

Bilton, A.M., Kelley, L.C., Dubowsky, S., 2011. Photovoltaic reverse osmosis—feasibility and a pathway to develop technology. Desalin. Water Treat. 31 (1-3), 24–34.

Birney, C.I., Jones, M.C., Webber, M.E., 2019. A Spatially resolved thermodynamic assessment of geothermal powered multi-effect brackish water distillation in Texas. Resources 8 (2), 65.

Budnitz, R.J., Rogner, H.H., Shihab-Eldin, A., 2018. Expansion of nuclear power technology to new countries–SMRs, safety culture issues, and the need for an improved international safety regime. Energy Policy 119, 535–544.

Bundschuh, J., Ghaffour, N., Mahmoudi, H., Goosen, M., Mushtaq, S., Hoinkis, J., 2015. Low-cost low-enthalpy geothermal heat for freshwater production: innovative applications using thermal desalination processes. Renew. Sustain. Energy Rev. 43, 196–206.

Burton, T., Sharpe, D., Jenkins, N., Bossanyi, E., 2011. Wind Energy Handbook, Second ed. John Wiley & Sons, United Kingdom.

Carta, J.G.P.C.J.A., 2019. Wind energy powered desalination systems. In: Kucera, J. (Ed.), Desalination: Water from Water. Second ed. John Wiley & Sons, NJ, pp. 567–646.

Chitgar, N., Emadi, M.A., Chitsaz, A., Rosen, M.A., 2019. Investigation of a novel multigeneration system driven by a SOFC for electricity and fresh water production. Energy Convers. Manage. 196, 296–310.

CRES, 1998. Desalination guide using renewable energies. THERMIE-DC XVII, European Commission Report, Greece.

Cuomo, M.A., Kool, E., Reddy, B.V., Rosen, M.A., 2018. Energy modelling and analysis of a renewable energy system for dairy farm applications. Biofuels 12 (3), 273–283.

Davies, J.H., 2013. Global map of solid earth surface heat flow. Geochem. Geophys. Geosyst. 14 (10), 4608–4622.

Delyannis, E.E., 1987. Status of solar assisted desalination: a review. Desalination 67, 3–19.

Dincer, I., Rosen, M.A., 2021. Thermal Energy Storage: Systems and Applications, Third ed. Wiley, London.

Dincer, I., Rosen, M.A., Ahmadi, P., 2017. Optimization of Energy Systems. Wiley, London.

DiPippo, R., 2012. Geothermal Power Plants: Principles, Applications, Case studies and Environmental Impact, Third ed. Elsevier, Amsterdam.

DOE, 2016. Solar Tower System. Solar Energy Technologies Office. U.S. Department of Energy, Washington DC. https://www.energy.gov/eere/solar/articles/skytrough-parabolic-solar-collector (Accessed 21 June 2021).

DOE, 2016. Sky-Trough Parabolic Solar Collector. Solar Energy Technologies Office. U.S. Department of Energy, Washington DC. https://www.energy.gov/eere/solar/articles/solar-two-tower-system (Accessed 21 June 2021).

DOE, 2021. Concentrating Solar-Thermal Power Basics. Solar Energy Technologies Office. U. S. Department of Energy, Washington DC. https://www.energy.gov/eere/solar/concentrating-solar-thermal-power-basics (Accessed 21 June 2021).

Duffie, J.A., Beckman, W.A., 2013. Solar Engineering of Thermal Processes, Fourth ed. John Wiley & Sons, NJ.

Eastman, 2020. Therminol, heat transfer fluid by Eastman. https://www.therminol.com/product/71093459 (Accessed 23 June 2021).

El Haj Assad, M., Rosen, M.A., 2021. Design and Performance Optimization of Renewable Energy Systems. Academic Press, Elsevier, New York.

El-Nashar, A.M., 1993. An optimal design of a solar desalination plant. Desalination 93 (1), 597–614.

Energy stage, 2021. Smarter energy decisions. https://news.energysage.com/what-are-the-most-efficient-solar-panels-on-the-market/ (Accessed 23 June 2021).

ESMAP, 2019. Global Solar Atlas. SOLARGIS. https://globalsolaratlas.info (Accessed 16 June 2021).

Fytikas, M., Andritsos, N., Dalabakis, P., Kolios, N., 2005. Greek geothermal update 2000–2004, Proc. 2005 World Geothermal Congress, 1–8.

Ghaffour, N., 2009. The challenge of capacity-building strategies and perspectives for desalination for sustainable water use in MENA. Desalin. Water Treat. 5 (1-3), 48–53.

Gocht, W., Sommerfeld, A., Rautenbach, R., Melin, T.H., Eilers, L., Neskakis, A., Herold, D., Horstmann, V., Kabariti, M., Muhaidat, A., 1998. Decentralized desalination of brackish water by a directly coupled reverse-osmosis-photovoltaic-system-a pilot plant study in Jordan. Renew. Energy 14 (1-4), 287–292.

Gowin, P.J., Konishi, T., 1999. Nuclear seawater desalination—IAEA activities and economic evaluation for southern Europe. Desalination 126 (1-3), 301–307.

Gregorzewski, A., Genthner, K., Zarza, E., Leon, J., de Gunzbourg, J., Alefeld, G., Scharfe, J., 1991. The solar thermal desalination research project at the Plataforma Solar de Almería. Desalination 82 (1-3), 142–144.

Habibollahzade, A., Ahmadi, P., Rosen, M.A., 2021. Biomass gasification using various gasification agents: optimum feedstock selection, detailed numerical analyses and tri-objective grey wolf optimization. J. Clean. Prod. 284, 124718.

Haghghi, M.A., Shamsaiee, M., Holagh, S.G., Chitsaz, A., Rosen, M.A., 2019. Thermodynamic, exergoeconomic, and environmental evaluation of a new multi-generation system driven by a molten carbonate fuel cell for production of cooling, heating, electricity, and freshwater. Energy Convers. Manage. 199, 112040.

Hamrang, F., Shokri, A., Mahmoudi, S.M.S., Ehghaghi, B., Rosen, M.A., 2020. Performance analysis of a new electricity and freshwater production system based on an integrated gasification combined cycle and multi-effect desalination. Sustainability 12 (19), 7996.

Hamut, H.S., Dincer, I., Rosen, M.A., 2014. Assessment of desalination technologies integrated with renewable energy sources in Turkey. In: Dincer, I., Midilli, A., Kucuk, H. (Eds.), Progress in Exergy, Energy, and the Environment. Springer, New York, pp. 279–292.

Hau, E., 2013. Wind Turbines: Fundamentals, Technologies, Application, Economics, Third translated ed. Springer, New York and London.

Herold, D., Neskakis, A., 2001. A small PV-driven reverse osmosis desalination plant on the island of Gran Canaria. Desalination 137 (1-3), 285–292.

Hoffman, D., 1992. The application of solar energy for large-scale seawater desalination. Desalination 89 (2), 115–183.

IAEA, 2001. Safety aspects of nuclear plants coupled with seawater desalination units. International Atomic Energy Agency. IAEA-TECDOC-1235, Vienna. https://www.iaea.org/publications/6275/safety-aspects-of-nuclear-plants-coupled-with-seawater-desalination-units (Accessed 01 December, 2021).

IAEA, 2002. Status of design concepts of nuclear desalination plants. International Atomic Energy Agency. IAEA-TECDOC-1326, Vienna. https://www.iaea.org/publications/6368/design-concepts-of-nuclear-desalination-plants (Accessed 01 December, 2021).

IAEA, 2007. Status of nuclear desalination in IAEA member states. International Atomic Energy Agency. IAEA-TECDOC-1524, Vienna. https://www.iaea.org/publications/7584/status-of-nuclear-desalination-in-iaea-member-states (Accessed 01 December, 2021).

IAEA, 2010. Environmental impact assessment of nuclear desalination. International Atomic Energy Agency. IAEA-TECDOC-1642, Vienna. https://www.iaea.org/publications/8262/environmental-impact-assessment-of-nuclear-desalination (Accessed on 29 November 2021).

IAEA, 2019. Preliminary Nuclear Power Facts and Figures for 2019. https://www.iaea.org/newscenter/news/preliminary-nuclear-power-facts-and-figures-for-2019 (Accessed 23 June 2021).

IEA, 2010. Technology Roadmap—Concentrating Solar Power, IEA, Paris. https://www.iea.org/reports/technology-roadmap-concentrating-solar-power (Accessed 23 June 2021).

ITA, 2018. International Technological Applications, Greece. https://ita-sa.gr/%ce%b1%cf%86%ce%b1%ce%bb%ce%ac%cf%84%cf%89%cf%83%ce%b7?lang=en (Accessed 22 June 2021).

Jalili, M., Chitsaz, A., Hashemian, M., Rosen, M.A., 2021. Economic and environmental assessment using emergy of a geothermal power plant. Energy Convers. Manage. 228, 113666.

Joyce, A., Loureiro, D., Rodrigues, C., Castro, S., 2001. Small reverse osmosis units using PV systems for water purification in rural places. Desalination 137 (1-3), 39–44.

Kalogirou, S.A., 2004. Solar thermal collectors and applications. Prog. Energy Combust. Sci. 30 (3), 231–295.

Kalogirou, S.A., 2005. Seawater desalination using renewable energy sources. Prog. Energy Combust. Sci. 31 (3), 242–281.

Kalogirou, S.A., 2013. Solar Energy Engineering: Processes and Systems, Second ed. Elsevier, Amesterdam.

Karytsas, K., Alexandrou, V., Boukis, I., 2002. The Kimolos geothermal desalination project, Workshop on Possibilities of Geothermal Energy Development in the Aegean Islands Region. Greece.

Kasaeian, A., Sharifi, S., Yan, W.M., 2018. Novel achievements in the development of solar ponds: a review. Sol. Energy 174, 189–206.

Keeper, B.G., Hembree, R.D., Schrack, F.C., 1985. Optimized matching of solar photovoltaic power with reverse osmosis desalination. Desalination 54, 89–103.

Kershman, S.A., Rheinländer, J., Neumann, T., Goebel, O., 2005. Hybrid wind/PV and conventional power for desalination in Libya—GECOL's facility for medium and small scale research at Ras Ejder. Desalination 183 (1-3), 1–12.

Keshavarzzadeh, A.H., Ahmadi, P., Rosen, M.A., 2020. Technoeconomic and environmental optimization of a solar tower integrated energy system for freshwater production. J. Clean. Prod. 270, 121760.

Khalid, F., Dincer, I., Rosen, M.A., 2016a. Analysis and assessment of a gas turbine-modular helium reactor for nuclear desalination. J. Nucl. Eng. Radiat. Sci. 2 (3), 031014.

Khalid, F., Dincer, I., Rosen, M.A., 2016b. Comparative assessment of CANDU 6 and sodium-cooled fast reactors for nuclear desalination. Desalination 379, 182–192.

Khan, S.U.D., Khan, S.U.D., Haider, S., El-Leathy, A., Rana, U.A., Danish, S.N., Ullah, R., 2017. Development and techno-economic analysis of small modular nuclear reactor and desalination system across Middle East and North Africa region. Desalination 406, 51–59.

Kilkis, S., Krajacic, G., Duic, N., Rosen, M.A., Al-Nimr, M.A., 2020. Advances in integration of energy, water and environment systems towards climate neutrality for sustainable development. Energy Convers. Manage. 225, 113410.

Koohi-Fayegh, S., Rosen, M.A., 2020a. A review of renewable energy options, applications, facilitating technologies and recent developments. Eur. J. Sustain. Dev. 4 (4), em0138.

Koohi-Fayegh, S., Rosen, M.A., 2020b. A review of energy storage types, applications and recent developments. J. Energy Storage 27, 101047.

Lai, W., Ma, Q., Lu, H., Weng, S., Fan, J., Fang, H., 2016. Effects of wind intermittence and fluctuation on reverse osmosis desalination process and solution strategies. Desalination 395, 17–27.

Lee, K., Lee, K.H., Lee, J.I., Jeong, Y.H., Lee, P.S., 2013. A new design concept for offshore nuclear power plants with enhanced safety features. Nucl. Eng. Des. 254, 129–141.

Li, C., Goswami, Y., Stefanakos, E., 2013. Solar assisted sea water desalination: a review. Renew. Sustain. Energy Rev. 19, 136–163.

Limberger, J., Boxem, T., Pluymaekers, M., Bruhn, D., Manzella, A., Calcagno, P., Beekman, F., Cloetingh, S., van Wees, J.D., 2018. Geothermal energy in deep aquifers: a global assessment of the resource base for direct heat utilization. Renew. Sustain. Energy Rev. 82, 961–975.

Locatelli, G., Bingham, C., Mancini, M., 2014. Small modular reactors: a comprehensive overview of their economics and strategic aspects. Prog. Nucl. Energy 73, 75–85.

Loutatidou, S., Liosis, N., Pohl, R., Ouarda, T.B., Arafat, H.A., 2017. Wind-powered desalination for strategic water storage: techno-economic assessment of concept. Desalination 408, 36–51.

Lu, H., Walton, J.C., Hein, H., 2002. Thermal desalination using MEMS and salinity-gradient solar pond technology. NASA Technical Report, 00498.

Lu, H., Walton, J.C., Swift, A.H., 2001. Desalination coupled with salinity-gradient solar ponds. Desalination 136 (1-3), 13–23.

Ma, Q., Lu, H., 2011. Wind energy technologies integrated with desalination systems: review and state-of-the-art. Desalination 277 (1-3), 274–280.

Mahmoudi, H., Ghaffour, N., Goosen, M.F., Bundschuh, J., 2017. Renewable Energy Technologies for Water Desalination. CRC Press. Taylor & Francis Group, London, UK.

Maleki, A., Gholipour Khajeh, M., Rosen, M.A., 2016. Weather forecasting for optimization of a hybrid solar-wind-powered reverse osmosis water desalination system using a novel optimizer approach. Energy 114, 1120–1134.

Maurel, A., 1991. Desalination by reverse osmosis using renewable energies (solar–wind) Cadarche centre experiment, Proc. Seminar on New Technologies for the Use of Renewable Energies in Water Desalination. Athens, 26–28.

McAdams, W.H., 1942. Heat Transmission, Second ed. McGraw-hill, New York and London.

Miranda, M.S., Infield, D., 2003. A wind-powered seawater reverse-osmosis system without batteries. Desalination 153 (1-3), 9–16.

Missimer, T.M., Choon Ng, K., Thuw, K., Wakil Shahzad, M., 2016. Geothermal electricity generation and desalination: an integrated process design to conserve latent heat with operational improvements. Desalin. Water Treat. 57 (48-49), 23110–23118.
Missimer, T.M., Ng, K.C., Thu, K., Kim, Y., 2014. Systems and methods for integrated geothermal electricity generation and water desalination. United States Provisional Patent Application, 6197081.
Mittal, G., Singh, D.B., Singh, G., 2021. Production of electric power from solar ponds using thermoelectric generator: a review. Mater. Today 43, 608–613.
Mohamed, A.M.I., El-Minshawy, N.A.S., 2009. Humidification–dehumidification desalination system driven by geothermal energy. Desalination 249 (2), 602–608.
Mohamed, E.S., Papadakis, G., 2004. Design, simulation and economic analysis of a stand-alone reverse osmosis desalination unit powered by wind turbines and photovoltaics. Desalination 164 (1), 87–97.
Mosaffa, A.H., Farshi, L.G., 2021. Thermodynamic feasibility evaluation of an innovative salinity gradient solar ponds-based ORC using a zeotropic mixture as working fluid and LNG cold energy. Appl. Therm. Eng. 186, 116488.
Moser, M., Trieb, F., Fichter, T., 2013. Potential of concentrating solar power plants for the combined production of water and electricity in MENA countries. J. Sustain. Dev. Energy Water Environ. Syst. 1 (2), 122–140.
Murray, R., Holbert, K.E., 2014. Nuclear Energy: An Introduction to the Concepts, Systems, and Applications of Nuclear Processes, Seventh ed. Elsevier, Amsterdam.
NREL, 2018. Photovoltaic Lifetime Project. National Renewable Energy Laboratory. https://www.nrel.gov/pv/lifetime.html (Accessed 23 June 2021).
Ntavou, E., Kosmadakis, G., Manolakos, D., Papadakis, G., Papantonis, D., 2016. Experimental evaluation of a multi-skid reverse osmosis unit operating at fluctuating power input. Desalination 398, 77–86.
Ophir, A., 1982. Desalination plant using low grade geothermal heat. Desalination 40 (1-2), 125–132.
Orhan, M.F., Dincer, I., Naterer, G.F., Rosen, M.A., 2010. Coupling of copper-chloride hybrid thermochemical water splitting cycle with a desalination plant for hydrogen production from nuclear energy. Int. J. Hydrogen Energy 35, 1560–1574.
Paulsen, K., Hensel, F., 2007. Design of an autarkic water and energy supply driven by renewable energy using commercially available components. Desalination 203 (1-3), 455–462.
Peng, W., Maleki, A., Rosen, M.A., Azarikhah, P., 2018. Optimization of a hybrid system solar-wind-based water desalination by reverse osmosis: comparison of approaches. Desalination 442, 16–31.
Peters, T., Pintó, D., 2008. Seawater intake and pre-treatment/brine discharge—environmental issues. Desalination 221 (1-3), 576–584.
Petersen, G., Fries, S., Mohn, J., Müller, A., 1979. Wind and solar-powered reverse osmosis desalination units-description of two demonstration projects. Desalination 31 (1-3), 501–509.
Popiel, C.O., Wojtkowiak, J., Biernacka, B., 2001. Measurements of temperature distribution in ground. Exp. Therm. Fluid Sci. 25 (5), 301–309.
Rezaie, B., Rosen, M.A., 2020. The energy-water-food nexus: a framework for sustainable development modeling. Energy Equip. Sys. 8 (2), 179–201.
Richards, B.S., Schäfer, A.I., 2003. Photovoltaic-powered desalination system for remote Australian communities. Renew. Energy 28 (13), 2013–2022.
Rosen, M.A., 2021. Nuclear energy: non-electric applications. Eur. J. Sustain. Dev. 5 (1), em0147.
Rosen, M.A., Koohi-Fayegh, S., 2016. Cogeneration and District Energy Systems: Modelling, Analysis and Optimization. Institution of Engineering and Technology, London.
Rosen, M.A., Koohi-Fayegh, S., 2017. Geothermal Energy: Sustainable Heating and Cooling Using the Ground. John Wiley & Sons, UK.

Salehi, S., Mahmoudi, S.M.S., Yari, M., Rosen, M.A., 2018. Multi-objective optimization of two double-flash geothermal power plants integrated with absorption heat transformation and water desalination. J. Clean. Prod. 195, 796–809.
Sinha, S., Chandel, S.S., 2015. Review of recent trends in optimization techniques for solar photovoltaic–wind based hybrid energy systems. Renew. Sustain. Energy Rev. 50, 755–769.
Soetedjo, A., Lomi, A., Mulayanto, W.P., 2011. Modeling of wind energy system with MPPT control, Proc. 2011 International Conference on Electrical Engineering and Informatics. IEEE, pp. 1–6.
Sorensen, B., 2017. Renewable Energy, Fifth ed. Academic Press, Elsevier, MA.
Stober, I., Bucher, K., 2013. Geothermal Energy: From Theoretical Models to Exploration and Development. Springer, New York and London.
Subiela, V.J., de la Fuente, J.A., Piernavieja, G., Peñate, B., 2009. Canary Islands Institute of Technology (ITC) experiences in desalination with renewable energies (1996–2008). Desalin. Water Treat. 7 (1-3), 220–235.
Tabor, H., 1981. Solar ponds. Sol. Energy 27 (3), 181–194.
Taslimi, M., Ahmadi, P., Ashjaee, M., Rosen, M.A., 2021. Design and mixed integer linear programming optimization of a solar/battery based conex for remote areas and various climate zones. Sustain. Energy Technol. Assess. 45, 101104.
Thomson, M., Infield, D., 2003. A photovoltaic-powered seawater reverse-osmosis system without batteries. Desalination 153 (1-3), 1–8.
Thomson, M., Miranda, M.S., Infield, D., 2003. A small-scale seawater reverse-osmosis system with excellent energy efficiency over a wide operating range. Desalination 153 (1-3), 229–236.
Trieb, F., 2007. Concentrating solar power for seawater desalination. AQUA-CSP Study Report, German Aerospace Center (DLR).
Trieb, F., Muller-Steinhagen, H., Kern, J., Scharfe, J., Kabariti, M., Al Taher, A., 2009. Technologies for large scale seawater desalination using concentrated solar radiation. Desalination 235 (1-3), 33–43.
Tzen, E., Theofilloyianakos, D., Sigalas, M., Karamanis, K., 2004. Design and development of a hybrid autonomous system for seawater desalination. Desalination 166, 267–274.
Voivontas, D., Misirlis, K., Manoli, E., Arampatzis, G., Assimacopoulos, D., 2001. A tool for the design of desalination plants powered by renewable energies. Desalination 133 (2), 175–198.
Weiner, D., Fisher, D., Moses, E.J., Katz, B., Meron, G., 2001. Operation experience of a solar-and wind-powered desalination demonstration plant. Desalination 137 (1-3), 7–13.
Weisser, D., 2007. A guide to life-cycle greenhouse gas (GHG) emissions from electric supply technologies. Energy 32 (9), 1543–1559.
Wenham, S.R., Green, M.A., Watt, M.E., Corkish, R., Sproul, A., 2011. Applied Photovoltaics, Third Ed. Taylor & Francis Group, Routledge. London and New York.
Wu, B., Maleki, A., Pourfayaz, F., Rosen, M.A., 2018. Optimal design of stand-alone reverse osmosis desalination driven by a photovoltaic and diesel generator hybrid system. Sol. Energy 163, 91–103.
Yan, H.M., Cao, C.Y., Bai, G., Bai, W., 2017. Seawater desalination technology route and analysis of production capacity for large commercial nuclear power plant, Proc. International Confernece Pacific Basin Nuclear Conference. Springer, Singapore, pp. 865–872.
Yu, X., Li, W., Maleki, A., Rosen, M.A., Komeili Birjandi, A., Tang, L., 2021. Selection of optimal location and design of a stand-alone photovoltaic scheme using a modified hybrid methodology. Sustain. Energy Technol. Assess. 45, 101071.
Zarzo, D., Prats, D., 2018. Desalination and energy consumption. What can we expect in the near future? Desalination 427, 1–9.

CHAPTER NINE

Economics of seawater desalination using sustainable energy technologies

9.1 Introduction

By 2050, it is expected that 5 billion people will be affected by water scarcity crises (GWI, 2020). Climate change has the effect of increasing the intensity of water scarcity, potentially placing all countries at risk. Global population growth combined with scarcities of local water resources make desalination systems important for many countries, especially those dealing with water shortages. Recent advances in seawater desalination technologies and reductions in their costs have expanded the utilization of desalination systems. Currently, over 19,000 desalination facilities are installed around the world, which produce almost 100 million m^3/day freshwater from brackish water and seawater (GWI, 2020). This rate of freshwater production can satisfy only 1% of the world's drinking water demand (GWI, 2020). The number of desalination projects worldwide is increasing at an average annual rate of 5.5%, which corresponds to an increase in the freshwater production rate by 3.5 million m^3/day for new installed facilities every year.

Desalination plants can be classified in terms of freshwater production capacity. Large plants (freshwater production capacity of over 10^6 m^3/day) and medium plants (freshwater production capacity between 10^4 and 10^6 m^3/day) contribute about 90% of the total freshwater production rate from installed desalination plants (Voutchkov, 2018). At present, due to the increased need for freshwater, the trend in desalination projects is the construction of large-scale plants. For instance, recent seawater desalination projects in Australia, Algeria, Israel, Singapore, and Spain aim to supply 25%–30% of their long-term freshwater demands (Voutchkov, 2018).

From 2018 to 2022, the largest seawater desalination projects are likely to be in the Middle East and North Africa (MENA), with the costs for all projects totaling 28.2 billion US$ (GWI, 2016). After these, the following are the rough project costs from 2018 to 2022 for large desalination projects

Sustainable Energy Technologies for Seawater Desalination
DOI: https://doi.org/10.1016/B978-0-323-99872-7.00011-5

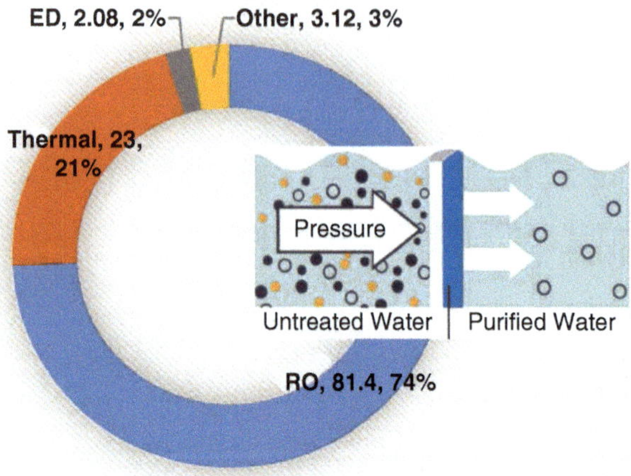

Fig. 9.1 ***Breakdown by industrial desalination system type of global freshwater production rate (in units of million m^3/day and percent of total) in 2020.*** The working principle for RO, the most widely used desalination technology, is shown as an overlay. *Data extracted from GWI (2020).*

worldwide: East Asia and the Pacific (9.6 billion US$), North America (5.1 billion US$), Latin America (4.9 billion US$), and Western Europe (2.9 billion US$) (GWI, 2016).

Fig. 9.1 provides a breakdown of industrial desalination technologies installed around the world. It is seen that reverse osmosis (RO) technology is the most commonly used desalination system, mainly due to its lower overall energy use and cost and higher efficiency compared to thermally driven desalination systems such as multieffect desalination (MED) and multistage flash (MSF) desalination. RO facilities are capable of producing 77.8 million m^3/day freshwater, which corresponds to 74% of total worldwide freshwater production (GWI. 2020).Table 9.1 lists the energy required

Table 9.1 Energy use and freshwater production cost for various industrial desalination facilities[a].

Energy type/medium	MSF	MED	SWRO[b]	BWRO[c]
Unit electrical energy equivalent of thermal energy (kWh/m^3)	9.5–11.0	4.5–6.0	-	-
Unit electrical energy use (kWh/m^3)	3.2–4.0	1.2–1.8	2.5–4.0	0.3–2.8
Unit overall energy use (kWh/m^3)	12.7–15	5.7–7.8	2.5–4.0	0.3–2.8
Unit cost of freshwater production (US$/m^3)	0.9–4.0	0.7–3.5	0.5–3.0	0.2–1.8

[a]Data extracted from GWI (2016).
[b]Seawater RO.
[c]Brackish water RO.

and water production cost for industrial desalination systems including MSF, MED, seawater RO, and brackish water RO (BWRO) desalination systems. RO desalination clearly is the preferred technology for producing freshwater around the world. In the MENA region, where over 50% of worldwide installed desalination systems are located; there has been a trend toward the use of RO seawater desalination over the past 8 years. In Saudi Arabia, the largest combined MSF–RO desalination system is established, with MSF producing 727,000 m^3/d freshwater and RO 310,000 m^3/d; but the same trend of using of sole RO seawater desalination for new projects is observed (WDR, 2017).

Generally, freshwater production costs depend on site-specific conditions such as local energy costs, level of freshwater quality required, and concentration of constituents in feed water. The latter is a particularly important factor in desalination system design and cost. The salinity of intake water can usually be found in the NCEI World Ocean Database (NECI, 2021). Low-salinity *brackish* water desalination systems are used for salinities between 500 ppm and 2500 ppm and have relatively simple configurations with relatively low initial and operating costs. For example, for desalting low-salinity BCRO, a single-stage RO system is sufficient to separate the feed water into freshwater and brine. In such systems, 5%–30% of the brackish water input commonly is bypassed and then mixed with the produced freshwater to yield the target freshwater quality. In some cases, however, when freshwater with a very low salt concentration is required, more than one stage for RO process may be required. For instance, a two-stage RO system is used in Florida and Texas to desalt low-salinity ground water; this approach results in a relatively higher freshwater cost compared to that for single-stage RO.

High-salinity brackish water desalination plants are commonly used for water salinities between 2500 ppm and 10,000 ppm. In these desalination facilities, all of the high-salinity brackish water input to the system is desalted. The produced permeate (freshwater) cost for such systems is typically 15%–35% higher than that for low-salinity brackish water desalination systems. The relatively higher cost is associated with the greater electricity requirement for separation of a relatively higher salt concentration from the feed stream, the complexities in the configuration of the RO set up, and the lower freshwater recovery in the operation of such systems.

Seawater desalination systems are commonly used to desalt seawater with salinities between 15,000 ppm and 46,000 ppm. These plants usually employ multistage, multipass RO modules. The relatively higher operating pressure in RO seawater desalination systems leads to the use of stronger membrane structures and vessels in such plants. Thus, for the same site-specific conditions (energy cost, etc.), the freshwater production cost via RO seawater desalination is higher than that for RO brackish water desalination.

The freshwater production cost for a BCRO desalination facility is less than that for a seawater RO system due to lower construction and operation costs. However, the limited nature of brackish water sources (1.1% of the worldwide water resource) restricts the utilization of such systems. Seawater RO desalination systems, on the other hand, have access to the largest sources of water, the seas and oceans (which account for 97.5% of the worldwide water resource), rendering them widely used as desalination plants.

Increasing environmental and ecological problems resulting from fossil fuel–based desalination systems, especially in coastal areas, is increasing the pressure on employing renewable energy resources. A competitive cost for freshwater production using renewable energy–driven seawater desalination systems can be attained in the case of optimum designs and proper selection and utilization of renewable energy sources available in a region, so as to maximize the synergy between diverse resources. With efficiency improvements and other advances in renewable energy technologies, the costs associated with such systems have decreased notably in recent years, as seen in Figs. 9.2–9.4. In these figures, the variation in unit capital costs is

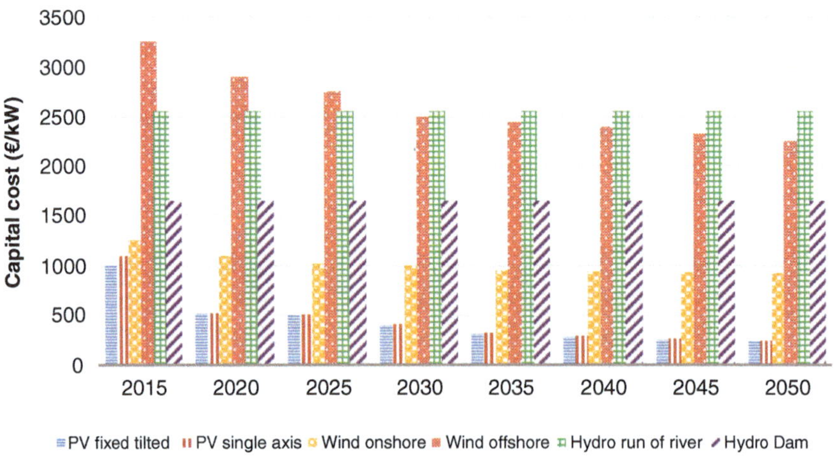

Fig. 9.2 ***Unit capital costs of various configurations of solar, wind, and hydraulic energy facilities for 2015 to 2050.*** For hydroelectric dams, *data are extracted from Lacal Arantegui (2014), Pleßmann et al. (2014)*. For PV systems, *data are extracted from Mayer (2015), Lacal Arantegui (2014), Pleßmann et al. (2014), Vartiainen (2017)*. For wind energy, *data are extracted from Lacal Arantegui (2014), Neij (2008), Pleßmann et al. (2014)*.

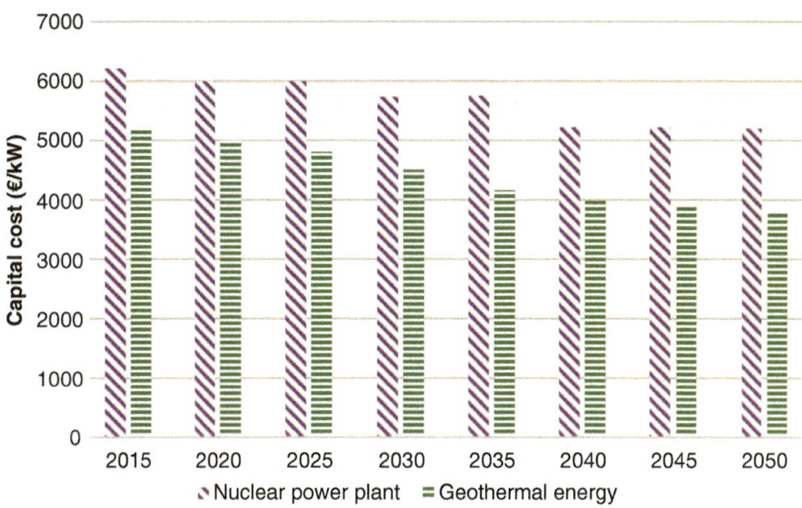

Fig. 9.3 ***Unit capital costs of nuclear and geothermal energy systems for 2015 to 2050.*** *Data extracted from Haysom et al. (2015), Lacal Arantegui (2014); Scott (2018); Pleßmann et al. (2014).*

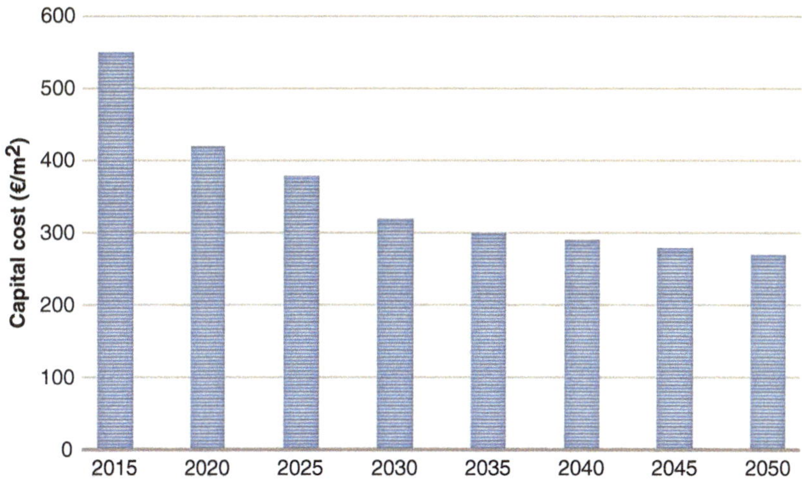

Fig. 9.4 ***Unit capital costs of concentrating solar power (CSP) plants for 2015 to 2050.*** *Data extracted from Haysom et al. (2015), Kutscher et al. (2010).*

shown/predicted for 2015 to 2050 for various renewable energy technologies. It is seen that, between 2015 and 2050, the capital cost of all renewable energy technologies is likely to decline. Further descriptions and cost estimations for renewable energy-based desalination systems are provided in subsequent sections.

In this chapter, an overview of desalination cost calculation is provided. Then an economic study of freshwater production is presented based on a technoeconomic analysis of desalination systems and renewable energy technologies. From the technoeconomic analysis, the cost of water produced from desalination facilities and the cost of energy from renewable energy technologies are predicted for systems, today and in the future. Next, exergy concepts are related to economic principles through exergoeconomic analysis, and suggestions for reducing the overall costs of freshwater and electricity are developed. Through exergoeconomic analyses of renewable energy–driven desalination systems, the proper allocation of economic resources can be determined to optimize the design and operation of such systems. Finally, an exergo-environmental analysis using life cycle assessment (LCA) is presented to evaluate the environmental impact of renewable and nonrenewable energy desalination systems.

It is noted that research and development in this and directly related fields has been ongoing, and applications have ensued. For instance, reports have been presented on the economics of renewable energy systems, using solar energy (Kool et al, 2018), geothermal energy (Jalili et al., 2021), biomass (Malfuzi et al., 2020), and renewable energy combined with fossil fuel (Sadeghi et al., 2020; Cai et al., 2020; Malfuzi et al., 2020). In addition, research has been reported on the application of exergy-based economic methods to renewable energy systems, including those driven by solar energy (Mosleh et al., 2020; Khoshgoftar Manesh et al., 2021) and geothermal energy (Ehyaei et al., 2020; Javanshir et al., 2020). For desalination systems, various types of economic assessments have been reported (Keshavarzzadeh et al., 2020; Haghghi et al., 2019; Chitgar et al., 2019), and many of these have involved optimizations based at least in part on economics for renewable energy forms such as wind (Peng et al., 2018; Maleki et al., 2016), solar (Wu et al., 2018), geothermal (Salehi et al., 2018), and combined renewable energy forms (Peng et al., 2018; Maleki et al., 2016).

9.2 Desalination project cost estimation and management

In a desalination project, the capital cost, the operation and maintenance (O&M) cost, and the water production cost are the main economic parameters.

The capital cost is associated with all expenses for implementation of the desalination project, from the initial time of conceptual development, through to design, construction, financing, commissioning, and testing. The construction cost (also known as direct capital cost) includes direct expenses required to construct a plant, the purchase and installation cost of all facility equipment, and connection costs of plant piping for delivering produced freshwater to final user(s). The direct capital cost of a seawater desalination facility is typically about 50%–85% of the total capital cost. The indirect capital cost, which accounts for the remaining 15%–50% of the total capital cost, includes funding, administrative, contractor, and engineering activities for the design and building of project (Voutchkov, 2018). The total capital cost is presented either as an annualized cost (i.e., monetary unit per year (US$/year)) or as a monetary value per unit of freshwater production capacity (US$/($m^3$/day)).

The operation cost includes expenses for operation of the desalination plant such as chemicals, electricity and other energy, labor, and replacement of membranes and filters. The operation cost also includes the compliance cost to adhere to applicable regulations such as finance, transport tax reporting, and environmental permit costs. Maintenance costs are related to the maintenance of building, equipment, and grounds. O&M costs are commonly presented on an annual basis (US$/year) or per unit volume of produced freshwater (US$/$m^3$).

O&M cost can be divided into fixed and variable costs. Fixed O&M costs are independent from the scale of the freshwater production process and often include labor cost, maintenance of equipment, performance and environmental monitoring, administrative costs, and insurance (Voutchkov, 2018). About 15%–50% of total annual O&M cost are usually associated with fixed O&M costs. Variable O&M costs, on the other hand, are a function of produced freshwater volume and include electricity and other energy costs, and expenses for replacement of filters and membranes, chemicals, and waste disposal. Around 50%–85% of total annual O&M cost is associated with variable O&M costs. Note that some labor costs, particularly those that depend on plant size, should be treated as variable O&M costs, whereas labor costs that are independent of plant size are counted as fixed O&M costs.

All capital and annual O&M costs are brought together to ascertain the total cost of freshwater production (cost of water). The cost of water can be expressed as the ratio of annual capital and O&M costs in US$/year to the target annual freshwater production rate in m^3/year. Then, the cost of water can be presented in monetary unit per volume of produced freshwater (US$/$m^3$).

Average costs of water production from large-scale high-salinity BCRO desalination, high-salinity BCRO desalination, and seawater RO desalination, respectively, reported as 0.7 US$/m^3, 0.9 US$/m^3 and 1.1 US$/m^3 (Voutchkov, 2018). These costs are based on information for more than 50 desalination projects around the world. The average cost of water produced via a brackish water desalination plant is 0.8 US$/m^3 and via a seawater desalination plant is 1.1 US$/m^3. It is seen that the average water production cost is 1.2–1.6 times lower for RO brackish water desalination plants than RO seawater desalination plants. However, comparisons of desalination systems must be carried out cautiously, as they only tend to be reasonable when approximately the same site-specific conditions exist.

A breakdown by economic parameter of the water production cost for a typical large/medium RO seawater desalination plant is shown in Fig. 9.5. The highest cost is associated with the direct capital cost (construction expenses), followed in order by the cost of electricity used in the plant, other O&M costs (membranes and chemicals replacement and maintenance), and indirect capital costs (project finance, engineering and development). Note that the cost associated with the electrical energy consumed by RO brackish water desalination systems (13%–27% of the total cost of water production) is less than that by RO seawater desalination systems. This means that, for RO brackish desalination systems, a greater portion of the total water cost is associated with the direct capital cost, which corresponds to 37%–57% of total cost of water production.

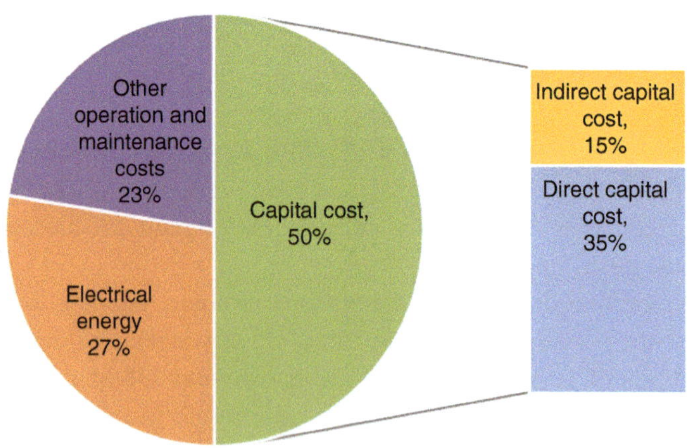

Fig. 9.5 ***Breakdown by economic parameter of the cost of freshwater production for a typical seawater RO desalination system.***

9.3 Technoeconomic analysis of combined sustainable energy and desalination technologies

A technoeconomic analysis examines the economic characteristics and technical performance of an energy technology by combining engineering design, process modeling, and economic evaluation. For a commercially available energy system, a reliable technoeconomic assessment requires recent data, for example, current costs of equipment, consumables, constructions, and labor. For energy technologies that are not commercialized, technoeconomic analyses can be speculative and subject to uncertainties. A technoeconomic analysis of a renewable energy–based desalination system examines the equipment costs and technical performance of such technologies. Such an analysis can also identify reasonable tradeoffs between cost and performance of renewable energy systems.

The viability of seawater or brackish water desalination driven by renewable energies to meet the freshwater demand in a region can be determined through economic assessments of electricity and water production costs and comparisons with the costs for fossil fuel–driven desalination plants. Such assessments help determine cost effectiveness.

Two important parameters are the levelized cost of water (*LCOW*) and the levelized cost of energy (*LCOE*). Each of these are now described.

9.3.1 Levelized cost of water

The LCOW or unit production cost of water in a region is usually expressed in dollars per cubic meter freshwater production (US$/m^3). The LCOW is as an important economic parameter for the analysis desalination plants. The LCOW can be expressed in terms of the capital cost and the O&M cost of a desalination facility. The LCOW can be written in terms of capital and operating costs of desalination and water storage, electricity costs, and costs of water transportation to users (industrial, municipal, or agricultural sectors). Consequently, the levelized cost of water produced from a desalination plant can be written as follows:

$$LCOW_D = \frac{\left[\left(\left(CC_D \cdot CRF_D\right) + \left(CC_{ws} \cdot CRF_{ws}\right)\right) + \left(OC_{fix,D} + OC_{fix,ws}\right)\right]}{V_{fw}} + \left(OC_{var,D} \cdot SEC_D\right) \quad (9.1)$$

where V_{fw} is the total volume of freshwater produced (m^3), CC_D and CC_{ws} are the capital costs of desalination and water storage units (US$), and $OC_{fix,D}$ and $OC_{fix,ws}$ are the fixed operational cost of desalination plant and water storage (US$). The plant capacity, V_{fw}, in the denominator of Eq. (9.1) reflects the unit cost of capital investment in US$/$m^3$. The term $OC_{var,D}$ is the variable operational cost of the desalination plant, which is equal to the *LCOE* in US$/kWh. Also, SEC_D is the specific energy consumption of the desalination plant in kWh/m^3. The product of $OC_{var,D}$ and SEC_D represents the energy cost of the desalination plant in US$/$m^3$.

The terms CRF_D and CRF_{ws} in Eq. (9.1) are the capital recovery factors of the desalination plant and the water storage. The annualized capital cost of component *j* can be specified by multiplying the corresponding capital cost (CC_j) by the corresponding capital recovery factor (CRF_j). The capital recovery factor is equal to the annualized factor in an economic analysis that describes the ratio of uniform payment made annually to the present value of the sum of the annuity over the given length of time (loan period). The capital recover factor can be expressed in terms of interest rate (*i*) and loan period of a component (*n*) as follows:

$$CRF = \frac{i\left(1+i\right)^n}{\left(1+i\right)^i - 1}. \tag{9.2}$$

The levelized cost of water transportation $LCOT_D$ (US$/$m^3$) from a desalination plant to user(s) can be written as

$$LCOT_D = \frac{\left[\left(\left(CC_p.CRF_p\right)+\left(CC_{pipe}.CRF_{pipe}\right)\right)+\left(OC_{fix,p}+OC_{fix,pipe}\right)\right]}{V_{fw}} + \left(OC_{var,pp}.SEC_{pp}\right) \tag{9.3}$$

where CC_p and CC_{pipe} are the capital costs of pumps (including horizontal and vertical) and pipes in US$; $OC_{fix,p}$ and $OC_{fix,pipe}$ are the fixed operating costs of pumps and pipes in US$; $OC_{var,pp}$ is the variable operational cost of pumps and pipes which is equal to *LCOE* (US$/kWh); and SEC_{pp} is the specific electrical energy consumption of pumps (kWh/m^3).

The levelized cost of water can be written as the summation of the levelized cost of water produced from a desalination plant and the levelized cost of water transportation from the desalination plant to the users, as

$$LCOW = LCOW_D + LCOT_D \tag{9.4}$$

9.3.2 Levelized cost of energy

The *LCOE* represents the total system lifetime costs, on an annualized basis. The levelized cost of electricity includes the expenditures made for the production of 1 kilowatt hour of electricity. It is usually expressed in dollars per kilowatt-hour (US$/kWh). Levelized costs of energy can be used to compare different energy production processes on a consistent basis. In renewable energy–based desalination systems, any action that results in a higher electrical energy production over a given period results in a lower total lifecycle cost and lowers the *LCOE* of the renewable energy–driven desalination system.

The *LCOE* of a plant in a region can be expressed in terms of the levelized cost of electricity for a primary generation source in the region ($LCOE_{pg}$), the levelized cost of curtailment (*LCOC*), the levelized cost for storage in the region (*LCOS*), and the levelized cost of transmission (*LCOT*), as follows:

$$LCOE = LCOE_{pg} + LCOC + LCOS + LCOT \tag{9.5}$$

where all of above terms are in US$/kWh.

In Eq. (9.5) both capital and operational costs contribute to *LCOE*. Both capital and operational costs can be further broken down into their corresponding components. The $LCOE_{pg}$ can be written as

$$LCOE_{pg} = \frac{\sum_{RE_j} (CC_j.CRF_j + OC_{fix,j})\dot{E}_{RE,j} + OC_{var,j}.E_{g,j}}{E_{Demand} + E_{exp} - E_{imp}} \tag{9.6}$$

where RE_j denotes renewable energy technology j used in the plant (e.g., solar or wind). In the numerator, CC_j and $OC_{fix,j}$, respectively, are the capital cost and fixed operational cost of the jth renewable energy technology (US$/kW) and $OC_{var,j}$ is the variable operational cost of the jth renewable energy technology (US$/kWh). The term $\dot{E}_{RE,j}$ is the installed capacity of the jth renewable energy technology in kiloWatt and $E_{g,j}$ is the annual electrical or thermal energy generated by the jth renewable energy technology (kWh). In the denominator, E_{Demand} is the annual electrical or thermal energy demand (kWh), which can be determined by multiplying the energy rate demand by the operational hours in a year. Also, E_{exp} is the annual exported energy (kWh) and E_{imp} is the annual imported energy to a region (kWh).

The levelized cost of curtailment accounts for the higher penetration of renewable energy when the electricity generation exceeds the plant load and overproduction may need to be curtailed. This results in an increase in the average unit cost of renewable energy. The levelized cost of curtailment (*LCOC*) in a region can be expressed as

$$LCOC = LCOE_{pg}.\frac{E_{curt}}{E_{Demand} + E_{exp} - E_{imp}} \tag{9.7}$$

Here, E_{curt} is the annual energy curtailed in the region (kWh).

The levelized cost of storage (*LCOS*) can be defined as the total lifetime cost of an energy storage technology divided by the cumulative energy delivered by the storage. The *LCOS* can be expressed as

$$LCOS = \frac{\sum_{storage_j} (CC_j.CRF_j + OC_{fix,j})E_{s,j} + OC_{var,j}.E_{disch,j}}{E_{Demand} + E_{exp} - E_{imp}} \tag{9.8}$$

where $storage_j$ denotes storage technology j used in the plant; Esj is the capacity of storage j (kWh), and $E_{disch,j}$ is the electrical energy received from discharging storage j (kWh). Also, CC_j, $OC_{fix,j}$, and $OC_{var,j}$, respectively, denote capital cost, fixed operational cost, and variable cost of storage j in US$/kWh.

The levelized cost of transmission (*LCOT*) is associated with transmitting high-voltage electricity from an electricity-generation plant through the transmission lines. Electricity transmission can be performed in two technology levels: distribution of power and transmission to subregions based on standard alternating current, and interregional transmission based on high voltage direct current grids (Bogdanov and Breyer, 2016). Using a high-voltage, direct current transmission grid can significantly reduce storage capacities and increase transmission capacities, which result in lower overall energy system costs. The *LCOT* can be expressed as follows:

$$LCOT = \frac{C_{total,tr}}{E_{Demand} + E_{exp} - E_{imp}} \tag{9.9}$$

where

$$\begin{aligned} C_{total,tr} &= \sum_{linej} (CC_{tl}.CRF_{tl} + OC_{tl}).E_{tl}.dl_{tl} + OC_{var,tl}E_{tr,j} \\ &+ \left(CC_{cs}CRF_{cs} + OC_{cs}\right)E_{cs} + OC_{var,cs}E_{tr,j} \end{aligned} \tag{9.10}$$

where $C_{total,tr}$ is the total unit cost of transmission (US$/kWh). Also, subscripts *tl* and *cs*, respectively, denote transmission lines and converter substation.

In general, the costs vary temporily, the degree to which depends on the reference year of the analysis, interest rates, and other economic factors. To ensure the data are appropriate and up to date, all cost data used in the economic analysis should be updated to the reference year, which is the year selected as the basis for cost calculations. A correction can be applied to costs that are for a different time than the reference year. The correction can be determined using appropriate cost index (CI) values as follows (CEPCI, 2021):

$$\text{Cost at reference year} = \text{Original cost} \times \frac{\text{Cost index for reference year}}{\text{Cost index for year when original cost was obtained}} \tag{9.11}$$

9.3.3 Technoeconomic performance of hybrid solar-wind driven seawater RO desalination plant

Determining the viability of RO seawater desalination driven by renewable energies to meet the freshwater demand in a region requires an economic study of electricity and water production costs and a comparison with the corresponding costs for a fossil-fueled RO desalination plant. This information helps determine the cost effectiveness of a plant.

Fig. 9.6 shows a renewable energy–based electricity and freshwater production plant. The electricity is produced from a hybrid solar-wind energy system comprised of photovoltaic (PV) panels and wind turbines. The electrical energy produced from the hybrid renewable energy system is used by the RO desalination unit for freshwater production. A water storage tank ensures continuous supply of feed water. When the electricity produced from the hybrid solar-wind field exceeds the electricity required by the RO plant, a portion of excess electricity can be stored in an electrical storage device such as a battery and the remainder can be converted to gaseous fuel in an electricity-to-gas system. The stored energy can be recovered when needed by chemical-to-electricity conversion in a battery and/or gas-to-electricity conversion in an electricity generation system.

In this section, the capital and operational costs associated with supplying freshwater in a region by a RO seawater desalination plant driven by a hybrid solar-wind energy system is presented. These costs can be used in the levelized cost of water relations in the previous section (Eqs. 9.1–9.11) to determine the unit cost of water for a renewable energy–driven RO plant.

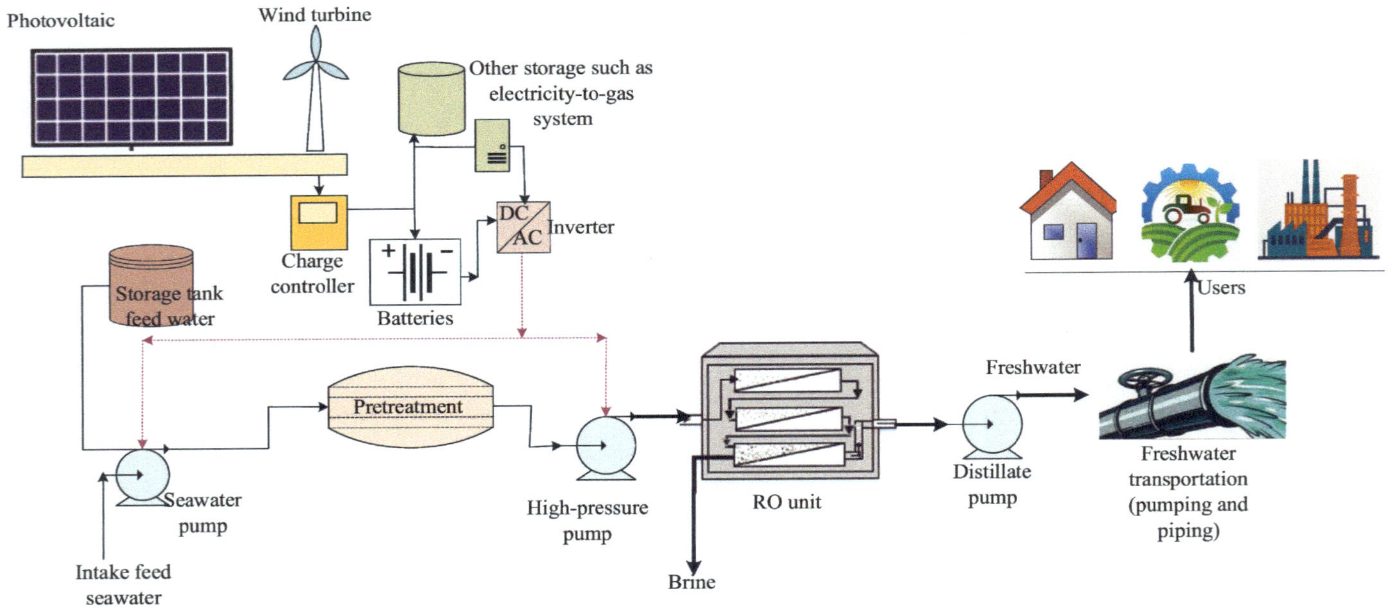

Fig. 9.6 *Freshwater supply from a hybrid solar-wind energy system including energy storages and an RO seawater desalination plant.*

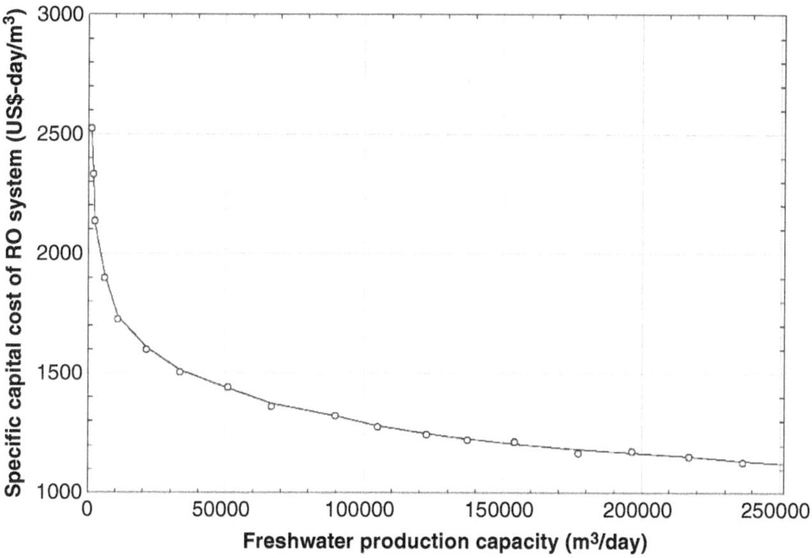

Fig. 9.7 ***Variation of specific capital cost of RO desalination system with freshwater production capacity.*** *Data extracted from DesalData (2017).*

9.3.3.1 RO seawater desalination plant cost

The capital cost of an RO seawater desalination plant can be found using the estimator tool at DesalData.com (DesalData, 2017). The estimator tool on that website can be used for various desalination plant capacities, salinities of intake seawater, and types of pretreatment. In Fig. 9.7 the variation of specific capital cost for RO in US$-day/$m^3$ with freshwater production capacity is shown. It is seen that the specific capital cost of an RO desalination unit varies logarithmically with the freshwater production capacity (V_{RO} (m^3/day)), as follows (DOW, 2016):

$$CC_{s,RO} = 3619 - 201.\ln\left(V_{RO}\right). \tag{9.12}$$

Therefore, the total capital cost of the RO system (CC_{RO} (US$)) can be determined by multiplying the specific capital cost in Eq. (9.12) by the volume flow rate of freshwater produced. Note that the total capital cost of RO after annualization can be obtained by multiplying CC_{RO} with capital recovery factor for RO (CRF_{RO}).

As mentioned before, the capital cost of components in an RO system are dependent on the freshwater production capacity. Based on data from

Table 9.2 Capital cost of components and processes in an RO desalination system with the freshwater production capacity of 150,000 m^3/day.

Component/process	Cost (US$-day/$m^3$)
Equipment and materials	247
Construction of intake/outfall structures	102
Installation services	109
Design	78
Pretreatment	96
Legal and professional tasks	16
Civil construction	241
Pumps	97
Piping	147
Pressure vessel	17
Membranes	59
Energy recovery devices	11
Total RO system	1219

Source: Data from (DesalData, 2017).

Desaldata.com, Table 9.2 shows a detailed breakdown of seawater RO costs, accounting for such items as design, material and equipment, pretreatment, pressure vessel, pumps, intake and outfall systems, piping, installation, engineering, legal and professional tasks.

There exist other methods for the capital costs of components in RO seawater desalination plants based on plant capacity, type of RO (BCRO or SWRO), region of plant, and initial year of construction and operation desalination project (Fulya, 2011; Loutatidou et al., 2014). Among these factors, plant capacity is the most influential, and the capital cost of a plant typically is inversely proportional to its capacity. So, it can usually be assumed that the capital cost of RO seawater desalination plant varies with freshwater production capacity. This assumption has been demonstrated to lead to be useful for economic analyses of RO desalination projects. For instance, Loutatidou et al. (2014) estimated the unit capital cost for a large-scale RO seawater desalination plant with the freshwater capacity of 20,000 m^3/day to be 3.7 US$/$m^3$ and 2.9 US$/$m^3$ in 2020 and 2030, respectively. They showed a trend of continual capital cost decreases over the period 2014–2030, corresponding to an average decrease of 2.23% per year (Loutatidou et al., 2014).

As mentioned before, the operational cost of an RO seawater desalination plant commonly includes fixed costs (labor, maintenance, etc.) and variable costs (replacements of membranes, consumable chemicals, energy use, etc.). The fixed operational cost of RO seawater plant can change with site-specific conditions. The fixed operational cost can be approximated as 4% of the total capital cost, according to Fichtner (Fulya, 2011). But DesalData (DesalData, 2017) reported constant specific costs, per cubic meter of freshwater produced by the RO system, for maintenance work, labor, and chemicals as 0.03, 0.08 and 0.07 US$/m^3, respectively.

The membrane replacement cost needs to be accounted for as an operating cost as the membrane needs to be replaced periodically. Seawater RO (SWRO) membranes are reported to last between 3 and 7 years (Avlonitis et al., 2003; Salgado et al., 2013). The annual operating cost associated with membrane replacement for an average lifetime of l_m years (l_m = 5) and project life of n can be expressed in terms of the total capital cost of the membrane (CC_m) and the rate of return on capital (r_{return}) as follows:

$$OC_m = CRF_m CC_m \begin{bmatrix} \left(1+r_{return}\right)^{-l_m} + \left(1+r_{return}\right)^{-2l_m} \\ + \cdots + \left(1+r_{return}\right)^{-((n/l_m)-1)} \end{bmatrix}. \quad (9.13)$$

Here, according to DesalData (DesalData, 2017), the capital cost of the membrane (CC_m) is between 5% and 6.5% of the total capital cost of the RO system (CC_{RO}) (Table 9.2). The rate of return on capital (or return on investment capital) is an economic parameter that indicates how effective a company is at turning capital into profits. This parameter typically is 5%–8% for RO desalination systems (DesalData, 2017).

The variable operating cost associated with input electrical energy is the greatest contributor to the total operational cost of RO plants. The annual energy cost for an RO desalination system can be determined from the net work rate input to the RO system (kW), the electricity cost (C_{elect} (US$/kWh)), and capacity factor (Cap_{fac}), as follows:

$$OC_{RO,\,energy} = \left(8760\ \text{h/year}\right) \cdot \dot{W}_{net,RO} \cdot C_{elect} \cdot Cap_{fac}. \quad (9.14)$$

The capacity factor basically indicates how often a plant is running at maximum energy output (i.e., ratio of actual energy output to the maximum energy output). For a desalination system it can be defined as the ratio of actual distillate production rate to the maximum distillate production rate. The net work rate input to an RO system ($\dot{W}_{net,RO}$) is the sum of the total work rate inputs for pumping, pretreatment (if used), and auxiliary devices, minus any electrical power produced from energy recover devices.

For a simple RO configuration, where there is only a circulation pump for pressurizing feed seawater, the work rate required by the RO pump can be expressed as

$$\dot{W}_{net,\,RO} = \dot{W}_{pump,\,RO} = \frac{\dot{V}_f \Delta P_{pump}}{\eta_{pump,RO}} \tag{9.15}$$

where $\dot{V}_f$ is the volumetric flow rate of feed seawater input to the RO system in m^3/s, ΔP_{pump} is the pressure difference between the outlet and inlet of pump, and $\eta_{pump,RO}$ is the isentropic efficiency of the pump.

The specific energy consumption for a seawater RO desalination plant is a function of feed temperature, seawater salinity, and the type of energy recovery devices (ERDs) used (if any). The specific energy consumption in an RO seawater desalination plant can be expressed as (Mistry et al., 2011; Thiel et al., 2015)

$$SEC_{RO} = \frac{\dot{W}_{net,RO}}{\dot{V}_{fw,h}} \tag{9.16}$$

where $\dot{V}_{fw,h}$ is the volumetric flow rate of produced freshwater for the RO desalination system (m^3/h). If a hydraulic turbine is used as the ERD, the specific power consumption (kWh/m^3) for an RO seawater desalination plant can be determined (Al-Zahrani et al., 2012) as:

$$SEC_D = 3\times10^{-4} w^2 + 1.8\times10^{-3} w + 2.605 \tag{9.17}$$

where w is the mass-based salinity of seawater in g/kg.

From above relations and descriptions, the total operating cost of an RO system can be written as

$$OC_{RO} = OC_{RO,energy} + OC_m + OC_{maintenance} + OC_{labor} + OC_{chemical} \tag{9.18}$$

9.3.3.2 Water transportation cost

The cost of water transportation from the desalination plant site to users (agriculture, domestic, industrial, etc.) is an important factor in the levelized cost of water. The cost of water transportation mainly consists of piping and pumping, over vertical and horizontal distances. The total capital and operational costs of water transportation are proportional to the RO desalination plant capacity. Missimer et al. (2014) estimated the capital and operational costs for a water transportation process. They reported a capital cost of 500,000 US$ for pumping 5000 m^3/day freshwater; by counting head loss, it was shown that the water can be transferred vertically 0.12 km

Table 9.3 Assumed and evaluated economic parameters for seawater RO desalination, water storage, and water transportation, for 2030[a].

Process and parameter	Value
Seawater RO desalination[b]	
Capacity	20,000 m^3/day
Capital cost	2.9 US$/$m^3$
Operational cost	0.148 US$/$m^3$ annually (assumed as 4% of capital cost)
Electricity consumption	2.6–3.3 kWh/m^3
Water storage[c]	
Capital cost	78.65 US$/$m^3$
Fixed operational cost	1.57 US$/$m^3$ annually (assumed as 4% of capital cost)
Lifetime	30 years
Water transportation[d]	
Capital cost of horizontal pipes	0.064 US$/($m^3$.km)
Fixed operational cost of horizontal pipes	2.8×10^{-4} US$/($m^3$.km)
Capital cost of vertical pipes	0.064 US$/($m^3$.km)
Fixed operational cost of vertical pipes	7.8×10^{-4} US$/($m^3$.km)
Horizontal pumping cost	1×10^{-4} US$/($m^3$.km)
Energy consumtion of horizontal pump	9.6×10^{-3} US$/($m^3$.km.day)
Vertical pumping cost	0.9 US$/($m^3$.km)
Energy consumption of vertical pump	86.4 US$/($m^3$.km.day)

[a]Lifetime is assumed to be 35 years and exchange rate is taken to be 1.21 US$/€.
[b]Data from Al-Zahrani et al. (2012), Fulya (2011), Loutatidou et al. (2014).
[c]Data from Missimer et al. (2014), UN (2009).
[d]Data from Al-Nory and El-Beltagy (2014), Newfoundland and Labrador (2011).

and horizontally 96 km. This means that the capital cost of horizontal pumps per m^3 water and per km would be (500,000 US$/(5000 m^3/day))/96 km = 1.04 US$/($m^3$/day) per km.

Table 9.3 lists capital and operational costs for water production by an RO seawater desalination plant, including water transportation and storage, for 2030.

9.3.3.3 Renewable energy plant cost

Hybrid renewable energy systems are often appropriate options for supplying electrical power in remote areas where electrical grid extensions are neither simple nor economical. Employing hybrid renewable energy facilities can potentially reduce or remove the dependency on fossil fuel–based generators. The availabilities of solar and wind resources depend greatly on weather and geographical conditions. The intermittent nature of both solar and wind energies often necessitates the use of energy storage to permit storage of the excess energy produced during periods of high solar insolation or wind and recovery of the energy during periods of low solar insolation or wind.

Generally, solar modules are responsible for about 30%–40% of the total installed cost of a solar facility (Hiendro et al., 2013). The cost of wind turbines is responsible for about 50% of the installed cost of wind power plants (Hiendro et al., 2013). Every year, the National Renewable Energy Laboratory (NREL) publishes annual technology baseline information that contains detailed cost data for renewable and conventional energy technologies (NREL, 2019). That data indicate that the prices of solar and wind modules have been continually declining from 2015 to 2020.

According to the International Energy Agency (IEA), the global total number of installed PV plants has increased sharply in recent years. The global PV installed capacity in 2020 reached 107 GW, representing a growth of 18% from the value in 2019 (IEA, 2020). The cumulative installed capacity of PV plants is anticipated to reach 117 GW in 2021, a nearly 10% rise from 2020 (IEA, 2020).

The total capital and operational costs of a hybrid solar-wind plant can be broken down by component, for example, PV modules, wind turbines, energy storages, and inverters. The levelized cost of energy for the hybrid solar-wind electricity generation plant can be defined as the average cost per kWh of electrical energy generated from the plant, and can be determined with Eqs. (9.5–9.11).

The electricity output from solar PV panels (i.e., the installed capacity of PV) can be expressed as (Hiendro et al., 2013)

$$\dot{W}_{PV} = \dot{W}_{PV,STC} f_{pv} f_T \left(\frac{I_T}{I_{T,STC}} \right) \tag{9.19}$$

Here, $\dot{W}_{PV,STC}$ is the capacity of PV modules under standard conditions (kW), f_{pv} is the derating factor of PV; f_{pv} is temperature derating factor; and I_T and $I_{T,STC}$, respectively, are the solar radiation incident on the PV panels (kW/m^2) and the solar radiation at standard test conditions (1 kW/m^2). The derating factor is the reduction in electricity production of PV modules due to shading, dust on PV surfaces, aging, elevated temperature, wiring losses, etc. The temperature derating factor can be determined based on the power temperature coefficient (α_P), PV cell temperature in each time step (T_C), and PV cell temperature at standard test conditions ($T_{C,STC}$ = 25 °C), as follows:

$$f_T = 1 + \alpha_P \left(T_C - T_{C,STC}\right) \tag{9.20}$$

The power temperature coefficient describes the sensitivity of electricity output from PV panels to the temperature of cells. This coefficient depends on the type of PV module. It has a negative value because the electricity output decreases as the cell temperature increases.

The actual electrical power produced from a wind turbine can be determined based on the wind turbine electricity production at standard temperature and pressure ($\dot{W}_{WT,STC}$), the actual air density (ρ_a), and air density at standard temperature and pressure ($\rho_{a,STC}$ = 1.225 kg/m^3), as follows (Hiendro et al., 2013; Saheb-Koussa et al., 2010):

$$\dot{W}_{WT} = \dot{W}_{WT,STC} \left(\frac{\rho_a}{\rho_{a,STC}} \right) \tag{9.21}$$

The capacity of wind power facilities has been growing every year. Globally, wind turbine stations could produce 65 GW of electrical power in 2020, an 8% increase over 2019 (IEA, 2020). The COVID-19 pandemic led to the decline in construction of wind power plants, especially onshore constructions (IEA, 2020). Nonetheless, the installed wind turbine capacity is anticipated to reach 68 GW in 2021, a rise 4.6% from 2020 (IEA, 2020). Kersten et al. (2011) and Liebreich (2016) predicted that capital costs of PV and wind power stations would continue descending to 2030.

The *lithium-ion battery* is a rechargeable type of battery that has seen increasing utilization in recent years, in part because of significant decreases in its capital cost (PV Magazine, 2016). Lithium-ion batteries have been applied in applications ranging from small applications such as electric vehicles to large applications such as aerospace and military technologies. The International Renewable Energy Agency (IRENA, 2017) investigated in detail electrical energy storages and their contributions to renewable energy applications, and predicted that the total electricity storage capacity in energy systems will triple by 2030 relative to 2017, if the share of renewable energy in energy systems doubles. In addition, NREL (Cole and Frazier, 2019) assessed capital cost predictions for lithium-ion battery energy storage systems for low, medium, and high cost projections, based on information in over 25 publications. These data are presented in Fig. 9.8, which suggests significant reductions are likely in the cost of batteries for the period 2020-2050

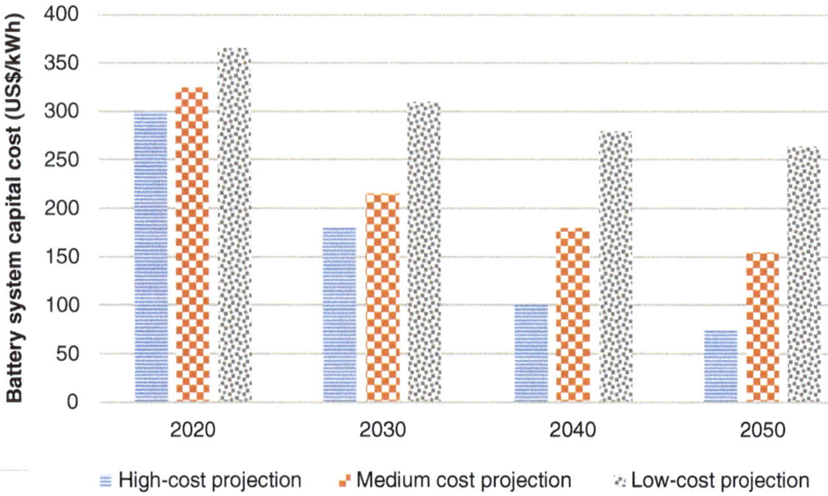

Fig. 9.8 ***Predicted capital costs of utility scale lithium-ion batteries with 4-h storage durations for 2020 to 2050.*** *Data from NREL (Cole and Frazier, 2019).*

Electricity-to-gas systems can be also used as energy storages, for storing excess electricity generated from renewable energy facilities in the form of gaseous fuel. This technology uses electrical energy to produce a gaseous fuel, typically hydrogen or methane. The produced fuel gas can be converted back to electricity by fuel cells or other conventional generators. The electricity-to-gas and gas-to-electricity efficiencies depend on the type of gaseous product, the technologies used for energy conversion (e.g., electrolyzers, fuel cells, conventional generators), and the gas storage efficiency. For example, the electricity-to-hydrogen efficiency for a water electrolysis system is about 65%–82% and the hydrogen-to-electricity efficiency for a fuel cell typically is about 40%–60% (Moseley and Garche, 2014). Further efficiency increases and cost reductions remain necessary for electricity-to-gas systems. Fig. 9.9 shows a typical electricity-to-gas and gas-to-electricity plant. It can be seen that the electricity generated from PV panels and wind turbines is used in the water electrolysis unit to produce hydrogen and oxygen gases. These gases are subsequently utilized in a fuel cell to generate electricity for a given load.

Table 9.4 provides typical capital and operational costs of a hybrid renewable energy facility consisting of wind turbines, PV fixed-tilt, and PV single-axis tracking systems. The battery and electricity-to-gas systems are also used as electrical energy storages to enable steady and continuous energy supply for an RO seawater desalination.

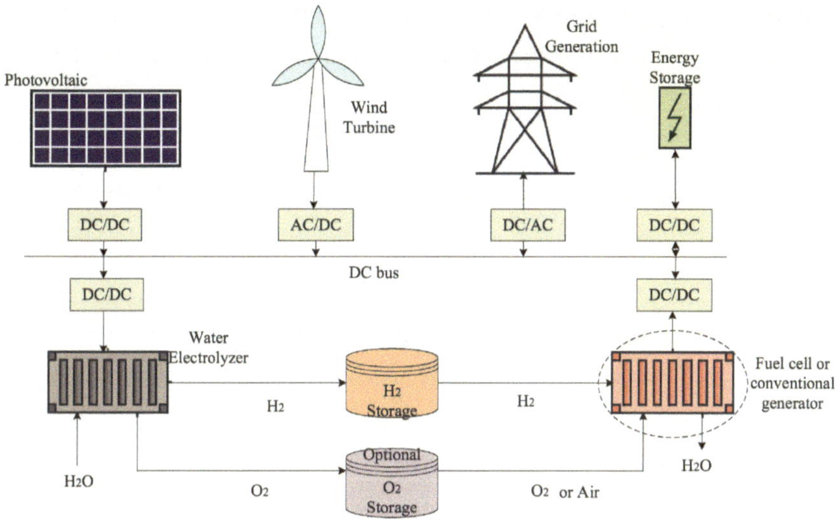

Fig. 9.9 ***Typical electricity-to-gas and gas-to-electricity plant for generating electricity, converting it to hydrogen gas with an electrolyzer for storage and then converting it back to electricity with a fuel cell or other generator.***

Table 9.4 Assumed and evaluated economic parameters for hybrid solar-wind electricity generation facility, for 2030[a].

System and parameter	Value
Fixed-tilted PV plant	
Capital cost	665.5 US$/kW
Operational cost	9.98 US$/kW annually (assumed as 1.5% of capital cost)
Single-axis tracking PV plant	
Capital cost	750 US$/kW
Operational cost	11.25 US$/kW annually (assumed as 1.5% of capital cost)
Wind power plant	
Capital cost	1210 US$/kW
Operational cost	24.2 US$/kW annually (assumed as 2% of capital cost)
Battery storage	
Capital cost	181.5 US$/kWh
Operational cost	12.1 US$/kWh annually
Efficiency	90%
Power-to-gas storage	
Capital cost	1173.7 US$/kW
Operational cost	39 US$/kW annually
Water electrolysis efficiency	82%

[a]Lifetime is assumed to be 35 years and exchange rate is taken to be 1.21 US$/€.
Data extracted from Bloomberg (2016), Breyer et al. (2015b), Kersten et al. (2011).

9.3.3.4 Levelized cost of energy and water in hybrid solar-wind driven seawater RO desalination plant

Using the assumptions and capital and operational costs in Tables 9.3 and 9.4 with Eqs. (9.1–9.21), the levelized cost of energy and water for a region can be determined. The hybrid solar-wind driven RO seawater desalination plant can be optimized to minimize its total annual cost. This involves the costs of the hybrid renewable energy and RO desalination plants, the cost of water transportation, and the cost of energy production. To achieve an optimal design, the plant can be modeled and analyzed to assess the energy conversion and storage efficiencies, and to determine the optimum distance from user(s). A model for use in conjunction with linear optimization was developed and is described elsewhere (Bogdanov and Breyer, 2016; Breyer et al., 2015a; Barbosa et al., 2017).

Table 9.5 shows a prediction of power/energy capacities of PV, wind turbine, battery, and power to gas subsystems in hybrid solar-wind driven RO seawater desalination plants in 2030. In this case, the PV solar plant is responsible for 85% of the energy supply, and 56% of this amount is attributable to single-axis and the remaining 29% to fixed tilted solar PV facilities. The battery bank capacity installed for supplying energy for desalination is estimated at 2011 TWh. Each battery is usually cycled every day, and there are around 360 cycles per year.

The levelized cost of energy from the plant varies between 0.072 US$/kWh and 0.18 US$/kWh depending on region. In addition, the levelized cost of water from the plant is determined to be between 1.21 US$/$m^3$ and 5.44 US$/$m^3$ in 2030 (Caldera et al., 2018a). The estimated levelized cost of water from a renewable energy–driven desalination plant is expected to be promising in the future compared to that from a fossil-fueled plant, which has been predicted to range between 0.78 US$/$m^3$ and 2.3 US$/$m^3$ in Saudi Arabia for the early 2020s (Caldera et al., 2018b; Fulya, 2011). Note that this value of *LCOW* from a fossil fuel–based desalination plant

Table 9.5 Predicted capacities of subsystems of a hybrid solar-wind driven seawater RO desalination plant for supplying a 2.6 billion m^3/day water global demand in 2030.

Subsystem	Capacity
PV units	3730 GW
Wind turbines	598 GW
Batteries	2011 TWh
Power to gas plant	316 GW

Data extracted from Caldera et al. (2018a).

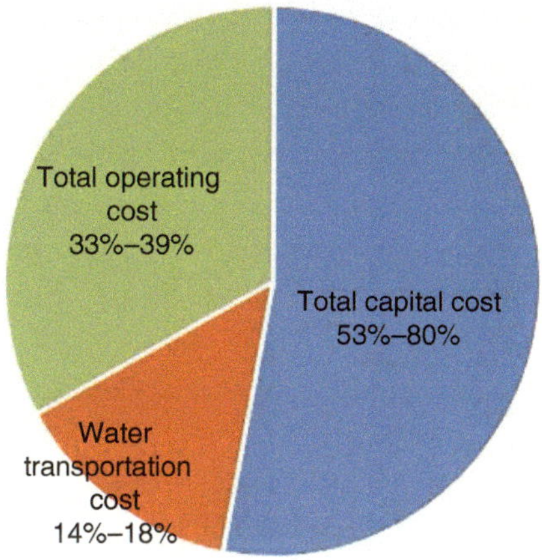

Fig. 9.10 ***Breakdown by capital, operational, and water transportation costs of the levelized cost of water for a hybrid solar-wind driven RO seawater desalination plant in 2030.***

excludes the cost associated with transporting produced water to the users. This suggests that utilizing renewable energy technologies with seawater desalination systems is a potentially advantageous way of responding to global water demands. Advances in renewable energy and desalination technologies along with reductions in their capital costs can further decrease the unit cost of water.

Fig. 9.10 provides a breakdown of the capital and operational costs associated with a hybrid solar-wind driven RO seawater desalination plant. It is seen that the capital cost of the plant has the greatest impact on the levelized cost of water for the plant, although the influence of the water transportation process also is noteworthy. This means that a significant amount of electricity is required for pumping the produced freshwater to users. So, it can make sense for the site of a renewable energy–based plant to be selected so that the distance from user(s) is not excessive and an appropriate balance is achieved for this and the other costs of the plant. The water transportation contribution increases with the installed desalination capacity. The costs of piping and pumping freshwater can be more reliably determined by considering ground conditions and location. The electricity required for pumping is higher when a desalination plant is installed far from a coastal area. The total pumping energy includes energy required for pumping intake seawater to the desalination system and for transporting the product freshwater from the desalination system to users. One advantage of RO desalination over MED and MSF desalination is that the electricity generation and RO facilities are not necessarily located in the same place, while MED and MSF systems are typically coupled with a steam generator unit (e.g., a natural gas back-up, steam power cycle) so as to use the heat of steam for the desalting process. Hence, by installing an RO system close to the sea and a solar power plant in an area with high levels of solar radiation, the electrical power required for pumping and transporting seawater to the RO module can be kept low.

A breakdown of the components of the levelized cost of water predicted for 2015–2050 is presented in Fig. 9.11. For 2015, it is assumed that there is no electricity cost associated with the cogeneration plant and that the only variable operating cost is the cost of gas. From 2015 to 2050, the levelized cost of water decreased from 3.8 $\$/m^3$ to 0.75 $\$/m^3$ for the following reasons: the replacement of natural gas fuel with free renewable energy resources (solar, wind, etc.); decreases in the electricity cost and increases in the efficiency for cogeneration plants; and increases in the efficiencies of MED and RO desalination plants. By 2050, the transportation infrastructure is expected to become one of the greatest contributions to the capital cost.

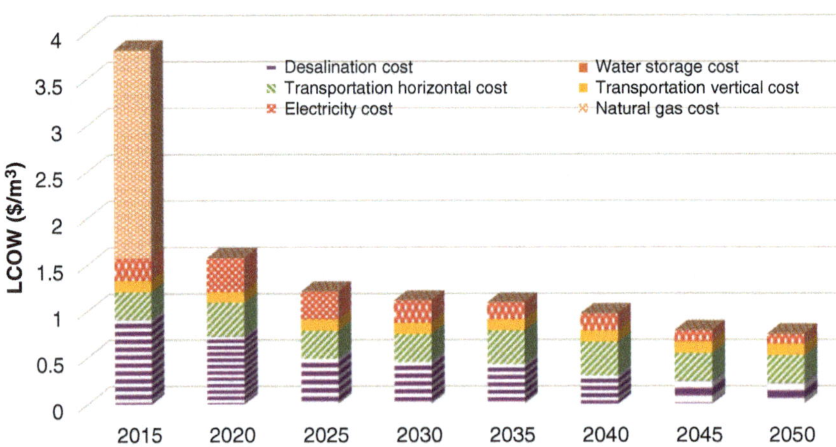

Fig. 9.11 ***Breakdown of components of the levelized cost of water (LCOW) for 2015 to 2050.*** *Data from Caldera et al. (2018b), GWID (2016), Reddy and Ghaffour (2007).*

9.4 Case study: technoeconomic assessment of CSP–MED and CSP–RO plant configurations

Concentrating solar power (CSP) plants and desalination systems have been successfully coupled in the past for large-scale cogeneration of electricity and freshwater, particularly in the MENA region (Trieb and Müller-Steinhagen, 2008). In this section, a case study is presented in which technoeconomic assessments are described of two renewable energy–based desalination systems: CSP–MED and CSP–RO. This case study is based on the work of Olwig et al. (2012). In that project, which was conducted for Israel (Ashdod) and Jordan (Aqaba), an Andasol-type CSP plant was combined with ultrafiltration RO desalination and MED plants, each with a freshwater production capacity of 24,000 m^3/day (Olwig et al., 2012). Andasol-type CSP units are comprised of parabolic trough solar collectors that use thermal oil as the heat transfer medium, thermal storage with molten salt utilized as the heat storage medium in two thermal storage tanks and a Rankine steam power cycle. Figs. 9.12 and 9.13, respectively, illustrate the CSP–MED and CSP–RO plant configurations. The CSP plant design is sized to be adequate to provide the steam needed for driving the MED plant. A deaerator is used to remove the dissolved gases from the water. Key design parameters for the CSP–MED and CSP–RO plants in Aqaba and Ashdod are presented in Tables 9.6–9.8.

The RO plant is installed by the sea away from the CSP plant, while the MED plant is located near the CSP plant. Consequently, the electrical power required for pumping seawater to the MED plant is significant, increasing the overall electricity and capital costs. The fact that, for the CSP–RO cogeneration plant, the individual CSP and RO plants are not necessarily located together is an advantage of CSP–RO over CSP–MED in terms of land costs and electricity for pumping seawater.

Direct normal irradiance (DNI) is one of the primary factors in CSP plant design. The hourly time series of DNI are obtained using data images from satellites collected for the period of 2001–2005. For 2005, the average DNI on the ground was 1984 kWh/(m^2year) at Ashdod and 2461 kWh/(m^2year) at Aqaba. In the MED system, the cooling seawater mass flow rate in each season is determined based on the corresponding intake seawater temperature (see Table 9.7).

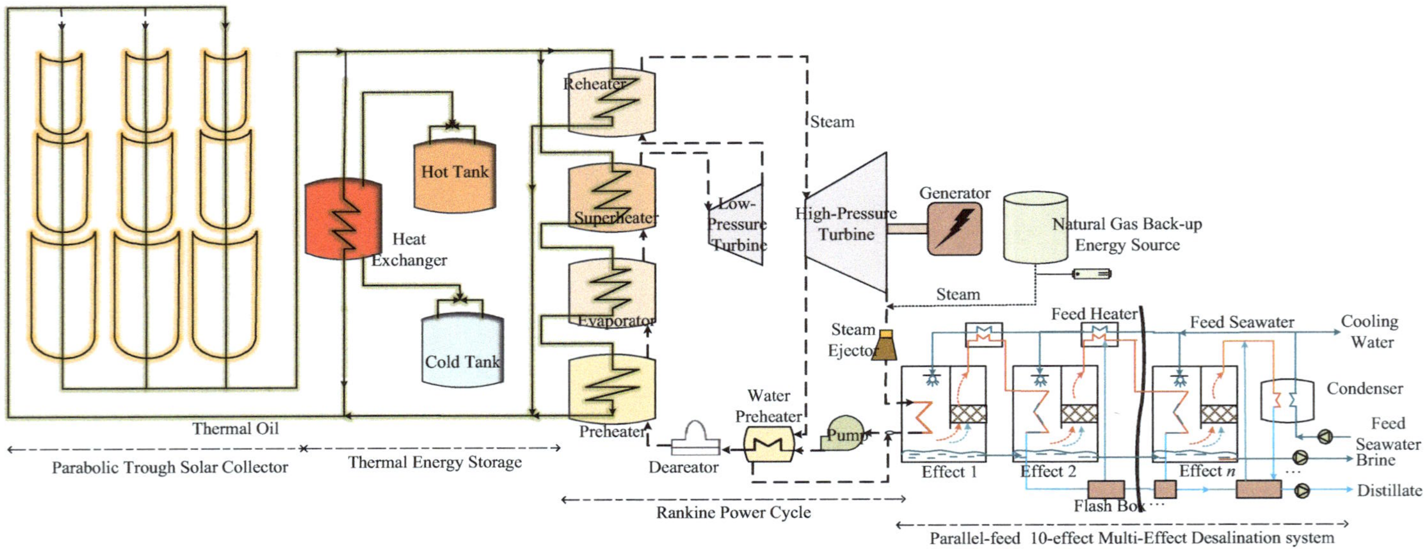

Fig. 9.12 ***Simplified diagram of concentrating solar power facility combined with parallel-feed MED plant for desalination (with only one MED unit shown).***

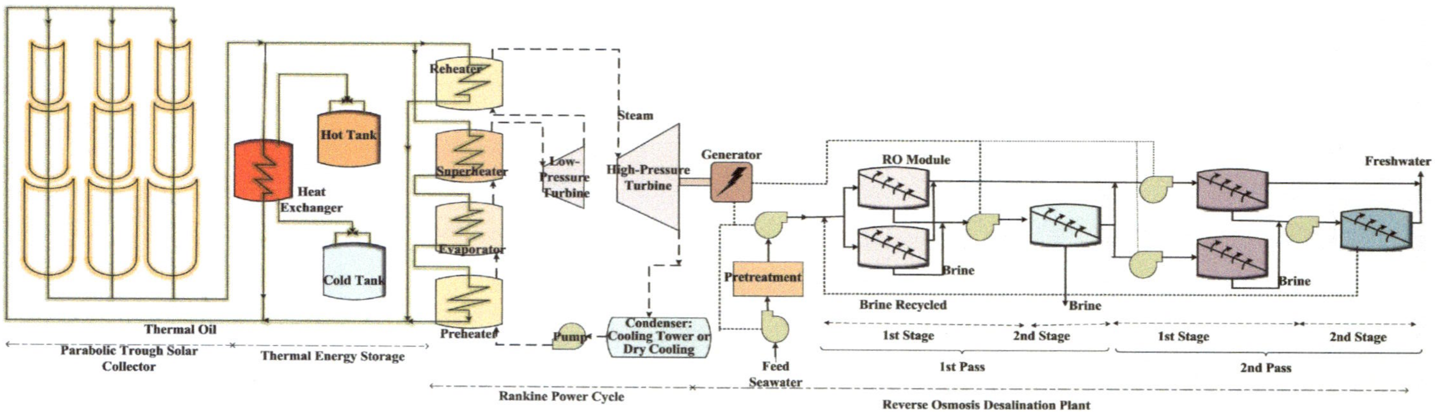

Fig. 9.13 *Simplified diagram of concentrating solar power facility combined with seawater two-pass RO desalination plant.*

Table 9.6 Design parameters for concentrating solar power system in the combined concentrated solar power (CSP)–MED and CSP–RO plants in Aqaba and Ashdod.

Parameter	CSP–MED	CSP–RO
Solar field		
Collector type	Eurotrough	Eurotrough
Absorber tube	PRT 70	PRT 70
Aperture area	Flexible	Flexible
Nominal inlet temperature, °C	293	293
Nominal outlet temperature, °C	393	393
Energy capacity of thermal storage, MWh	123	109
Power cycle		
Turbine type	Backpressure turbine with controlled extraction	Condensing turbine
Gross capacity, MW	42	42
Gross efficiency of steam turbine, %	34.1	38.5

Source: Data from Olwig et al. (2012).

Table 9.7 Design parameters for MED desalination system in the combined CSP-MED plant in Aqaba and Ashdod.

Parameter	Aqaba		Ashdod	
	Summer	*Winter*	*Summer*	*Winter*
Seawater temperature, °C	28	20	32	18
Seawater salinity, g/kg	42.5		40.5	
Freshwater mass flow rate, ton/day	12,000		12,000	
Distillate salinity, ppm	<10		<10	
Number of stages	10		10	
Gained output ratio (GOR)	8.35		8.52	
Top brine temperature, °C	65		65	
Heat transfer area of one stage, m^2	4626		4893	
Heating steam mass flow rate, kg/s	16.63		16.31	
Heating steam pressure at first effect, bar	0.278		0.276	
Temperature of heating steam condensate, °C	67.3		67.2	
Specific heat consumption, kWh/t	77.8		76.3	
Cooling water mass flow rate, kg/s	751	142	1539	94
Total seawater mass flow rate to MED, kg/s	1178	569	1918	472
Specific electrical power consumption in MED located at coast, kWh/t	1.7	1.2	2.4	1.0
Steam pressure at turbine outlet, bar	0.35			
Motive steam pressure at steam ejector inlet, bar	4.0			

Source of data Olwig et al. (2012).
Data for MED are based on an MED system with a freshwater production capacity of 12,000 m^3/day.

Table 9.8 Design parameters for RO desalination system in the combined CSP-RO plant in Aqaba and Ashdod.

Parameter	Aqaba		Ashdod	
	First pass	*Second pass*	*First pass*	*Second pass*
Feed pressure, bar	68.88	4.9	65.62	5.48
Concentrate pressure, bar	66.66	4.97	63.4	3.58
Concentrate flow rate, m^3/h	1010	32	1020	32
Specific energy consumption without ERD, kWh/m^3	4.78	0.22	4.56	0.24
Specific energy consumption, kWh/m^3	3.16		3.05	
Specific energy consumption with ERD, kWh/m^3	3.04		2.92	
Overall specific energy consumption, kWh/m^3	3.16		3.05	

Source of data Olwig et al. (2012).
ERD: energy recovery device

CSP–RO plant: Intake seawater is pumped into the pretreatment unit of the RO plant. In the pretreatment unit, a Zenon ZeeWeed ultrafiltration membrane is used which is capable of desalting seawater well. The flow rate per unit area in the RO module determines the number of membranes. A two-pass RO module is utilized here, in which each pass is comprised of two stages (Fig. 9.13). In the first pass, brine from the first stage enters the second stage where its salt concentration is reduced. The permeate from the first pass is introduced to the first stage of the second pass for further reduction in salt concentration. Then the brine stream enters the second stage of the second pass to yield the final permeate and brine streams. The staging ratio between the first and second stages is obtained based on the recovery ratio of the considered pass (r), as follows:

$$R_{RO} = \sqrt{\frac{1}{1-r}} \tag{9.22}$$

where r is defined as the ratio of the mass flow rate of permeate exiting a pass to the mass flow rate of feed entering the pass. The second pass, which operates at a relatively higher salt concentration, is designed with a high recovery ratio of 80%. An ERD is also employed, with an isobaric pressure exchanger efficiency of 96%. The electrical energy consumed by the RO plant per unit of produced freshwater is calculated based on the volumetric flow rate of feed ($\dot{V}_f$), the pressures of feed (P_f) and brine (P_b), the pump isentropic efficiency (η_{pump}), and the isobaric pressure exchange efficiency of the ERD (η_{ERD}), as follows (Oh et al., 2009):

$$\dot{W}_{RO} / \dot{V}_{fw} = \left(\frac{P_f \dot{V}_f}{\eta_{pump}} - P_b \dot{V}_b \eta_{ERD} \right) \tag{9.23}$$

The economic parameters of the ultrafiltration RO unit in the combined CSP–RO plant for Ashdod and Aqaba are listed in Table 9.9. It is seen that the first pass of the RO module is the most expensive component. The electricity and maintenance costs are the main contributors to the O&M costs.

CSP–MED plant: Natural gas is used as a back-up energy source for the MED plant to obtain full-load operation during periods when no heat (direct or stored) is available from the solar collectors. Two units of parallel-feed MED plants are employed, with each MED unit having 10 effects. Also, each unit has a freshwater production capacity of 12,000 m^3/day. The specific energy consumption in the MED plant can be determined using thermodynamic modeling. The overall temperature difference between the steam heat source and the feed seawater determines the optimum number of stages. Having more stages in the MED facility results in a higher desalination efficiency, but also increases the heat transfer area by decreasing the driving temperature difference for each stage. A larger heat transfer area correlates with a higher capital cost for the MED plant. The steam temperature of is set several degrees higher than the top brine temperature *TBT* (i.e., the brine temperature in the first effect). The *TBT* should be chosen conservatively to avoid the scaling problem, at less than 70 °C.

Applying low-pressure steam with a high flow rate is considered more advantageous for combining with a CSP plant with MED compared to applying high-pressure steam (up to 16 bar) with a relatively low flow rate (Olwig et al., 2012). Mass flow rates and temperatures in all effects are taken to be constant in both winter and summer operation, except for the cooling water mass flow rate, which varies seasonally based on the seawater temperature. Therefore, the electrical energy required for pumping varies with ambient temperature. A vacuum system is employed to eliminate the non-condensable gas formed from the evaporation of brine, which can penetrate the evaporators. In the CSP–MED plant, a back-pressure turbine is utilized with a constant steam exhaust pressure of 0.35 bar. In the nominal load condition for the CSP–MED cogeneration plant design, the turbine exhaust steam can fully supply the thermal energy required for the MED plant.

Table 9.9 Economic parameters for CSP–MED and CSP–RO cogeneration plant in Aqaba and Ashdod.

Parameter	Aqaba		Ashdod	
	CSP–MED	CSP–RO	CSP–MED	CSP–RO
Capital costs				
CSP solar field, US\$/m^2	455	455	455	455
Power plant, US\$/kWh	1092	1092	1092	1092
TES, \$/kWh	72.8	72.8	72.8	72.8
Desalination plant, coast, US \$/m^3/d	1667	1705	1667	1705
First pass in RO, kUS\$/year		3703		3740
Second pass in RO, kUS\$/year		45		47
Posttreatment plant in RO, kUS\$/year		132		133
Two MED units, \$/m^3/day	1394		1394	
Posttreatment plant in MED, \$/m^3/d				
Open seawater intake in MED, US\$/m^3/d	217		217	
Water transmission system to inland, US\$/m^3/d	292		152	
Operation and maintenance (O&M) costs				
Solar field, US\$/m^2/year	112	112	112	112
Power block, US\$/kWh/year	4.2	4.2	4.2	4.2
Thermal storage, US\$/MWh/year	70	70	70	70
Desalination plant, inland, no cofiring for MED, US\$/m^3	0.126	0.200	0.151	0.235
Energy costs				
Electricity price from grid, US\$/kWh	0.069	0.069	0.115	0.115
Fossil fuel price (natural gas), US\$/kWh	0.020		0.010	
Economic parameters				
Project lifetime, year	20	20	20	20
Investment interest rate, %	8	8	8	8

Data from Olwig et al. (2012).

The *LCOW* for the CSP–MED and CSP–RO plants can be determined using Eqs. (9.1–9.4). The *LCOW* for the CSP–MED and CSP–RO plants is a function of freshwater production capacity, total capital investment, cost of fuel used for cofiring the MED plant, costs of chemical, electricity costs for RO or MED desalination, personnel costs, replacement costs, maintenance costs, administrative and general costs, and revenues from electricity sales.

The capital, operating and maintenance costs and the energy cost resulting from a techno-economic analysis of the CSP–MED and CSP–RO combined plants in Aqaba and Ashdod are listed in Table 9.9. The *LCOW* is determined with Eq. (9.23). For operation of the RO and MED plants, it is assumed that electricity from the grid is used. The *LCOW* is highly sensitive to the price for selling electricity generated from solar facilities.

Fig. 9.14 shows values of *LCOW* for CSP–MED and CSP–RO plants in Aqaba and Ashdod for several feed-in tariff rates. The feed-in tariff is a policy instrument used to promote investment in renewable energy technologies. The feed-in tariff guarantees energy producers a price for electricity above the market price for electricity delivered to the grid. A feed-in tariff commonly is offered to producers using renewable energy sources. For the expensive electricity prices in Aqaba, the CSP–MED plant has economic advantages relative to the CSP–RO plant, while the CSP–RO configuration is economically advantageous over the CSP–MED configuration for other electricity prices. In Ashdad, the CSP–RO plant yields freshwater at a lower cost than the CSP–MED plant. For both CSP–MED and CSP–RO plants, the combined system configuration has a lower cost of water when thermal energy storage (TES) is not utilized. The levelized cost of water production is lower in Jordan (Aqaba) than Israel (Ashdod), mainly due to the lower wages for personnel and poorer irradiation conditions in Aqaba.

Despite the result regarding thermal storage in the preceding paragraph, which indicates that the use of TESs leads to an increase in the final cost of water, thermal storage systems are important for enabling solar energy technologies to contribute to greater percentage of the energy supply for desalination. Usually, it is necessary to incorporate TES systems in solar energy facilities. TES systems are required to operate at relatively high temperatures in high-temperature solar facilities such as CSP systems, and it is highly desired to avoid thermal losses. Efforts are being made to lower the capital cost of TES. The target cost for a TES system depends on the cost and performance factors of the overall solar plant. When the temperature drop through the electricity generation plant is relatively high, the stored

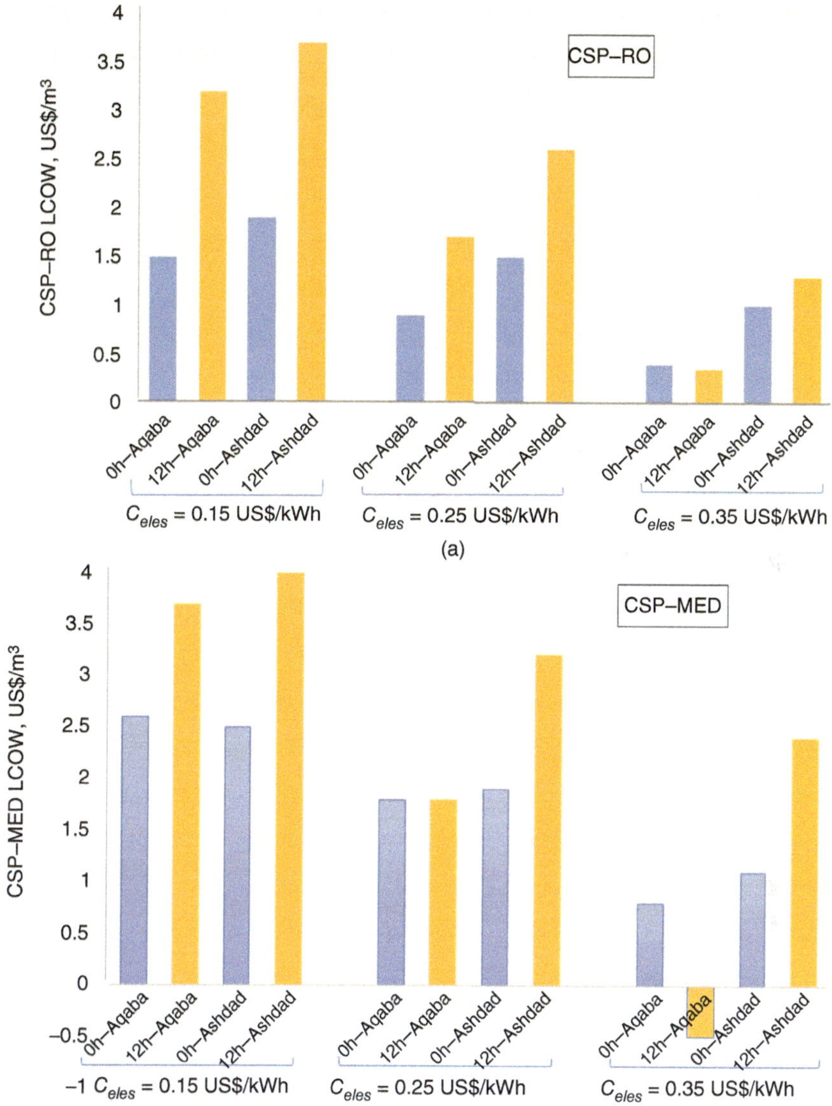

Fig. 9.14 Levelized cost of water for (A) CSP–RO and (B) CSP–MED cogeneration plants in Aqaba and Ashdod for storage capacities of 0 h (blue bars) and 12 h (yellow bars). Three feed-in tariff values are considered for electricity production for the grid (i.e., C_{elec}). *Data extracted from Olwig et al. (2012).*

energy density associated with the sensible part of a TES increases, which helps lower the storage inventory and capital cost. For instance, the cost of molten for salt TES linked to trough solar systems operating at 390 °C is around 2.5 times greater than the cost of molten salt for solar towers operating at 565 °C (Glatzmaier, 2011).

Note that the technoeconomic results in this case study are sensitive to the economic and technical input parameters used in the analysis. Using up-to-date data and considering relevant influences are both important for technoeconomic system analyses. The results of this case study suggest that the combination of a CSP plant and a seawater desalination system can be economically viable for meeting freshwater demands in the MENA region.

9.5 Exergoeconomic analysis of seawater desalination systems using renewable energy resources

Exergoeconomic analysis is an economic analysis method based on exergy. Exergoeconomic analysis combines thermodynamic and economic principles to understand better system and process economics, and to determine optimum designs. Often the results of exergoeconomic analyses cannot be obtained with conventional thermodynamic and cost analyses, which are usually based on energy. Results of exergoeconomic analyses of sustainable energy technologies and desalination systems provide useful information regarding potential modifications of components, cost-effectiveness, and optimum designs. Such information is important in making economic and investment decisions. By using the results of exergoeconomic analyses, measures for reducing the overall cost of the product(s) (freshwater, electricity, heat, etc.) can be proposed. Exergoeconomic analyses of renewable energy–based desalination systems permit recommendations that draw on the second law of thermodynamics to be developed for process designs.

To evaluate an energy system with exergoeconomic analysis, an exergy analysis of the system is normally carried out first. The exergy-relevant relations are presented in detail in Chapters 3, 6, and 7. The fundamental energy and exergy balances can be written, respectively, as (Dincer and Rosen, 2021)

$$\text{Energy input} - \text{Energy output} = \text{Energy accumulation} \tag{9.24}$$

$$\begin{aligned}\text{Exergy input} - \text{Exergy output} - \text{Exergy consumption} \\ = \text{Exergy accumulation}\end{aligned} \tag{9.25}$$

Eqs. (9.24) and (9.25) indicate that energy can be neither consumed nor generated as it is a conserved quantity, whereas exergy can be consumed in a process due to irreversibilities. Energy and exergy outputs can be written, respectively, as

$$\text{Energy output} = \text{Product energy output} + \text{Waste energy output} \quad (9.26)$$

$$\text{Exergy output} = \text{Product exergy output} + \text{Waste exergy output.} \quad (9.27)$$

A general exergy rate balance under steady-state conditions for component (or device) k when kinetic and potential energy values are treated as negligible can be expressed as follows:

$$\dot{E}x_{Des,k} + \dot{E}x_{loss,k} = \dot{E}x_{P,k} - \dot{E}x_{F,k} \quad (9.28)$$

Here, $\dot{E}x_{Des,k}$ and $\dot{E}x_{loss,k}$ denote the exergy destruction rate and exergy loss rate, which are measure of thermodynamic inefficiencies occurring in the component. The destroyed exergy represents the amount of available work that is lost due to system irreversibilities, while the exergy loss represents the amount of available work that is lost with outgoing waste streams. The exergy rates of product ($\dot{E}x_{P,k}$) and fuel ($\dot{E}x_{F,k}$) for component k can be specified based on the purpose of the component. Eq. (9.28) is used when applying the specific exergy costing (SPECO) method, which is a useful type of exergy-based economic analysis.

The exergy rates associated with work rate ($\dot{W}$) and heat transfer rate ($\dot{Q}$), respectively, can be written as

$$\dot{E}x^{\dot{Q}} = \left(1 - \frac{T_0}{T}\right)\dot{Q} \quad (9.29)$$

$$\dot{E}x^{\dot{W}} = \dot{W} \quad (9.30)$$

where T_0 and T, respectively, are the environment temperature (in K) and the temperature at which the heat transfer occurs (in K). The exergy rate of a material stream can be written as

$$\dot{E}x = \dot{m}ex \quad (9.31)$$

where $\dot{m}$ is the mass flow rate and ex is the specific exergy. The total specific exergy of a flow can be written in terms of its physical exergy (ex_{ph}) and chemical exergy (ex_{chem}) as follows:

$$ex = \underbrace{h - h^* - T_0(s - s^*)}_{ex_{ph}} + \underbrace{\sum_{i=1}^{n} (\mu_i^* - \mu_{o,i})\, w_i}_{ex_{chem}} \quad (9.32)$$

where h and s, respectively, denote specific enthalpy and specific entropy, and w_i is the mass fraction of constituent i. The superscript * used for h, s, and μ indicates that the properties are those at the environmental temperature and pressure. The term μ_i* denotes specific chemical potential of constituent i at environmental temperature and pressure, and $\mu_{0,i}$ denotes chemical potential of constituent i at the environmental temperature, pressure, and constituent concentrations. Note that the chemical exergy term is zero for a substance that forms a part of the environment such as liquid water.

The exergy efficiency of component k in a system can be defined as the ratio of the exergy rate of product to the exergy rate of fuel, as follows:

$$\Psi_k = \frac{\dot{E}x_{P,k}}{\dot{E}x_{F,k}} \tag{9.33}$$

SPECO method: The SPECO method is used here in the exergoeconomic analysis of an energy system. The SPECO method assesses the effects of design parameters in an energy system on the costs, and identifies possible improvements for more cost-effective approaches (Lazzaretto and Tsatsaronis, 2006). In this method, cost rates associated with each stream are determined based on the corresponding exergy rate and unit cost. A general balance for cost which is an increasing and nonconserved quantity can be written as

$$\text{Cost input} + \text{Cost generation} - \text{Cost output} = \text{Cost acculülation} \tag{9.34}$$

where the cost generation in a system can be expressed as

$$\begin{aligned} \text{Cost generation} = {} & \text{Capital cost of equipment} \\ & + \text{All other creation and maintennce costs} \end{aligned} \tag{9.35}$$

Then, under steady-state conditions, the cost balance rate can be written for each component of a system in terms of fuel cost rate ($\dot{C}_F$) and product cost rate ($\dot{C}_P$) as follows (Lazzaretto and Tsatsaronis, 2006):

$$\underbrace{\sum_{out}(C_{out} \cdot \dot{E}x_{out})k + c_{\dot{w},k}\dot{E}x_{\dot{w},k}}_{\dot{C}_{P,k}} = \underbrace{C_{\dot{Q},k}Ex_{\dot{Q},k} + \sum_{in}(C_{in} \cdot \dot{E}x_{in})k}_{\dot{C}_{F,k}} + \underbrace{\dot{Z}_k}_{\dot{Z}_k} \tag{9.36}$$

where c is cost per unit exergy (\$/kJ), $\dot{E}x$ is exergy rate of a material stream (kW), and $\dot{E}x_{\dot{Q}}$ and $\dot{E}x_{\dot{w}}$ are the exergy rates associated with the heat and work rates. The terms $c_{\dot{Q}}$ and $c_{\dot{w}}$ are the unit costs of electricity and heat, and their values depend on the conversion efficiency of generator and the type of energy used to generate the electricity and heat (e.g., natural gas, renewable energy, coal). The unit costs of electricity and heat vary by region and country, sometimes significantly. In Eq. (9.36) it is seen that the summation of cost rates of all exergy streams entering a component ($\dot{C}_{F,k}$) and the cost rates of capital and O&M of component k ($\dot{Z}_k$) are equal to the summation of cost rates of exergy streams leaving that component ($\dot{C}_{P,k}$).

The component cost rate of capital investment and O&M cost, $\dot{Z}_k$, can be written as

$$\dot{Z}_k = \dot{Z}_{capital,k} + \dot{Z}_{O\&M,k}. \tag{9.37}$$

In the economic analysis, annualized values are commonly used, in part because they undergo a change over their economic life. Then, the annual levelized capital cost rate for component k in \$/h can be expressed using the capital recover factor (CRF) and annual operation time of system in hours (τ) as

$$\dot{Z}_{capital,k} = \frac{\left(Z_{capital,k}.CRF\right)}{\tau} \tag{9.38}$$

where $\dot{Z}_{capital,k}$ is the capital cost for component k. As mentioned before, cost data that are not up to date need to be updated, and this can be done using the Marshall and Swift equipment cost index. The economic parameter CRF is the capital recovery factor, defined in Eq. (9.2). We can also write

$$\dot{Z}_k = CRF.\frac{\varphi.Z_{capital,k}}{\tau} \tag{9.39}$$

where ϕ is the O&M factor. Recalling that the exergy destruction rate for a component refers to all exergy destroyed in the component, and that the exergy loss in for component refers to the exergy released to the environment, the cost rate of exergy destruction, and exergy loss, respectively, can be written as

$$\dot{C}_{Des,k} = c_{f,k}\dot{E}x_{Des,k} \tag{9.40}$$

$$\dot{C}_{loss,k} = c_{f,k}\dot{E}x_{loss,k} \tag{9.41}$$

Here, $\dot{E}x_{Des,k}$ is the exergy destruction rate of component k and $\dot{E}x_{loss,k}$ the exergy loss rate for component k, while c_f is the average unit cost of fuel flow in component k, which can be determined by summing the unit cost rates of fuels of the component and their corresponding exergy rates. That is,

$$c_{f,k} = \frac{\dot{C}_{F,k}}{\dot{E}x_{F,k}} \quad (9.42)$$

The total cost rate of component k ($\dot{C}_{k,Total}$) is an important exergo-economic parameter. It can be defined as the summation of cost rates associated with the destruction and loss of exergy ($\dot{C}_{Des,k}$, $\dot{C}_{loss,k}$) and the total cost rate of component k ($\dot{Z}_k$), as follows:

$$\dot{C}_{k,Total} = \dot{C}_{Des,k} + \dot{C}_{loss,k} + \dot{Z}_k \quad (9.43)$$

The terms $\dot{C}_{Des}$ and $\dot{Z}$ in Eq. (9.43) reflect the importance of a component in the optimum operation of a system from both thermodynamic and economic viewpoints.

The sources of cost in a component can be from nonexergy-related costs (capital investment and operating and maintenance expenses) and exergy destruction and exergy loss. In evaluating the performance of a component and knowing the relative significance of each cost source, the component cost effectiveness, often called the exergo-economic factor, can be expressed. That is, for component k,

$$f_k = \frac{\dot{Z}_k}{\dot{C}_{k,Total}} = \frac{\dot{Z}_k}{\dot{Z}_k + \dot{C}_{Des,k} + \dot{C}_{loss,k}}. \quad (9.44)$$

The exergoeconomic factor of component k, f_k, describes the relation between thermodynamic losses of a component and its capital cost. A low value of exergoeconomic factor of a component indicates that the efficiency of component merits increasing to reduce the exergy destruction rate within the component, even at the expense of an increase in the capital cost for the component. A high value of exergoeconomic factor of a component indicates that the investment cost for the component is worth decreasing, even if the exergy efficiency of the component decreases.

The relative cost difference (r) is another important exergoeconomic variable. This variable is useful in optimizing a system component and it describes the relative increase in the average cost per unit exergy between fuel and product of the component. Thus, the relative cost difference of component k can be defined as the difference between the average cost per unit exergy of product and fuel divided by the cost per unit fuel exergy. That is,

$$r_k = \frac{c_{p,k} - c_{f,k}}{c_{f,k}} \tag{9.45}$$

where the average cost per unit exergy of product can be expressed as

$$c_{p,k} = \frac{\dot{C}_{P,k}}{\dot{E}x_{P,k}}. \tag{9.46}$$

Generally, when N exergy streams exit a component, there are N unknowns and only one equation (i.e., the cost balance rate). Thus, auxiliary equations using F and P rules can be employed to formulate $N - 1$ equations, thereby permitting the equations to be solved. By solving all cost rate balances and auxiliary equations simultaneously, the specific cost (c_j) of each exergy stream can be determined. Next the method for employing F and P rules is described for a general case.

Application of SPECO method using F and P rules: When the cost rate balance is applied for a component with N exergy streams, the number of unknowns is N while the number of equations is one (the cost rate balance). With the aid of F and P rules, $N - 1$ auxiliary equations can be formulated. According to this rule, fuel (F) can be defined as the summation of all exergy values at the inlet plus all decreasing inlet and outlet exergy flows minus all exergy increases that are not associated with the purpose of component. Similarly, product (P) can be defined as the summation of all exergy values at the outlet plus all increasing inlet and outlet exergy flows that are aligned with the purpose of component (Lazzaretto and Tsatsaronis, 2006). Fig. 9.15 shows a generalized component in a thermal system and its interacting streams to define fuel, product, and axillary equations.

With the SPECO method, the components can be ordered based on decreasing cost importance (i.e., $\dot{Z} + \dot{C}_{Des}$). As mentioned before, the exergoeconomic factor (f) for each system component represents the relation between the cost associated with the exergy destruction and capital cost. A low value of f for a component suggests that some improvements are required for enhancing the efficiency of the component at the expense of increasing the component capital cost. But a high value of f for a component indicates that it is more cost effective to decrease the capital cost of component at the expense of decreasing its efficiency.

Several studies have been conducted using exergoeconomic analysis on desalination plants. Lawal et al. (2018) carried out an exergoeconomic study of a closed-air open-water (CAOW) humidification–dehumidification

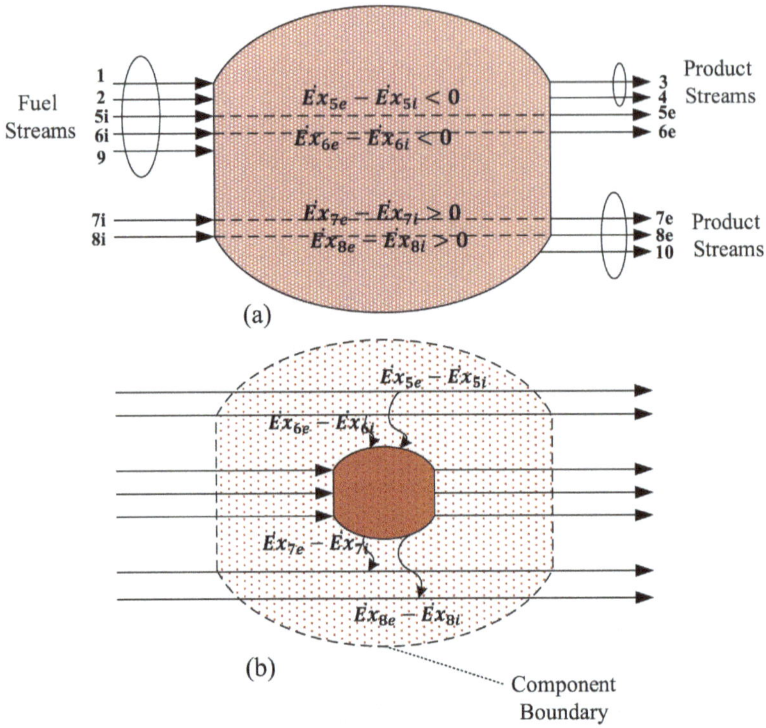

Fig. 9.15 Interacting exergy flow rates ($\dot{E}x$), describing fuel and product principles for a general component in (a) a thermal system and (b) a thermal system with a productive unit. *Modified from Lazzaretto and Tsatsaronis (2006).*

(HDH) desalination system driven by a vapor compression heat pump. CAOW water-heated and air-heated HDH systems are shown in Fig. 9.16. In the vapor compression heat pump, the refrigerant in the condenser supplies the thermal energy required for the desalination system. The input feed seawater vaporizes the refrigerant in the evaporator of the heat pump system. The reduced temperature in the feed seawater makes the condensation process in the dehumidifier more efficient. In the CAOW water-heated configuration, seawater is preheated in the condenser of the heat pump before entering the humidifier while, in the air-heated configuration, air is preheated. In the humidifier, a portion of the water is vaporized and absorbed by air. The moist air is condensed in the dehumidifier, producing freshwater.

Exergy-based cost equations for the system components are presented by Lawal (2018), and used to determine the product costs and exergy destruction costs associated with the desalting process. It is shown that, when the isentropic efficiency of compressor is low, the exergy destruction

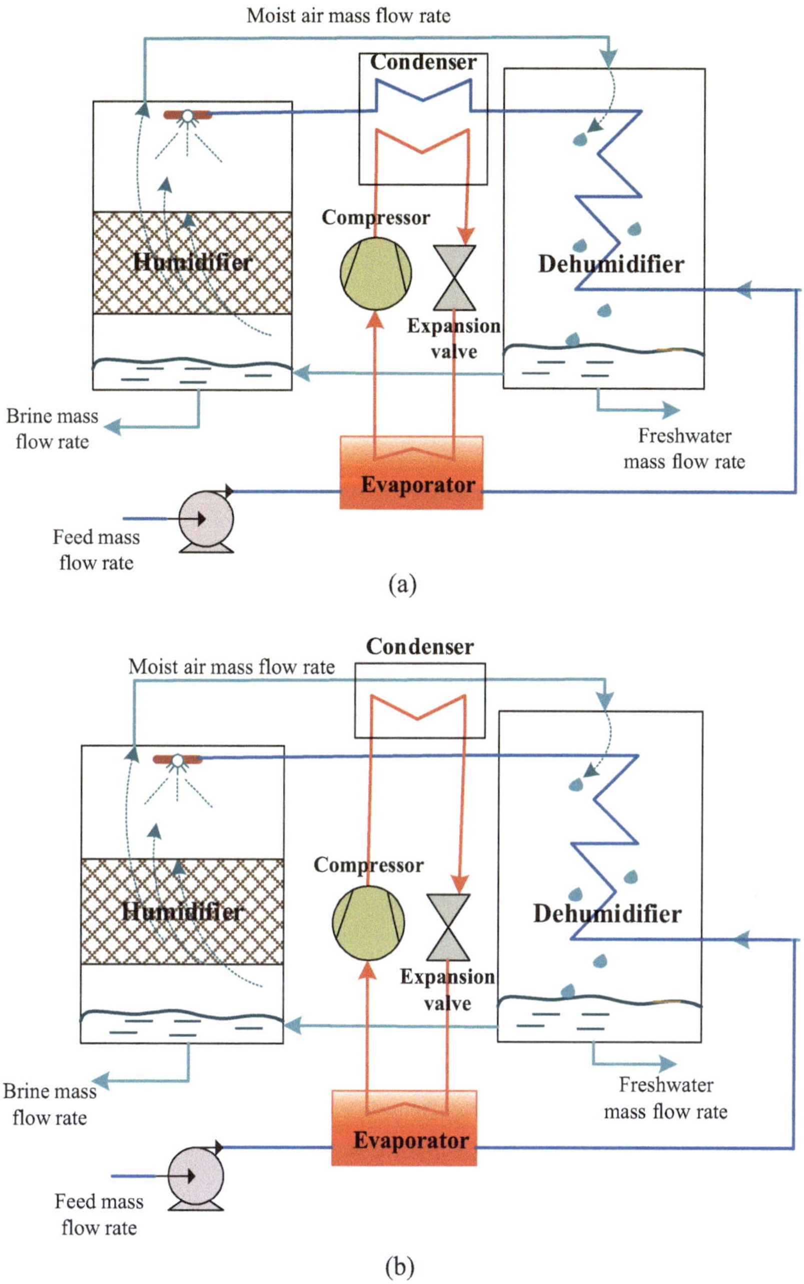

Fig. 9.16 Closed air, open water (CAOW) humidification–dehumidification (HDH) desalination systems (A) water-heated, (B) air-heated.

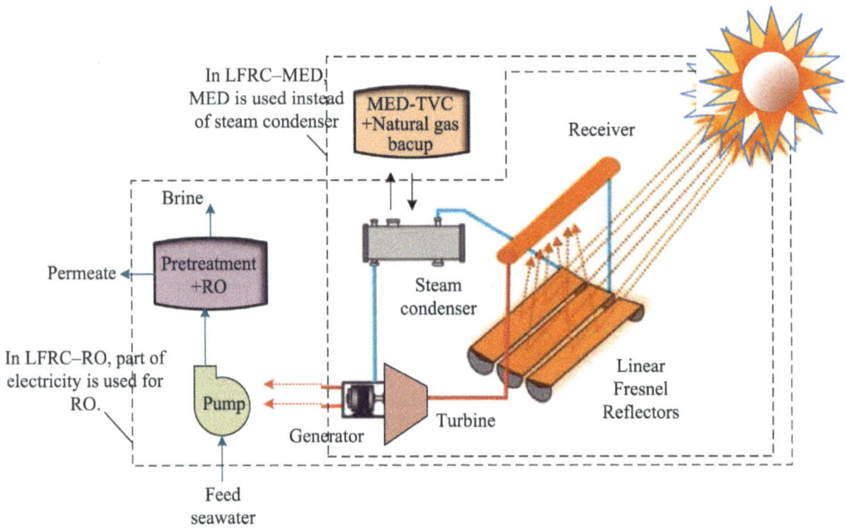

Fig. 9.17 ***Simplified schematic of linear Fresnel Rankine cycle (LFRC) integrated with an RO system and an MED system.***

rate is larger in this component than all others. After compressor, the greatest exergy destruction occurs in the evaporator. Also, the greatest cost rate is associated with the compressor (0.0122 US\$/h) in both CAOW air-heated and CAOW water-heated systems. Next, the condenser and evaporator have the greatest cost rates, at 0.0061 US\$/h for both components. The freshwater production cost for the CAOW air-heated and CAOW water-heated desalination systems are found to vary between 4.6 US\$/m^3 and 5.5 US\$/m^3, and between 6.0 US\$/m^3 and 7.1 US\$/m^3, respectively.

Baniasad Askari et al. (2018) performed comprehensive energy, exergy, and exergoeconomic analyses for a Linear Fresnel Rankine Cycle (LFRC) integrated with an RO system and MED with and without thermal vapor compressor (TVC. Fig. 9.17 shows a simplified schematic diagram of the LFRC–MED/RO system. A natural gas backup system is used when thermal energy production from the solar field is insufficient. The exergy analysis results for the LFRC–MED/RO plant indicate that the Linear Fresnel (LF) solar field is responsible for the greatest contribution to the exergy destruction rate, at 78%–90%. The exergy destruction contributions for the MED and MED–TVC systems, respectively, are 5% and 7%. The unit cost of freshwater production is determined for each plant, and it is shown that the cost of water is more sensitive to solar radiation than seawater temperature.

Baniasad Askari et al. (2018) show that the freshwater cost for the single-purpose solar desalination (LF–MED/RO) system is higher than for the solar-based electricity-water cogeneration system (LFRC–MED/RO). The

Table 9.10 Cost of water (US$/m^2) for various solar-based desalination systems[a].

Configuration	**Design parameters**		
	DNI = 2700 kWh/m^2/year, T_{sw} = 31 °C, Freshwater production rate = 38,100 m^3/day	**DNI = 2300 kWh/m^2/year, T_{sw} = 36 °C, Freshwater production rate = 34,460 m^3/day**	**DNI = 1990 kWh/m^2/year, T_{sw} = 26 °C, Freshwater production rate = 40,970 m^3/day**
RO-LFRC with electricity consumption of 3.5 kWh/m^3	1.4	1.4	1.49
RO-LFRC with electricity consumption of 5.5 kWh/m^3	1.7	1.7	1.75
MED-LFRC with electricity consumption of 1.5 kWh/m^3	1.3	1.45	1.5
MED-TVC-LFRC with motive steam pressure of 362 kPa and electricity consumption of 1.5 kWh/m^3	1.4	1.72	1.75
MED-TVC-LFRC with motive steam pressure of 4540 kPa and electricity consumption of 1.5 kWh/m^3	1.7	2.4	2.5
MED-TVC-LF with motive steam pressure of 4540 kPa and electricity consumption of 1.5 kWh/m^3	2.1	2.4	2.7
MED-TVC-LF with motive steam pressure of 362 kPa and electricity consumption of 1.5 kWh/m^3	2.25	2.6	2.9
MED-LF with electricity consumption of 1.5 kWh/m^3	2.45	3.1	3.35

[a]Data extracted from Askari et al. (2018).
DNI, direct normal irradiance; RO-LFRC, reverse osmosis-Linear Fresnel Rankine Cycle; MED-TVC-LF, Multieffect desalination-thermal vapor compressor-Linear Fresnel; LF, Linear Fresnel.

freshwater cost from the solar-based electricity-water cogeneration system is equal to that of the fossil fuel–based system, when the fuel price is 0.23 \$/m^3. Also, for regions with relatively high solar radiation levels (2700 kWh/m^2/year), the capital cost associated with the solar field and Rankine cycle for the LFRC–MED configuration is lower than that for LFRC-RO configuration (with electricity consumption of 5.5 kWh/m^3). In fact, higher solar radiation results in a lower cost for the thermal energy used in the MED system compared to the electricity cost used by the RO unit. Table 9.10 presents the cost of water for several system configurations for various values of DNI, intake seawater temperature (T_{sw}), and freshwater production capacity.

9.5.1 Example: exergoeconomic analysis of a geothermal steam turbine combined with MED and RO desalination plant

Consider a combined geothermal electrical power-freshwater production system, as shown in Fig. 9.18. It consists of a geothermal steam turbine in which 140 kg/s saturated liquid geo-fluid (water) at a temperature of 230 °C and a pressure of 2.8 MPa passes through an expansion valve, where its pressure is dropped to 1.5 MPa. Then the saturated vapor-liquid mixture enters a separator in which the vapor and liquid phases are separated. The vapor portion (stream 4) enters the steam turbine and expands to 0.08 MPa with isentropic efficiency of 80%, while the liquid portion (stream 3) is returned to the ground. About 20% of the electricity generated by the steam turbine is used in an RO desalination system to drive RO pump, which pressurizes feed seawater to 6 MPa. Distillate and brine leave the RO membrane module at the low-pressure and high-pressure sides, respectively. An expansion valve is used to reduce the pressure of the high-pressure brine stream to atmospheric pressure. The geothermal saturated vapor leaving the steam turbine is transferred to the first effect of an MED system, where it condenses and provides thermal energy for freshwater production. Then, the geothermal fluid, which is now a saturated liquid, is further cooled to 47 °C in a heat exchanger before being discharged into the injection well.

Design data are given in Table 9.11 for the combined geothermal power cycle and MED–RO desalination plant. Economic parameter values considered for the geothermal-based combined power-freshwater production system are listed in Table 9.12.

Assuming no pressure drops and thermal losses in the system components other than those stipulated above:

1. Develop mass, energy, exergy, and cost rate balances for each component of the geothermal-based combined system.
2. Determine the exergy rate (kW), cost rate ($/h), and specific cost of exergy (or unit cost of exergy) ($/kWh) for each stream in the geothermal-based combined system. Do calculations for both high-interest rate countries (assume an interest rate of 14%) and low-interest rate countries (assume an interest rate of 6%).
3. Determine the cost rate of exergy destruction ($/h), average unit cost of fuel flow ($/kWh), total cost rate ($/h), and exergo-economic factor (%) for all components of the system. Do calculations for both high-interest rate countries (assume an interest rate of 14%) and low-interest rate countries (assume an interest rate of 6%).

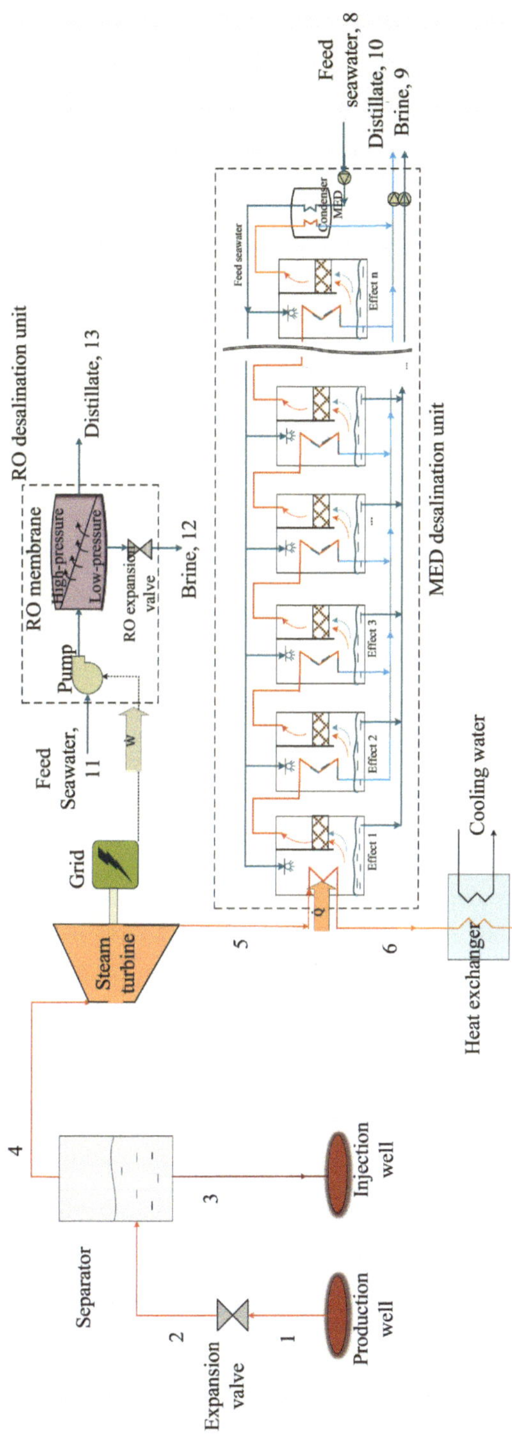

Fig. 9.18 *Geothermal-based steam turbine combined with MED and RO desalination system.*

Table 9.11 Design characteristics of geothermal-based combined power-freshwater production system.

Parameter	Value
Geothermal steam turbine	
Environment temperature, T_o	25 °C
Environment pressure, P_o	0.101 MPa
Geo-fluid temperature from production well, T_1	230 °C
Quality of geo-fluid at production well, x_1	0
Geo-fluid pressure at expansion valve outlet, P_2	1.5 MPa
Turbine isentropic efficiency	80%
Turbine outlet steam pressure, P_5	0.08 MPa
Heat exchanger outlet geo-fluid temperature, T_7	47 °C
MED desalination	
Intake feed seawater temperature	25 °C
Intake feed seawater pressure	0.101 MPa
Intake feed seawater salinity	42 g/kg
Brine salinity	70 g/kg
Distillate salinity	0 g/kg
Number of effects	6
First effect temperature (i.e., top brine temperature)	60 °C
Temperature difference between effects	3.2 °C
RO desalination	
Intake feed seawater temperature	25 °C
Intake feed seawater pressure	0.101 MPa
Intake feed seawater salinity	42 g/kg
Brine salinity	70 g/kg
Distillate salinity	0 g/kg
Pressure of RO module high pressure side	6 MPa

Table 9.12 Economic parameters for the geothermal-based combined power-freshwater production system.

Parameter	Value
Annual operation hours, τ	7000 h
Lifetime of system, n	20 years
Annual interest rate, i	14% for high-interest rate countries and 6% for low-interest rate countries
Maintenance factor, ϕ	1.06
Unit cost of intake geo-fluid exergy, c_1	1.3 $/GJ (0.0046 $/kWh)

Solution

For conducting an exergoeconomic analysis of a system, a cost rate balance should be written for each component. This balance, combined with appropriate auxiliary equations, results in a system of linear algebraic equations that can be solved for the unknown values of cost rate ($\dot{C}$) or unit cost of exergy (c). In this example, mass, energy, and exergy rate balances are first written for the components of the geothermal-based combined system to determine the thermodynamic properties of each state. The cost rate balance in addition to auxiliary equations is also written for each component to determine the cost rates of streams in the system. The information on stream costs is used to calculate exergoeconomic variables ($\dot{C}_{Des}$, $\dot{C}_{total}$, f, etc.) associated with each component.

We assume generally that:

- The system operates at steady-state conditions.
- The startup period can be neglected.
- Kinetic and potential energy values are negligible.
- The steam turbine is adiabatic.
- Pressure drops and heat losses in geothermal steam turbine and MED unit are neglected.

1. General expression for mass, energy, exergy, and cost rate balances neglecting kinetic and potential energy values under steady-state conditions for component k can be written, respectively, as:

$$\sum_{in} \dot{m} = \sum_{out} \dot{m} \tag{9.47}$$

$$\dot{Q} + \sum_{in} \dot{m}h = \dot{W} + \sum_{out} \dot{m}h \tag{9.48}$$

$$\dot{E}x^{\dot{W}}{}_{k} + \sum_{out} \left(\dot{m}ex\right)_k + \dot{E}_{Des,k} = \dot{E}_k^{\dot{Q}} + \sum_{in} \left(\dot{m}ex\right)_k \tag{9.49}$$

$$\underbrace{\sum_{out} (c_{out} \cdot \dot{E}x_{out})k + c_{\dot{w},k}}_{\dot{C}_{P,k}} = \underbrace{C_{\dot{Q},k}\dot{E}x_{\dot{Q},k} + \sum_{in} (C_{in}\dot{E}x_{in})_k + \dot{Z}_k}_{\dot{C}_{F,k}} \tag{9.50}$$

where $\dot{m}$, h, and ex, respectively, denote mass flow rate (kg/s), specific enthalpy (kJ/kg), and specific exergy (kJ/kg). The definitions of exergy and cost of fuel ($\dot{E}x_{F,k}$ and $\dot{C}_{F,k}$) and exergy and cost rate of product ($\dot{E}x_{P,k}$ and $\dot{C}_{P,k}$) for each component of the system are listed in Table 9.13.

Table 9.13 Definition of fuel and product exergy rates of geothermal-based system shown in Fig. 9.18, based on describing fuel and product principles.

Component	**Definition of fuel and product exergy and cost rates**
Expansion valve 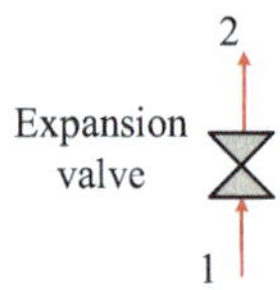	$\dot{E}x_F = \dot{m}_1 ex_1$ $\dot{E}x_P = \dot{m}_2 ex_2$ $\dot{C}_F = \dot{C}_1$ $\dot{C}_P = \dot{C}_2$
Separator 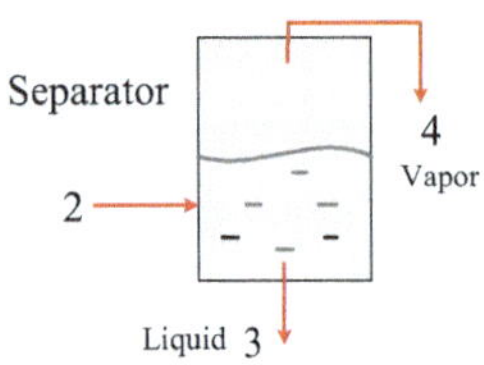	$\dot{E}x_F = \dot{m}_2 ex_2$ $\dot{E}x_P = \dot{m}_3 ex_3 + \dot{m}_4 ex_4$ $\dot{C}_F = \dot{C}_2$ $\dot{C}_P = \dot{C}_3 + \dot{C}_4$
Steam turbine 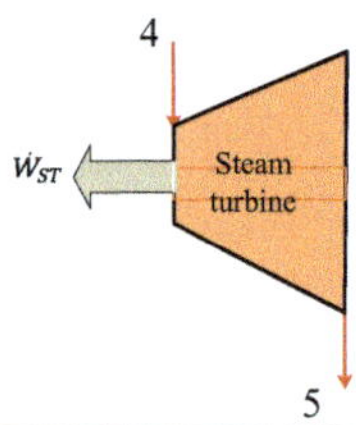	$\dot{E}x_F = \dot{m}_4 ex_4 - \dot{m}_5 ex_5$ $\dot{E}x_P = \dot{W}_{ST}$ $\dot{C}_F = \dot{C}_4 - \dot{C}_5$ $\dot{C}_P = \dot{C}_{\dot{W}_{ST}}$
MED unit 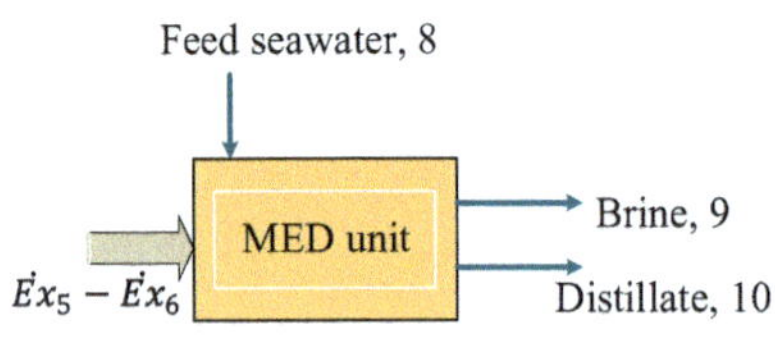	$\dot{E}x_F = \dot{m}_5 ex_5 - \dot{m}_6 ex_6 + \dot{m}_{f,MED} ex_{f,MED} - \dot{m}_{B,MED} ex_{B,MED}$ $\dot{E}x_P = \dot{m}_{D,MED} ex_{D,MED}$ $\dot{C}_F = \dot{C}_5 - \dot{C}_6 + \dot{C}_{f,MED} - \dot{C}_{B,MED}$ $\dot{C}_P = \dot{C}_{D,MED}$
Heat exchanger 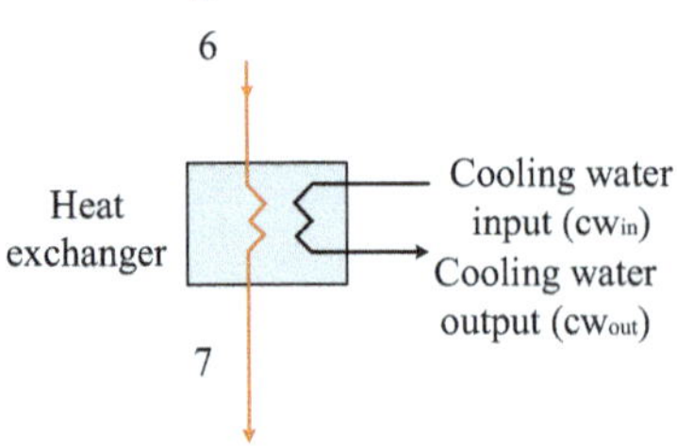	$\dot{E}x_F = \dot{m}_{cw} ex_{cw,out} - \dot{m}_{cw} ex_{cw,in}$ $\dot{E}x_P = \dot{m}_6 ex_6 - \dot{m}_7 ex_7$ $\dot{C}_F = \dot{C}_{cw,out} - \dot{C}_{cw,in}$ $\dot{C}_P = \dot{C}_7 - \dot{C}_6$

Component	Definition of fuel and product exergy and cost rates
RO unit 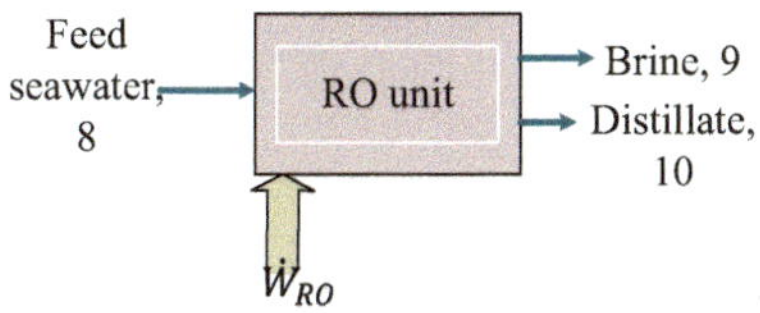	$\dot{E}x_F = \dot{W}_{RO} + \dot{m}_{f,RO}ex_{f,RO}$ $\dot{E}x_P = \dot{m}_{D,RO}ex_{D,RO} + \dot{m}_{B,RO}ex_{B,RO}$ $\dot{C}_F = \dot{C}_{\dot{W}_{RO}} + \dot{C}_{f,RO} - \dot{C}_{B,RO}$ $\dot{C}_P = \dot{C}_{D,MED}$

The mass, energy, exergy, and cost rate balances and auxiliary equations (based on F and P rules) for the various system components can be written as follows:

Expansion valve

$$\dot{m}_1 = \dot{m}_2 \tag{9.51}$$

$$\dot{m}_1 h_1 = \dot{m}_2 h_2 \tag{9.52}$$

$$\dot{m}_2 ex_2 + \dot{E}x_{Des,exp.valve} = \dot{m}_1 ex_1 \tag{9.53}$$

$$c_2 \dot{E}x = c_1 \dot{E}x_1 + \dot{Z}_{exp.valve} \tag{9.54}$$

$$c_1 = c_{Geo.fluid} \tag{9.55}$$

Note that the unit cost of exergy of the geothermal source ($c_{Geo.fluid}$) can be determined based on the exergy rate of the geothermal fluid and the cost rate of the production well, as $c_{Geo.fluid}\dot{E}x_1 = \dot{Z}_{well}$. Akbari et al. (2014) calculated $c_{Geo.fluid}$ as 1.3 \$/GJ which is used in this example.

Separator

$$\dot{m}_2 = \dot{m}_3 + \dot{m}_4 \tag{9.56}$$

$$\dot{m}_3 = x_{fg}\dot{m}_2 \tag{9.57}$$

where x_{fg} is the quality of saturated water mixture, expressible as

$$x_{fg} = \frac{h_2 - h_{f,2}}{h_{fg,2}} \tag{9.58}$$

Here, h_2, $h_{f,2}$, and $h_{fg,2}$, respectively, denote specific enthalpy of saturated water mixture at temperature T_2, specific enthalpy of saturated liquid at T_2, and enthalpy of evaporation at T_2. It is assumed that the separation process of water vapor from liquid water occurs at constant pressure. The exergy and cost rate balances and auxiliary equations for the separator are

$$\dot{m}_3 ex_3 + \dot{m}_4 ex_4 + \dot{E}x_{Des,sep} = \dot{m}_2 ex_2 \quad (9.59)$$

$$c_3 \dot{E}x_3 + c_4 \dot{E}x_4 = c_2 \dot{E}x_2 + \dot{Z}_{sep} \quad (9.60)$$

$$c_3 = c_4 \quad (9.61)$$

Steam turbine

$$\dot{m}_5 = \dot{m}_6 \quad (9.62)$$

$$\dot{m}_5 h_5 + \dot{W}_{ST} = \dot{m}_4 h_4 \quad (9.63)$$

$$\dot{m}_5 ex_5 + \dot{W}_{ST} + \dot{E}x_{Des,ST} = \dot{m}_4 ex_4 \quad (9.64)$$

$$\dot{W}_{ST} = \dot{m}_5 ex_5 \quad (9.65)$$

$$c_5 \dot{E}x_5 + c_{\dot{W}_{ST}} \dot{W}_{ST} = c_4 \dot{E}x_4 + \dot{Z}_{ST} \quad (9.66)$$

$$c_4 = c_5 \quad (9.67)$$

Here $c_{\dot{W}_{ST}}$ is the unit cost of exergy of the electricity generated by the steam turbine.

MED desalination unit

Detailed energy and exergy analyses of an MED desalination system are presented in Chapters 6–8. In this example, the work rates for pumping feed, distillate, and brine are neglected in the MED. Also, the entire MED unit shown in Fig. 9.18 is considered as a single control volume and the following balance equations are written for it:

$$\dot{m}_6 = \dot{m}_5 \quad (9.68)$$

$$\dot{m}_{D,MED} + \dot{m}_{B,MED} = \dot{m}_{f,MED} \quad (9.69)$$

$$\dot{m}_{D,MED} w_{D,MED} + \dot{m}_{B,MED} w_{B,MED} = \dot{m}_{f,MED} w_{f,MED} \quad (9.70)$$

where w is the salinity (or salt concentration) in g/kg. Based on Table 9.11, the salt concentration for feed, distillate, and brine streams are 42 g/kg, 0 g/kg, and 70 g/kg, respectively. The temperature difference between the MED effects is the same at 3.2 °C. Therefore the temperature of the last effect is 44 °C (60 °C – 5 × 3.2°C). It is assumed that the brine and distillate streams exit the MED unit at the average temperature of the MED unit as $T_{ave, MED} = (T_{effect1} + T_{effect6})/2 = 60°C+44°C)/2)$. The energy, exergy, and cost rate balances and auxiliary equations for MED follow:

$$\dot{m}_6 ex_6 + \dot{m}_{D,MED} ex_{D,MED} + \dot{m}_{B,MED} ex_{B,MED} + \dot{E}x_{Des,MED} = \dot{m}_5 ex_5 + \dot{m}_{f,MED} ex_{f,MED} \tag{9.71}$$

$$c_6 \dot{E}x_6 + c_{D,MED} \dot{E}x_{D,MED} + c_{B,MED} \dot{E}x_{B,MED} = c_5 \dot{E}x_5 + c_{f,MED} \dot{E}x_{f,MED} + \dot{Z}_{MED} \tag{9.72}$$

$$c_{f,MED} = 0 \tag{9.73}$$

$$c_{B,MED} = 0 \tag{9.74}$$

$$c_5 = c_6 \tag{9.75}$$

where subscripts *f*, *B*, and *D*, respectively, denote feed, brine, and distillate. Here, the unit cost of exergy of the intake feed seawater $c_{f,MED}$ is zero because there is no preprocessing for using it. Also, the unit cost of exergy of the brine stream $c_{B,MED}$ is zero because it is treated as a loss which is rejected into the sea.

Heat exchanger

$$\dot{m}_6 = \dot{m}_7 \tag{9.76}$$

$$\dot{Q}_{HEX} = \dot{m}_{cw} ex_{cw,out} - \dot{m}_{cw} ex_{cw,in} = \dot{m}_6 h_6 - \dot{m}_7 h_7 \tag{9.77}$$

$$\dot{m}_7 ex_7 + \dot{m}_{cw} ex_{cw,out} + \dot{E}x_{Des,cond} = \dot{m}_6 ex_6 + \dot{m}_{cw} ex_{cw,in} \tag{9.78}$$

$$c_7 \dot{E}x + c_{cw,out} \dot{E}x_{cw,out} = \dot{Z}_{cond} + c_6 \dot{E}x_6 + c_{cw,in} \dot{E}x_{cw,in} \tag{9.79}$$

$$c_{cw,in} = 0 \tag{9.80}$$

$$c_6 = c_7 \tag{9.81}$$

Here, the unit cost of exergy of the incoming cooling water ($c_{cw,in}$) is considered zero for the same reason as for the intake feed seawater.

RO desalination unit

Rate balance equations and auxiliary equations are written for the overall RO desalination system shown in Fig. 9.18, as

$$\dot{m}_{D,RO} + \dot{m}_{B,RO} = \dot{m}_{f,RO} \tag{9.82}$$

$$\dot{m}_{D,RO} w_{D,RO} + \dot{m}_{B,RO} w_{B,RO} = \dot{m}_{f,RO} w_{f,RO} \tag{9.83}$$

$$\dot{m}_{D,RO} h_{D,RO} + \dot{m}_{B,RO} h_{B,RO} = \dot{W}_{RO} + \dot{m}_{f,RO} h_{f,RO} \tag{9.84}$$

$$\dot{m}_{D,RO}ex_{D,RO} + \dot{m}_{B,RO}ex_{B,RO} + \dot{E}x_{Des,RO} = \dot{W}_{RO} + \dot{m}_{f,RO}ex_{f,RO} \quad (9.85)$$

$$c_{D,RO}\dot{E}x_{D,RO} + c_{B,RO}\dot{E}x_{B,RO} = c_{\dot{W}_{RO}}\dot{W}_{RO} + c_{f,RO}\dot{E}x_{f,RO} + \dot{Z}_{RO} \quad (9.86)$$

$$c_{f,RO} = 0 \quad (9.87)$$

$$c_{B,RO} = 0 \quad (9.88)$$

$$c_{\dot{W}_{RO}} = c_{\dot{W}_{ST}} \quad (9.89)$$

Again, the unit costs of exergy of intake feed seawater ($c_{f,RO}$) and brine ($c_{D,RO}$) are considered to be zero. In addition, the unit cost of exergy of electricity consumed by the RO pump ($c_{\dot{W}_{RO}}$) is taken to be equal to the unit cost of exergy of electricity produced by steam turbine ($c_{\dot{W}_{ST}}$), as a part of electricity produced by steam turbine is used to drive RO pump.

From the design characteristics of the system listed in Table 9.11 and the mass, energy, and exergy rate balances written for each component, the thermodynamic properties of each stream in the geothermal-based system can be specified. Engineering equation software is used solve the equations for the thermodynamic model of the system. Table 9.14 presents the thermodynamic properties and mass flow rates evaluated at each state of the geothermal-based combined power-freshwater production system.

2. High-interest rate countries (interest rate is 14%)

The component cost rate $\dot{Z}_k$, which is a function of component capital investment cost (Z_k), is calculated based on Eq. (9.39). Table 9.12 lists the assumptions used in the economic analysis of the geothermal-based system. The capital and investment cost function for the components of the geothermal steam power production system are now presented.

The investment cost equation for calculating the capital cost of the steam turbine (in \$) can be written as (Baghernejad and Yaghoubi, 2011)

$$Z_{ST} = 6000\left(\dot{W}_{ST}\right)^{0.7} \quad (9.90)$$

The capital cost for the expansion valve is taken to be \$200 (Ghazizade-Ahsaee et al., 2019) and for separator unit is considered to be 7% of the capital cost of the steam turbine (Bina et al., 2018). The investment cost equation for the heat exchanger is taken to be (Rodríguez et al., 2012)

$$Z_{HEX} = 588\left(A_{HEX}\right)^{0.8} \quad (9.91)$$

Table 9.14 Thermodynamic quantities evaluated of streams in the geothermal-based combined power-freshwater production system.

State	Temperature (°C)	Pressure (MPa)	Specific enthalpy (kJ/kg)	Specific entropy (kJ/kg.K)	Specific exergy (kJ/kg)	Quality (-)	Mass flow rate (kg/s)
1	230	2.80	990.1	2.61	216.9	0	140
2	198.3	1.5	990.1	2.62	212.9	0.07	140
3	198.3	1.5	844.5	2.31	159.4	0	129.5
4	198.3	1.5	2791	6.44	875.5	1	10.47
5	93.49	0.08	2400	6.71	404.7	0.88	10.47
6	93.49	0.08	391.7	1.23	28.77	0	10.47
7	47	0.08	196.9	0.66	3.26	0	10.47
8	25	0.101	98.71	0.34	0	-	188.8
9	52	0.101	202.2	0.66	5.38	-	113.3
10	52	0.101	221.9	0.74	8.31	-	75.52
11	25	0.101	98.71	0.34	0	-	283.5
12	25	0.101	94.31	0.32	0.64	-	170.1
13	25	0.101	104.9	0.37	3.14	-	113.4

where A_{HEX} is the heat transfer area of the heat exchanger. The heat transfer area of heat exchanger is calculated through modeling of heat transfer process in the heat exchanger as

$$\dot{Q}_{HEX} = UA_{HEX}\Delta T_{LMTD} \tag{9.92}$$

where $\dot{Q}_{HEX}$ is the heat transfer rate, U overall heat transfer coefficient, $\dot{m}$ mass flow rate, and ΔT_{LMTD} is the log mean temperature difference (LMTD). For the condenser using cooling water to cool the geo-fluid (water) to 47 °C, the overall heat transfer coefficient, U, is assumed to be 2 kW/m^2 °C (EDGO, 2021). The LMTD can be expressed as

$$\Delta T_{LMTD} = \frac{\Delta T_1 - \Delta T_2}{\ln\left(\Delta T_1 / \Delta T_2\right)} \tag{9.93}$$

where ΔT_1 is the maximum temperature difference between two fluids at each side of the heat exchanger, and ΔT_2 is the temperature difference between the two fluids at the other side of heat exchanger. It is assumed that the temperature of cooling water changes from 25 °C at the inlet to

35 °C at the outlet of the heat exchanger (i.e., $\Delta T_2 = 10°C$). Also, ΔT_1 is the difference between the geo-fluid temperature leaving the MED (i.e., T_6) and the inlet cooling water temperature (i.e., $T_{cw,in} = 25$ °C).

Based on Energy Net (2012) and Loutatidou and Arafat (2015), the main investment cost for MED and RO desalination systems are 1500 $\$/m^3{}_{\text{freshwater}}$ /day and 900 $\$/m^3{}_{\text{freshwater}}$/day, respectively. A sample calculation for determination of the cost rate for the MED unit follows:

$$\dot{Z}_{MED} = CRF.\frac{\varphi.Z_{capital,MED}}{\tau} \tag{9.94}$$

$$CRF = \frac{i(i+1)^n}{(i+1)^n - 1} = \frac{0.14(0.14+1)^{20}}{(0.14+1)^{20} - 1} = 0.151 \tag{9.95}$$

$$\begin{aligned} Z_{capital,MED} &= 1500\,(\$/(\text{m}^3/\text{day})) \times 75.52\,(= \dot{m}_{D,MED})(\text{kg/s}) \\ &\times \frac{1}{997(=\rho_D)}(\text{m}^3/\text{kg}) \times \frac{3600 \times 24\text{s}}{1\text{ day}} \\ &= 9{,}817{,}000\ \$ \end{aligned} \tag{9.96}$$

$$\dot{Z}_{MED} = 0.151 \times \frac{1.06 \times 9{,}817{,}000\$}{7000\text{ h}} = 224.47\ \$/\text{h} \tag{9.97}$$

By solving the exergy rate balance equations for each system component from section *(a)*, the exergy rate of each stream ($\dot{E}x_i$) and the exergy destruction rate ($\dot{E}x_{Des}$) associated with each component can be determined. Also, from the cost rate balance applied for each component and auxiliary equations, the unknown unit cost of exergy (*c*) can be determined for the streams. In this example, there are six components (expansion valve, separator, steam turbine, MED, condenser, and RO), so six cost rate balance equations can be written. Then 12 auxiliary equations are written for the system components to reduce the unknown parameters from 12 to 6. Therefore, the final number of unknown parameters is six including $c_2, c_3, c_{\dot{W}_{ST}}, c_{D,MED}, c_{cw,out}$, and $c_{D,RO}$, and this is equal to the number of cost rate balance equations.

Table 9.15 provides the exergy rate, unit cost of exergy, and cost rate of each stream in the geothermal-based system. Note that the highest exergy unit cost (*c*) values are associated with streams 10 and 13. These streams are associated with the RO and MED units, which are the most expensive units in the geothermal-based combined system. Also, the exergy-based unit cost of electrical power ($c_{\dot{W}_{ST}}$) produced from the steam turbine is found to be 4.72 $/GJ.

Table 9.15 Values of exergy rate ($\dot{E}x$), unit cost of exergy (c), and cost rate ($\dot{C}$) for streams in the geothermal-based system.

State	$\dot{E}x$ (kW)	c (\$/GJ)	$\dot{C}$ (\$/h)
1	30,361	1.30	139.7
2	29,813	1.31	139.7
3	20,645	1.33	98.96
4	9168	1.33	43.95
5	4238	1.33	20.31
6	301.3	1.33	1.44
7	34.12	1.33	0.16
8	0	0	0
9	609.7	0	0
10	627.3	105.52	243.3
11	0	0	0
12	109.4	0	0
13	356.1	166.62	216.2

3. High-*interest rate countries (interest rate is* 14%*)*

The cost rate associated with the exergy destruction of each component can be determined using Eq. (9.40) (i.e., $\dot{C}_{Des,k} = c_{f,k}\dot{E}x_{Des,k}$). To specify the average unit cost of fuel flow (c_f) for each component, Eq. (9.42) is used (i.e., $c_{f,k} = \dot{C}_{F,k} / \dot{E}x_{F,k}$). Table 9.15 tabulates the values of $c, \dot{C}$, and $\dot{E}x$ for all streams. Also, $\dot{E}x_{F,k}$ and $\dot{C}_{F,k}$ are written for each component in Table 9.13.

To illustrate the calculations involved in obtaining these values, consider MED desalination system as an example. The unit costs c_f for the MED desalination system can be calculated as

$$c_{f,MED} = \frac{\dot{C}_{F,k}}{\dot{E}x_{F,k}} = \frac{\dot{C}_5 - \dot{C}_6 + \dot{C}_8 - \dot{C}_9}{\dot{m}_5 ex_5 - \dot{m}_6 ex_6 + \dot{m}_8 ex_8 - \dot{m}_9 ex_9} = \frac{20.31 - 1.44 + 0 - 0}{4238 - 301.3 + 0 - 609.7}$$
$$= 0.0056\,\$/\text{kWh} = 1.55\,\$/\text{GJ}\$ \tag{9.98}$$

Then,

$$\dot{C}_{Des,MED} = c_{f,MED}\dot{E}x_{Des,MED} = 0.0056\ \$/\text{kWh} \times 2699\ \text{kW}$$
$$= 15.11\,\$/\text{h} \tag{9.99}$$

Also, the cost rate of exergy loss for the MED unit can be calculated as

$$\dot{C}_{loss,MED} = c_{f,MED}\dot{E}x_{loss,MED} = c_{f,MED} \times \dot{E}x_9$$
$$= 0056\ \$/\text{kWh} \times 609.7\ \text{kW} = 3.41\,\$/\text{h} \tag{9.100}$$

Table 9.16 Exergoeconomic parameters for components of the geothermal-based combined power-desalination system for high-interest rate countries.

Component	$\dot{E}x_{Des}$ (kW)	$\dot{Z}$ ($/h)	c_f ($/GJ)	$\dot{C}_{Des}$ ($/h)	$\dot{C}_{loss}$ ($/h)	$\dot{C}_{Total}$ ($/h)	f (%)
Expansion valve	548.6	0.004	1.27	2.52	-	2.52	0.2
Separator	0	3.24	1.27	0	96.72	99.96	3.2
Steam turbine	832.6	46.35	1.3	3.99	-	50.34	92
Heat exchanger	207.3	0.22	6.9	5.21	1.5	5.43	4.1
MED unit	2699	224.4	1.55	15.11	3.41	242.92	92.3
RO unit	354	202.2	5.27	6.98	2.15	211.4	95

Finally, the total cost rate of the MED unit is

$$\begin{aligned}\dot{C}_{Total,MED} &= \dot{C}_{Des,MED} + \dot{C}_{loss,MED} + \dot{Z}_{MED} \\ &= 15.11\,\$/\text{h} + 3.41\,\$/\text{h} + 224.4\,\$/\text{h} = 242.92\,\$/\text{h}\end{aligned} \tag{9.101}$$

It is important to note that $\dot{Z}$ is the cost rate term that is not related to exergy, while $\dot{C}_{Des,MED}$ and $\dot{C}_{loss,MED}$ are the cost rates terms related to exergy destruction and exergy loss rates. In evaluating the performance of a component, the exergoeconomic factor reveals the relative significance of each category and is thus an informative parameter. The exergoeconomic factor of the MED unit can be determined as follows:

$$f_k = \frac{\dot{Z}_k}{\dot{C}_{k,Total}} = \frac{224.4\,\$/\text{h}}{242.92\,\$/\text{h}} = 92.3\% \tag{9.102}$$

Table 9.16 presents, for all components of the geothermal-based combined power-desalination system, calculated values of exergy destruction rate, exergy loss rate, component cost rate, average unit cost of fuel flow, cost rate of exergy destruction, cost rate of exergy loss, and exergoeconomic factor.

Low-interest rate countries (interest rate is 6%)

For countries such as the United States, Canada, Japan, Europe, and China, the interest rate is about 6%. The general calculation procedure for determining exergoeconomic parameters is the same as for high-interest rate countries. The only difference between them is the interest rate value

Table 9.17 Exergoeconomic parameters for components of the geothermal-based combined power-desalination system for low-interest rate countries.

Component	$\dot{Z}$ (\$/h)	c_f (\$/GJ)	$\dot{C}_{Des}$ (\$/h)	$\dot{C}_{loss}$ (\$/h)	$\dot{C}_{Total}$ (\$/h)	f (%)
Expansion valve	0.0026	1.27	2.52	-	2.52	0.1
Separator	1.87	1.30	0	97.72	99.59	1.9
Steam turbine	26.76	1.31	3.95	-	30.72	87
Heat exchanger	0.129	6.4	4.84	1.39	4.97	2.5
MED unit	129.6	1.55	15.16	3.42	148.2	87
RO unit	116.8	3.92	5.0	1.54	123.3	94

that is used in the calculations. Thermodynamic properties of streams interacting within the overall geothermal-based steam turbine and desalination system are identical for high and low interest countries, as they are not dependent on interest rate. Table 9.17 shows the exergoeconomic parameters such as $\dot{Z}$ (\$/h), c_f (\$/GJ), $\dot{C}_{Des}$ (\$/h), $\dot{C}_{loss}$ (\$/h), $\dot{C}_{Total}$ (\$/h), and f (%) for the components of the geothermal-based combined power-desalination system in low-interest rate countries.

Discussion: From Tables 9.16 and 9.17, it can be concluded that a higher interest rate negatively affects the cost rate of the system. With a rise of interest rate, the component cost rate, cost rate of destroyed exergy, and total cost rate of the components in the geothermal-based system increase.

Also, Tables 9.16 and 9.17 show that the highest values of exergoeconomic factor are associated with the RO, MED, and steam turbine units. This indicates that the capital investment costs are dominant in these units. Therefore, it is cost effective to reduce the capital cost for the MED and RO units at the expense of their efficiency. The high capital cost of the RO unit is mainly associated with the purchase and replacement costs of membranes and the cost of the high pressure pump. For the MED unit, the large heat transfer area required is the main reason of the high capital cost. The capital investment cost of the steam turbine depends on the temperature T_4, the pressure ratio P_5/P_4, and the isentropic efficiency of the steam turbine. To reduce the high capital cost of the steam turbine, a reduction in the value of at least one of these parameters merits consideration. The low value of the exergoeconomic factor for the expansion valve, separator and heat exchanger suggests that the efficiency of these component can be improved by increasing their capital investment.

According to Table 9.16, MED unit has the highest exergy destruction rate. To reduce the exergy destruction rate in the MED unit, any subprocesses that increase the exergy destruction rate or exergy loss rate can be eliminated without contributing to the reduction of capital cost or of fuel costs for other components. By considering measures for reducing the high cost rate of exergy destruction in the MED, two key design variables can be identified, the operating temperatures of the effects and the intake feed salinity. An increase in the first effect temperature (i.e., an increase in the average temperatures of the brine and distillate) reduces the exergy destruction rate and cost associated with the exergy destruction rate in the MED unit. Also, a reduction in the intake feed seawater salinities for both the MED and RO units reduces the exergy destruction rates in these units. Therefore a pretreatment process for the intake feed seawater to reduce its salinity before entering the MED and RO modules merits consideration. Moreover, reducing the high-pressure side pressure (P_{12}) of the RO unit can help reduce the exergy destruction rate of the RO unit.

9.6 Exergo-environmental analysis of seawater desalination systems using renewable energy

Exergo-environmental analysis combines environmental assessments with exergy analysis following the principles of exergoeconomic analysis. To evaluate an energy system with exergo-environmental analysis, first an exergy analysis of the system is carried out. Then, an LCA of the system is performed to determine the environmental impacts, considering all relevant input and output streams for the system and all relevant system components. Next, the environmental impacts are applied to the exergy streams of the system to determine the environmental impacts per unit of exergy. Auxiliary equations are also required, as explained subsequently. Through an exergo-environmental analysis of a system, the components with the highest and lowest environmental impacts can be identified. Improvements in components with high environmental impacts typically lead to increases in monetary costs. The holistic information provided with exergo-environmental analysis makes it useful for decision makers.

Life cycle assessment: LCA is a standardized method used for evaluating and assessing the environmental impact of a system, product, or process over all stages of its life cycle. For example for a product manufactured by an energy system, environmental impacts are evaluated from extraction of raw materials, production processes, transportation, installation, operation, maintenance, and disposal steps (ISO, 2006). The LCA concept for freshwater production via desalination is illustrated in Fig. 9.19.

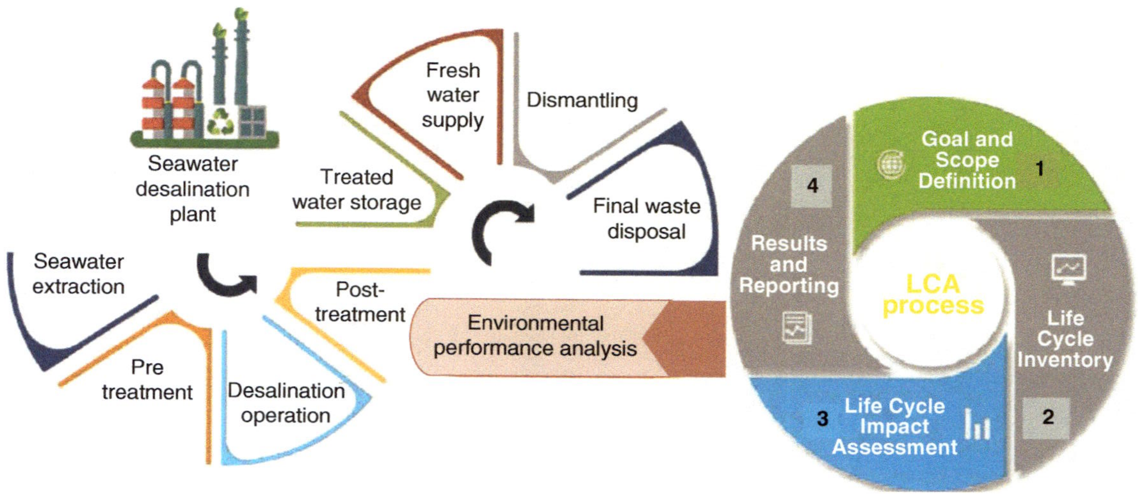

Fig. 9.19 ***Components of life cycle assessment for a freshwater production process.***

An LCA is comprised in general of four stages:

1. *Goal and scope definition:* The goal and scope of the assessment are defined to specify boundaries for the product life cycle to be applied in the assessment, if any, and also on the extent of the assessment.
2. *Inventory analysis:* The material and energy flows associated with the product or process of a system and their corresponding interactions with the environment are described.
3. *Impact assessment:* The impacts of the flows identified in the inventory analysis are assessed. Assessments of the values of indicators for all environmental impact categories are part of in this stage. Normalization and weighting can be used for assessing the importance of each impact category.
4. *Interpretation:* To complete the assessment, a critical review of the results over the life cycle is carried out to interpret the environmental impacts of greatest significance and requiring the most attention. It is often necessary to evaluate data sensitivity as part of this stage.

Usually software can be used for LCA and the calculations involved. One such common software is Package SimaPro (SimaPro, 2020). It contains the fundamental tools for LCA and a database with the required information about wastes and materials processes. For evaluation of the impact, the Eco-indicator 99 method can be applied in LCA (Goedkoop and Spriensma, 2000). This method utilizes average European data and considers many environmental aspects (Goedkoop and Spriensma, 2000). There are other evaluation methods, such as CML 2 baseline 2000, and Eco-Points 97, for environmental assessments of a system. LCA methods and tools have evolved over time and continue to do so.

Eco-indicator 99 is a life cycle environmental impact evaluation method used for LCA of a product that estimates eco-indicator scores for processes and materials used. The score, which can be determined using a weighting method, indicates the potentials for improvements of the process and product. The eco-indicator can be divided into three parts: production of raw materials, processing, and manufacture; transportation of product energy in use and consumables in use; and disposal. The Eco-indicator 99 describes the "environment damage" in three categories: damage to human health, damage to ecosystem quality, and damage to resources.

The emissions from seawater desalination and renewable energy systems can be assessed by LCA. The environmental impact from those systems can be compared with conventional desalination systems using fossil fuels. The main potential environmental impacts of seawater desalination systems are typically as follows:

- Impingement and entrainment of organisms due to the intake seawater for desalting and cooling processes.
- Airborne emissions of carbon dioxide and pollutants during generation of the heat and electricity needed to drive the desalination facilities.
- Adversely affecting organisms by discharging brine to the surroundings, mainly because there may be biocides and chemical additives in the brine to avoid corrosion, fouling, and scaling of seawater desalination systems.
- Adversely affecting the local species and coastal environment due to the discharge of high-salinity and high-temperature brine to the sea.

Table 9.18 presents unit airborne emissions (including CO_2, NO_x, NMVOC, and SO_x) and the Eco-indicator 99 score for various renewable energy–based desalination systems. From this table, it can be seen that renewable energies can significantly reduce the environmental impacts of producing water with desalination plants. An increase in the production capacities of both solar and wind energies commonly decreases the Eco-indicator 99 score and many airborne emissions (Raluy et al., 2005). Thus, using renewable energy resources in desalination facilities provides promising alternatives in terms of environmental impact.

9.6.1 Exergo-environmental balances, auxiliary equations, and variables

Fig. 9.20 illustrates the environmental impact flows for component k of a system. An environmental impact balance for component k can be written based on the sum of the environmental impacts associated with all input streams (y=1:n) plus the component-related environmental impact minus the sum of the environmental impacts associated with all output streams (y=*1*:m). That is, on a rate basis:

$$\underbrace{\sum_{y=1}^{n} \dot{B}_{y,k,in} + \dot{Y}_k}_{\dot{B}_{F,k}} \underbrace{-\sum_{y=1}^{m} \dot{B}_{y,k,out}}_{\dot{B}_{P,k}} = 0 \tag{9.103}$$

Here $\dot{B}_{F,k}$ and $\dot{B}_{P,k}$, respectively, are the environmental impact rates associated with fuel and product and can be expressed in Eco-indicator 99 points per time unit (Pts/s). Each term in Eq. (9.103) is described now.

The environmental impact rate of material stream y can be written in terms of the specific environmental impact and its exergy rate, as follows:

Table 9.18 Airborne emissions and Eco-indicator 99 score for various renewable energy–driven desalination systems[a].

Desalination system	Unit emission of CO_2 (kg/m³ product water)	Unit emission of NO_x (g/m³ product water)	Unit emission of NMVOC (g/m³ product water)[b]	Unit emission of SO_x (g/m³ product water)	Eco-indicator 99 score
MSF driven by solar thermal energy[c]	10.915	9.819	4.210	19.858	0.741
MSF driven by waste heat and wind energy (150 kW)	0.369	0.826	0.340	6.173	0.028
MSF driven by waste heat and PV (100 kW)	0.683	1.165	0.501	8.841	0.044
MSF cogeneration with steam power cycle using natural gas	17.24	19.65	5.67	16.17	1.226
MSF cogeneration with steam power cycle using coal	26.10	63.62	3.35	154.89	0.898
MED driven by solar thermal energy[c]	8.164	6.803	2.954	20.107	0.576
MED driven by waste heat and wind turbines (150 kW)	0.309	0.688	0.151	11.811	0.0314
MED driven by waste heat and PV (100 kW)[c]	0.363	0.902	0.196	14.992	0.403
MED cogeneration with steam power cycle using natural gas	12.91	14.77	4.15	19.34	0.924
MED cogeneration with steam power cycle using coal	19.48	47.41	2.54	122.25	0.68
RO-wind energy (150 kW)	0.170	0.412	0.22	2.05	0.136
RO-PV (100 kW)[c]	0.900	2.105	0.726	7.809	0.0534
RO-PV (500 kW)[c]	0.626	1.816	0.697	16.155	0.0624
RO with steam power cycle using natural gas	2.79	3.38	0.93	3.25	0.199

[a]Data sources Raluy et al. (2005, 2004).
[b]NMVOC denotes nonmethane volatile organic compound.
[c]Data extrapolated from a plant in Spain Raluy et al. (2005, 2004).

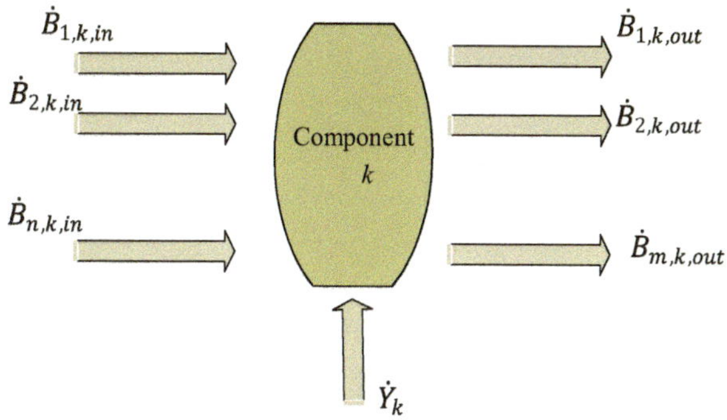

Fig. 9.20 ***Environmental impact flows for kth component in a system.***

$$\dot{B}_y = b_y \dot{E}x_y \tag{9.104}$$

Here, $\dot{B}_y$ denotes the environmental impact rate (in Eco-indicator points per unit of time, b_y the specific environmental impact (or specific environmental cost), and $\dot{E}x_y$ the exergy rate of material stream y in GJ/s. The specific environmental impact, b_y, also known as specific environmental cost, is the average environmental impact associated with the production of stream y per unit exergy (in Pts/GJ exergy). As mentioned before (Eq. 9.32), the specific exergy contains physical exergy and chemical exergy. Therefore, Eq. (9.104) can be broken down into components as follows

$$\dot{B}_y = b_{ph,y} \dot{E}x_{ph,y} + b_{chem,y} \dot{E}x_{chem,y} \tag{9.105}$$

where the subscripts *ph* and *chem,* respectively, denote physical and chemical. Splitting exergy into chemical and physical parts is necessary in desalination systems because the chemical exergy of seawater changes through the desalting process. For example, in the membrane of an RO desalination system, the chemical exergy of seawater changes while passing through the membrane.

The environmental impact rate associated with the work rate and heat rate, respectively, can be written as follows:

$$\dot{B}_{\dot{W}} = b_{\dot{W}} \dot{W} \tag{9.106}$$

$$\dot{B}_{\dot{Q}} = b_{\dot{Q}} \left(1 - \frac{T_0}{T}\right) \dot{Q} \tag{9.107}$$

Here, T_0 is the environment temperature (in K) and T the temperature at which the heat transfer process occurs (in K).

The component-related environmental impact rate, $\dot{Y}_k$, can be expressed based on three life cycle phases: construction (consumption of resources and pollutant emissions during manufacturing, transport, and installation) $\dot{Y}_{CO,k}$, O&M $\dot{Y}_{O\&M,k}$, and disposal, $\dot{Y}_{Disp,k}$. That is,

$$\dot{Y}_k = \dot{Y}_{CO,k} + \dot{Y}_{O\&M,k} + \dot{Y}_{Disp,k}. \tag{9.108}$$

All the environmental impacts values can be quantified with the aid of LCA for the aforementioned life cycle phases. Note that, for a conventional electricity-generation plant, the operation phase normally is responsible for the highest contribution to the overall environmental impact of the system. This is mainly due to the use of fossil fuels for electricity generation and the pollutant emissions from combustion.

Similar to exergoeconomic analysis, when writing the environmental impact balance for a component with N outlet streams, there are N unknowns and one equation (i.e., the environmental impact balance). So, auxiliary equations can be used to formulate N - 1 equations. Auxiliary equations can be written in a manner analogous to the F and P rules in exergoeconomic analysis. However, instead of the cost rates in exergoeconomic analysis, environmental impact rates are used when applying F and P rules. By solving the environmental impact rate balance and auxiliary equations simultaneously, the environmental impact rate ($\dot{B}_{y,out}$) or the specific environmental impact ($b_{y,out}$) associated with all outlet streams can be calculated. For example, the exergo-environmental balance rate and auxiliary equations for a typical RO-pump system, in which a feed stream is split into permeate and brine, can be written as follows:

$$\dot{B}_{F,Pump} = \dot{B}_{elect,Pump} \tag{9.109}$$

$$\dot{B}_{P,Pump} = \dot{B}_{outlet} - \dot{B}_{inlet} \tag{9.110}$$

$$\dot{B}_{F,RO} = \dot{B}_{feed,Ph} - \left(\dot{B}_{permeate,Ph} + \dot{B}_{Brine,Ph}\right) \tag{9.111}$$

$$\dot{B}_{P,RO} = \dot{B}_{feed,Chem} - \left(\dot{B}_{permeate,Chem} + \dot{B}_{Brine,Chem}\right) \tag{9.112}$$

$$b_{feed,Ph} = b_{permeate,Ph} = b_{brine,Ph} \tag{9.113}$$

$$\frac{\dot{m}_{permeate}b_{permeate,Chem}ex_{permeate,Chem} - \dot{m}_{feed}b_{feed,Chem}ex_{feed,Chem}}{\dot{m}_{permeate}ex_{permeate,Chem} - \dot{m}_{feed}ex_{feed,Chem}} = \frac{\dot{m}_{brine}b_{brine,Chem}ex_{brine,Chem} - \dot{m}_{feed}b_{feed,Chem}ex_{feed,Chem}}{\dot{m}_{brine}ex_{brine,Chem} - \dot{m}_{feed}ex_{feed,Chem}} \quad (9.114)$$

As for exergoeconomic analysis, in which the specific cost of product and fuel is defined, the average specific environmental impact of the product and fuel in exergo-environmental analysis can be expressed as

$$b_{P,k} = \frac{\dot{B}_{P,k}}{Ex_{P,k}} \quad (9.115)$$

$$b_{F,k} = \frac{\dot{B}_{F,k}}{Ex_{F,k}} \quad (9.116)$$

The environmental impact rate associated with the destruction of exergy in component k can be written as

$$\dot{B}_{Des,k} = b_{F,k}\dot{E}x_{Des,k} \quad (9.117)$$

The total environmental impact rate associated with component k can thus be expressed in terms of the component-related environmental impact rate ($\dot{Y}_k$) plus the environmental impact rate associated with the exergy destruction rate within component k ($\dot{B}_{Des,k}$), as follows:

$$\dot{B}_{total,k} = \dot{Y}_k + \dot{B}_{Des,k} \quad (9.118)$$

The exergo-environmental factor of component k ($f_{b,k}$) relates the component environmental impact ($\dot{Y}_k$) to the sum of environmental impacts associated with the component k ($\dot{Y}_k + \dot{B}_{Des,k}$), as follows:

$$f_{b,k} = \frac{\dot{Y}_k}{\dot{Y}_k + \dot{B}_{Des,k}} \quad (9.119)$$

With the exergo-environmental variable $f_{b,k}$, the sources of environmental impact formation in a component can be compared. For instance, this variable shows how much of the impact is due to the exergy destruction within a component and how much is due to other causes. High values of $f_{b,k}$ (usually more than 0.7) indicate that $\dot{Y}_k$ is dominant, while low values of $f_{b,k}$ (less than 0.3) indicate that $\dot{B}_{Des,k}$ is dominant.

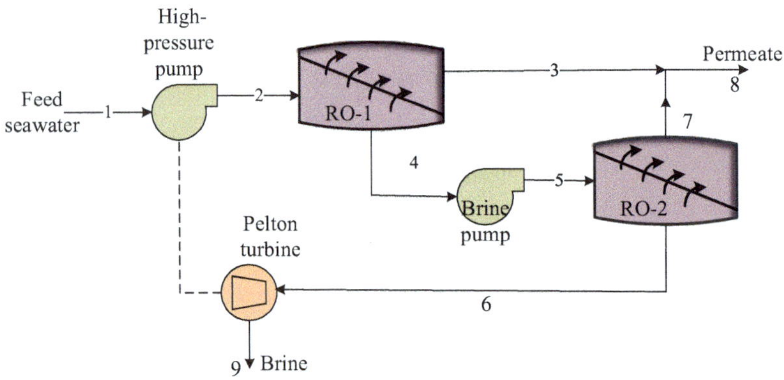

Fig. 9.21 ***Two-stage RO desalination plant using a Pelton turbine for energy recovery.***

The relative difference of specific environmental impact of component k ($r_{b,k}$) is an exergo-environmental variable that indicates the potential for reducing the environmental impact of component k. It can be written based on the difference between the average specific environmental impact of the product ($b_{p,k}$) and the fuel ($b_{f,k}$), as follows:

$$r_{b,k} = \frac{b_{P,k} - b_{F,k}}{b_{F,k}} \tag{9.120}$$

A high value of $r_{b,k}$ suggests that the environmental impact of component k can be reduced with a relatively smaller effort compared to a low value of $r_{b,k}$.

The results obtained from an exergo-environmental analysis of a system can provide useful information for identifying the component(s) with highest potential for improvement. The improvement can be achieved by reducing thermodynamic inefficiencies in the component or by reducing material consumption during operation or construction of the component.

Numerous exergo-environmental analyses of desalination systems have been reported. Blanco-Marigorta et al. (2014) analyzed a two-stage RO system (Fig. 9.21) from an exergo-environmental perspective. The main RO plant processes consist of pretreatment and filtration units, high-pressure pump, membranes, and posttreatment units. The study uses actual operating data from the Las Palmas III plant located in Gran Canaria, Spain in the thermodynamic analysis of the RO desalination plant. The freshwater production capacity of the plant is 82,000 m^3/day. A Pelton turbine is used for energy recovery in the plant. The mean salinity of the feed seawater is 37,000 ppm and its temperature varies during the year between 19 °C and 26 °C.

Table 9.19 Exergy destruction rate and exergo-environmental parameters for two-stage RO system with a Pelton turbine.

Component	$\dot{E}x_{Des}$ (kW)	$b_{F,k}$ (mPt/kJ)	$b_{P,k}$ (mPt/kJ)	$\dot{Y}_k$ (mPt/s)	$\dot{B}_{Des,k}$ (mPt/s)
Pelton turbine	246	12.9	36	0.10	2948
High-pressure pump	229	38	68.3	0.10	2247
RO-1	201	34.1	49.4	0.09	9031
RO-2	66	3.7	5.5	0.12	7645
Brine pump	66	36.7	60.5	0.10	245

Data extracted from Blanco-Marigorta et al. (2014).

Table 9.19 lists the exergo-environmental variables of the RO system components examined by Blanco-Marigorta et al. (2014). It is seen that the environmental impact rate associated with the exergy destruction rate of each component ($\dot{B}_{Des,k}$) in the system is the greatest contributor to the total environmental impact, rather than the component-related environmental impact rate ($\dot{Y}_k$). Further, the first stage of the RO membrane and the high-pressure pump are seen to have the highest exergy destruction rates. In addition, these components have high component-related environmental impact rates. Therefore, the thermodynamic efficiency of the first stage RO membrane and high-pressure pump can be improved to reduce the overall environmental impact.

9.6.2 Discussion

It is important to carry out a comprehensive environmental impact analysis for desalination plants before implementation of any large-scale freshwater production project. Several quick and effective methods exist for diluting and cooling a brine stream before discharging it to the sea, which can potentially reduce the direct and indirect impacts of desalination systems on the local species and coastal environment. One practical method for prefiltering, precooling, and distribution of a brine stream involves installing horizontal drain tubes beneath the seabed. In addition, using nanofiltration can potentially improve the prefiltering process, which partially or totally removes chemical additives and biocides such as antiscaling and antifouling substances. However, applying such prefiltering increases the energy input, while significantly reducing emissions and chemical additives.

In the future, the use of advanced renewable energy technologies in conjunction with seawater desalination plants is likely to provide the promise of lower environmental impacts compared to conventional desalination

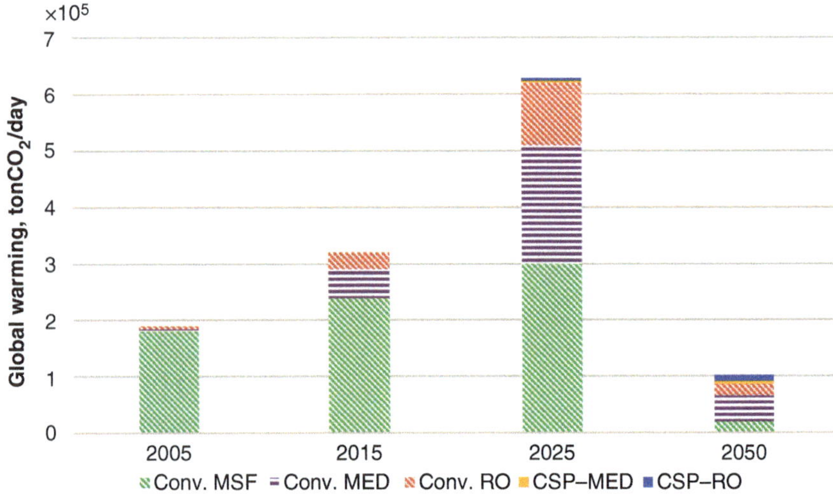

Fig. 9.22 ***Variations in greenhouse emissions from 2005 to 2050 as the share of renewable energies in desalination systems increases.*** *Source: Data from (DLR, 2007).*

plants. However, using renewable energy facilities in desalination systems increases the capital cost. Thus, improvements in renewable energy technologies that reduce their capital costs and the use of renewable energy sources that are significantly cheaper than fossil fuels can make renewable-based desalination plants more advantageous than conventional desalination plants. It is predicted that the share of solar desalination systems will increase significantly after 2025 and helping to mitigate greenhouse gas emissions considerably (DLR, 2007). Fig. 9.22 shows the carbon dioxide emissions from the AQUA-CSP desalination project, based on an LCA in the MENA region when the share of renewable energies increases, for the period 2015 to 2050.

9.7 Closing remarks

In this chapter, the economics of seawater desalination systems using sustainable energy technologies are described. First, an overview of desalination cost projects is provided. Then, the technoeconomic analysis of renewable energy–based desalination systems is presented to examine the economic and technical performances of those systems. The cost of water generally depends on site-specific conditions such as energy cost, target of produced freshwater quality, concentration of constituents in feed water, and distance from the users. All potentially influential conditions need to be considered. The viability of freshwater supply in a region from a seawater desalination plant when integrated with one or more renewable energy resources is assessed via an economic study of the electricity and water production costs and a comparison with fossil fuel–based desalination plants. It is shown that it is possible to meet the worldwide water demand with SWRO desalination plants driven by electricity production plants based on renewable energy. The *LCOW* from hybrid solar-wind electricity generation plants combined with RO desalination plants is found to be between 1.21 US$/$m^3$ and 5.44 US$/$m^3$ in 2030. Compared to fossil-fueled desalination plants, for which *LCOW* is typically between 0.72 US$/$m^3$ and 2.3 US$/$m^3$, it is evident that using renewable energy sources in seawater desalination systems is a potentially advantageous method for addressing global water demands. It is also shown that the water transportation costs are one of the greatest contributions to the final *LCOW*.

In a case study, technoeconomic analyses of CSP–MED and CSP–RO plant configurations are compared in terms of cost of water, for two regions. It is shown that the CSP–MED configuration is economically advantageous to the CSP–RO plant when electricity is expensive, while the CSP–RO configuration is advantageous for regular electricity prices.

Finally, exergoeconomic and exergo-environmental analyses for renewable energy–driven desalination systems are provided and discussed. Through exergoeconomic analysis, the relations between thermodynamic losses (exergy destructions and wastes) in system components and their capital costs can be identified. The SPECO approach is used here; it involves utilizing the specific exergy and the cost per unit of exergy to define the exergy efficiencies and auxiliary costing equations for system components. An exergoeconomic analysis of sustainable desalination technologies reveals useful information about appropriate modifications in component cost-effectiveness and optimum designs. A case study is also provided to clarify how to implement the principles of exergoeconomic analysis in a desalination system driven by renewable energy resources. Exergo-environmental analysis measures the environmental impacts associated with sustainable desalination technologies by combining exergy concepts and LCA. This analysis follows the approach of exergoeconomic analysis, but with environmental impacts related to the exergy streams of a system. A quantitate evaluation of sustainability and the greatest potentials for improvement of environmental performance can be achieved with the exergo-environmental approach. It is shown that using renewable energies for driving desalination plants can significantly reduce the environmental impacts of freshwater production.

Nomenclature

Symbols

b	specific environmental impact (Pts/kJ)
$\dot{B}$	environmental impact rate (Points/s or Pts/s)
c	cost per unit exergy ($/kJ)
C	cost ($)
$\dot{C}$	cost rate ($/s)
C_{elec}	unit electricity cost ($/kWh)
Cap_{fac}	capacity factor (-)
CC	capital cost ($)
CRF	capital recovery factor (-)
E	energy (kWh)
$\dot{E}$	installed capacity (kW)
ex	specific exergy (kJ/kg)
Ex	exergy rate
f	derating factor (-)
f_k	exergoeconomic factor of component k (-)
$f_{b,k}$	exergo-environmental factor of component k (-)
GOR	gained output ratio (-)
h	specific enthalpy (kJ/kg)
h_{fg}	enthalpy of evaporation (kJ/kg)
I	solar radiation incident per unit area (kW/m^2)
l_m	average lifetime (year)
$LCOC$	levelized cost of curtailment ($/m^3)
$LCOE$	levelized cost of electricity ($/kWh)
$LCOS$	levelized cost of storage ($/m^3)
$LCOT$	levelized cost of transmission ($/m^3)
$LCOT_D$	levelized cost of water transportation ($/m^3)
$LCOW$	levelized cost of water ($/m^3)
$\dot{m}$	mass flow rate (kg/s)
n	loan period, project lifetime (year)
OC	operation cost ($)
P	pressure (kPa)
$\dot{Q}$	heat rate (kW)
r_k	relative cost difference of component k (-)
$r_{b,k}$	relative difference of specific environmental impact of component k (-)
r_{return}	rate of return on capital (%)
R_{RO}	staging ratio in RO (-)
s	specific entropy (kJ/kg.K)
SEC	specific energy consumption (kWh/m^3)
T	temperature (°C, K)
TBT	top brine temperature (°C, K)
V	volume (m^3)
$\dot{V}$	volumetric flow rate (m^3/s)
w	mass fraction (g/kg or kg/kg)
$\dot{W}$	work rate (kW)

$Z_{capital}$	capital investment cost ($)
$\dot{Z}$	capital cost and operation and maintenance cost rate ($/s)

Greek symbols

a_p	power temperature coefficient (-)
η	efficiency
μ	chemical potential (kJ/kg)
ρ	density (kg/m^3)
τ	annual operation time of system (h)
ϕ	operation and maintenance factor (-)
ψ	exergy efficiency (%)

Subscripts

a	air
A&G	administrative and general
b	brine
c	photovoltaic cell
chem	chemical
conv	conventional
curt	curtailed
D	desalination
Des	destruction
disp	disposal
disch	discharge
exp	export
F	fuel
f	feed
Fix	fixed
HEX	heat exchanger
imp	import
inv	investment
isen	isentropic
m	membrane
M	maintenance
O&M	operation and maintenance
p	pump
P	product
ph	physical
RE	renewable energy
Rev,elec	revenue from electricity sale
sep	separator
ST	steam turbine
STC	standard test conditions
tr	transmission
var	variable

ws	water storage
wt	wind turbine

Acronyms

BWRO	Brackish water reverse osmosis
CAOW	Closed air, open water
CSP	Concentrating solar power
DNI	Direct normal irradiance
ERD	Energy recovery device
HDH	Humidification–dehumidification
LCA	Life cycle assessment
LF	Linear Fresnel
LFRC	Linear Fresnel Rankine cycle
LMTD	Log mean temperature difference
MED	Multieffect desalination
MENA	Middle East and North Africa
MSF	Multistage flash
O&M	Operation and maintenance
PV	Photovoltaic
RO	Reverse osmosis
SPECO	Specific exergy costing
SWRO	Seawater reverse osmosis
TES	Thermal energy storage
TVC	Thermal vapor compressor

References

Akbari, M., Mahmoudi, S., Yari, M., Rosen, M.A., 2014. Energy and exergy analyses of a new combined cycle for producing electricity and desalinated water using geothermal energy. Sustainability 6 (4), 1796–1820.

Al-Nory, M., El-Beltagy, M., 2014. An energy management approach for renewable energy integration with power generation and water desalination. Renew. Energy 72, 377–385.

Al-Zahrani, A., Orfi, J., Al-Suhaibani, Z., Salim, B., Al-Ansary, H., 2012. Thermodynamic analysis of a reverse osmosis desalination unit with energy recovery system. Procedia Eng. 33, 404–414.

Avlonitis, S.A., Kouroumbas, K., Vlachakis, N., 2003. Energy consumption and membrane replacement cost for seawater RO desalination plants. Desalination 157 (1-3), 151–158.

Baghernejad, A., Yaghoubi, M., 2011. Exergoeconomic analysis and optimization of an integrated solar combined cycle system (ISCCS) using genetic algorithm. Energy Convers. Manage. 52 (5), 2193–2203.

Baniasad Askari, I., Ameri, M., Calise, F., 2018. Energy, exergy and exergo-economic analysis of different water desalination technologies powered by linear Fresnel solar field. Desalination 425, 37–67.

Barbosa, L.S.N.S., Bogdanov, D., Vainikka, P., Breyer, C., 2017. Hydro, wind and solar power as a base for a 100% renewable energy supply for South and Central America. PLoS One 12 (3), e0173820.

Bina, S.M., Jalilinasrabady, S., Fujii, H., 2018. Exergoeconomic analysis and optimization of single and double flash cycles for Sabalan geothermal power plant. Geothermics 72, 74–82.

Blanco-Marigorta, A.M., Masi, M., Manfrida, G., 2014. Exergo-environmental analysis of a reverse osmosis desalination plant in Gran Canaria. Energy 76, 223–232.
Bogdanov, D., Breyer, C., 2016. North-East Asian Super Grid for 100% renewable energy supply: optimal mix of energy technologies for electricity, gas and heat supply options. Energy Convers. Manage. 112, 176–190.
Breyer, C., Bogdanov, D., Komoto, K., Ehara, T., Song, J., Enebish, N., 2015. North-east Asian super grid: renewable energy mix and economics. Japan. J. Appl. Phys. 54 (8S1), 08KJ01.
Breyer, C., Tsupari, E., Tikka, V., Vainikka, P., 2015. Power-to-gas as an emerging profitable business through creating an integrated value chain. Energy Procedia 73, 182–189.
Cai, W., Li, X., Maleki, A., Pourfayaz, F., Rosen, M.A., Alhuyi Nazarie, M., Bui, D.T., 2020. Optimal Sizing and location based on economic parameters for an off-grid application of a hybrid system with photovoltaic, battery and diesel technology. Energy 201, 117480.
Caldera, U., Bogdanov, D., Afanasyeva, S., Breyer, C., 2018b. Role of seawater desalination in the management of an integrated water and 100% renewable energy based power sector in Saudi Arabia. Water 10 (1), 3.
Caldera, U., Bogdanov, D., Breyer, C., 2018. Desalination costs using renewable energy technologies. In: Gude, V.G. (Ed.), Renewable Energy Powered Desalination Handbook. Butterworth-Heinemann, MA, pp. 287–329.
CEPCI, 2021. Chemical engineering plant cost index. Chemical Engineering, Access Intelligence LLC, New York (USA). https://www.chemengonline.com/pci (Accessed 26 May 2021).
Chitgar, N., Emadi, M.A., Chitsaz, A., Rosen, M.A., 2019. Investigation of a novel multi-generation system driven by a SOFC for electricity and fresh water production. Energy Convers. Manage. 196, 296–310.
Cole, W.J. and Frazier, A., 2019. Cost projections for utility-scale battery storage (Report no. NREL/TP-6A20-73222). National Renewable Energy Lab. (NREL), Golden, CO (United States).
DesalData, 2017. DesalData, Cost estimator. https://www.desaldata.com/cost_estimator (Accessed 30 May 2021).
Dincer, I., Rosen, M.A., 2021. Exergy: Energy, Environment and Sustainable Development, Third ed. Elsevier, Oxford, UK.
Dow, 2016. The DOW Chemical Company, FILMTEC Membranes, DOW XUS180808 Reverse Osmosis Element. http://msdssearch.dow.com (Accessed 30 May 2021).
EDGO, 2021. Engineers EDGO, Overall Heat Transfer Coefficient Table Chart and Equation, https://www.engineersedge.com/thermodynamics/overall_heat_transfer-table.htm (Accessed 30 May 2021).
Ehyaei, M.A., Ahmadi, A., El Haj Assad, M., Rosen, M.A., 2020. Investigation of an integrated system combining an organic Rankine cycle and absorption chiller driven by geothermal energy: energy, exergy, and economic analyses and optimization. J. Clean. Prod. 258, 120780.
EnergyNet, 2012. Technology data for energy plants, https://ens.dk/sites/ens.dk/files/Analyser/c_teknologikatalog_for_individuelle_varmeanlaeg_og_energitransport_2012.pdf (Accessed 30 May 2021).
Fulya, V., 2011. MENA Regional Water Outlook: Part II Desalination Using Renewable Energy. Fichtner. Final report. www.dlr.de/tt/Portaldata/41/Resources/dokumente/institut/system/projects/MENA_REGIONAL_WATER_OUTLOOK.pdf (Accessed 31 May 2021).
German Aerospace Center (DLR), 2007. Aqua-CSP Study Report, Concentrating Solar Power for Seawater Desalination. http://www.dlr.de/tt/aqua-csp (Accessed 30 May 2021).
Ghazizade-Ahsaee, H., Ameri, M., Askari, I.B., 2019. A comparative exergo-economic analysis of four configurations of carbon dioxide direct-expansion geothermal heat pump. Appl. Therm. Eng. 163, 114347.

Glatzmaier, G., 2011. Developing a cost model and methodology to estimate capital costs for thermal energy storage. Report No. NREL/TP-5500-53066. National Renewable Energy Lab (NREL), Golden, CO (United States).
Goedkoop, M., Spriensma, R., 2000. The Eco-indicator 99: A Damage Oriented Method for Life Cycle Impact Assessment Methodology Annex. PRé Consultants B.V. Amersfoort, Netherland.
GWI, 2016. IDA Desalination Yearbook. Global Water Intelligence (Media Analytics), Oxford; UK.
GWI, 2020. The IDA Water Security Handbook 2020-2021. Media Analytics Ltd., Oxford, UK.
GWID, 2016. Global Water Intelligence Desal Data. www.desaldata.com/projects (Accessed 30 May 2021).
Haghghi, M.A., Shamsaiee, M., Holagh, S.G., Chitsaz, A., Rosen, M.A., 2019. Thermodynamic, exergoeconomic, and environmental evaluation of a new multi-generation system driven by a molten carbonate fuel cell for production of cooling, heating, electricity, and freshwater. Energy Convers. Manage. 199, 112040.
Haysom, J.E., Jafarieh, O., Anis, H., Hinzer, K., Wright, D., 2015. Learning curve analysis of concentrated photovoltaic systems. Prog. Photovoltaics Res. Appl. 23 (11), 1678–1686.
Hiendro, A., Kurnianto, R., Rajagukguk, M., Simanjuntak, Y.M., 2013. Techno-economic analysis of photovoltaic/wind hybrid system for onshore/remote area in Indonesia. Energy 59, 652–657.
IEA, 2020. Renewables 2020, International Energy Agency, Paris. https://www.iea.org/reports/renewables-2020 (Accessed 30 May 2021).
IRENA, 2017. Electricity Storage and Renewables: Costs and Markets to 2030. International Renewable Energy Agency, Abu Dhabi.
ISO, 2006. ISO 14044: 2006Environmental Management – Life Cycle Assessment – Requirements and Guidelines, First ed. International Organization for Standardization, Geneva, Switzerland.
Jalili, M., Chitsaz, A., Hashemian, M., Rosen, M.A., 2021. Economic and environmental assessment using emergy of a geothermal power plant. Energy Convers. Manage. 228, 113666.
Javanshir, N., Seyed Mahmoudi, S.M., Akbari Kordlar, M., Rosen, M.A., 2020. Energy and cost analysis and optimization of a geothermal-based cogeneration cycle using an ammonia-water solution: thermodynamic and thermoeconomic viewpoints. Sustainability 12, 484.
Kersten, F., Doll, R., Kux, A., Huljic, D.M., Gorig, M.A., Breyer, C., Müller, J.W., Wawer, P., 2011. PV learning curves: past and future drivers of cost reduction, Proc. 26th European Photovoltaic Solar Energy Conference and Exhibition. Hamburg, Germany, 4697–4702.
Keshavarzzadeh, A.H., Ahmadi, P., Rosen, M.A., 2020. Technoeconomic and environmental optimization of a solar tower integrated energy system for freshwater production. J. Clean. Prod. 270, 121760.
Khoshgoftar Manesh, M.H., Abdolmaleki, M., Vazini Modabber, H., Rosen, M.A., 2021. Dynamic advanced exergetic, exergoeconomic and environmental analyses of a hybrid solar city gate station. J. Energy Res. Technol. 143 (10), 102105.
Kool, E., Cuomo, M.A., Reddy, B.V., Rosen, M.A., 2018. Multi-generation renewable energy system for dairy farms: Exergy analysis. Eur. J. Sustain. Dev. Res. 2 (3), 37.
Kutscher, C., Mehos, M., Turchi, C., Glatzmaier, G., Moss, T., 2010. Line-focus solar power plant cost reduction plan (Milestone Report). Report No. NREL/TP-5500-48175. National Renewable Energy Lab (NREL), Golden, CO (United States).
Lacal Arantegui, R., Jaeger-Waldau, A., Vellei, M., Sigfusson, B., Magagna, D., Jakubcionis, M., Perez Fortes, M., Lazarou, S., Giuntoli, J., Weidner Ronnefeld, E., De Marco, G., Spisto, A., Gutierrez Moles, C., Carlsson, J., 2014. ETRI 2014 - Energy Technology

Reference Indicator projections for 2010-2050. EUR 26950. Luxembourg: Publications Office of the European Union; JRC92496.
Lawal, D.U., Zubair, S.M., Antar, M.A., 2018. Exergo-economic analysis of humidification-dehumidification (HDH) desalination systems driven by heat pump (HP). Desalination 443, 11–25.
Lazzaretto, A., Tsatsaronis, G., 2006. SPECO: a systematic and general methodology for calculating efficiencies and costs in thermal systems. Energy 31 (8-9), 1257–1289.
Liebreich, M., 2016. BLOOMBERG new energy finance summit: in search of the miraculous. Bloomberg New Energy Finance. http://www.bbhub.io/bnef/sites/4/2016/04/BNEF-Summit-Keynote-2016.pdf (Accessed 31 May 2021).
Loutatidou, S., Arafat, H.A., 2015. Techno-economic analysis of MED and RO desalination powered by low-enthalpy geothermal energy. Desalination 365, 277–292.
Loutatidou, S., Chalermthai, B., Marpu, P.R., Arafat, H.A., 2014. Capital cost estimation of RO plants: GCC countries versus southern Europe. Desalination 347, 103–111.
Maleki, A., Gholipour Khajeh, M., Rosen, M.A., 2016. Weather forecasting for optimization of a hybrid solar-wind-powered reverse osmosis water desalination system using a novel optimizer approach. Energy 114, 1120–1134.
Malfuzi, A., Mehr, A.S., Rosen, M.A., Alharthi, M., Kurilova, A.A., 2020. Economic viability of bitcoin mining using a renewable-based SOFC power system to supply the electrical power demand. Energy 203, 117843.
Mayer, J.N., 2015. Current and future cost of photovoltaics: Long-term scenarios for market development, system prices and LCOE of utility-scale PV systems. Fraunhofer-Institute for Solar Energy Systems, Agora Energiewende 82. https://www.ise.fraunhofer.de/content/dam/ise/de/documents/publications/studies/AgoraEnergiewende_Current_and_Future_Cost_of_PV_Feb2015_web.pdf (Accessed 31 May2021).
Missimer, T.M., Maliva, R.G., Ghaffour, N., Leiknes, T., Amy, G.L., 2014. Managed aquifer recharge (MAR) economics for wastewater reuse in low population Wadi communities, Kingdom of Saudi Arabia. Water 6 (8), 2322–2338.
Mistry, K.H., McGovern, R.K., Thiel, G.P., Summers, E.K., Zubair, S.M., Lienhard, J.H., 2011. Entropy generation analysis of desalination technologies. Entropy 13 (10), 1829–1864.
Moseley, P.T., Garche, J., 2014. Electrochemical Energy Storage for Renewable Sources and Grid Balancing. Newnes. Elsevier, AE Amsterdam, Netherlands.
Mosleh, H.J., Behnam, P., Kamazani, M.A., Mohammadi, O., Kavian, S., Ahmadi, P., Rosen, M.A., 2020. A comprehensive comparative investigation on solar heating and cooling technologies from a thermo-economic viewpoint: a dynamic simulation. Energy Sci. Eng. 9 (5), 724–742.
NECI, 2021. National Center for Environmental Information. https://www.ncei.noaa.gov/ (Accessed 30 May 2021).
Neij, L., 2008. Cost development of future technologies for power generation: a study based on experience curves and complementary bottom-up assessments. Energy Policy 36 (6), 2200–2211.
Newfoundland & Labrador, 2011. Evaluation of Potable Water Storage Tanks in Newfoundland and Labrador and their Effect on Drinking Water Quality. Government of Newfoundland & Labrador, Department of Environment and Conservation, Water Resources Management Division, St. John's, NL, Canada. https://www.gov.nl.ca/eccm/files/waterres-reports-drinking-water-tank-report-july-12-2011.pdf (Accessed 30 May 2021).
NREL, 2019. Electricity Annual Technology Baseline (ATB) Data Download. https://atb.nrel.gov/electricity/2020/data.php (Accessed 30 May 2021).
Oh, H.J., Hwang, T.M., Lee, S., 2009. A simplified simulation model of RO systems for seawater desalination. Desalination 238 (1-3), 128–139.
Olwig, R., Hirsch, T., Sattler, C., Glade, H., Schmeken, L., Will, S., Ghermandi, A., Messalem, R., 2012. Techno-economic analysis of combined concentrating solar power and desalination plant configurations in Israel and Jordan. Desalin. Water Treat. 41 (1-3), 9–25.

Peng, W., Maleki, A., Rosen, M.A., Azarikhah, P., 2018. Optimization of a hybrid system solar-wind-based water desalination by reverse osmosis: comparison of approaches. Desalination 442, 16–31.

Pleßmann, G., Erdmann, M., Hlusiak, M., Breyer, C., 2014. Global energy storage demand for a 100% renewable electricity supply. Energy Procedia 46, 22–31.

Pothecary, S., 2016. PV Magazine, Energy storage market to grow to USD250 billion by 2040. https://www.pv-magazine.com/2016/06/13/energy-storage-market-to-grow-to-usd250-billion-by-2040_100024952/ (Accessed 30 May 2021).

Raluy, R.G., Serra, L., Uche, J., 2005. Life cycle assessment of desalination technologies integrated with renewable energies. Desalination 183 (1-3), 81–93.

Raluy, R.G., Serra, L., Uche, J., Valero, A., 2004. Life-cycle assessment of desalination technologies integrated with energy production systems. Desalination 167, 445–458.

Reddy, K.V., Ghaffour, N., 2007. Overview of the cost of desalinated water and costing methodologies. Desalination 205 (1-3), 340–353.

Rodríguez, C.E.C., Palacio, J.C.E., Sotomonte, C.R., Leme, M., Venturini, O.J., Lora, E.E.S., Cobas, V.M., dos Santos, D.M., Dotto, F.R.L., Giallucca, V., 2012. Exergetic and economic analysis of Kalina cycle for low temperature geothermal sources in Brazil, Proc. 25th International Conference on Efficiency, Cost, Optimization, Simulation and Environmental Impact of Energy (ECOS), 2012, 167–179.

Sadeghi, S., Ghandehariun, S., Rosen, M.A., 2020. Comparative economic and life cycle assessment of solar-based hydrogen production for oil and gas industries. Energy 208, 118347.

Saheb-Koussa, D., Haddadi, M., Belhamel, M., Hadji, S., Nouredine, S., 2010. Modeling and simulation of the fixed-speed WECS (wind energy conversion system): application to the Algerian Sahara area. Energy 35 (10), 4116–4125.

Salehi, S., Mahmoudi, S.M.S., Yari, M., Rosen, M.A., 2018. Multi-objective optimization of two double-flash geothermal power plants integrated with absorption heat transformation and water desalination. J. Clean.Prod. 195, 796–809.

Salgado, B., Gasia-Bruch, E., García-Molina, V., Casañas, A., Coker, S., 2013. Long-term benefits of DOW FILMTEC™ RO membranes. Desalin.Water Treat. 51 (1-3), 333–342.

Scott, M., 2018. Nuclear power outlook, Annual Energy Outlook 2018. Report No. DOE/EIA-0383(2016). https://www.eia.gov/outlooks/aeo/npo.php (Accessed 30 May 2021).

SimaPro, 2020. SimaPro database manual. Methods library. Amersfoort. https://simapro.com/wp-content/uploads/2020/10/DatabaseManualMethods.pdf (Accessed 30 May 2021).

Thiel, G.P., Tow, E.W., Banchik, L.D., Chung, H.W., 2015. Energy consumption in desalinating produced water from shale oil and gas extraction. Desalination 366, 94–112.

Trieb, F., Müller-Steinhagen, H., 2008. Concentrating solar power for seawater desalination in the Middle East and North Africa. Desalination 220 (1-3), 165–183.

Vartiainen, E., Masson, G., Breye, B., 2017. The true competitiveness of solar PV - A European case study, ETIP-PV, European Technology and Innovation Platform. Solar Industry Report.

Voutchkov, N., 2018. Desalination Project Cost Estimating and Management, First ed. CRC Press, Boca Raton, USA.

WDR, 2017. Saudi Arabia Desal market re-ignited, Water Desalination Report (WDR). https://www.desalination.com/publications/desalination-report-wdr (Accessed 30 May 2021).

Wu, B., Maleki, A., Pourfayaz, F., Rosen, M.A., 2018. Optimal design of stand-alone reverse osmosis desalination driven by a photovoltaic and diesel generator hybrid system. Sol. Energy 163, 91–103.

CHAPTER TEN

Optimization of seawater desalination systems

10.1 Introduction

Growing freshwater demands around the world have led to an increasing need for new and improved desalination processes and also for a better understanding of currently used desalination technologies. The performance of desalination systems of various configurations can be evaluated through a thorough understanding of desalination processes. With the help of mathematical models, new and improved designs for desalination processes can be developed and assessed, resulting in a deeper understanding of desalination processes and the ability to predict performance more effectively. Improvements and advances in computational speed and capacity have enabled researchers to study the feasibility of various desalination plants, whether they utilize renewable or nonrenewable energy resources. Mathematical models of performance parameters are used in optimization procedures for desalination plants, for example to design plants that minimize cost and maximize energy efficiency.

Generally, modeling can be viewed as describing a real-world problem via mathematical equations and then using suitable approaches to solve those equations. Optimization is a mathematical procedure that is commonly employed. In optimization, a minimization and/or maximization approach is applied to single or multiple objective functions, based on a set of input values, while satisfying relevant constraints, to compute values of objective functions.

In prior decades, there has been an increasing interest in optimization of desalination technologies. Desalination techniques are basically classified as phase change processes (e.g., multieffect desalination (MED), multistage flash (MSF), mechanical vapor compression, humidification–dehumidification (HDH), and solar distillation) or membrane-based processes (e.g., reverse osmosis (RO) and membrane distillation). MED and MSF desalination techniques are commonly used industrial thermally driven desalination systems and RO desalination is the most widely used work-driven desalination system.

Sustainable Energy Technologies for Seawater Desalination
DOI: https://doi.org/10.1016/B978-0-323-99872-7.00006-1

Improvements in configurations of desalination technologies can lead to a significant reduction in cost and specific energy requirement of those systems. For instance in an MSF desalination system, brine recovery process can potentially improve the energy efficiency of the system. Also, using long tube evaporator bundle in each stage lowers the heat transfer area, which results in reduction in the capital cost of system (Mabrouk, 2013). Reducing seawater temperature increases freshwater production capacity in MSF system (Ali and Kairouani, 2016). In addition, increasing top brine temperature (maximum operating temperature of MSF) can increase freshwater production, lower specific energy consumption (SEC), lower reduced specific area, and increase performance ratio (Al-Sofi, 1999; Roy et al., 2017). It is important to note that scale/fouling formation in the MSF system should be controlled when increasing the top brine temperature (i.e., beyond 100 °C). In RO desalination systems, some improvements in system performance can be applied, including optimizing pretreatment and posttreatment (Mousa et al., 2013, Shaffer et al., 2012), configuration of RO membranes in pressure vessels (Geise et al., 2011), membrane selectivity and permeability considerations (Geise et al., 2011; Kim and Hong, 2018; Mousa et al., 2013; Shaffer et al., 2012), multi-stage pressurization process (Jeong et al., 2019), and employing energy recovery devices (ERDs) (Kim and Hong, 2018).

This chapter presents a structure for optimization of stand-alone or combined desalination systems operating with renewable energy resources. This allows one to determine the optimum design parameters for each component based on the objective functions considered. To formulate an appropriate optimization approach for desalination plants, the following questions should be addressed: (1) What equipment/processes should be selected for a desalination process? (2) How should the components/parts be arranged for the best outcome? (3) What are the best characteristics (e.g., capacity, size, cost) for the components/parts? (4) What are the best values of process parameters (e.g., pressure, temperature, composition, flow rate) for each stream interacting with the desalination system?

To answer these questions, optimization elements need to be defined, including optimization criteria, system boundaries, objective functions, and decision variables. In this chapter, various optimization methods are described, and a step-by-step procedure for the optimization of renewable energy–based desalination systems is presented and discussed. Also, several case studies are provided to show how optimization algorithms can be applied to desalination systems that use one or more renewable energy resources.

10.2 Optimization

Optimization is one of the most practical approaches in engineering and other fields for achieving the best design and operation relative to a set of prioritized criteria. In engineering, parameters are adjusted to optimize the performance of designs. In manufacturing, the efficiency in design and the operation of production processes are generally to be maximized. In physical systems the energy use tends toward a minimum. In an isolated chemical system, the molecules react with each other to reach the minimum overall potential energy for their electrons.

Mathematically, optimization is the process of finding values of a function that maximize or minimize its value subject to several constraints. Usually, there are a number of variables, sometimes a great many. For each of these variables a range is defined based on relevant conditions. If there is only one objective function in an optimization problem, the process of maximizing or minimizing the objective function is called single-objective optimization. The single-objective optimization has been used for a long time for many applications. The need for more than one objective function when other objectives are also important at the same time leads to the need for multiobjective optimization. In most cases involving multiobjective optimization, the objective functions are contradictory to each other. Also, a solution that is proper for one objective function may not work for another objective function(s). With multiobjective optimization problems, therefore, it is not straightforward—and often not possible—to search for one solution that satisfies all objective functions. Rather, a set of feasible solutions usually exists in this condition.

Important features for all optimization algorithms follow (Nocedal and Wright, 2006):

- *Robustness:* Optimization methods should exhibit good performance for a broad variety of problems in their class and for all sensible values of initial estimates.
- *Efficiency:* Optimization methods should not require extremely high computer storage and time.
- *Accuracy:* Optimization methods should yield accurate solutions.

These features may conflict. For instance, a robust method may be the slowest, while a method that achieves rapid convergence during iterations may need excessive computer storage. Tradeoffs between the aforementioned optimization features are one of the main challenges in numerical optimization.

10.3 Optimization terminology

In the following, to aid in comprehension of this chapter, several optimization terms are now presented and described:

System boundaries: Defining system boundaries is the first step in an optimization problem. In this step, any subsystem that might affect the performance of system is required to be included. When the system is comprised of several subsystems, the process of optimization can be performed on each subsystem independently. In other words, suboptimization of the subsystems is carried out (Ahmadi, 2013). The principle of suboptimization does not generally lead to global optimization, but properly deriving the requirements for the design of the subsystems from the system requirements may lead to an optimum system design (Wymore, 1998). So optimization of the overall system may be necessary after optimization of subsystems.

Objective functions: Another important step in solving an optimization problem is the definition of the objective function(s). An objective function can be related to efficiency (such as energy and exergy efficiencies), economics (such as levelized cost, capital cost, cost of exergy destruction and environmental impacts), technological factors (such as production time and rate), environmental impact, and other factors. Basically, minimization or maximization of objective functions is based on the goal of the user or decision maker. For instance, in almost all cases, it is desirable to maximize the energy/exergy efficiency of an energy system, while minimizing its cost and environmental impacts.

Decision variables: An important step in an optimization problem is the selection of decision variables. The decision variables are independent variables (i.e., their values cannot be determined from other variables) that can be used to specify the possible design options for a system. Decision variables should be selected so as to include all important variables that affect the objective function(s) (e.g., influence the performance and cost effectiveness of the system). Considering the variables with minor importance may increase the complexity and time required for the solution of an optimization problem.

Constraints: In the optimization process of a system, a range is often specified for the decision variables, usually including upper and lower limits, based on physical factors, conservation principles, and other restrictions. In a thermal system, some of the constraints may arise from mass, energy, and momentum conservation laws. For example, the mass balance principle for a thermal system indicates that the input mass flow rates are equal to the output mass flow rates under steady-state conditions. This conservation principle restricts the values of relevant design variables that might be applied in the search for an optimum solution. In addition, temperature, heat flux, and dimension ranges can be restricted by applying the energy balance equations for a thermal system (Dincer and Rosen, 2021).

There are many examples of constraints for decision variables in optimization of a thermal system. For instance, the maximum allowable temperature for a component due to physical limitations, the minimum pressure required to provide adequate ventilation, and restrictions on physical dimensions of a system due to space limitations. Thus, the optimum values of objective functions need to be determined with respect to these constraints (i.e., while satisfying them).

10.4 Optimization methods

The optimization methodology is described as the process of identifying variables, objective functions, and constraints for a given problem. Building a proper model is an important first step in an optimization problem. If the model is too complex, it might be too difficult to solve the optimization problem and if the model is too simple, it may not yield useful or practical results.

When the system model is developed, an optimization algorithm can be employed to determine its solution(s), typically using a computer. There are many optimization algorithms, which have different attributes and which can be tailored to a specific kind of optimization problem. The selection of an appropriate optimization algorithm by the user is important, in part because the speed of solving for the solutions of a problem is highly dependent on the type of algorithm selected. After an optimization algorithm is chosen and applied to the model, it is important to ascertain whether it has succeeded in finding one or more solutions or not.

10.4.1 Primary techniques for optimization

Various optimization methods can be used to achieve optimal systems. The three primary techniques for optimization are classical, numerical, and evolutionary, and each is now described.

Classical optimization methods: These methods can be employed to find the optimal solution of problems involving continuous and differentiable functions. Classical optimization methods basically are analytical techniques to find maximum and minimum points for unconstrained and constrained continuous objective functions (Reed and Marks, 1999). Such methods can be used in some optimization problems with: (1) single variable functions, (2) multivariable functions with no constraints, and (3) multivariable functions with both inequality and equality constraints. For problems with inequality constraints, the Kuhn–Tucker conditions is usually employed to determine the optimum solution, while for problems with equality constraints, the Lagrange multiplier technique is typically used (Dincer and Rosen, 2021). One of the drawbacks of classical optimization methods is that they can yield unsatisfactory results when used to find multiple solutions. This is because they can obtain different solutions for different starting points in multiple runs of the algorithm. In addition, this technique cannot be widely applied in practical applications, as it commonly involves objective functions that are not continuous or differentiable.

Numerical optimization methods: These optimization methods are usually iterative, beginning with an initial guess of a variable. The methods then produce a sequence of improved iterates (i.e., estimates) until they approach a solution. Such optimization methods can be distinguished from each other based on the strategy used to move from one iterate to the next. For instance, some methods store and use the information from previous iterations.

There are several algorithms for numerical optimization methods (Dincer et al., 2017):

- *Linear programming:* This method can be used for a linear objective function and the set design variable space is determined using only linear inequalities and equalities.
- *Quadratic programming:* In this method, the objective functions are allowed to have quadratic terms, whereas the set design variable space is determined with linear inequalities and equalities.
- *Nonlinear programing:* This method can be used for cases in which the constraints, objective functions, or both contain nonlinear parts.
- *Integer programming:* This method can be applied to linear programs in which all or some variables are constrained to integer values.
- *Dynamic programing:* This method can be applied to the case in which the optimization problem is divided into smaller subproblems.
- *Combinatorial optimization:* This method can be applied in cases in which there is a set of discrete feasible solutions or a single discrete one.
- *Stochastic programming:* In this method some of the constraints are dependent on random variables.

Evolutionary algorithms: Evolutionary optimization algorithms are optimization techniques that mimic biological evolution, including processes such as reproduction, recombination, natural selection, and mutation. Such algorithms have developed over the last few decades and many techniques exist under the class of evolutionary algorithm. The candidate solutions in this type of algorithm can be treated as individuals in a population. Also, a fitness function is often applied, which specifies the environment in which the solutions live. The main evolutionary optimization techniques are genetic algorithm (GA), fuzzy logic, and artificial neural network (ANN). All of these are available in toolboxes developed by Math Works in the Matlab software. These techniques differ in terms of the presentation of solutions, implementation details, and how a given problem is applied. The general procedure for an evolutionary algorithm is shown in Fig. 10.1 and can be described as follows (Sangaiah, 2019):

- Select an initial population in the search space.
- Determine the value of the objective function for each population member.
- Carry out for each iteration until the termination condition is satisfied the following steps:
  - Perform selection;
  - Perform crossover with a probability;
  - Perform mutation with a probability;
  - Determine the new population and fitness function;
  - Replace if the new members are better in terms of fitness
  - Else, retain the same members and proceed with iterations.

10.4.2 Types of evolutionary optimization algorithms

Genetic algorithm: The GA is a search heuristic-based evolutionary technique that is commonly used to generate high-quality solutions in optimization problems. It mimics Darwinian evolution by following only the strongest survival strategy. Natural selection is the fundamental principle used in the GA.

Inspired by evolutionary biology and involving inheritance, natural selection, learning, and mutation, the process begins with an initial random population of candidate solutions, called individuals. The algorithm next progresses through iterations (also called generations), in which the fitness of each individual is evaluated, based on objective function. The properties of each individual can be mutated and altered. The individuals having better fitness are selected from the current generation and modified to form a new population. The next iteration employs this new population. By

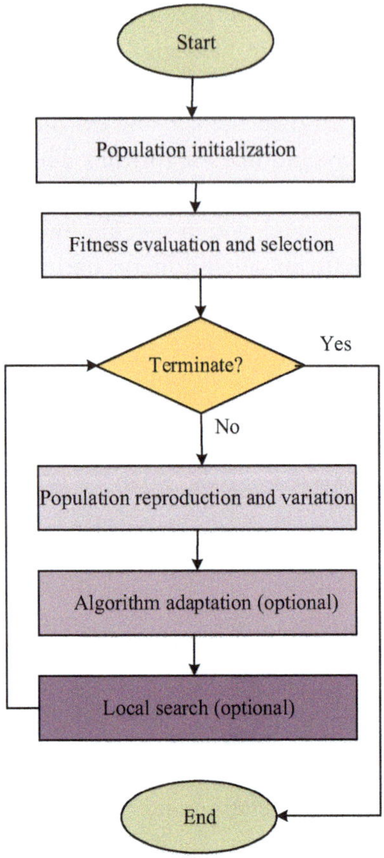

Fig. 10.1 ***General procedure for an evolutionary algorithm.***

iteration of this procedure, the fitness of each individual in the population is evaluated, and the algorithm evolves toward the desired optimal point (Goldberg, 1989). The evolution process is stopped when the maximum number of generations has been produced or a desired fitness level has been achieved. Fig. 10.2 illustrates the concept of evolutionary computation in a GA. A GA can be used for a wide range of optimization problems, including those that are multidimensional, nondifferential, noncontinuous, and nonparametrical. Other benefits of GAs for optimization problems are: (1) they can yield multiple solutions for problems, (2) they are easily understandable and thus do not require prior high-level knowledge of mathematics and (3) they can be applied easily to simulations and models.

Artificial neural network: The ANN method is based on the structure and function of the human brain for processing information. Similar to neurons

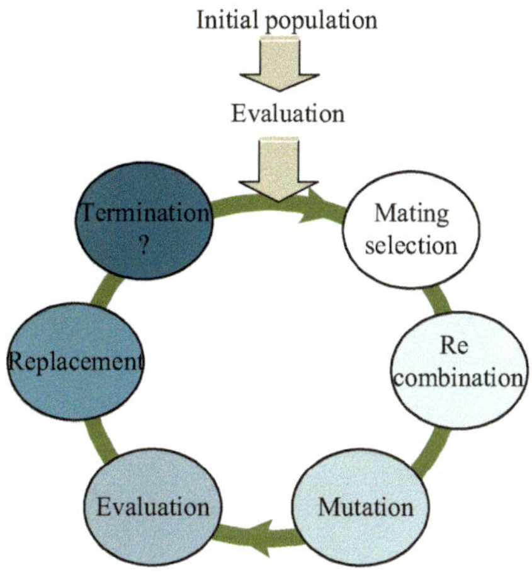

Fig. 10.2 ***General concept of a genetic algorithm (GA).***

in the brain, an ANN is comprised of neurons arranged in various layers. Finding a relation between the input and output parameters analytically or using process simulator software with many components and streams is complex. Simulating the deep learning in a human brain to extract the best function to connect the objectives and variables can be achieved by machine learning, such as by using an ANN (Farsi et al., 2020). A commonly used ANN method is the forward-feed neural network, which involves an input layer (which receives external data), an output layer (which gives the problem solution), and an intermediate hidden layer (which separates the other layers). Nodes from adjacent layers have connections (or edges) with other nodes. All these connections have weights that indicate the strength of input influence on the output. The ANN employs a back-propagation training algorithm to teach datasets, which modifies the weights of neurons based on the error rate between the target and the actual output. Fig. 10.3 shows the general concept of an ANN. The ANN approach may lead to nonlinear, parallel, adaptive, distributed, and local processing representations of systems.

Fuzzy optimization: Fuzzy logic optimization is a search-based strategy that solves optimization problems based on simple if-then rules. Fuzzy logic mimics human reasoning in terms of linguistic variables based on if-then mathematical rules. In fuzzy logic, linguistic variables are natural language variables that humans usually employ to specify semantic rules from several processes (Cuevas et al., 2020).

Fuzzy logic is a form of many-valued logic, a propositional calculus in which there are more than two truth values (Novák et al., 2012, Suganthi

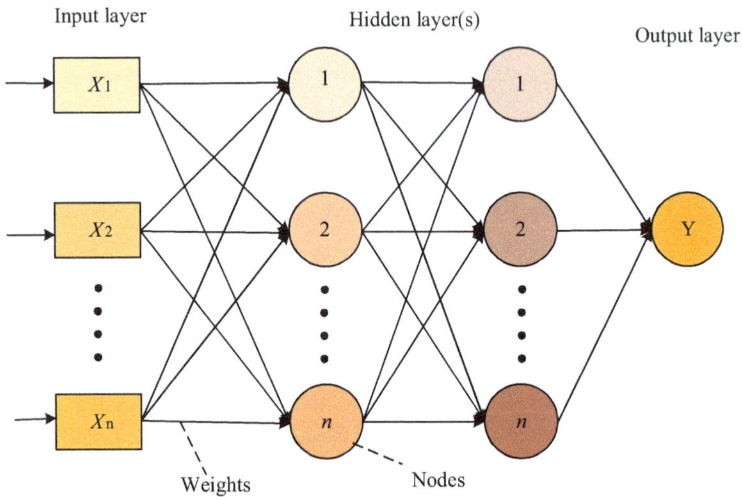

Fig. 10.3 ***General concept of an artificial neural network (ANN).***

et al., 2015). The fuzzy logic method uses approximate reasoning that has linguistic values (i.e., nonnumeric values). Fuzzy logic deals with the concept of a truth value that can be any real number between 0 and 1. With this method, the concept of partial truth can be handled, where the truth value may range between completely true and completely false (Novák et al., 2012). Fuzzy logic together with probability, which uses the concept of subjective probability, provides another means of expressing uncertainty. Fuzzy logic–based models in energy and thermal systems can be varied from the most simple to the most complex. Some of the classes of fuzzy optimization are fuzzy particle swarm optimization (PSO), fuzzy honey bee optimization, fuzzy cuckoo search optimization, fuzzy quantum PSO, and fuzzy ant colony optimization (Suganthi et al., 2015).

It is important to develop increasingly realistic models for energy systems to enhance systems for providing reliable, efficient, environmentally friendly energy in future years. Fuzzy logic assists in conceptualizing the fuzziness in an energy system into a crisp quantifiable parameter. Then, fuzzy logic–based models can be adopted for effective energy planning, by yielding realistic solutions. Niknam et al. (2011) conducted a fuzzy honey bee optimization for renewable energy resources. In that study, the objective functions of the problem were the minimization of electrical active power losses, voltage deviations, total electrical energy costs, and total emissions of renewable energy resources. A fuzzy clustering technique was used to control the nondominated solutions stored in the repository. The superiority of the modified fuzzy honey bee optimization method was shown with respect to other optimization methods.

10.4.3 Single-objective optimization

Optimization problems that have only one objective function only require single-objective optimization, a process by which the best solution for that sole objective function is found. In other words, in single-objective optimization, only one criterion is considered in the solution of the problem.

Although single-objective optimization has been applied for decades to a wide range of applications, most real-world optimization problems involve more than one objective function. This led to the advent of multiobjective optimization. In fact, in many practical problems when applying an optimization method, more than one objective function merits consideration. In the following section, multiobjective optimization is described.

10.4.4 Multiobjective optimization

In multiobjective optimization, there is not a single solution, but rather there a set of solutions. In defining different types of optimization objective functions, $f(x)$, both minimum and maximum functions can be transferred to each other via the following expression:

$$\max\{f(x)\} \dot{\Leftrightarrow} \min\{-f(x)\} \tag{10.1}$$

Then, any multiobjective optimization problem can be written as the following mathematical expression (Cui et al., 2017):

$$\min(y) = f(x) = [f_1(x), f_2(x), \dots, f_m(x)]^T \tag{10.2}$$

Subject to

$$g_j(x) \le 0 \ (j = 1, 2, 3, .., p)$$

$$h_k(x) = 0 \ (k = 1, 2, 3, .., q)$$

$$x_i^{min} \le x_i \le x_i^{max} \ (i = 1, 2, 3, \dots, n) \tag{10.3}$$

$$x = [x_1, x_2, \dots, x_n]^T \in \Theta$$

$$y = [y_1, y_2, \dots, y_m]^T \in \Psi$$

where m is the number of optimized objective functions. Also, Θ is the n-dimensional search space which can be determined through the upper bound $x^{max} = \left[x_1^{max}, x_2^{max}, \ldots, x_n^{max}\right]^T$ and the lower bound $x^{min} = \left[x_1^{min}, x_2^{min}, \ldots, x_n^{min}\right]^T$ of decision variables $x_i (i = 1, \ldots, n)$. Ψ is the m-dimensional vector space of objective functions and can be specified by Θ and the objective function $f(x)$. The expressions $g_j(x) \leq 0$ ($j = 1, 2, 3,..,$ p) and $h_k(x) = 0$ ($k = 1, 2, 3,.., q$) represent p inequality constraints and q equality constraints. If $p = q = 0$, the multiobjective optimization problem is simplified to a unconstrained problem (Cui et al., 2017).

The set of nondominated solutions is called the Pareto frontier, which represents the collection of optimal solutions. In fact, multiobjective optimization aims to search for such nondominated solutions to find real optimal solutions. When the Pareto-optimal set is found, decision makers select the final solution from these optimal solution based on supplemental criteria or preferences. The decision space is the n-dimensional space formed by the variables, and the objective space is the m-dimensional space formed by the objectives.

Fig. 10.4 shows the relation between the design space, the objective space, and Pareto-optimal solutions for a problem with two variables and two objectives. For a solution vector, $x \in \Theta$, it is defined as a feasible solution provided it satisfies both equality and inequality constraints for all j ($= 1, 2, 3,..p$) and k ($= 1, 2, 3, ..q$). Otherwise it is an infeasible solution. In Fig. 10.4, the infeasible regions are also depicted out of the feasible solutions. There are six points in the decision space which are placed into the objective function space. If points a and b are dominated by c, a is dominated by b, and points b and d are dominated by f, then, points c and f are nondominated solutions in the Pareto front. There are no dominance

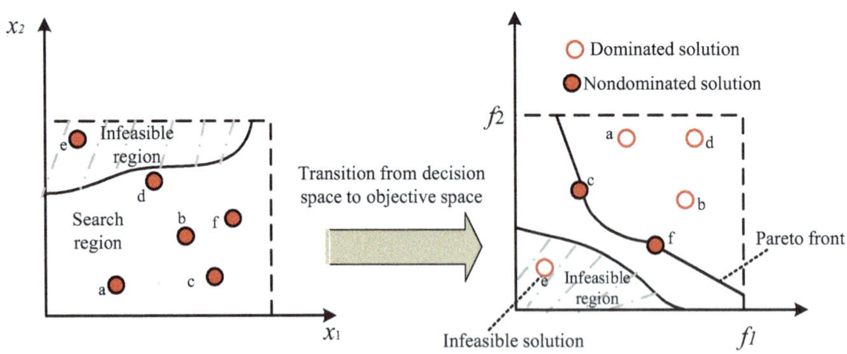

Fig. 10.4 ***Design space, objective space, and Pareto optimal solutions for a two-objective optimization problem.***

relations between c and f and also between b and d. A multiobjective optimization method aims to find the solutions that are as close as possible to the Pareto-optimal front.

In the desalination industry, much attention is being directed toward environmental performance and sustainable development. Efficiencies, pollutant emissions, economic benefits, and production rates are important considerations in multiobjective optimization in real-world problems. For instance, in the design of heat/work-driven desalination systems, it is usually more meaningful to consider more than one objective function. As mentioned before, the chosen objective functions are often conflicting. For example, in an MED desalination system, some objective functions such as energy and exergy efficiencies, heat transfer rate, production rate, and quality are to be maximized, while others such as cost, environmental impact, and pump work are usually aimed to be minimized. The structural optimization of a seawater desalination system is detailed in the next section.

10.5 Structural optimization of seawater desalination

Desalination technologies are usually complex and operate based on various physical laws and principles. The behavior of a desalination system can be described by a mathematical model, which is usually based on thermodynamics, heat and mass transfer, and fluid dynamics. The resulting model equations need to be solved simultaneously. A set of reasonable assumptions are also needed to eliminate unnecessary variables and equations and thus to improve the mathematical model. Desalination systems generally consist of thermal components forming an open cycle in which feed seawater is an input stream and brine and freshwater are the output streams. It is important to know for each component in a system characteristics such as efficiency and performance details.

For instance in an MED system, heat is the primary energy source used to drive successive evaporation/condensation processes in heat exchangers to produce freshwater. Pumps are used to transport fluid streams (feed, distillate, and brine) within the system. When the heat exchangers (which act as effects) are integrated, one of the potentially most significant problems is thermal loss during the heat exchange process. Thus, optimization of these components can be very beneficial for minimizing the thermal losses and consequently maximizing the efficiency of the MED desalination system. It is be noted that enhancing the efficiency of a desalination system is often at the expense of higher capital and operating cost, through employing better materials, designs, and components. Therefore, multiobjective optimization, considering both efficiency and cost as objective functions, can potentially assist in identifying the best compromise between higher efficiency and lower cost.

A generalized structure for optimization a stand-alone, hybrid desalination plant can be developed and evaluated by considering technical (e.g., energy requirement, water recovery rate), economic (e.g., investment and operating costs), and environmental performance indicators. By developing this structure for a desalination process, the conceptual design considerations can be used to reduce its complexity. A flexible structure of a desalination plant includes a distribution box and a box for each possible unit that might be present in the system. In the distribution box, the feed and product streams for the system units are connected, according to the system design and characteristics. Fig. 10.5 shows a generalized structure for a desalination system. Considering a hybrid MED–RO desalination plant as an example, the unit boxes can be an MED unit, an RO unit, an ERD, a pressure exchanger (PX), a high-pressure pump, a booster pump, etc. It is seen that intake feed seawater is the only system input, while the final produced freshwater and brine streams are the system outputs.

In the optimization of desalination systems, the objectives are typically minimization of SEC, system cost, and environmental impacts, and maximization of energy and/or exergy efficiencies and freshwater production rate. Based on the objective functions chosen by decision makers, a corresponding

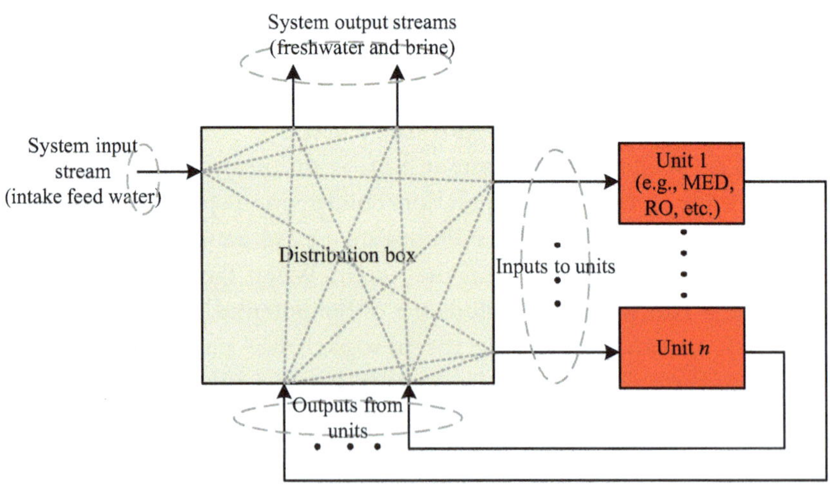

Fig. 10.5 ***General structure of a desalination plant.***

mathematical model needs to be developed for the system. For instance, when considering SEC and cost of freshwater product as the objective functions, thermodynamic and heat and mass transfer models need to be developed, as does an economic model. Then, optimization is applied to the models. When multiobjective optimization is utilized to optimize several objective functions that are conflicting, there typically ends up being a set of optimal tradeoffs between the objective functions. Multiobjective optimization algorithms usually proceed to determine the Pareto-optimal solutions. Therefore, there is a set of nondominated (i.e., Pareto frontier) solutions that give the decision maker the flexibility to select the final solution from these optimal solutions based on designated preferences.

In the design and optimization of desalination systems, it is necessary to distinguish decision variables and dependent variables. Decision variables are a set of independent variables that may vary in the optimization process. The value of dependent variables can be calculated from independent variables using mathematical equations.

10.5.1 Mathematical model

A mathematical model can be developed for the distribution box and each possible unit in desalination plant. Reasonable assumptions and simplifications are used in the mathematical modeling and optimization. These assumptions are based on the design and specifications of the system, and often are used to eliminate unnecessary variables and equations and thus simplify the mathematical modeling and optimization. For instance, if there is no restriction for the brine concentration discharged from the desalination plant, mixing the final brine stream with the feed is unnecessary. Also, it can be assumed that the intake feed seawater enters the system at atmospheric pressure, and that the permeate stream leaving the RO modules is at atmospheric pressure.

In the distribution box shown in Fig. 10.5, all streams can be characterized by their mass flow rate $\dot{m}$, salt concentration w (g/kg), and pressure P (bar). The splitting of each inlet stream and the mixing for each outlet stream, respectively, can be expressed as

$$\dot{m}_i = \sum_{k=1}^{n_{out}} \dot{m}_{i,k} \tag{10.4}$$

$$\dot{m}_k = \sum_{i=1}^{n_{in}} \dot{m}_{i,k} \tag{10.5}$$

where $\dot{m}_i$ and $\dot{m}_k$, respectively, are the mass flow rate of input stream i and output stream k; and $\dot{m}_{i,k}$ denotes the kth split of stream $\dot{m}_i$. The salt concentration of each split $w_{i,k}$ is the same as that for its input stream w_i (i.e., $w_{i,k} = w_i$; $k = 1, 2, \ldots n_{out}$). Also, the salt mass balance rate for each outlet stream can be written as

$$\dot{m}_k w_k = \sum_{i=1}^{n_{in}} \dot{m}_{i,k} w_i \tag{10.6}$$

In addition, all n_{in} inputs can be split and distributed to n_{out} outputs after a possible isobaric mixing process, that is,

$$P_{i,k} = P_i k = 1, 2, \ldots, n_{out} \tag{10.7}$$

$$\dot{m}_{i,k} P_i - P_k) = 0 i = 1, 2, \ldots, n_{in} \tag{10.8}$$

Then, a mathematical model is required to develop for each unit in the design of the desalination plant (e.g., MED, RO, PX, ERD, high-pressure pump, thermal vapor compressor). The models for commonly used desalination techniques are provided in Chapter 5.

10.5.2 Objective functions

As noted earlier, the objectives in the optimization of desalination processes are typically minimization of SEC, costs, and environmental impacts, and maximization of energy and/or exergy efficiencies and freshwater production rate. Each of these objective functions is now described and formulated.

Energy efficiency: The energy efficiency can be an objective function in the optimization of a desalination system. The energy efficiency of thermal desalination process is commonly expressed by the gained output ratio (GOR). This parameter is defined as the ratio of the enthalpy of vaporization (or latent heat of vaporization) of pure water at the inlet water temperature to the thermal energy input to the desalination system. Expressing terms on a rate basis, we can write

$$GOR = \frac{\dot{m}_D \lambda}{\dot{Q}_D} \tag{10.9}$$

where $\dot{m}_D$ is the mass flow rate of freshwater (kg/s), λ is the latent heat of vaporization (kJ/kg), and $\dot{Q}_D$ is the input heat rate to the desalination system (kW). The higher the GOR of a thermal desalination system, the better is its efficiency.

For a work-driven desalination system, which uses electrical energy, the SEC can be used as a performance parameter. SEC can be expressed as the ratio of the work rate required for desalination ($\dot{W}_D$) to the mass flow rate of produced freshwater ($\dot{m}_D$):

$$SEC = \frac{\dot{W}_D}{\dot{m}_D} \tag{10.10}$$

Although in a desalination system energy efficiency is often used to evaluate the performance of system, energy analysis does not distinguish between work and heat thermodynamically and does not specify the sources of irreversibility or dissipation. Energy assessments therefore do not offer complete information related to optimum energy conversion. As an alternative, applying exergy analysis to a desalination process permits the extent of waste exergy emissions and the location of exergy destructions and irreversibilities to be identified. Reducing the sources of irreversibilities for a desalination system helps to improve it thermodynamically.

Exergy efficiency: If exergy efficiency is considered as one of the objective functions of an optimization problem, an exergy analysis needs to be performed for the overall system and each unit comprising it. In Chapters 5–7, energy and exergy analyses of commonly used desalination technologies are presented and discussed. The exergy rate balance equation under steady-state conditions for an open system in a control volume can be written as

$$\dot{E}x^{\dot{Q}} - \dot{E}x^{\dot{W}} + \sum_{in} \dot{m}_{in} ex_{in} - \sum_{out} \dot{m}_{out} ex_{out} - \dot{E}x_{Des} = 0 \tag{10.11}$$

Consider a generalized desalination process in a black box, as shown in Fig. 10.6. It is seen that the process input and output streams for a desalination system can in general be in the form of heat, work, and material (chemical fuel, feed, brine, distillate), or a combination thereof. The exergy rates

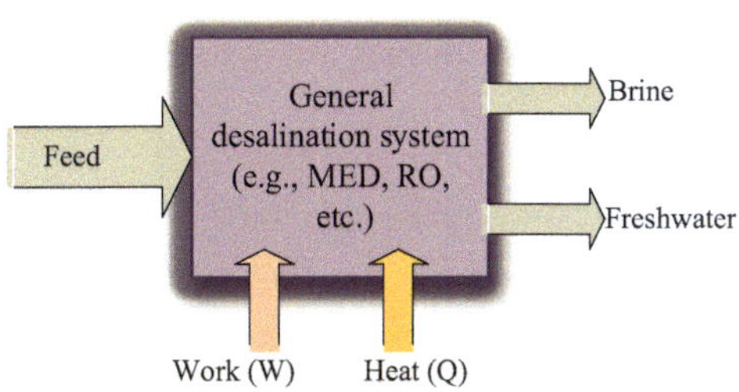

Fig. 10.6 ***A general desalination system treated as a black box.***

associated with the heat transfer rate and the work rate in Eq. (10.11) can be expressed as

$$\dot{E}x^{\dot{W}} = \dot{W} \tag{10.12}$$

$$\dot{E}x^{\dot{Q}} = \left(1 - \frac{T_0}{T}\right)\dot{Q} \tag{10.13}$$

where T and T_0, respectively, are the temperature at which the heat transfer occurs (in K) and the reference environment temperature (in K). The exergy rate of a material stream (e.g., chemical fuel, feed, brine, freshwater) in Eq. (10.11) can be written as follows:

$$\dot{E}x = \dot{m}ex \tag{10.14}$$

where $\dot{m}$ is mass flow rate (in kg/s) and ex specific exergy (in kJ/kg). The specific exergy of a flow comprises its specific physical exergy (ex_{ph}) and specific chemical exergy (ex_{chem}). That is,

$$ex = \underbrace{h - h* - T_0(s - s*)}_{ex_{ph}} + \underbrace{\sum\nolimits_{i=1}^{n} (\mu_i * - \mu_{0,i}) w_i}_{exc_{hem}} \tag{10.15}$$

where h denotes specific enthalpy (in kJ/kg), s specific entropy (in kJ/kg.K), w_i mass fraction of the ith constituent, and μ chemical potential (in kJ/kg). The superscript * used for h, s, and μ indicates that the properties are those at the environmental pressure and temperature. The term $\mu_{0,i}$ is the chemical potential of constituent i at the reference environment pressure, temperature, and constituent concentrations.

In general, exergy efficiency can be expressed as the exergy rate of useful process outputs ($\dot{E}x_{out,\ useful}$) to the exergy rate of the process inputs ($\dot{E}x_{in}$), that is,

$$\Psi = \frac{\dot{E}x_{out,useful}}{\dot{E}x_{in}} \tag{10.16}$$

The exergy rate of useful process outputs can be written as to the input exergy rate ($\dot{E}x_{in}$) less the exergy destruction rate ($\dot{E}x_{Des}$) and the exergy loss rate ($\dot{E}x_{loss}$), as follows:

$$\dot{E}x_{out,useful} = \dot{E}x_{in} - \left(\dot{E}x_{Des} + \dot{E}x_{loss}\right) \tag{10.17}$$

Then, the exergy efficiency can be expressed as

$$\Psi = 1 - \frac{\dot{E}x_{Des} + \dot{E}x_{loss}}{\dot{E}x_{in}} \tag{10.18}$$

The destroyed exergy is the exergy that is lost due to system irreversibilities, while the exergy loss indicates the exergy that is lost with outgoing waste streams. This latter term is often equivalent to the exergy loss due to bringing the brine and freshwater streams to the reference environment temperature, pressure, and constituent concentrations.

Cost: Another objective function that can be considered in the optimization of desalination systems is total annualized cost of the system. This can be written as the summation of its components cost rates. Although the optimization process can be simplified by considering only SEC or exergy efficiency as a single-objective function, total annualized cost is an alternative and realistic objective function needed to quantify the performance of a desalination system. Thus, in addition to the thermodynamic and heat and mass transfer models for desalination systems, economic models/analyses are also required.

In Chapter 9, economic analyses of desalination systems driven by renewable energy facilities are provided and discussed. The component cost rate of capital investment and operation and maintenance, $\dot{Z}_k$, can be expressed as

$$\dot{Z}_k = \dot{Z}_{capital,k} + \dot{Z}_{O\&M,k} \tag{10.19}$$

Here, $\dot{Z}_{capital,k}$ and $\dot{Z}_{O\&M,k}$, respectively, are the component cost rates for capital investment and operation and maintenance. In the economic analysis, annualized values are often used, as they undergo a change over their economic life. Thus, the annual levelized capital cost rate for component k (in \$/h) can be written as

$$\dot{Z}_{capital,k} = \frac{\left(Z_{capital,k}.CRF\right)}{\tau} \tag{10.20}$$

where CRF is the capital recover factor and τ the annual operation time of the system (in hours). $Z_{capital,k}$ is the capital cost for component k (in \$). Cost data that are not up to date need to be updated. This can be accomplished using a correction that can be applied to costs that are for a different time than the reference year. That is (CEPCI, 2021),

$$\text{Cost at reference year} = \text{Original cost} \times \frac{\text{Cost index for reference year}}{\text{Cost index for year when original cost was obtained}} \tag{10.21}$$

In addition, the CRF can be expressed in terms of interest rate (i) and loan period of a component (n) as follows:

$$CRF = \frac{i\left(1+i\right)^n}{\left(1+i\right)^i - 1} \tag{10.22}$$

We can also write Eq. (10.19) based on the operation and maintenance factor ϕ, as

$$\dot{Z}_k = CRF.\frac{\varphi.Z_k}{\tau} \tag{10.23}$$

Environmental impact: The environmental impacts associated with the desalination industry are mainly concerned with energy consumption and brine discharge into local water resources. The environmental impacts of desalination systems can be reduced by several treatment processes such as disposal treatment and concentrate dilution. In addition, the environmental impacts can be effectively mitigated by minimizing the upstream impact sources via process optimization. To do so, environmental indicators for a desalination system need to be defined and integrated with the process design of the system.

One useful approach for evaluating the environmental impacts of desalination processes is life cycle assessment. This method is based on the life cycle inventory of the inputs, such as raw materials, chemicals, and electricity, and outputs derived from the process operation and decommissioning and construction. With this information, the environmental impacts associated with a process can be determined. In other words, each process configuration can be characterized by a specific inventory and consequently specific impacts. For instance, in an RO desalination process, it has been shown that more than 90% of total plant life cycle impacts are associated with the electricity consumed in the desalination process (Raluy et al., 2005). The impact of discharging concentrated brine into local water resources is usually another measure of the environmental impacts of RO and other desalination techniques. The environmental impacts associated with membrane renewal, construction, and decommissioning are often negligible when compared with those for electrical energy consumption and concentrate discharges.

Normally, reducing SEC of a desalination system through optimization and brine discharge management appear to be the main two actions available to reduce the environmental impacts of desalination processes. Furthermore, by using renewable energy resources and nuclear energy, carbon dioxide emission can be reduced more than 10 times compared to the case when fossil fuels such as coal are consumed to generate electricity.

Other approaches to evaluate the environmental impacts associated with desalination exist. For instance, exergo-environmental analysis, which combines environmental and exergy assessments, can be very useful. With this analysis, the components with the highest and lowest environmental impacts can be identified. In Section 9.6 of Chapter 9, the exergo-environmental assessment of desalination systems driven by renewable energies is discussed.

10.5.3 Pareto frontier for optimization of a desalination process

To provide insights into the optimization of desalination systems, consider the general desalination process shown in Fig. 10.6. A general process design of the desalination process can be developed through several steps (Fig. 10.7). In the process synthesis step in Fig. 10.7, system process configurations are systematically generated. Then through process characterization, the performance of each of the process configurations generated in the previous step is evaluated. Finally, in the optimization step, the best configurations in terms of the specified objective functions are selected. As a consequence, three main components of desalination process design are technical modeling of the desalination system components, meaningful performance indicators, and useful optimization procedures.

If we assume that the cost and SEC of freshwater production are considered as objective functions, a general Pareto diagram obtained from

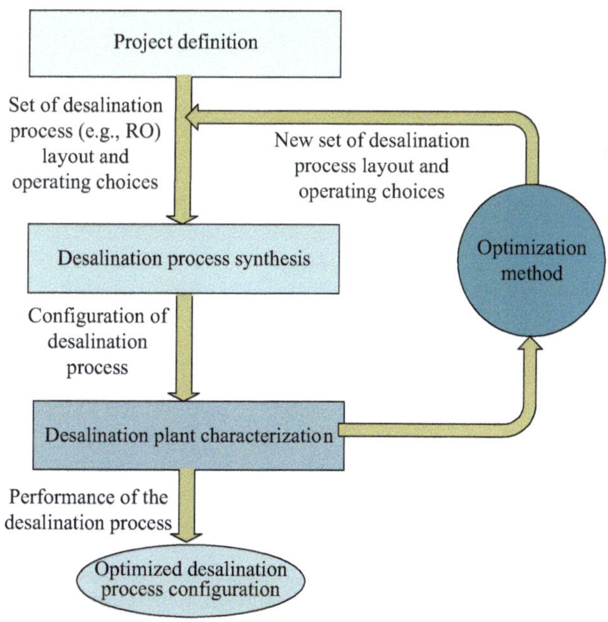

Fig. 10.7 ***General desalination process design method.***

optimization of the desalination process can be developed, as shown in Fig. 10.8. In this figure, three process configurations (A, B, and C) are shown. Each configuration is represented by its corresponding total water cost (TWC) and SEC. The characteristics of each process configuration (e.g., water recovery rate, permeate and brine salinities, SEC, TWC) can be realized in the process characterization step described previously.

For multiobjective optimization, a decision-making procedure can be used to select the favorable solution from the optimal solutions in the Pareto frontier. This can be carried out by the Utopian method, in which a hypothetical ideal point (point P in Fig. 10.8) is specified in a way that all objective functions are at their optimum values independently. The Utopian method minimizes the distance from points on the Pareto frontier to the ideal point (also called Utopia point). The Utopia point is not itself an optimal point, as it is not possible to have both objective functions possess their maximum/minimum values simultaneously. Thus, the Utopia point is not located on the Pareto frontier of optimal solutions.

Fig. 10.8 also shows the relationship between two typical objective functions in the optimization of a desalination process. For process configuration A, which is a Pareto-optimal solution, for any TWC, it is not technically feasible to achieve a lower SEC than SEC_A. For process configuration B, another Pareto-optimal solution, for any SEC it is not technically feasible

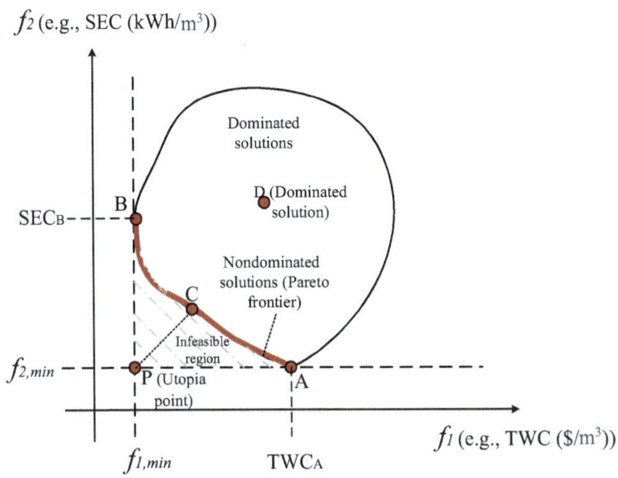

Fig. 10.8 ***Acceptable solutions for minimization of two objective functions (specific energy consumption (SEC) and total water cost (TWC)) with Pareto frontier and Utopia point specified.***

to achieve a lower cost than TWC_B. Therefore, the Utopia point (P) can be specified at the intersection of two dashed lines containing points A and B, as shown in Fig. 10.8. The closer is the solution of the Pareto frontier to the hypothetical ideal point (P), the more optimal is the solution in terms of both objective functions (i.e., point C). This method is reliable when the variations in objective functions are almost identical at the optimum region (i.e., for this case, when small increases/decreases in SEC do not result in a significant increase/decrease in TWC).

10.6 Optimization of renewable energy–driven desalination systems

Many remote and arid areas suffer from shortages of freshwater resources. The combination of renewable energy resources with desalination systems as a "green" approach to freshwater production is a promising and reliable alternative for supplying freshwater demand. In Fig. 10.9, various possible combinations of renewable energy–driven desalination systems are shown.

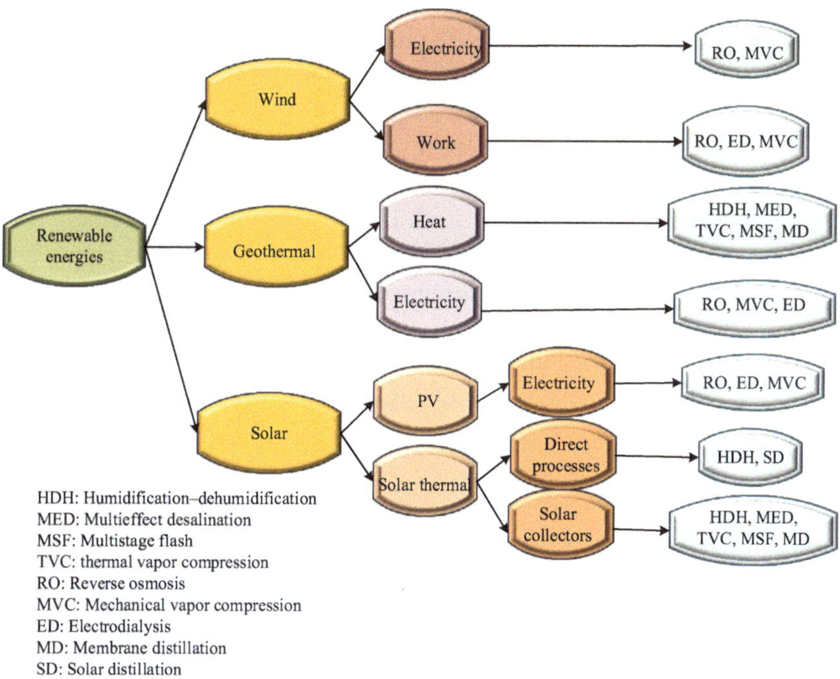

Fig. 10.9 ***Various combinations of renewable energy sources with desalination technologies.***

In this section, a methodology for selecting the most appropriate combined system involving renewable energy–driven desalination is presented. A site is specified with specific characteristics (feed water salinity, energy prices, etc.).

The selection of an appropriate combination of renewable energy and desalination systems for a given location normally requires an iterative approach in which each possible combination is assessed for its ability to meet the freshwater demand in that location and its economic viability. Information on renewable energy and desalination plants is required by the decision maker before deciding on and starting any construction of project. This information can include the salinity and production rate of freshwater, the type and capacity of desalination system, available renewable energy systems, and the commercial maturity of the candidate technologies. Such information is needed to determine operational specifications. Then, from the candidate options deemed viable, the optimum solution is selected based on the desired objectives (e.g., economic benefit) and the characteristics of location or region. For example, for the installation of a photovoltaic (PV)-RO desalination system in a region, the solar radiation in that region and PV efficiency are important inputs for calculation of the energy flows between solar PV system and the RO module.

In Fig. 10.10, a general algorithm for designing a desalination process integrated with a renewable energy system in a region is shown. In the first step, the desalination system is designed and modeled based on the freshwater demand in the region. From this step, the energy required for desalination system can be determined. There should be a balance between the energy supplied from renewable energy sources and backup systems and the energy required for the desalination system. The second step is to design and model the energy supply systems (e.g., renewable energy systems and backup units such as batteries, diesel generators, and natural gas combustors). The potential renewable energy sources in the region need to be identified, and then the corresponding electricity provision configurations (e.g., PV panels, wind turbines, batteries) need to be examined. The last step is the calculation of the objective function(s) (e.g., system cost and efficiency) for each case, so that the best objective quantities can be obtained. The comparison of the values of the objective functions for different configurations of renewable energy-desalination systems can reveal the optimum solution the specific region.

In the next few subsections, several case studies are presented involving optimization of renewable energy systems combined with desalination

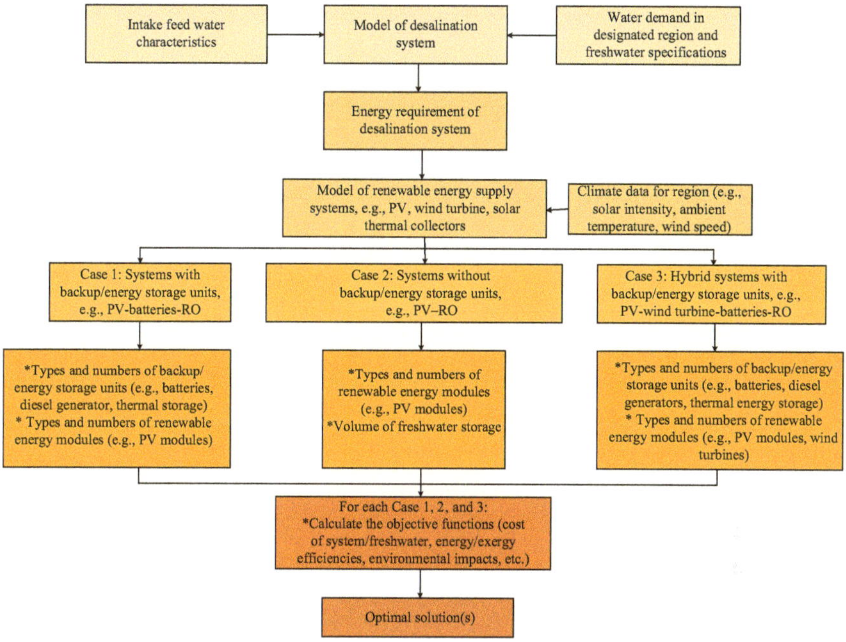

Fig. 10.10 ***Flowchart of the steps for designing a renewable energy–driven desalination plant.***

technologies. In these case studies, different optimization algorithms are used based on the objective functions and other specifications of the relevant problem.

10.7 Case study 1: optimization of a renewable energy–driven RO system using a GA

This section draws extensively from a study by Bourouni et al. (2011) in which GA optimization was applied to an RO desalination driven by PV arrays, wind turbines, and battery storages. The objective function for this study was minimization of the total cost of the system. In addition, a real case study in Ksar Ghilène, a village in Tunisia, is considered in which a PV-battery-RO system was installed in 2007. The design methodology and simulation of the PV-battery-RO desalination system were performed for this real case to find the best cost of freshwater from the optimum configuration of the system. Various configuration scenarios were considered including PV–RO, PV-battery-RO, wind–RO, wind-battery-RO and PV-wind turbine-battery-RO. Each scenario was modeled and the

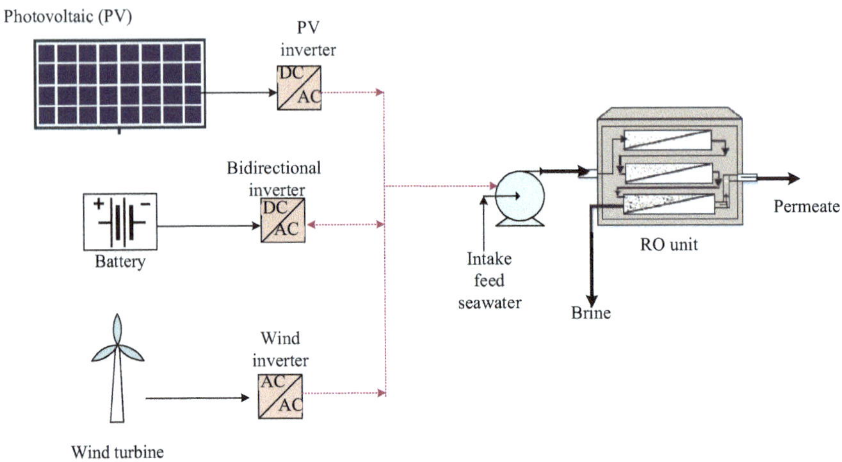

Fig. 10.11 ***Integration of photovoltaic-wind turbine-battery technologies with an RO desalination unit.***

corresponding cost of water was determined for each. Fig. 10.11 shows the PV-wind turbine-battery-RO configuration scenario.

10.7.1 Design and optimization methodology

In the *first step* of the design and optimization methodology for the system, the water demand in the village of Ksar Ghilène was estimated. The village has 300 inhabitants with an average daily water consumption of 50 L/day per inhabitant for primary uses, such as cooking and drinking. Thus, the freshwater required for all inhabitants of the village was determined to be 15 m^3/day. This rate was assumed to be constant throughout the year.

10.7.2 RO model

In the *second step*, the RO unit was modeled to optimally design it using climatic data for Ksar Ghilène, feed water characteristics, and product freshwater quantity and quality requirements. Table 10.1 presents the technical characteristics of the installed unit in the village of Ksar Ghilène. Also in Table 10.2, the model equations for the RO unit are presented. The type of RO membranes was selected based on the feed water characteristics. From Eq. (10.24) in Table 10.2, the number of membranes, which is a function of membrane surface area, stream flow, and unit capacity can be determined. Using Eqs. (10.26)–(10.31), the concentration and flow rate of freshwater and brine streams in the first element of RO can be determined.

Table 10.1 Photovoltaic-reverse osmosis (PV–RO) system characteristics in the village of Ksar Ghilène.

Characteristic	Value
Freshwater salinity	138.86 ppm
Feed water salinity	3500 ppm
Operation pressure	6 bar
Recovery ratio (ratio of freshwater flow rate to feed water flow rate)	32.33%
Freshwater flow rate	15 m^3/day
Feed water flow rate	54 m^3/day
Design temperature	35 °C

From Bourouni et al. (2011).

The produced freshwater is collected from the central tube and the brine is transferred to the second element. This process is continued until the last element of the RO unit is reached. Fig. 10.12 shows the model structure developed for the RO unit.

10.7.3 PV model

In the *third step*, the electrical power required for the system was calculated. Two cases of batteries and no batteries were considered in the model. For the case using batteries, the daily operation of the system was assumed to be 24 h, while for the case without batteries, the daily operation time varies with solar radiation availability. The batteries are in a charging mode when the power produced by the PV modules exceeds the RO power requirement and in a discharging mode when the power produced by PV is lower than the RO power requirement. In the case in which batteries are used in the system, the total energy required for the RO plant was calculated as 558.03 Wh. For the case without batteries, the capacity of the RO unit was overestimated.

In the *fourth step*, PV modules with and without batteries were modeled and simulated based on the RO energy requirement. Table 10.3 provides the model equations for the PV and battery storage systems. From the energy model of the PV–RO system, the electrical power delivered by the PV module can be determined based on the climate data (including hourly radiation and temperature variations) for Ksar Ghilène village. Here, data are used for the year 1990. The PV model begins from the average number of the PV modules required for the system (Eq. 10.38) and average daily radiation G_j used to determine daily power P_j delivered by the PV system

Table 10.2 Equation set for modeling of the reverse osmosis process.

Quantity	Equation		Description
Number of membranes in RO unit	$N_{mb} = \frac{\dot{V}_p}{fA_{mb}}$	(10.24)	$\dot{V}_p$ is permeate (freshwater) volumetric flow rate (m^3/s), f pure water transport coefficient (m^3/m^2 min) and A_{mb} surface area of membrane (m^2).
Number of pressure tube vessels in RO unit	$N_{pt} = \frac{N_{mb}}{N_t}$	(10.25)	N_t is the number of membranes in a pressure vessel.
Permeate product flow rate	$\dot{V}_p = k_{mb}.A_{mb}.TCF.FF(\Delta P - \Delta\pi)$	(10.26)	k_{mb} is membrane pure water permeability at 25 °C (kg/m^2/s/Pa), TCF temperature correction factor (%), FF membrane fouling factor, ΔP applied transmembrane pressure (Pa), and $\Delta\pi$ osmotic pressure (Pa).
Osmotic pressure	$\pi = 0.002654\ T.w.\frac{1}{1000 - \frac{w}{1000}}$	(10.27)	w is salt concentration (ppm), and T temperature (K).
Pressure drop between feed and brine	$\Delta P = P_f - 0.5\ \Delta P_{fs}$	(10.28)	P_f is concentration polarization factor, and ΔP_{fs} transmembrane pressure drop (Pa) in the membrane.
Pressure drop in membrane	$\Delta P_{fs} = 0.01\ \overline{V}_{fb}^{\ 1.7}$	(10.29)	$\overline{V}_{fc}$ is average volumetric flux between feed and brine (m/s).

*k*th membrane efficiency	$Y_{mb,k} = 1-(1-Y)^{1/N_{mb}}$	(10.30)	$Y_{mb,k}$ is recovery fraction for membrane *k* (assumed the same for all membranes).
Permeate salt concentration	$w_p = (1-R_{mb})w_{fb}P_f.TCF.\frac{A_{mb}}{\dot{V}_p}$	(10.31)	R_{mb} is salt rejection fraction of membrane, w_{fb} average salt concentration of feed and brine (ppm), P_f concentration polarization factor, and *TCF* temperature correction factor (%).
Mass rate balance for salt through membrane	$\dot{V}_f w_f = \dot{V}_p w_p + \dot{V}_b w_b$	(10.32)	$\dot{V}$ is flow rate (m^3/s) and *w* salt concentration (ppm). Subscripts, *f*, *p*, and *b* denote feed, permeate, and brine, respectively.
Total volume flow rate of permeate produced by RO unit	$\dot{V}_{p,T} = \sum_{k=1}^{N_{mb}} \dot{V}_{p,k}$	(10.33)	
Salt concentration of permeate produced by RO unit	$w_{p,T} = \frac{\sum_{k=1}^{N_{mb}} \dot{V}_{p,k} w_{p,k}}{\dot{V}_{p,T}}$	(10.34)	

Modified from Bourouni et al. (2011), Rheinländer et al. (2003), Weiner et al. (2001).

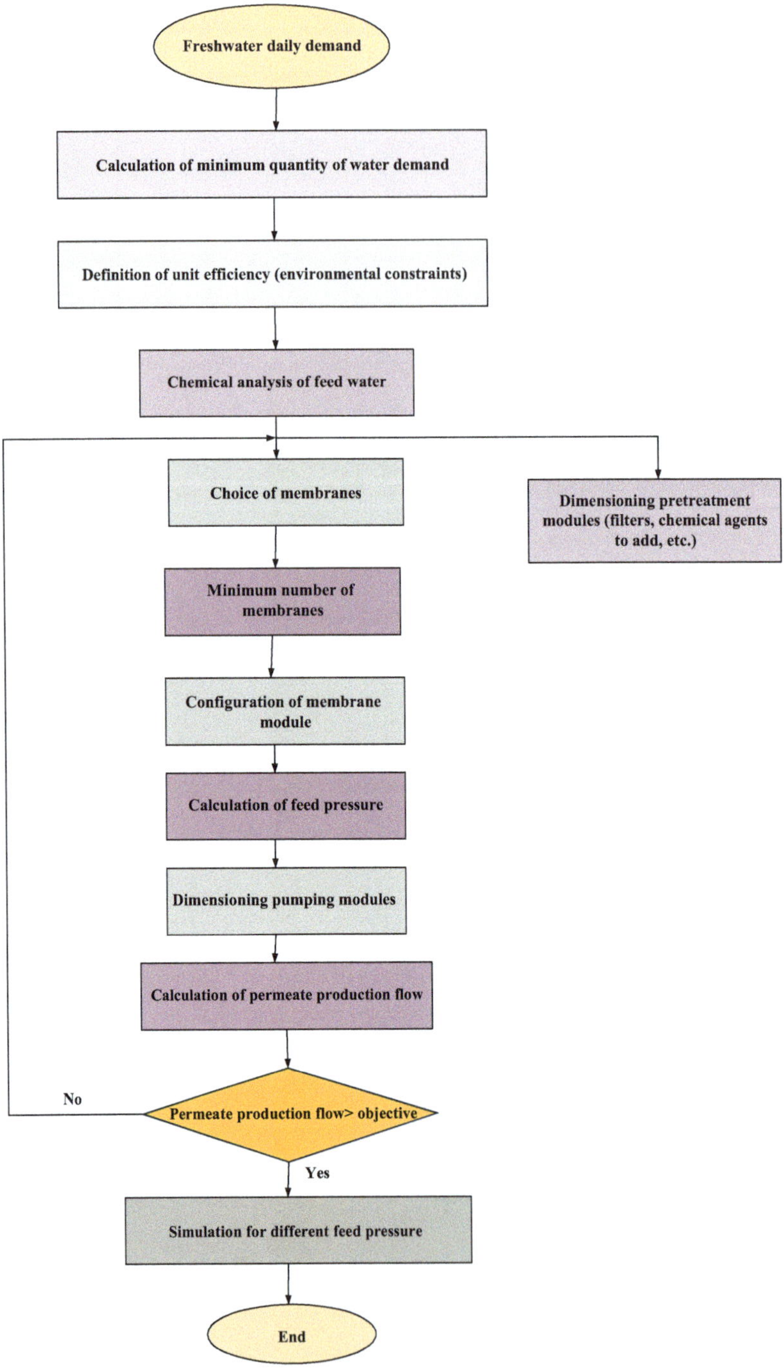

Fig. 10.12 ***Flowchart of reverse osmosis design model.*** *Modified from Bourouni et al. (2011).*

Table 10.3 Equation set for modeling the photovoltaic and battery storage systems.

Quantity	Equation		Description
Photovoltaic power	$P_i = \eta_{PV}(T_{a,i})N_{PV}A_{PV}G_i$	(10.35)	η_{PV} is photovoltaic efficiency, $T_{a,i}$ ambient temperature, N_{PV} number of PV modules, A_{PV} surface area of PV, and G_i solar radiation (W/m^2).
Efficiency of PV system	$\eta_{PV}(T_{a,i}) = \eta_{PV}(1 - \beta)(T_{PV,i} - T)$	(10.36)	β is temperature coefficient of efficiency of PV module
Daily state of storage	$E_c(j+1) = E_c(j) + \Pr(j) - B(j)$	(10.37)	Pr denotes permeate production (m^3) and B freshwater demand (m^3)
Average number of PV modules	$N_{PV} = \dfrac{\overline{P}_{RO}}{\overline{P}_{PV}}$	(10.38)	$\overline{P}_{RO}$ is average power requirement for RO and $\overline{P}_{PV}$ average power produced by PV.
Maximum volume of the storage	$V_c = max_{j=1,..,365}\left(\Pr(j) - B(j)\right)$	(10.39)	
Average number of batteries	$N_B = \dfrac{E(RO).A_t}{V_B \eta_B DOD}$	(10.40)	$E(RO)$ denotes energy requirement of RO, A_t system autonomy (ratio of the storage bank size to the electric load) in (h), V_B battery voltage (V), η_B battery efficiency, and DOD maximum permissible depth of battery discharge (%).
Charging equation for the system	$E_B(t) = \eta_B(E_{Ch,B}(1 - \alpha)) + (Pr(t) - B(t))$	(10.41)	$E_{Ch,B}$ is hourly state of battery tank, and a self-discharge rate of the battery bank.
Discharging equation for the system	$E_B(t) = (E_{Ch,B}(1 - \alpha)) - (Pr(t) - B(t))$	(10.42)	

Modified from Bourouni et al. (2011).

(Eqs. 10.35 and 10.36). Then, P_j is used in the RO model to determine the daily freshwater production. From this calculation, the volume of water storage can be evaluated.

Fig. 10.13 shows the model structure developed for the PV system with and without batteries. From Fig. 10.13, the algorithm developed for designing the PV system tries increasing the number of PV modules to obtain a positive volume of water storage over a year. From this model structure, the number of PV modules, volume of water storage, and minimum electricity requirement for RO can be determined. In the PV-battery-RO configuration, to track the hourly permeate production of the RO unit and charge–discharge state of batteries, the simulation was performed hourly. To achieve a feasible solution, it is presumed that, if the permeate production rate surpasses freshwater demand, the charge state of the battery remains between its nominal and minimum values. Otherwise, the simulation procedure increases the number of PV modules to obtain the minimum number of PV modules that ensure nonstop operation of the RO process. From the simulation results, the average electrical power delivered over a year by 320 PV modules with a nominal power of 230 W each was found to be 47.5 kW.

In the next step, an alternative design was proposed in which wind turbines were also included to form a hybrid PV-wind turbine energy system for the RO unit at the Ksar Ghilène site. HOMER software was used to simulate the hybrid system. Table 10.4 provides simulation results for several combination scenarios of PV, wind turbine, and battery technology with an RO system in the village of Ksar Ghilène.

10.7.4 System optimization

Finally, the hybrid system was optimized using a GA to determine the optimum configuration based on the total net cost of the system, which was considered as the objective function. The economic model was applied to the considered combinations of PV, wind turbine, and battery technology with the RO system (Table 10.4), for a freshwater production capacity of 15 m^3/day in Ksar Ghilène. The total net present cost of the hybrid system includes the capital cost of system components (RO unit, PV units, wind turbines, batteries, etc.), the replacement cost of membranes, the pretreatment cost in the RO installation, the replacement cost of batteries and their charger, the cost of the charge regulator and inverter, and cost of maintenance of the membranes. The system life time was assumed to be 20 years, which corresponds to the typical lifetime of a PV system.

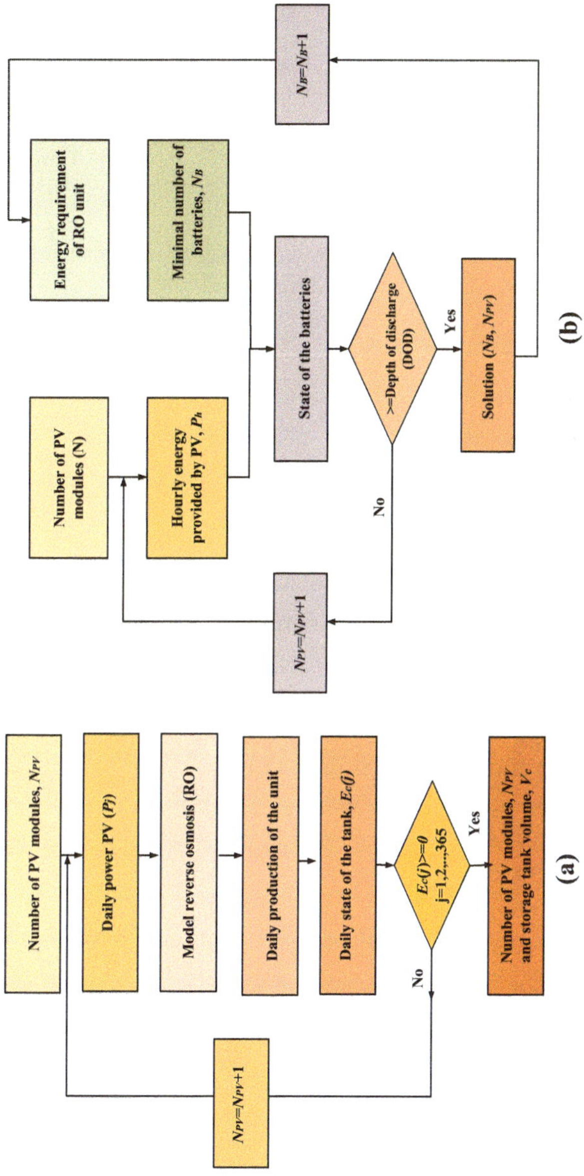

Fig. 10.13 Flowchart of model structure for photovoltaic system (A) without batteries and (B) with batteries. *Modified from Bourouni et al. (2011).*

Table 10.4 Characteristics of various combination desalination scenarios, from simulation with HOMER software.

Scenario	Photovoltaic (PV) panels	Wind turbines (TB)	Batteries (B)
PV–RO	$N_{PV} = 206$ Power ratings = 230 W	-	-
PV-batteries-RO	$N_{PV} = 48$ Power ratings = 230 W	-	$N_B = 26$ Capacity = 200 Ah/12 V (dc)
Wind-batteries-RO		$N_{TB} = 2$ Power = 8 kW Wind speed = 5 m/s	$N_B = 30$
PV-wind turbine-batteries-RO	Power of PV = 1 kW	$N_{TB} = 2$	$N_B = 20$

Source: Bourouni et al. (2011).

A possible solution is represented by a vector, in which the components represent the parameters of the system. The developed algorithm for the hybrid system consists of a vector with the following parameters: type and number of RO membranes, recovery ratio of RO, type and number of PV modules, type and number of wind turbine, type and number of batteries, and the total cost of freshwater produced. The modeling and optimization principles of the hybrid solar PV-wind turbine-batteries-RO system are presented in Fig. 10.14.

The optimization process begins from initial parents corresponding to wind-battery-RO and PV-battery-RO systems using the same type of battery. The optimization proceeds based on the elimination of a wind turbine from wind-battery-RO configuration and the addition of PV modules to the PV-battery-RO system. The possible solutions are obtained from the first case (i.e., first wind turbine replaced by PV modules). If the cost of freshwater produced from the hybrid system is not desirable compared to that obtained by the parents, the process is repeated (i.e., the second wind turbine is replaced by PV modules). If not, it returns to the initial solutions corresponding to the parents (wind-battery-RO and PV-battery-RO). Table 10.5 provides the characteristics of various system components.

The simulation results indicate that four PV modules with nominal power of 230 W per module combined with two wind turbines, each of which can produce 8 kW (at a wind speed of 5 m/s) and 20 batteries (type 6FM200, with a storage capacity per module of 200 Ah at 12 V (dc)) are required to supply 15 m^3/day freshwater. The cost of electricity produced from PV or wind turbine units is the main contributor to the freshwater

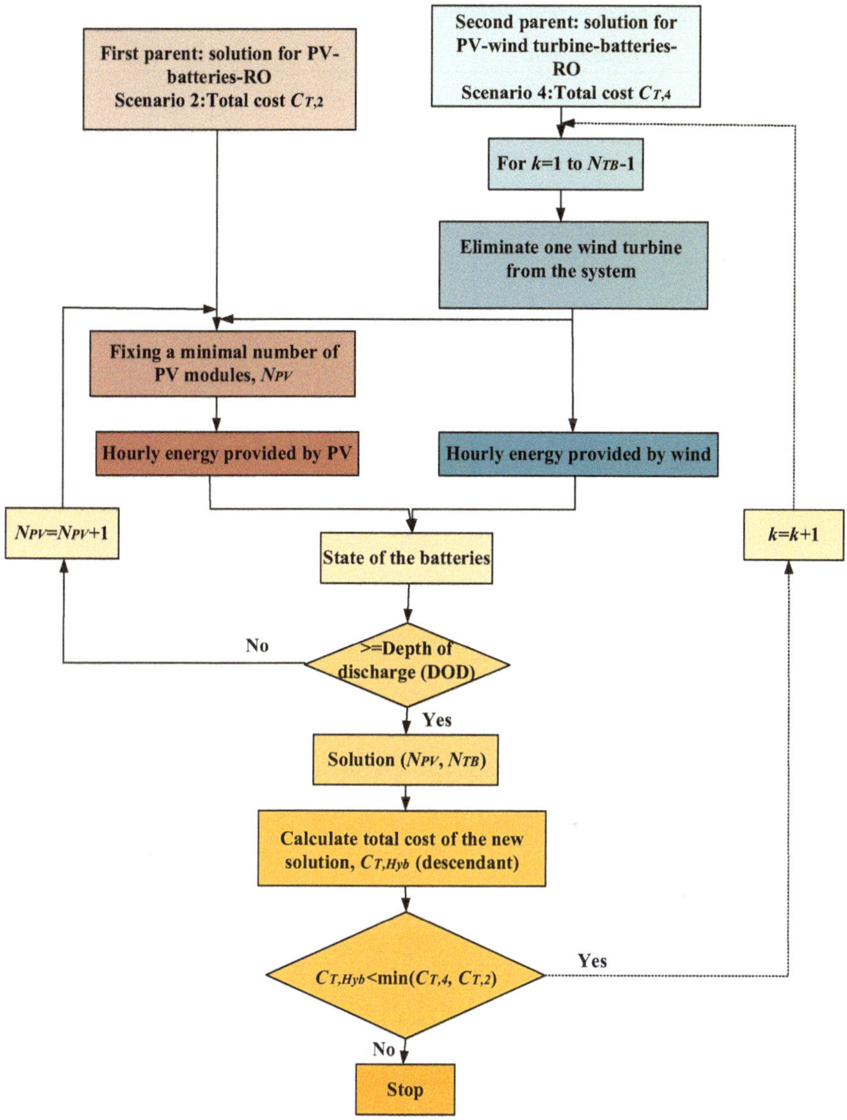

Fig. 10.14 ***Flowchart of modeling and optimization principles of the hybrid solar PV-wind turbine-batteries-RO system.*** *Modified from Bourouni et al. (2011).*

production cost. For better understanding, the cost associated with electricity generation by renewable energy sources (solar PV and wind turbine) is compared. The highest cost of electricity generated from solar and wind was associated with the PV-battery-RO system at 0.912 \$/kW, while the lowest electricity cost was associated with the wind-battery-RO system at 0.253 \$/kW. Therefore, the wind-battery-RO configuration achieves

Table 10.5 Characteristics of units used in the system model.

RO membrane	Photovoltaic	Wind turbine	Battery
Diameter = 20.32 cm (8 in)	Nominal power = 230 W	Rotor diameter = 6.7 m	Nominal capacity = 200 Ah
Active surface area = 37 m^2	Maximum allowable current = 7.2 mA	Height = 18–43 m	Nominal voltage = 12 V
Recovery-15%	Module area = 1.18 m^2	Rated power = 7.5–10 kW	DOD_{max} = 40%
Salt rejection = 99.5%	Acquisition cost = 4.7 $/W		Efficiency = 80%
			Acquisition cost = 760 $

Source: Bourouni et al. (2011).

the lowest cost for freshwater production, indicating it to be the optimum configuration as the objective function was minimization of the total cost of the system.

10.8 Case study 2: thermodynamic optimization of MED system using thermal vapor compressor combined with RO system

This section is adapted from the study conducted by Sadri et al. (2017), in which a mathematical and optimization was developed for an MED-thermal vapor compression (TVC) desalination system installed in Qeshm Island in the southern part of Iran. In the MED–TVC system, the motive (or primary) steam and a portion of the vapor generated in the last effect of MED are compressed via TVC. In fact, through using a TVC, the latent heat of the low pressure steam in the last effect of the MED system is recovered, which reduces the boiler load that supplies primary steam to drive the system. The compressed vapor is transferred to the evaporating tubes of the first effect and provides heat in the first effect on release of the latent heat of condensation. The generated vapor enters the second effect, vaporizing a part of the feed stream by releasing its latent heat of condensation. This process is repeated in subsequent effects. In the last effect, the generated vapor is condensed at the external condenser unit and preheats the feed seawater input to the system

To improve the performance of the system, it was proposed that an RO desalination unit with seven membrane modules in one stage be combined with the MED–TVC system (Fig. 10.15). The intake feed seawater is seen in Fig. 10.15 to be split in two parts, one that enters the RO unit, while the second part is mixed with the brine stream leaving the RO unit. The mixture then is fed to the MED–TVC unit for desalination purposes. A multiobjective optimization using a GA was used to determine the best tradeoff between the exergy efficiencies of MED–TVC and RO desalination systems, with the aim of maximizing the overall exergy efficiency of the combined system.

10.8.1 System model

Energy and exergy analyses were performed of the combined system, while the diffusion and convection transport mechanisms in the RO membrane were assessed. Table 10.6 shows the characteristics of the MED–TVC combined with RO system. Also, Table 10.7 presents the equation set for modeling the MED–TVC system based on salt mass and energy balances for each system component (e.g., effects, thermal vapor compressor, flash boxes, and condenser). Several simplifying assumptions are made in the mathematical model of the MED–TVC system: (1) the temperature difference between effects can be treated as equal, (2) the product freshwater is salt free, (3) steady-state conditions apply, and (4) the input feed flow rate into each effect is identical.

The exergy balance (Eq. 10.11) was written for each component of the MED–TVC–RO combined system. The exergy destruction and exergy loss rates in each of the MED–TVC and RO systems were calculated to determine the inefficiencies. The exergy efficiency of the system was determined based on deducting the relative irreversibilities from the reversible operation exergy efficiency, that is, unity (i.e., Eq. 10.18).

10.8.2 System optimization

After developing a mathematical model for the MED–TVC and RO systems with Matlab simulation software, the MED–TVC–RO combined system was optimized using a GA to maximize the exergy efficiencies of the MED–TVC and RO systems. The settings for the GA for the multiobjective optimization of the combined system are presented in Table 10.8. The decision variables of the optimization problem were feed seawater temperature, feed salinity, motive steam flow rate, pressure and temperature, top brine and bottom brine temperatures, inlet steam temperature, number of effects, and reference environment temperature. Nonfeasible solutions were reduced by setting lower and upper bounds (constraints) for the decision variables. The range of decision variables is listed in Table 10.8. The

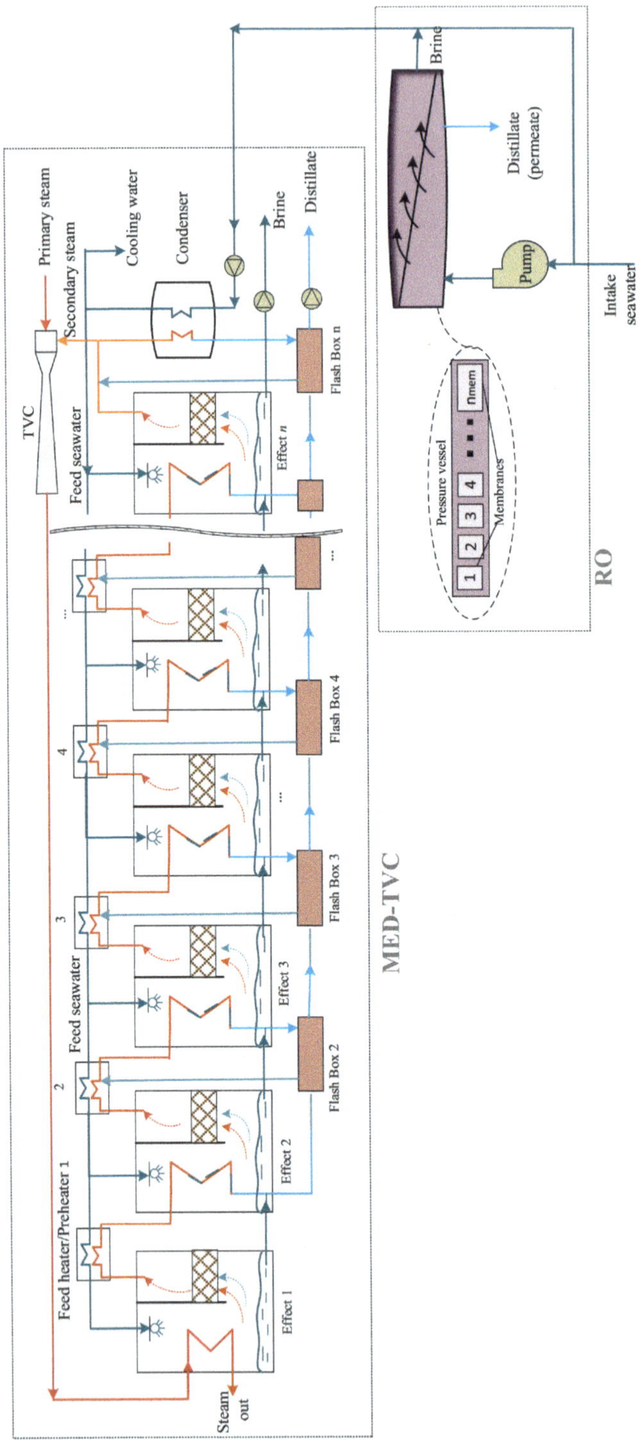

Fig. 10.15 ***Multieffect desalination-thermal vapor compressor (MED–TVC) combined with reverse osmosis (RO) desalination system.***

Table 10.6 Modeling input parameters of multieffect desalination-thermal vapor compressor (MED–TVC) combined with reverse osmosis (RO) desalination system.

MED-TVC[a]	
Primary steam flow rate	6.44 kg/s
Motive steam pressure	1290 kPa
Feed water salinity	42,000 ppm
Feed seawater temperature	44 °C
Top brine temperature	64.2 °C
Bottom brine temperature	48.9 °C
Cooling water temperature	35 °C
RO[b]	
Feed flow rate	31 kg/s
Feed concentration	42,000 ppm
Water permeability	1.2×10^{-9} m^2/s
Membrane length	1 m
Membrane area	37.2 m^2
Height of spacer channel	2.965×10^{-4} m
Number of membranes	7

[a]Data from Sadri et al. (2017).
[b]Data from Abbas (2007).

exergy efficiencies of the MED–TVC and RO systems were considered as objective functions to be maximized and the decision-making method was used to select the optimum point (Fig. 10.16).

The objective functions were normalized to allow for fair comparisons and decision making. Generally, normalization of objective functions allows for comparing and making decisions about different objective functions that have different units and/or considerably different orders of magnitude. With normalization, the objective functions can be transformed such that they have similar orders of magnitude. A robust approach for transforming objective functions is to normalize each objective function by dividing by its representative absolute function values for the initial (or current) designs. There are several approaches for normalization of objective functions, and descriptions of these can be found elsewhere (Chang, 2014; Marler and Arora, 2004). One of the simplest approaches is first to optimize each of the objectives individually, and then to divide each objective by those optimum values and then sum up all normalized terms as one objective.

Fig. 10.16 shows the normalized Pareto frontier for the multiobjective optimization of the MED–TVC–RO combined system. The Utopian method (described in Section 10.3.3) was used to select the optimum

Table 10.7 Equation set for mathematical modeling of multieffect desalination-thermal vapor compressor (MED–TVC) system in Fig. 10.15.

Quantity	Equation		Description
Temperature difference between effects	$\Delta T = \dfrac{T_1 - T_n}{n-1}$	(10.43)	T_l and T_l are the first and last effect temperatures; n is the number of effects.
Compressed steam temperature	$T_s = T_l - \Delta T$	(10.44)	
Last effect vapor temperature	$T_{v,n} = T_n - BPE$	(10.45)	BPE is boiling point elevation (°C) which is related to dissolved salt value in the water.
Compressed steam pressure, P_s, and entrained steam pressure, P_{ev}	$P_s = 1000 \times \exp\left(\dfrac{-3892.7}{T_s + 273.15 - 42.67} + 9.5\right)$	(10.46)	
	$P_{ev} = 1000 \times \exp\left(\dfrac{-3892.7}{T_{vn} + 273.15 - 42.67} + 9.5\right)$	(10.47)	
Expansion ratio, ER, and compression ratio, CR	$ER = \dfrac{P_m}{P_{ev}}$	(10.48)	P_m denotes motive (or primary) steam pressure (kPa).
	$CR = \dfrac{P_s}{P_{ev}}$	(10.49)	
Entrainment ratio	$R_a = 0.235 \dfrac{P_s^{1.19}}{P_{ev}^{1.04}} . ER^{0.015}$	(10.50)	

Mass flow rate of entrained vapor	$\dot{m}_{ev} = \frac{\dot{m}_{ms}}{R_a}$	(10.51)	$\dot{m}_{ms}$ denotes motive steam flow rate (kg/s).
Brine and vapor temperature in each effect	$T_{i+1} = T_i - \Delta T$ $T_{v,i} = T_i - BPE$	(10.52)	
Feed seawater flow rate in each effect	$\dot{m}_f w_f = \dot{m}_b w_b + \dot{m}_d w_d$	(10.53)	Subscripts *f*, *b* and *d*, respectively, denote feed, brine and distillate; *w* is salt concentration.
	$\dot{m}_f = \dot{m}_b + \dot{m}_d$	(10.54)	
	$\dot{m}_{f_i} = \frac{\dot{m}_f}{n} \quad n = 1, 2, \ldots n$	(10.55)	
Mass balance in each effect	$\dot{m}_{b_1} = \dot{m}_{f_1} - \dot{m}_{d_1}$	(10.56)	
	$\dot{m}_{b_i} = \dot{m}_{f_i} - \dot{m}_{d_i} + \dot{m}_{b_{i-1}} \quad i = 2, 3, \ldots n$	(10.57)	
Salinity of brine at outlet of each effect	$w_1 = \frac{\dot{m}_{f_1}}{\dot{m}_{b_1}} w_F$	(10.58)	
	$w_i = \frac{\dot{m}_{f_i}}{\dot{m}_{b_i}} w_F + \frac{\dot{m}_{b_{i-1}}}{\dot{m}_{b_i}} w_{i-1} \quad i = 2, 3, \ldots n$	(10.59)	

(Continued)

Table 10.7 (Cont'd)

Quantity	Equation		Description
Energy balance for each effect	$\dot{m}_{d_1} = \frac{1}{\lambda_1}\left[\dot{m}_s\lambda_s - \dot{m}_{f_1}c_p\left(T_1 - T_f\right)\right]$	(10.60)	$\dot{m}_s$ denotes mass flow rate of steam, λ latent heat of evaporation/condensation, c_p specific heat at constant pressure, T_i temperature of flashing brine (K),*NEA* nonequilibrium allowance in flashing box i (K) and *Ti* boiling temperature of brine in effect i. Note: Temperature of vapor formed by flashing is less than the effect boiling temperature T_i by the boiling point elevation *BPE* and the nonequilibrium allowance *NEA*.
	$\dot{m}_{d_i} = \frac{1}{\lambda_i}\left[\left(\dot{m}_{d_{i-1}} + \dot{m}'_{d_{i-1}}\right)\lambda_{i-1} - \dot{m}_{f_i}c_p\left(T_{i-1} - T_i\right)\right]$	(10.61)	
	$\dot{m}'_{d_{i-1}} = \dot{m}_{d_{i-1}}c_p\frac{T_{v,i-1}}{T'_i}$	(10.62)	
	$T_{v,i-1} = T_i - BPE$	(10.63)	
	$T'_i = T_{v,i-1} - NEA_i$	(10.64)	
	$NEA_i = 33.\frac{(T_{i-1} - T_i)^{0.55}}{T_{v,i}}$	(10.65)	
Condensing vapor flow rate	$\dot{m}_{d_c} = \dot{m}_{d_n} - \dot{m}_{d_{ev}}$	(10.66)	Vapor generated in the last effect is divided into two parts: one enters the TVC and the other the MED condenser.
Overall heat transfer coefficient	$U_i = \left(1939.4 + 1.40562T_i - 0.0207525T_i^2 + 0.0023186T_i^3\right)/1000$	(10.67)	

Heat transfer area for each effect	$A_1 = \frac{(\dot{m}_s + \dot{m}_{ev})\lambda_s}{U_1(T_s - T_1)}$	(10.68)
	$A_i = \frac{\dot{m}_{d_i}\lambda_i}{U_i(T_{c,i} - T_i)} \quad i = 2,3,..,n$	(10.69)
Logarithmic mean temperature difference, $LMTD_C$, and overall heat transfer coefficient for condenser, U_C	$LMTD_C = \frac{(T_f - T_{cw})}{\ln\left[\frac{T_{v,n} - T_{cw}}{T_{c,n} - T_f}\right]}$	(10.70)
	$U_c = 1.7194 + 3.2063\times10^{-2}T_{v,n} - 1.5971\times10^{-5}T_{v,n}^2 + 1.9918\times10^{-7}T_{v,n}{}^3$	(10.71)
Cooling seawater flow rate	$\dot{m}_{cw} = \frac{\dot{m}_c\lambda_n}{c_p(T_f - T_{cw})}$	(10.72)
Condenser heat transfer area	$A_c = \frac{\dot{m}_c\lambda_n}{U_c LMTD_c}$	(10.73)
Gained output ratio (GOR) and specific heat consumption, q	$GOR = \frac{\sum_{i=1}^{n}\dot{m}_{d_i}}{\dot{m}_{ms}}$	(10.74)
	$q = \frac{\dot{m}_{ms}\lambda_{ms}}{\sum_{i=1}^{n}\dot{m}_d}$	(10.75)

Equations from El-Dessouky (2002) and Sadri et al. (2017).

Table 10.8 Settings for genetic algorithm and range of decision variables in the optimization of multieffect desalination-thermal vapor compressor (MED–TVC) combined with RO system.

Settings for GA

Population size	400	
Probability crossover	0.7	
Number of crossover points	2	
Probability of mutation	0.01	
Selection process	Tournament	
Tournament size	2	
Range for decision variables	**Lower bound**	**Upper bound**
Reference environment temperature	25 °C	40 °C
Feed concentration	35,000 ppm	45,000 ppm
Feed seawater temperature	38 °C	45 °C
Motive steam pressure	1100 kPa	2500 kPa
Motive steam flow rate	5 kg/s	15 kg/s
Motive steam temperature	175 °C	210 °C
Number of effects	4	10
Bottom brine temperature	48 °C	50 °C
Top brine temperature	63 °C	66 °C
Steam inlet temperature	65 °C	75 °C

From Sadri et al. (2017).

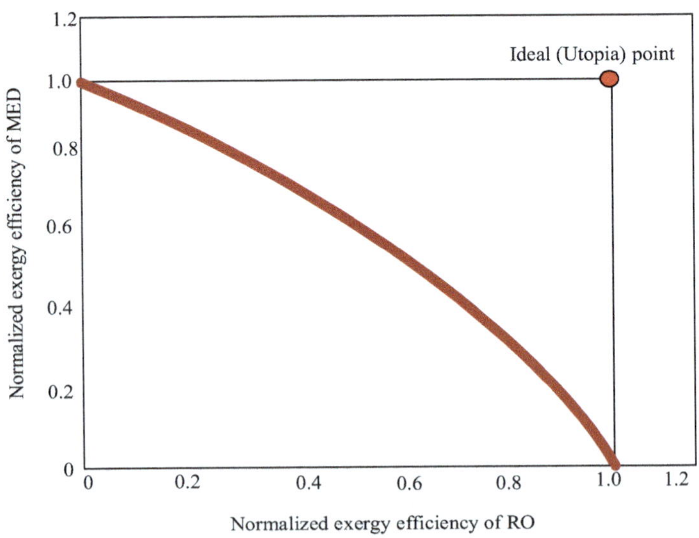

Fig. 10.16 ***Pareto frontier for multiobjective optimization of multieffect desalination-thermal vapor compressor (MED–TVC) combined with RO system.*** *Modified from Sadri et al. (2017).*

Table 10.9 Optimum values of parameters obtained with GA optimization for the multieffect desalination-thermal vapor compressor (MED–TVC) combined with RO system.

Parameter	Value
RO exergy efficiency	12.8%
MED exergy efficiency	8.6%
Reference environment temperature	32 °C
Intake seawater salt concentration	35,000 ppm
Feed seawater temperature	45 °C
Motive steam pressure	2500 kPa
Motive steam flow rate	15 kg/s
Motive steam temperature	175 °C
Number of membranes	9
Bottom brine temperature	49.9 °C
Top brine temperature	63.7 °C
Steam inlet temperature	75 °C

Modified from Sadri et al., (2017).

exergy efficiency values for the MED–TVC and RO systems. Table 10.9 lists the optimal values of parameters in the MED–TVC–RO system. The optimum exergy efficiency of the MED–TVC system at the optimal design parameters provided in Table 10.9 was found to be 8.63%, which is 3.43% higher than the exergy efficiency for the base case design characteristics listed in Table 10.6. Furthermore, the optimum exergy efficiency of the RO system was found to be 12.84%, which is 2.84% higher than that of the base case condition presented in Table 10.6.

10.9 Case study 3: optimization of combined flash-binary geothermal and HDH desalination systems

This section is extensively based on a study by Kolahi et al. (2020) in which two configurations of combined flash-binary geothermal systems for electricity and freshwater production were modeled, analyzed, optimized, and compared. The two configurations include an organic Rankine cycle (ORC) and a HDH desalination system, combined in series and parallel configurations (Fig. 10.17). In the ORC, zeotropic mixtures (combination of two organic fluids) are used as the working fluids. The two configurations were optimized using multiobjective PSO (MOPSO) in terms of maximization of electricity and freshwater production from the ORC and HDH systems.

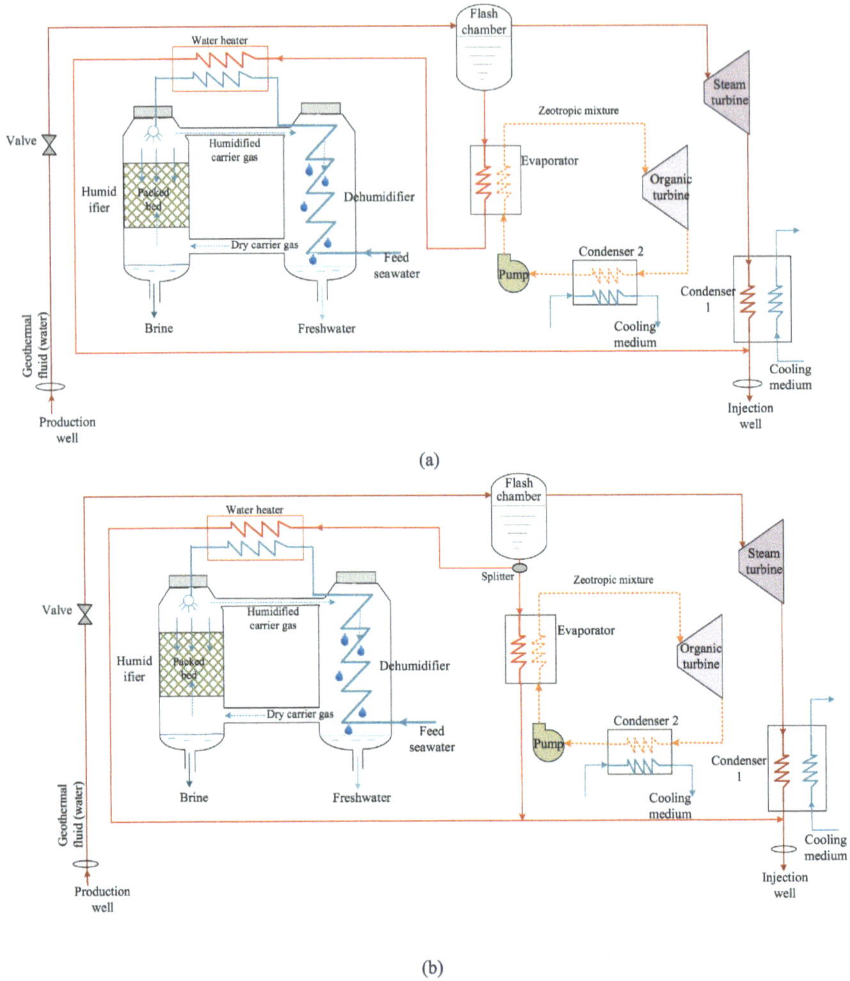

Fig. 10.17 ***Flash-binary geothermal system combined with an organic Rankine cycle-humidification dehumidification (ORC-HDH) system.*** (a) Series configuration, (b) parallel configuration. *Modified from Kolahi et al. (2020).*

In the parallel configuration, the flash chamber outlet stream, which is saturated liquid water, is divided into two streams with a splitter. These streams heat the ORC and HDH systems separately. In the series configuration, however, the saturated liquid first passes through the evaporator of the ORC, and then enters to the heater of the HDH system. In the HDH system, the intake brackish water is preheated in the dehumidifier using the latent heat of condensation of the water vapor (distillate) content in the moist air. Then the brackish feed water enters the heater, where it is further heated and increased in temperature by the high-temperature geothermal fluid, before passing to the humidifier. In the humidifier, the hot feed stream is sprayed to heat and humidify the dry air leaving the dehumidifier unit. This procedure is repeated to yield freshwater from the dehumidifier condensed stream.

In the ORC system, two organic fluids were incorporated in the zoetrope mixture to enhance the characteristics of the working fluid in terms of thermodynamic performance, thermal conductivity, cost, global warming potential, availability, etc. As a result of the investigation, isopentane and pentane were selected to be mixed with butane, trans-2-butene, and cis-2-butene (Kolahi et al., 2020; Lecompte et al., 2015; Lemmon et al., 2018).

10.9.1 System modeling and analysis

The modeling and analysis of the combined system were performed using the first and second laws of thermodynamics and other relations. The combined system was simulated with Matlab software. The thermodynamic properties of the working fluids (brackish water, pure water, humid air, and the zeotropic mixture) were obtained using correlations from Sharqawy et al. (2010), CoolProp (Bell et al., 2014), and REFPROP 8.0 (Lemmon et al., 2018).

The mass and energy rate balances were written for each component of the combined flash-binary geothermal-ORC–HDH system, for steady-state conditions, using the following general relations:

$$\sum \dot{m}_{in} - \sum \dot{m}_{out} = 0 \tag{10.76}$$

$$\dot{Q} - \dot{W} + \sum \dot{m}_{in} h_{in} - \sum \dot{m}_{out} h_{out} = 0 \tag{10.77}$$

where $\dot{m}$, $\dot{Q}$, $\dot{W}$, *0,* and *ex,* respectively, denote mass flow rate (kg/s), heat rate (kW), work rate (kW), and specific enthalpy (kJ/kg). Detailed mass and energy rate balances for each component are available elsewhere (Kolahi et al., 2020). The overall energy and exergy efficiencies of the combined flash-binary geothermal ORC–HDH system were defined as

$$\eta_{overall} = \frac{\text{energy rate of products}}{\text{total input energy rate}} = \frac{\dot{W}_{total} + \dot{m}_D \lambda}{\dot{m}_{Geo}\left(h_{Geo} - h_{Geo,0}\right)} \tag{10.78}$$

$$\varphi_{overall} = \frac{\text{exergy rate of products}}{\text{total input exergy rate}} = \frac{\dot{W}_{total} + \dot{m}_D ex_D}{\dot{m}_{Geo}\left(ex_{Geo}\right)} \tag{10.79}$$

where h_0 and ex, respectively, are the specific enthalpy of the geothermal fluid (water) at the dead state condition and the specific exergy (kJ/kg); $\dot{W}_{total}$ is the total electrical power produced by the ORC and the steam turbine of the flash-binary geothermal system. The term $\dot{m}_D$ is total mass flow rate of distillate (freshwater) produced by the HDH desalination system and λ is the latent heat of vaporization, calculated at the average partial pressure of water vapor. In the HDH desalination system, the effectiveness of freshwater production was evaluated in terms of the gained output ratio, defined as

$$GOR = \frac{\dot{m}_D \lambda}{\dot{W}_{heater}} \tag{10.80}$$

10.9.2 Optimization methodology

After developing the mathematical model for the combined system, it was optimized based on the MOPSO method to determine the optimum total electricity and freshwater production rates (i.e., $\dot{W}_{total}$ and $\dot{m}_D$). These are taken to be the two objective functions. With this optimization method, every particle moves during each iteration through a trajectory which is obtained from its previous velocity. This procedure allows particles to return to the former search space, thereby remaining the best position discovered in the past iteration (Sharaf and Elgammal, 2018).

The general steps in standard PSO are shown in Fig. 10.18. Also, the steps and procedures for applying MOPSO to the combined system are illustrated in Fig. 10.19.

The variables of the optimization problem were taken to be mass flow rate ratio of HDH (defined as mass flow rate of water ($\dot{m}_w$) divided by mass flow rate of air ($\dot{m}_a$) in the humidifier), mass fraction of splitter in parallel configuration (mf_{sp}), ORC top and low temperatures (T_{top}, T_{low}), and mass fraction of zeotropic mixtures (mf_{ORC}). Therefore, the general mathematical relation between the objective functions and decision variables can be written as

$$g\left(\dot{W}_{total}, \dot{m}_D\right) = f\left(\frac{\dot{m}_w}{\dot{m}_a}, mf_{sp}, T_{top,ORC}, T_{low,ORC}, mf_{ORC}\right) \tag{10.81}$$

The bounds on variables and the constraints were taken to be

$$T_{low} < T_{top} < T_{cr} \tag{10.82}$$

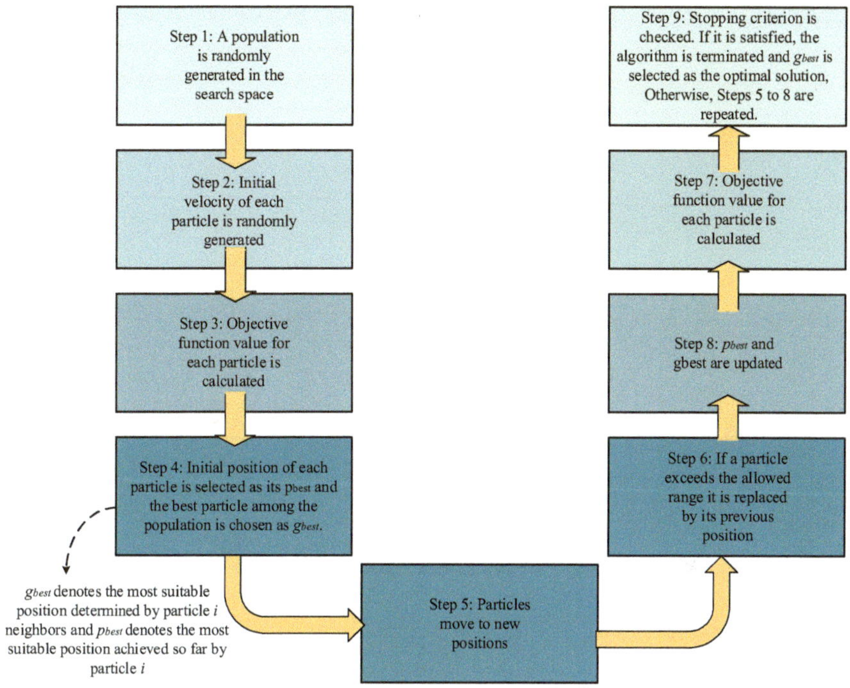

Fig. 10.18 ***Steps in standard particle swarm optimization.*** *Modified from Peng et al. (2018).*

$$T_0 < T_{low} < T_{top} \tag{10.83}$$

$$1 \le \frac{\dot{m}_w}{\dot{m}_a} \le 3 \tag{10.84}$$

$$\text{Temperature difference at pinch point in the evaporator of ORC } \left(\Delta T_{pp,ev}\right) \ge 15\ \text{K} \tag{10.85}$$

$$\text{Temperature difference of pinch point in the condenser of ORC } \left(\Delta T_{pp,co}\right) \ge 10\ \text{K} \tag{10.86}$$

The effect of varying the two objective functions (i.e., $\dot{W}_{total}$, $\dot{m}_D$) simultaneously, in both parallel and series configurations, can be demonstrated by the Pareto frontier diagrams in which the nondimensional solutions are shown (Fig. 10.20). From that figure, it can be seen that the value of $\dot{W}_{total}$ increases as the freshwater production rate decreases and vice versa. This is

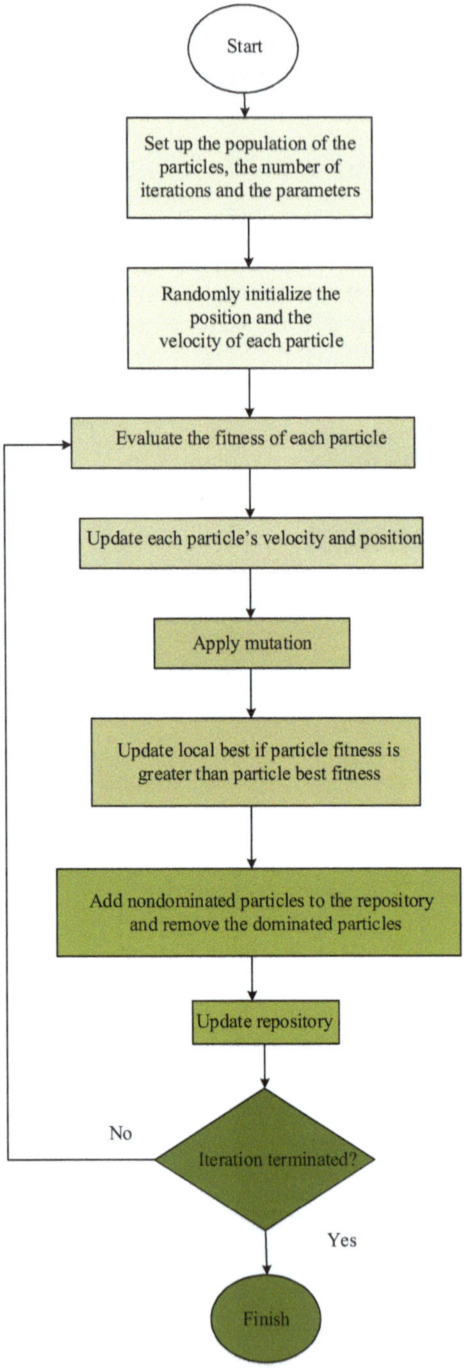

Fig. 10.19 ***Flowchart of multiobjective particle swarm optimization algorithm.*** *Modified from Kolahi et al., (2020).*

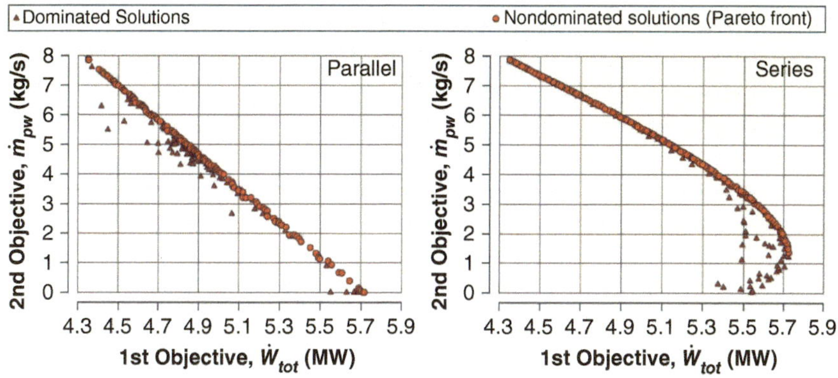

Fig. 10.20 ***Dominated and nondominated solutions for optimization of two configurations of combined flash-binary geothermal and HDH desalination system.*** Here, the effectiveness of the humidification and dehumidification units are both 80%, the air humidity ratio is 95%, and the flash chamber pressure is 5.5 bar. Pentane/butane is the working fluid of the ORC. *From Kolahi et al. (2020).*

mainly due to the fact that, as the ORC system is increasingly heated, the HDH desalination unit receives less energy for freshwater purification. It can also be seen in Fig. 10.20 that the maximum and minimum values of $\dot{W}_{total}$ in both configurations are 5.67 kW and 3.8 kW, respectively, under the optimal operating conditions. Furthermore, the maximum value of $\dot{m}_D$ in both configurations was found to be 11.42 kg/s while the minimum value of $\dot{m}_D$ was determined to be zero for the parallel system and 1.16 kg/s for the series system.

The effect of varying several parameters on the optimization objective functions was also examined (Kolahi et al., 2020). It was observed that a higher freshwater production rate results from raising the effectiveness of the humidification and dehumidification heat exchangers and the relative humidity of air, as well as reducing the flash chamber pressure. It was also found that the series configuration exhibits better thermodynamic performance than the parallel configuration, in large part because for the same value of $\dot{W}_{total}$ the series system configuration generates more freshwater.

As an example, the effect of a typical parameter, flash chamber pressure, on $\dot{W}_{total}$ and $\dot{m}_D$ for series and parallel configurations is shown in Fig. 10.21, based on (Kolahi et al., 2020). In that figure, the dominated solutions are excluded for better clarity. It can be seen that flash chamber pressure has a significant effect on both objective functions ($\dot{W}_{total}$ and $\dot{m}_D$). Specifically, as the flash chamber pressure increases, the maximum values of $\dot{W}_{total}$ and $\dot{m}_D$ decrease and increase, respectively.

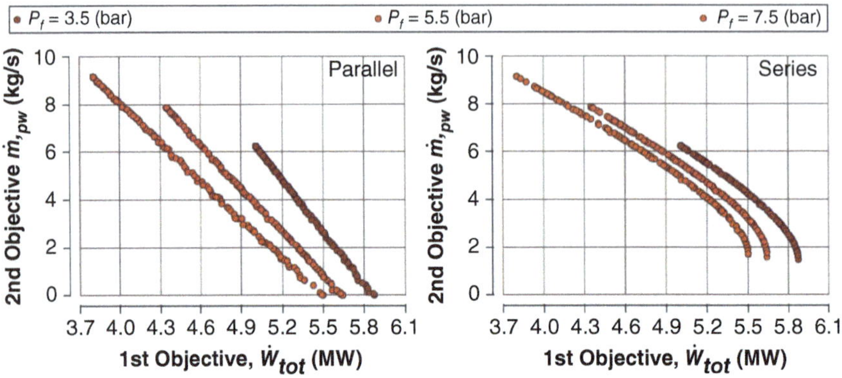

Fig. 10.21 ***Effect of flash chamber pressure on Pareto frontiers for optimization of two configurations of combined flash-binary geothermal and HDH desalination system.*** Here, the effectiveness of the humidification and dehumidification units are both 80%, and the air humidity ratio is 95%. Isopentane/trans-2-butene is the working fluid of the ORC. *From Kolahi et al. (2020).*

10.10 Future directions

Among the renewable energy resources and facilities used for driving the various desalination technologies, PVPV panels combined wiwth RO desalination systems appear to have been more extensively studied. Further studies are merited for combining other desalination technologies with one or more renewable energy sources. Numerous recent studies have been reported on the modeling, analysis, and optimization of renewable energy resources used in various desalination technologies, (Keshavarzzadeh et al., 2020; Klaimi et al., 2019; Okampo and Nwulu, 2021; Peng et al., 2018; Wu et al., 2018). In addition, research is needed on applying artificial intelligence algorithms to renewable energy–driven desalination systems. Developing and applying models to describe fouling and scaling phenomenon in membranes and heat exchangers, and employing optimization processes to minimize the presence of those phenomena can potentially provide attractive research benefits.

10.11 Closing remarks

In this chapter, the modeling and optimization of widely used desalination technologies coupled with renewable energy resources is described. A general structure for optimization of stand-alone and hybrid desalination plants is developed and evaluated by technical (energy requirement, water recovery rate, etc.), economic (investment and operating costs), and environmental performance indicators. To achieve this, each system component is modeled by considering reasonable simplifying assumptions. Next, the objective functions are defined for optimization purposes. Then, reasonable constraints are established for the design parameters, to ensure realistic optimization results. Finally, an appropriate optimization algorithm is applied to the system from which the Pareto diagram of optimization results is produced. This chapter helps in developing optimal renewable energy–driven desalination systems that operate at the optimal design parameter values. Several case studies are presented to demonstrate practical applications of various optimization algorithms for several renewable energy–based desalination plants.

Nomenclature

Symbols

A	surface area of membrane (m^2)
B	freshwater need (m^3)
BPE	boiling point elevation (°C, K)
c_p	specific heat at constant pressure (kJ/kg.K)
CR	compression ratio
CRF	capital recovery factor
DOD	depth of discharge for a battery (%)
E	energy (kWh, kJ)
$E_{Ch,B}$	hourly state of battery charge
ER	expansion ratio
ex	specific exergy (kJ/kg)
$\dot{E}x$	exergy rate (kW)
Ex_{Des}	exergy destruction rate (kW)
Ex_{loss}	exergy loss rate (kW)
f	pure water transport coefficient (m^3/m^2 min)
FF	fouling factor
G	solar radiation (W/m^2)
GOR	gained output ratio
h	specific enthalpy (kJ/kg)
i	interest rate
k_{mb}	membrane pure water permeability ($kg/m^2/s/Pa$)
$\dot{m}$	mass flow rate (kg/s)
n	loan period of a component (year)
NEA	non-equilibrium allowance (°C, K)
P	pressure (bar, kPa)
$\overline{P}$	average power requirement (kW)
P_f	concentration polarization factor
Pr	permeate production (m^3)

q specific heat consumption (kJ/kg)
$\dot{Q}$ heat rate (kW)
R_a entrainment ratio
R_{mb} salt rejection fraction of membrane
s specific entropy (kJ/kg.K)
SEC specific energy consumption (kWh/m^3)
T temperature (°C, K)
T' temperature of flashing brine (K)
TCF temperature correction factor (%)
TWC total water cost ($\$/m^3$)
U overall heat transfer coefficient ($W/m^2.K$)
V voltage (V)
$\dot{V}$ volumetric flow rate (m^3/s)
w salt concentration (kg/kg, ppm)
$\dot{W}$ work rate (kW)
X decision variable vector, solution vector
x_i *i*th decision variable, *i*th solution
Y_{mb} recovery fraction of membrane
$\dot{Z}$ component cost rate ($/h)
Z_k cost of component k ($)

Greek symbols

Θ n-dimensional search space
Ψ m-dimensional vector space of objective functions
μ specific chemical potential (kJ/kg)
τ annual operation time of system (h)
ϕ operation and maintenance factor
$\Delta\pi$ osmotic pressure (Pa)
η efficiency
β temperature coefficient of efficiency of PV module
α self-discharge rate of battery bank
λ latent heat of evaporation/condensation (kJ/kg)

Subscripts

a air
b brine
B battery
cw cooling water
c condenser, condensing
D distillate, desalination
ev entrained vapor
f feed
mb membrane
ms motive steam
p permeate

pp	pinch point
s	steam
v	vapor
w	water

Acronyms

ANN	artificial neural network
ERD	energy recovery device
GA	genetic algorithm
HDH	humidification–dehumidification
MOPSO	multi-objective particle swarm optimization
MED	multi-effect desalination
MSF	multistage flash
ORC	organic rankine cycle
PSO	particle swarm optimization
PV	photovoltaic
PX	pressure exchanger
RO	reverse osmosis
TVC	thermal vapor compression

References

Abbas, A., 2007. On the performance limitation of reverse osmosis water desalination systems. Int. J. Nucl. Desalin. 2 (3), 205–218.

Ahmadi, P., 2013. Modeling, analysis and optimization of integrated energy systems for multigeneration purposes, Doctoral dissertation. University of Ontario Institute of Technology, Oshawa, Canada. https://ir.library.dc-uoit.ca/bitstream/10155/338/1/Ahmadi_Pouria.pdf (Accessed 29 November 2021).

Ali, M.B., Kairouani, L., 2016. Multi-objective optimization of operating parameters of a MSF-BR desalination plant using solver optimization tool of Matlab software. Desalination 381, 71–83.

Al-Sofi, M.A.K., 1999. Fouling phenomena in multi stage flash (MSF) distillers. Desalination 126 (1-3), 61–76.

Bell, I.H., Wronski, J., Quoilin, S., Lemort, V., 2014. Pure and pseudo-pure fluid thermophysical property evaluation and the open-source thermophysical property library CoolProp. Ind. Eng. Chem. Res. 53 (6), 2498–2508.

Bourouni, K., M'Barek, T.B., Al Taee, A., 2011. Design and optimization of desalination reverse osmosis plants driven by renewable energies using genetic algorithms. Renew. Energy 36 (3), 936–950.

CEPCI, 2021. Chemical engineering plant cost index, Chemical Engineering, Access Intelligence LLC, New York (USA). https://www.chemengonline.com/pci (Accessed 26 May 2021).

Chang, K.H., 2014. Design Theory and Methods Using CAD/CAE: The Computer Aided Engineering Design Series. Academic Press and Elsevier, Oxford, UK.

Cuevas, E., Gálvez, J., Avalos, O., 2020. Recent Metaheuristics Algorithms for Parameter Identification. Springer, Cham.

Cui, Y., Geng, Z., Zhu, Q., Han, Y., 2017. Multi-objective optimization methods and application in energy saving. Energy 125, 681–704.

Dincer, I., Rosen, M.A., 2021. Exergy: Energy, Environment and Sustainable Development, Third ed. Elsevier, Oxford, UK.

Dincer, I., Rosen, M.A., Ahmadi, P., 2017. Optimization of Energy Systems. Wiley, New York.
El-Dessouky, H.T., Ettouney, H.M., 2002. Fundamentals of Salt Water Desalination. Elsevier, Amsterdam, The Netherlands.
Farsi, A., Dincer, I., Naterer, G.F., 2020. Multi-objective optimization of an experimental integrated thermochemical cycle of hydrogen production with an artificial neural network. Int. J. Hydrogen Energy 45 (46), 24355–24369.
Geise, G.M., Park, H.B., Sagle, A.C., Freeman, B.D., McGrath, J.E., 2011. Water permeability and water/salt selectivity tradeoff in polymers for desalination. J. Membr. Sci. 369 (1-2), 130–138.
Golberg, D.E., 1989. Genetic Algorithms in Search, Optimization, and Machine Learning. Addison-Wesley, Reading, MA.
Jeong, K., Park, M., Chong, T.H., 2019. Numerical model-based analysis of energy-efficient reverse osmosis (EERO) process: performance simulation and optimization. Desalination 453, 10–21.
Keshavarzzadeh, A.H., Ahmadi, P., Rosen, M.A., 2020. Technoeconomic and environmental optimization of a solar tower integrated energy system for freshwater production. J. Clean. Prod. 270, 121760.
Kim, J., Hong, S., 2018. Optimizing seawater reverse osmosis with internally staged design to improve product water quality and energy efficiency. J. Membr. Sci. 568, 76–86.
Klaimi, R., Alnouri, S.Y., Al-Hindi, M., Azizi, F., 2019. Optimization techniques for coupling renewable/hybrid energy options with desalination systems for carbon footprint reduction. Chem. Eng. Res. Des. 151, 270–290.
Kolahi, M.R., Amidpour, M., Yari, M., 2020. Multi-objective metaheuristic optimization of combined flash-binary geothermal and humidification dehumidification desalination systems. Desalination 490, 114456.
Lecompte, S., Huisseune, H., van den Broek, M., De Paepe, M., 2015. Methodical thermodynamic analysis and regression models of organic Rankine cycle architectures for waste heat recovery. Energy 87, 60–76.
Lemmon, E.W., Bell, I.H., Huber, M.L., McLinden, M.O., 2018. NIST Standard Reference Database 23: Reference Fluid Thermodynamic and Transport Properties-REFPROP, Version 10.0, National Institute of Standards and Technology. Standard Reference Data Program.
Mabrouk, A.N.A., 2013. Technoeconomic analysis of once through long tube MSF process for high capacity desalination plants. Desalination 317, 84–94.
Marler, R.T., Arora, J.S., 2004. Survey of multi-objective optimization methods for engineering. Struct. Multidiscipl. Optim. 26 (6), 369–395.
Mousa, K., Diabat, A., Fath, H., 2013. Optimal design of a hybrid solar-wind power to drive a small-size reverse osmosis desalination plant. Desalin. Water Treat. 51 (16-18), 3417–3427.
Niknam, T., Taheri, S.I., Aghaei, J., Tabatabaei, S., Nayeripour, M., 2011. A modified honey bee mating optimization algorithm for multiobjective placement of renewable energy resources. Appl. Energy 88 (12), 4817–4830.
Nocedal, J., Wright, S., 2006. Numerical Optimization, Second ed. Springer, New York.
Novák, V., Perfilieva, I., Mockor, J., 2012. Mathematical Principles of Fuzzy Logic, First ed. Springer, US.
Okampo, E.J., Nwulu, N., 2021. Optimisation of renewable energy powered reverse osmosis desalination systems: a state-of-the-art review. Renew. Sustain. Energy Rev. 140, 110712.
Peng, W., Maleki, A., Rosen, M.A., Azarikhah, P., 2018. Optimization of a hybrid system for solar-wind-based water desalination by reverse osmosis: comparison of approaches. Desalination 442, 16–31.
Raluy, R.G., Serra, L., Uche, J., Valero, A., 2005. Life cycle assessment of water production technologies-Part 2: reverse osmosis desalination versus the Ebro river water transfer. Int. J. Life Cycle Assess. 10 (5), 346–354.

Reed, R., Marks, II,R.J., 1999. Neural Smithing: Supervised Learning in Feedforward Artificial Neural Networks. (A Bradford Book), MIT Press, MA.
Rheinländer, J., Perz, E.W., Goebel, O., 2003. Performance simulation of integrated water and power systems—software tools IPSEpro and RESYSpro for technical, economic and ecological analysis. Desalination 157 (1-3), 57–64.
Roy, Y., Thiel, G.P., Antar, M.A., 2017. The effect of increased top brine temperature on the performance and design of OT-MSF using a case study. Desalination 412, 32–38.
Sadri, S., Ameri, M., Khoshkhoo, R.H., 2017. Multi-objective optimization of MED-TVC-RO hybrid desalination system based on the irreversibility concept. Desalination 402, 97–108.
Sangaiah, A.K., 2019. Deep Learning and Parallel Computing Environment for Bioengineering Systems, First ed. Academic Press, MA.
Shaffer, D.L., Yip, N.Y., Gilron, J., Elimelech, M., 2012. Seawater desalination for agriculture by integrated forward and reverse osmosis: improved product water quality for potentially less energy. J. Membr. Sci. 415, 1–8.
Sharaf, A.M., Elgammal, A.A., 2018. Novel AI-based soft computing applications in motor drives. In: Rashid, M.H. (Ed.), Power Electronics HandbookFourth ed. Butterworth-Heinemann, MA, pp. 1261–1302.
Sharqawy, M.H., Lienhard, J.H., Zubair, S.M., 2010. Thermophysical properties of seawater: a review of existing correlations and data. Desalin. Water Treat. 16 (1-3), 354–380.
Suganthi, L., Iniyan, S., Samuel, A.A., 2015. Applications of fuzzy logic in renewable energy systems–a review. Renew. Sustain. Energy Rev. 48, 585–607.
Weiner, D., Fisher, D., Moses, E.J., Katz, B., Meron, G., 2001. Operation experience of a solar-and wind-powered desalination demonstration plant. Desalination 137 (1-3), 7–13.
Wu, B., Maleki, A., Pourfayaz, F., Rosen, M.A., 2018. Optimal design of stand-alone reverse osmosis desalination driven by a photovoltaic and diesel generator hybrid system. Sol. Energy 163, 91–103.
Wymore, A.W., 1998. Subsystem optimization implies system suboptimization: Not!. The University of Arizona, USA. http://sysengr.engr.arizona.edu/wymore/optimal.html (Accessed 27 May 2021).

Index

Page numbers followed by "*f*" and "*t*" indicate, figures and tables respectively.

FSC
www.fsc.org
MIX
Papier | Fördert
gute Waldnutzung
FSC® C083411

Kolibri 360 GmbH
Ferdinand-Jühlke-Straße 7
99095 Erfurt, Deutschland
produktsicherheit@kolibri360.de